Environmental Chemistry for a Sustainable World

Volume 61

Series Editors
Eric Lichtfouse, Aix-Marseille University, CNRS, IRD, INRAE, Coll France, CEREGE, Aix-en-Provence, France
Jan Schwarzbauer, RWTH Aachen University, Aachen, Germany
Didier Robert, CNRS, European Laboratory for Catalysis and Surface Sciences, Saint-Avold, France

Other Publications by the Editors

Books
Environmental Chemistry
http://www.springer.com/978-3-540-22860-8

Organic Contaminants in Riverine and Groundwater Systems
http://www.springer.com/978-3-540-31169-0

Sustainable Agriculture
Volume 1: http://www.springer.com/978-90-481-2665-1
Volume 2: http://www.springer.com/978-94-007-0393-3

Book series
Environmental Chemistry for a Sustainable World
http://www.springer.com/series/11480

Sustainable Agriculture Reviews
http://www.springer.com/series/8380

Journals
Environmental Chemistry Letters
http://www.springer.com/10311

More information about this series at http://www.springer.com/series/11480

Lucas Meili • Guilherme Luiz Dotto
Editors

Advanced Magnetic Adsorbents for Water Treatment

Fundamentals and New Perspectives

Springer

Editors
Lucas Meili
Center of Technology
Federal University of Alagoas
Maceió, Alagoas, Brazil

Guilherme Luiz Dotto
Department of Chemical Engineering
Universidade Federal de Santa Maria
Santa Maria, Rio Grande do Sul, Brazil

ISSN 2213-7114 ISSN 2213-7122 (electronic)
Environmental Chemistry for a Sustainable World
ISBN 978-3-030-64091-0 ISBN 978-3-030-64092-7 (eBook)
https://doi.org/10.1007/978-3-030-64092-7

This Springer imprint is published by the registered company Springer Nature Switzerland AG
The registered company address is: Gewerbestrasse 11, 6330 Cham, Switzerland

Preface

It is a great pleasure and honour to be asked to write the preface for this important book covering such a significant theme dealing with the treatment of water pollution by the use of magnetic sorbents. By way of emphasizing the significant benefits and importance of this book, we should remember the thousands of negative impacts from contaminating our water resources. From the massive oil spillages such as the Deepwater Bay incident and the tanker spillages from Atlantic Express, Amoco Cadiz and Exxon Valdez, all making world news headlines and having long lasting adverse effects on the environment and damage to our flora and fauna ecosystems. The serious detrimental chemical spills in the River Rhine from Sandoz; the selenium poisoning of fish and widlife from farm runoff in Kesterton National Wildlife Refuge and wetland; the Minimata neurological disease for mercury poisoning and the itai-itai disease from cadmium poisoning; the release of cyanide, heavy metals and acid into the Alamosa River in Colorado, killed all aquatic life for 17 miles downstream. There are thousands more incident, like heavy metal containing acid leachate spillages from the mining industries and accidental releases of radioactive containing liquids into our water courses; all sources of contamination and damages to beaches, aquatic life, aesthetic sites and the enormous economic costs involved to both the companies involved and the rejuvenation of affected communities; amounting to thousands, millions and even billions of dollars. Due to these devastating impacts, the search for contaminated water and wastewater treatment systems has been extensive and includes various technologies, including:- chemical precipitation and coagulation, oxidation processes and chlorination, membrane technology, dispersants, demulsifiers, bioremediation chemicals – but adding additional active chemical compounds to the water; booms and skimmers can help in controlling the spread of insoluble organics and oil slicks by physically containing them by forming a barrier around the spill and then the oil can be picked up by skimmers (mechanical devices), however this is expensive for large spillages due to the requirement much labour and equipment and speed of implementation; burning-in-situ was popular, but in recent years the technique has almost been abandoned due to atmospheric pollution. For several years now there has been recognition that the process of sorption can offer a clean pollutant treatment technology and the field has become a highly popular and highly significant research area worldwide. Various hydrophobic sorbent materials have been developed and tested for the removal of contaminants from water and wastewater sources. The range of these sorbents can be either man made or synthetic. In terms of natural sorbents, many vegetable products such as rice straw, cotton, wood fiber or wool have been tested, but most have a relatively poor adsorption capacity and have only been tested on a small scale. Mineral products including zeolites and hydrophilic natural clays, with surface modification, can adsorb certain contaminants well. The sorbents are mostly in the form of particulates, flakes or powders and in recent years, nanomaterials have been used for water pollution treatment. The application of sorbents for water and wastewater treatment has tremendous potential from the available literature results but the main criticism has been with regard to the harvesting and safe handling of the contaminant laden sorbents after application for regeneration. In recent years, the advent of magnetic sorbents has opened up a novel solution to this dilemma with the potential to control, collect, concentrate and move a mass of magnetic sorbent particles in a safe manner by virtue of magnetism – applied from outside the sorbent system itself. Consequently, the topic of this book focuses on reviewing the recent research on the development of magnetic sorbents and offers a tremendous potential move-forward solution to this field of polluted water treatment application.

It is most appropriate that this book is edited by two of the world's most eminent researchers in the field of adsorption and in the development of magnetic sorbents for water pollution treatment. Professor Guilherme L. Dotto, is a Professor in Department of Chemical Engineering in the Universidade Federal de Santa Maria in Brazil. He has published extensively in the field of adsorption and his work in the field of Magnetic Sorbents for water remediation will be demonstrated in chapters in both volumes of the book. His research is focused on the following areas: transport phenomena, unit operations, wastewater treatment, physicochemical treatments for wastewater and separation processes. In specific, has specialty in adsorption/biosorption of contaminants from aqueous solutions, preparation and characterization of biomaterials and nanobiomaterials, wastewater treatment, wastes management and reuse, drying of biomaterials, statistical optimization, experimental design, response surface methodology, linear and non-linear regression analysis.

The co-editor Dr Lucas Meili is from the Center of Technology, Federal University of Alagoas, Brazil. He has published widely in the field of adsorption and water treatment and recovery using magnetic sorbents; he is also a contributor to chapters in both volumes of this book. His areas of interest are focused in separation processes, water and wastewater treatment, and synthesis of materials. Particularly he has interest in processes of adsorption of dyes and pharmaceuticals and synthesis of biochars, clays and carbon quantum dots.

The editors have solicited chapters from the world renowned experts in the field producing an excellent book, comprising updated literature in the field of magnetic sorbents and their application for oil spill remediation. The book is a pathway for the development of a major breakthrough in the treatment technology for water pollution.

Ar-Rayyan, Qatar Gordon McKay

Contents

About the Editors

Lucas Meili is working as Full Professor in the Center of Technology at Federal University of Alagoas (UFAL), Maceió, Alagoas, Brazil. Prof. Meili graduated in Chemical Engineering at Federal University of Rio Grande (Brazil) and obtained his Doctor degree in Chemical Engineering at Federal University of São Carlos (Brazil) in 2009. He is permanent member of Chemical Engineering post-graduation program and also of Materials post-graduation program. His areas of interest are focused in separation processes, water and wastewater treatment, and synthesis of materials. Particularly he has interest in adsorption and advanced oxidation processes. He has several international publications in Chemical Engineering, focusing in adsorption, development of adsorbents and advanced oxidation processes.

Guilherme Luiz Dotto is professor in Chemical Engineering Department at Federal University of Santa Maria (UFSM). Is a permanent member of the Chemical Engineering post-graduation program and also of the Chemistry post-graduation program. Is B.Sc. in Food Engineering, M.Sc. in Food Engineering and Ph.D. in Food Engineering. His research is focused on the following areas: transport phenomena, unit operations, wastewater treatment, physicochemical treatments for wastewater and separation processes. In specific, has specialty in adsorption/biosorption of contaminants from aqueous solutions, preparation and characterization of biomaterials and nanobiomaterials, wastewater treatment, wastes management and reuse, drying of biomaterials, statistical optimization, experimental design, response surface methodology, linear and non-linear regression analysis. In these research areas, has more than 160 published articles and is reviewer of more than 100 international journals. Its H-index is 25.

Contributors

Ioannis Anastopoulos Department of Chemistry, University of Cyprus, Nicosia, Cyprus

Julia Resende de Andrade School of Chemical Engineering, Department of Processes and Products Design, University of Campinas, Campinas, Sao Paulo, Brazil

Eva Baldikova Department of Nanobiotechnology, Biology Centre, ISB, CAS, Ceske Budejovice, Czech Republic

Alireza Bazargan School of Environment, College of Engineering, University of Tehran, Tehran, Iran

Giani de Vargas Brião School of Chemical Engineering, Department of Processes and Products Design, University of Campinas, Campinas, Sao Paulo, Brazil

Tito R. S. Cadaval Jr School of Chemistry and Food, Federal University of Rio Grande–FURG, Rio Grande, RS, Brazil

Yaned Milena Correa-Navarro Facultad de Ciencias, Departamento de Química, Grupo de Investigación en Sólidos Porosos y Calorimetría, Universidad de los Andes, Bogotá, Colombia
Facultad de Ciencias Exactas y Naturales, Departamento de Química, Universidad de Caldas, Manizales, Colombia

Talles Barcelos da Costa School of Chemical Engineering, Department of Processes and Products Design, University of Campinas, Campinas, Sao Paulo, Brazil

Parmila Devi Department of Chemical and Biological Engineering, College of Engineering, University of Saskatchewan, Saskatoon, SK, Canada

Guilherme L. Dotto Chemical Engineering Department, Federal University of Santa Maria–UFSM, Santa Maria, RS, Brazil

José Leandro da Silva Duarte Laboratório de Processos, Centro de Tecnologia, Universidade Federal de Alagoas, Maceió-AL, Brazil

Alessandro Erto Dipartimento di Ingegneria Chimica, dei Materiali e della Produzione Industriale, Università di Napoli Federico II, Napoli, Italy

Leticia B. Escudero Laboratory of Environmental Biotechnology (BioTA), Interdisciplinary Institute of Basic Sciences (ICB), UNCUYO – CONICET, Faculty of Natural and Exact Sciences, National University of Cuyo, Mendoza, Argentina

L. Carolina Espinoza Laboratorio de Electroquímica del Medio Ambiente (LEQMA), Universidad de Santiago de Chile, Santiago, Chile
Centro de Investigación de Procesos Redox (CIPREx), Facultad de Ciencias Químicas y Farmacéuticas, Universidad de Chile, Independencia, Chile

Bruna S. Farias School of Chemistry and Food, Federal University of Rio Grande–FURG, Rio Grande, RS, Brazil

Emiliano F. Fiorentini Laboratory of Analytical Chemistry for Research and Development (QUIANID), Interdisciplinary Institute of Basic Sciences (ICB–UNCUYO–CONICET), Faculty of Natural and Exact Sciences, National University of Cuyo, Mendoza, Argentina

Alejandra García García Laboratorio de síntesis y modificación de nanoestructuras y materiales bidimensionales, Centro de Investigación en Materiales Avanzados S.C. (CIMAV), Apodaca, NL, Mexico

Liliana Giraldo Facultad de Ciencias, Departamento de Química, Universidad Nacional de Colombia, Bogotá, Colombia

Josephine M. Hill Department of Chemical and Petroleum Engineering, University of Calgary, Calgary, AB, Canada

Wan Aini Wan Ibrahim Department of Chemistry, Faculty of Science, Universiti Teknologi Malaysia, Johor Bahru, Johor, Malaysia

Andrei V. Igansi School of Chemistry and Food, Federal University of Rio Grande–FURG, Rio Grande, RS, Brazil

Gordon McKay Division of Sustainable Development, College of Science and Engineering, Hamad Bin Khalifa University, Doha, Qatar

Lucas Meili Laboratório de Processos, Centro de Tecnologia, Universidade Federal de Alagoas, Maceió-AL, Brazil

Juan Carlos Moreno-Piraján Facultad de Ciencias, Departamento de Química, Grupo de Investigación en Sólidos Porosos y Calorimetría, Universidad de los Andes, Bogotá, Colombia

Bin Mu Key Laboratory of Clay Mineral Applied Research of Gansu Province, Center of Eco-material and Green Chemistry, Lanzhou Institute of Chemical Physics, Chinese Academy of Sciences, Lanzhou, People's Republic of China

Henrique Santana de Carvalho Neves School of Chemical Engineering, Department of Processes and Products Design, University of Campinas, Campinas, Sao Paulo, Brazil

María N. Oviedo Laboratory of Analytical Chemistry for Research and Development (QUIANID), Interdisciplinary Institute of Basic Sciences (ICB–UNCUYO–CONICET), Faculty of Natural and Exact Sciences, National University of Cuyo, Mendoza, Argentina

Ioannis Pashalidis Department of Chemistry, University of Cyprus, Nicosia, Cyprus

Luiz A. A. Pinto School of Chemistry and Food, Federal University of Rio Grande–FURG, Rio Grande, RS, Brazil

Kristyna Pospiskova Regional Centre of Advanced Technologies and Materials, Palacky University, Olomouc, Czech Republic

Jitka Prochazkova Department of Nanobiotechnology, Biology Centre, ISB, CAS, Ceske Budejovice, Czech Republic

Pamela Y. Quintas Laboratory of Analytical Chemistry for Research and Development (QUIANID), Interdisciplinary Institute of Basic Sciences (ICB–UNCUYO–CONICET), Faculty of Natural and Exact Sciences, National University of Cuyo, Mendoza, Argentina

Hamid Rashidi Nodeh Food Technology and Agricultural Products Research Centre, Standard Research Institute, Karaj, Iran

Mir Amir Mohammad Reshadi Department of Civil Engineering, K. N. Toosi University of Technology, Tehran, Iran

Paola Rodríguez-Estupiñán Facultad de Ciencias, Departamento de Química, Grupo de Investigación en Sólidos Porosos y Calorimetría, Universidad de los Andes, Bogotá, Colombia

Ivo Safarik Department of Nanobiotechnology, Biology Centre, ISB, CAS, Ceske Budejovice, Czech Republic
Regional Centre of Advanced Technologies and Materials, Palacky University, Olomouc, Czech Republic
Department of Magnetism, Institute of Experimental Physics, SAS, Kosice, Slovakia

Ricardo Salazar Laboratorio de Electroquímica del Medio Ambiente (LEQMA), Universidad de Santiago de Chile, Santiago, Chile

Juliana M. N. dos Santos Chemical Engineering Department, Federal University of Santa Maria–UFSM, Santa Maria, RS, Brazil

Anil K. Saroha Department of Chemical Engineering, Indian Institute of Technology, Delhi, New Delhi, India

Pamela Sepúlveda Facultad de Química y Biología, Universidad de Santiago de Chile, Santiago, Chile
Departamento de Física, Universidad de Santiago de Chile, Santiago, Chile
Centro para el Desarrollo de la Nanociencia y la Nanotecnología, CEDENNA, Santiago, Chile

Meuris Gurgel Carlos da Silva School of Chemical Engineering, Department of Processes and Products Design, University of Campinas, Campinas, Sao Paulo, Brazil

Thiago Lopes da Silva School of Chemical Engineering, Department of Processes and Products Design, University of Campinas, Campinas, Sao Paulo, Brazil

Melissa Gurgel Adeodato Vieira School of Chemical Engineering, Department of Processes and Products Design, University of Campinas, Campinas, Sao Paulo, Brazil

Aiqin Wang Key Laboratory of Clay Mineral Applied Research of Gansu Province, Center of Eco-material and Green Chemistry, Lanzhou Institute of Chemical Physics, Chinese Academy of Sciences, Lanzhou, People's Republic of China

Ye Xiao Department of Chemical and Petroleum Engineering, University of Calgary, Calgary, AB, Canada

Hui Yu Key Laboratory of Clay Mineral Applied Research of Gansu Province, Center of Eco-material and Green Chemistry, Lanzhou Institute of Chemical Physics, Chinese Academy of Sciences, Lanzhou, People's Republic of China
Center of Materials Science and Optoelectronics Engineering, University of Chinese Academy of Sciences, Beijing, People's Republic of China

Yongfeng Zhu Key Laboratory of Clay Mineral Applied Research of Gansu Province, Center of Eco-material and Green Chemistry, Lanzhou Institute of Chemical Physics, Chinese Academy of Sciences, Lanzhou, People's Republic of China

Chapter 1
Fundamentals of Adsorption in Liquid Phase

Pamela Sepúlveda, Alessandro Erto, José Leandro da Silva Duarte, and Lucas Meili

Contents

P. Sepúlveda
Facultad de Química y Biología, Universidad de Santiago de Chile, Santiago, Chile

Departamento de Física, Universidad de Santiago de Chile, Santiago, Chile

Centro para el Desarrollo de la Nanociencia y la Nanotecnología, CEDENNA, Santiago, Chile

A. Erto
Dipartimento di Ingegneria Chimica, dei Materiali e della Produzione Industriale, Università di Napoli Federico II, Napoli, Italy

J. L. da S. Duarte · L. Meili (✉)
Laboratório de Processos, Centro de Tecnologia, Universidade Federal de Alagoas, Maceió-AL, Brazil
e-mail: Lucas.meili@ctec.ufal.br

L. Meili, G. L. Dotto (eds.), *Advanced Magnetic Adsorbents for Water Treatment*, Environmental Chemistry for a Sustainable World 61,
https://doi.org/10.1007/978-3-030-64092-7_1

Abstract This chapter covers the fundamental concepts of adsorption in liquid phase. Aspects of the adsorption mechanism and operation characteristics are discussed. Kinetic and equilibrium aspects, both theoretical and experimental, are presented and reviewed. Thermodynamic adsorption properties are also commented and considered in this chapter.

Keywords Adsorption · Fundamentals · Pollutants · Water

1.1 Introduction

The widespread contamination of water from different pollutants has stimulated the development of effective technologies able to remove them or, at least, to significantly reduce their impact. Nowadays, water resources are threatened by many xenobiotic substances, commonly referred as micropollutants, including both organic (chlorinated, aromatic, etc.) and inorganic (mainly heavy metals) compounds. These substances are usually characterized by a high degree of toxicity and recalcitrance; for this reason, they are typically resistant to classical biological treatments and are not effectively removed by physicochemical methods, as those commonly adopted in municipal wastewater plants. In this context, adsorption has attracted the attention of several researchers due to its high efficiency, low cost, high selectivity at the molecular level, besides presenting a low energy consumption. Adsorption phenomena are spontaneous and occur when a solid surface is exposed to a gas or liquid stream in which the target molecule (adsorbate) is present, resulting in an increase of molecule density in the proximity of the material surface (adsorbent). Being a mass transfer phenomenon that involves the outer surface of solid materials, porous solids are commonly employed such as activated carbons, silica gel, some synthetic polymers, alumina, natural materials (e.g., zeolite, clays, etc.), or waste materials (e.g., lignocellulosic wastes, and red mud) (Babel and Kurniawan 2003; Bertocchi et al. 2006; Iovino et al. 2015; Sellaoui et al. 2018; Toumi et al. 2018; Li et al. 2019). The optimal choice of the adsorbent derives from a thorough evaluation of the nature of the pollutant and of the physical and chemical properties of the polluted stream.

From a practical point of view, the process configuration commonly adopted for the treatment of polluted water provides for the passage of the liquid to be purified through a column in which is present a fixed bed of porous material, capable of retaining pollutants on its surface. During the operation, the concentration of pollutants on the outer surface of the adsorbent solid increases, until a complete saturation, that is, exhaustion of the adsorption capacity. In order to be reused, these materials must be replaced or treated, aiming at restoring, albeit in part, the ability to bind pollutants on their surface. Regeneration consists in the desorption of pollutants, which is its removal from the solid matrix that can be achieved in different ways: by desorption with a (hot) inert gas passing through the bed of adsorbent; by thermal treatment, using a stream of water vapor or microwaves; or by using an acid/basic/saline solution for the extraction of the pollutant (generally for heavy metals) (Di Natale et al. 2013; Marques et al. 2017; Oladejo et al. 2020). From an economic point of view, it is worth using a nondestructive desorption agent, with high pollutant removal efficiency, not toxic to humans and to environment, and that does not alter the adsorption characteristics of the material. At the end of the regeneration treatment, the adsorbing capacity is not fully restored, so that after a series of cycles of adsorption and regeneration the material must be disposed of. In order to ensure a continuous operation of the system, the polluted stream is treated in two or more alternate treatment units, so that the exhausted one can be replaced, possibly reusing the adsorbent after a proper regeneration step.

Meantime, adsorption process presents some problems regarding sustainability. The main costs associated with this treatment rely on the adsorbent acquisition, and on the efficiency of regeneration techniques, while the operational costs are less intensive. For this reason, adsorbents deriving from low-cost sources capable to efficiently remove specific hazardous substances and to assure a cost-effective water treatment represent one of the most important challenges in adsorption intensification. Moreover, the optimization of the operational parameters for both thermodynamic and kinetic aspects of the adsorption process is the starting point for an effective application of this important technology for water treatment.

In this context, this chapter aims to address the fundamental aspects of adsorption in liquid phase in order to guide the reader to a better understanding of subsequent chapters of this book.

1.2 Adsorption Operation

For a thorough adsorption design and optimization process, a complete and accurate understanding of the effects of the main operational parameters must be accomplished. In general, the performance of an adsorption treatment depends on the thermodynamic aspects of solute–solvent–sorbent interactions and on the diffusive–convective transport phenomena involving the adsorbent (Ruthven, 1984; Vocciante et al. 2014). Adsorption equilibrium defines the limits of the applicability of the process and of a particular adsorbent, while the kinetic and transport phenomena allow defining the real efficiency and the overall extent of the intervention. To this aim, a thorough knowledge of the adsorption mechanism occurring at liquid–solid interface, which comprises the interactions between the adsorbent, the solvent (water), and adsorbate, is indispensable. The mass transfer phenomena can occur chemically, in which electrons are shared or exchanged between solute (adsorbate) and adsorbent; it is commonly referred as chemisorption. Alternatively, adsorption can involve weak physical forces with very low binding energy, such as Van der Waals forces or dispersion forces, and it is called physisorption.

In the case of adsorption in liquids, and in particular in electrolytic solutions, the description of adsorption phenomena is more difficult. In fact, while in gas systems the chemical form in which the molecule is present and, consequently, also adsorbed is known *apriori*, in aqueous solutions this is not true, because the chemical species can vary (e.g., for acid/base dissociation, hydrolysis reactions, or oxidation/reduction phenomena on the adsorbent surface) (Benjamin 2002). The phenomenon is dependent on the composition of the solution and on typical parameters such as pH, ionic strength, and temperature, which define the relative concentrations of ionic species in solution. Furthermore, the specific interactions between the solid surface and each of the ionic species present can determine significant modifications of the overall adsorption performances (Stumm and Morgan 1996).

The adsorption of organic compounds from aqueous solutions presents peculiar characteristics, which differentiate from inorganic compounds because, especially for nonionic species, some parameters have less influence (e.g., ionic strength and pH).

In general, adsorption from water streams is closely related to the following factors (Ruthven 1984; Do 1998):

- Physical characteristics of the adsorbent: specific surface area, total pore volume, structure, size and distribution of pores, etc.
- Chemical characteristics of the adsorbent: composition, presence of functional groups superficial, pH at point of zero charge (pH_{PZC}), hydrophilicity/hydrophobicity, etc.
- Chemical characteristics of the pollutant: nature, concentration, water solubility, molecular weight, etc.
- Characteristics of the aqueous matrix: temperature, pH, salinity, presence of natural organic matter (NOM), presence of other micropollutants, etc.

As adsorption is a surface phenomenon, the importance of a well-developed porous structure of the adsorbents is of utmost importance and the distribution and size of the porous matrix influence the adsorption mechanism occurring. Simultaneously, chemical interactions can occur on the surface of the solid due to the presence of surface functional groups variously reactive. In liquid adsorption, the chemistry of the adsorbent can be also characterized by the pH at point of zero charge (pH_{PZC}). It is defined as the pH value of a water/solid suspension in correspondence of which an equal adsorption of H^+ and OH^- ions is observed. Consequently, for solution pH below the pH_{PZC} value, the outer surface of the adsorbent will be positively charged; thus, it will tend to attract anions. On the contrary, for solution pH higher than pH_{PZC} the outer surface will be negatively charged, then attracting cations (Noh and Schwarz 1990).

The composition of the liquid phase also plays a very important role in the adsorption of a stated compound or for its adsorption at the maximum possible extent, which can be evaluated from the specific solute–adsorbent interactions (Nascimento et al. 2014).

From a thermodynamic point of view, adsorption is a spontaneous process ($\Delta G < 0$) and is characterized by a decrease in the entropy of the adsorbed molecule/ion, which is included into the adsorbent matrix ($\Delta S < 0$). Since, at constant temperature and pressure (Ruthven 1984):

$$\Delta G = \Delta H - T\Delta S \tag{1.1}$$

It results that adsorption is an exothermic process, that is, characterized by decrease of enthalpy, also referred as heat of adsorption, ($\Delta H < 0$) and, therefore, it is favored by low temperatures. However, in the presence of water, the overall phenomenon can be coupled by a water desorption step, for to allow the target-compound to be adsorbed. As desorption is an endothermic process, in these cases the overall process can result as endothermic.

The main thermodynamic parameter, expressing the maximum adsorption amount of a pollutant that an adsorbent can bind on its surface at equilibrium, is the adsorption capacity. The evaluation of the thermodynamic characteristics of the phenomenon (i.e., adsorption capacity) and the influence that the characteristic parameters have on the phenomenon itself are independent of the particular plant configuration chosen for tests; therefore, for the sake of simplicity, the experiments aimed at investigating these properties are conducted in batch mode.

The adsorption capacity (q) is expressed as moles or mass of pollutant per mass of adsorbent (mol/g or mg/g) and can be calculated by using a material balance on the pollutant, as in Eq. (1.2):

$$q = \frac{(C_0 - C_{\text{eq}})}{m} * V \tag{1.2}$$

where C_0 is the initial adsorbate concentration (mg/L); C_{eq} is the final (equilibrium) adsorbate concentration (mg/L); m is the mass of adsorbent (mg); and V is the volume of batch adsorbate solution (L).

From a kinetic point of view, the mass transfer from the solution to the surface of the adsorbent solid can be schematized with a series of successive steps, each of which representing one of the contributions to the global resistance. In particular, adsorption mechanism (Fig. 1.1) is rate-controlled by one of these steps or by their combination (Sanghi and Bhattacharya 2002; Plazinski et al. 2009; Vocciante et al. 2014):

1. Solute transportation in the bulk of the solution
2. Solute diffusion through the liquid film surrounding the solid particles
3. Solute diffusion inside the pores (intraparticle diffusion)
4. Adsorption/desorption equilibrium (reaction)

Similarly, in a batch adsorption test, the evolution of the adsorbed amount on the adsorbent surface (q_t) as a function of contact time with the adsorbent can be expressed by Eq. (1.3):

$$q_t = \frac{(C_0 - C_t)}{m} * V \tag{1.3}$$

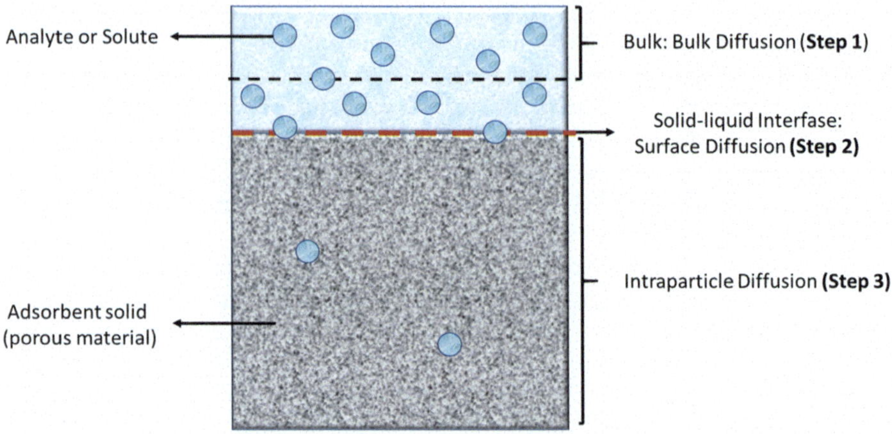

Fig. 1.1 Adsorption mechanism in terms of mass transfer phenomena

where C_0 is the initial adsorbate concentration (mg/L); C_t is the adsorbate concentration at time t (mg/L); m is the mass of adsorbent (mg); and V is the volume of adsorbate solution (L).

1.3 Adsorption Kinetics in Liquid Phase: Theoretical Aspects

In a heterogeneous system such as liquid–solid mixtures, the knowledge of adsorption–desorption kinetic is necessary to evaluate the rate of the process, the limiting steps of the mass transfer, and the influence of the textural properties of adsorbent material (Azizian 2004). Besides, this study provides useful information about the specific adsorption mechanism between adsorbent and adsorbate, considering that the adsorption process follows three main steps, as previously observed:

1. Transport of analyte from the bulk to the surface of the materials and diffusion in the liquid film surrounding the solid particle
2. Diffusion of analyte in the pores of the solid particle(pore diffusion/surface diffusion)
3. Formation of physical or chemical interaction between surface atoms of the adsorbent and the adsorbate

In this context, in order to understand the solid–liquid interactions as well as adsorbate–adsorbent complex dynamics, different mathematical models have been developed, which consider the dependence of adsorption pseudo-reaction rates with the concentration of pollutant either on the adsorbent or in the liquid contacted. In turn, the best fitting to one of the existing models provides information about the most probable reaction mechanism (Largitte and Pasquier 2016). The diverse kinetic mathematical models that have been formulated in the literature are summarized in the following.

1.3.1 Crank Model

This model assumes that the internal diffusion is the rate-limiting step and the adsorbent is considered as a sphere, assuming a constant surface diffusivity for all the adsorbent particles. The General Crank equation (Eq. 1.4) is described as (Largitte and Pasquier 2016; Westwater and Drickamer 1957):

$$(q_t/q_e) = 1 - (6/\pi^2)\sum\nolimits_{n-1}^{\infty}(1/n^2)\exp\left(\mathrm{D}_S n^2\pi^2 t/R^2\right) \quad (1.4)$$

where q_t is the amount of adsorbate adsorbed at time t (mg/g), q_e is the maximum amount of analyte adsorbed at equilibrium (mg/g), n is a fitting parameter (−), D_S is a surface diffusion coefficient (cm^2/s), t is time (s), and R (cm) is the radius of the particle of adsorbent, assumed as spherical.

1.3.2 Weber and Morris Model

In this model, the limiting step of the mass transfer is the intraparticle diffusion, and the plot of the amount of analyte (adsorbate) adsorbed at any time (mg/g) *vs* the square root of time ($t^{1/2}$) provides information about the rate of the process. This model is given by Eq. (1.5) (Yousef et al. 2011):

$$q_t = k_{id}t^{1/2} + C_i \quad (1.5)$$

where k_{id} is the intraparticle diffusion rate constant (mg/g $h^{0.5}$), t is time (h), and C_i corresponds to the boundary layer thickness (Zhou et al. 2017).

However, the plot of this model shows three curves, the first stage associated with faster or instantaneous process, followed by intra-particle diffusion stage, and finally equilibrium stage (Naseem et al. 2019). In this sense, if in the plot of q_t *vs* $t^{1/2}$ is a line passing through the origin, the more important and limiting process is the intraparticle diffusion but if the line does not pass through the origin, this stage is not the determining step (Yousef et al. 2011).

1.3.3 Pseudo First-Order Model (PFO)

In this model, also known as Lagergren model, the adsorption pseudo-reaction is the rate-limiting step or the slowest ones. This model is described by a nonreversible equation (Largitte and Pasquier 2016), assuming that the relationship between the adsorbate adsorbed onto the solid and the active site is 1:1 (Boparai et al. 2011). Moreover, adsorption does not involve the interaction between adsorbate–adsorbent and the process occurs on local sites. Adsorption bond takes place by means of the formation of monolayer on solid surface, where the energy of this process is independent on surface coverage (Largitte and Pasquier 2016).

The mathematical representation of the PFO model is shown in Eq. (1.6) (Azizian 2004; Sen Gupta and Bhattacharyya 2011; Yousef et al. 2011):

$$\frac{d_{q_t}}{d_t} = k_1(q_e - q_t) \tag{1.6}$$

Integrating this equation with the boundary conditions ($q_t = 0$ at $t = 0$; and $q_t = q_{eq}$ at $t = t_{eq}$), it is obtained:

$$q_t = q_e\left(1 - e^{-\mathrm{k_1 t}}\right) \tag{1.7}$$

which can be rewritten in linear form as:

$$\ln(q_e - q_t) = \ln q_e - k_1 t \tag{1.8}$$

where q_t and q_e (mg/g) are the amounts of analyte adsorbed at any times t (min) and at equilibrium, respectively, and k_1 is the rate constant of PFO reaction (1/min), which depends on the initial concentration of adsorbate. It is opportune to mention that Lagergren equation correlates better with the kinetic data for the adsorption systems that are not far from equilibrium.

1.3.4 Pseudo Second-Order (PSO) Model

In this case, the model assumes that adsorbate adsorption takes place by a second-order pseudo-reaction, that is, two active sites bind one adsorbate molecule. Moreover, the pollutant concentration is constant, the total number of active site depends on the ions adsorbed at equilibrium and considered that the mechanism follows a chemisorption kinetic, expressed as (Ho and McKay 1998, 1999):

$$\frac{d_{q_t}}{d_t} = k_2(q_e - q_t)^2 \tag{1.9}$$

Considering the boundary conditions ($q_t = 0$ at $t = 0$ and $q_t = q_{eq}$ at $t = t_{eq}$), the equation has the form:

$$\frac{t}{q_t} = \left(\frac{t}{q_e}\right) + \frac{1}{(k_2 q_e^2)} \tag{1.10}$$

or the linear formulation:

$$q_t = \frac{(k_2 q_e^2)\ t}{(1 + k_2 q_e t)} \tag{1.11}$$

The initial adsorption rate is obtained by defining $h = k_2 q_e^2$ directly from the intercept of the curve. Furthermore, q_t and q_e (mg/g) are the amount of analyte adsorbed at any time t (min) and at equilibrium, respectively, and k_2 is the overall rate constant for PSO reaction and depends on the operating conditions, such as analyte concentration, pH, and temperature.

1.3.5 Elovich Model

This mathematical model considers that in the adsorption process the adsorbent surface has a heterogeneous energy (Sen Gupta and Bhattacharyya 2011). The adsorption mechanism involves interactions between ions and adsorbent and there is a linear relationship between the energy of adsorption and the surface coverage (Largitte and Pasquier 2016). The equation has the form:

$$\frac{d_{q_t}}{d_t} = \alpha \exp(-\beta q_e) \tag{1.12}$$

Assuming $q_t = 0$ at $t = 0$ and $t = t$ and $q_t = q_{eq}$ and $\alpha\beta t >> 1$, the linear expression is:

$$q_t = \beta \ln(\alpha\beta) + \beta \ln t \tag{1.13}$$

And the nonlinear form is (Largitte and Pasquier 2016):

$$q_t = \left(\frac{1}{\beta}\right) \ln(\alpha\beta t) \tag{1.14}$$

where α and β are the Elovich coefficients that correspond to the initial adsorption rate (g/mg/min^2) and desorption coefficient (mg/g/min), respectively, and q_t is the amount of analyte adsorbed at any time t (min).

1.3.6 Experimental Procedure

The effect of contact time (kinetic study) can be performed by batch adsorption technique, using different glass flasks, capped polyethylene bottles, or Falcon tubes. Each flask consists in a volume containing a defined concentration of the pollutant in aqueous solution and a specific mass of adsorbent. They are put in contact and agitated in a thermostat shaker at constant temperature and, periodically, the supernatant is sampled at different time intervals, to quantify the analyte present in the solution and the adsorbed amount according to Eq. (1.3) (AL-Othman and Naushad 2012).

After the experimentation, the amount of adsorbate adsorbed per unit mass of adsorbent (q_t) can be plotted *vs* the adsorption time (t). In order to assess the best fitting model of the experimental data, a fitting procedure can be performed and different statistical parameters can be calculated for a fair evaluation of the fitting quality. For example, R^2, chi-square test (χ^2), root mean square error (RMSE), and standard deviation Δq (%) are excellent statistical parameters to individuate and validate the best fitting model. These parameters are calculated as (Harja and Ciobanu 2018; Wang et al. 2010):

$$\chi^2 = \sum_{i=1}^{N} \frac{\left(q_{\mathrm{e,\ exp}} - q_{\mathrm{e,cal}}\right)^2}{q_{\mathrm{e,cal}}} \tag{1.15}$$

$$R^2 = 1 - \frac{\sum_{\mathrm{i}=1}^{N} \left(q_{e,\mathrm{exp}} - q_{e,\mathrm{cal}}\right)^2}{\sum_{\mathrm{i}=1}^{N} \left(q_{e,\mathrm{exp}} - \overline{q_{e,\mathrm{exp}}}\right)^2} \tag{1.16}$$

$$\mathrm{RMSE} = \sqrt{\frac{1}{N-2} \sum_{1=1}^{N} \left(q_{e,\mathrm{exp}} - q_{e,\mathrm{cal}}\right)^2} \tag{1.17}$$

$$\Delta q(\%) = 100\sqrt{\frac{\sum\left(\left(q_{e,\mathrm{exp}} - q_{e,\mathrm{cal}}\right)/q_{e,\mathrm{exp}}\right)^2}{N-1}} \tag{1.18}$$

where $q_{e,\mathrm{exp}}$ and $q_{e,\mathrm{cal}}$ correspond to the experimental and calculate amount of analyte adsorbed on the adsorbent (mg/g) and N is the number of measurements. A low value of Δq suggests that the model has a good mathematical fitting and, therefore, that the parameters deriving from the model adequately describe the experimental evidences.

In Table 1.1, different kinetic experiments and the related best-fitting models are resumed as retrieved from the literature and taken as example.

1.4 Adsorption Equilibrium: Theoretical Aspects of Adsorption Isotherms

The study of adsorption equilibrium defines the potentiality of the specific application and it also helps to characterize the kinetic aspects of the process, as reported in the previous section.

The main instrument for to assess the thermodynamic properties of an adsorption system is the experimental retrieving of adsorption isotherms, which represent a correlation between the mass of the analyte (adsorbate) bond per unit mass of the adsorbent and its concentration in the solution, at constant temperature (Do 1998).

A modeling analysis of the experimental adsorption isotherms can be carried out in order to define a theoretical interpretation of the observed evidences.

In the case of solid–liquid systems, the models deriving from Langmuir theory have been extensively used to study the adsorption phenomenon for their wide applicability and undoubtable success (Chung et al. 2015). The use has been related to the interpretation of the adsorption of both organic and inorganic compounds. As an example, Hu et al. (2014) observed that by the study of the isotherms it is possible determining the distribution of ions of a given analyte between liquid and solid phase, at equilibrium. In order to characterize this distribution, different mathematical models have been proposed, which are summarized below.

1.4.1 Langmuir Model

This model assumes that the surface of adsorbent is homogeneous, all the active sites present in the substrate have similar energy or affinity for the investigated analyte and, therefore, the adsorption is homogeneous and occurs without any interaction among adsorbed analytes on the adsorbent surface (Ghosal and Gupta 2017). Besides, the Langmuir isotherm considers the formation of a monolayer and adsorption is assumed as reversible (Naseem et al. 2019). The mathematical expression is (Uslu et al. 2016):

$$q_e = q_{\max} \frac{K_L\, C_e}{1 + K_L\, C_e} \tag{1.19}$$

Table 1.1 Kinetic models and parameters determined for different analyte adsorption

Analyte	Adsorbent	Kinetic model	Kinetics parameters	References
Arsenate (V)	Biochar/γ-Fe_2O_3 composite	PSO	$q_e = 3.525$(mg/g) $k_2 = 15.1$ (*10^{-3} kg/mg h)	Zhang et al. (2013)
Lead	Nano-scale zero valent iron	PSO	$q_e = 67.99$ (mg/g) $k_2 = 2.4$ (*10^{-3} g/mg min)	Arancibia-Miranda et al. (2014)
Chromium (VI)	Magnetic multi-wall carbon nanotubes	PSO	$C_{inicial} = 25$ mg/L $q_e = 11.571$ (mg/g) $k_2 = 5.27$ (*10^{-3} g/mg min)	Huang et al. (2015)
Fluoride	γ-AlOOH @CS magnetic nanoparticle	PSO	$C_{initial} = 4 -$ 30 mg/L $q_e = 3.33 - 23.64$ (mg/g) $k_2 = 0.110 - 0.025$ (*10^{-3} g/mg min)	Wan et al. (2015)
Tetracycline (TC) Oxytetracycline (OTC) Chlortetracycline (CTC)	Nanoscale Cu/Fe bimetallic particle	PSO	**TC** $q_e = 312.15$ (mg/g) $k_2 = 1.0$ (*10^{-3} g/mg min) **OTC** $q_e = 85.47$ (mg/g) $k_2 = 4.0$ (*10^{-3} g/mg min) **CTC** $q_e = 82.64$ (mg/g) $k_2 = 4.0$ (*10^{-3} g/mg min)	Aslan et al. (2016)
Chromium (VI)	Magnetic biochar derived from peanut hull on	Elovich	$\alpha = 78{,}872$ mg/kg $\beta = 0.078$ kg/mg	Han et al. (2016)
Phosphate	Magnetic Fe_3O_4/MgAl-NO3 layered double hydroxide	PSO	$q_e = 33.90$ (mg/g) $k_2 = 0.0129$ (g/mg min)	Koilraj and Sasaki (2016)
Arsenate (V)	FeCu bimetallic nanoparticles	PSO	$q_e = 60.179$ (mg/g) $k_2 = 64.584$ (*10^{-3} g/mg min)	Sepúlveda et al. (2018))
Chromium (VI)	Magnetic biochar derived from cotton stalks	PSO	$q_e = 2.81$ (mg/g) $k_2 = 6.72$ (kg/mg h)	Ma et al. (2019)
Phenol	Magnetic chitosan	PFO	$K_p = 1.281 -$ 4.962 (mg/gmin$^{1/2}$)	Salari et al. (2019)
Methylene blue	Magnetic alginate/rice husk bio-composite	PSO	$q_e = 90.91 - 36.90$ (mg/g) $k_2 = 0.34 - 9.90$ (*10^{-3} g/mg min)	Alver et al. (2020)

which can be also linearized as:

$$\frac{C_e}{q_e} = \frac{1}{q_{\max} K_{\mathrm{L}}} + \frac{C_e}{q_{\max}} \tag{1.20}$$

where C_e is the concentration of adsorbate at equilibrium (mol/L), q_e is the analyte mass adsorbed per unit of adsorbent mass at equilibrium (adsorption capacity) (mg/g), $q_{\max}$ is the maximum adsorption capacity (mg/g), and K_L is the Langmuir constant, which is related to the affinity between adsorbate and adsorption sites or the binding energy between adsorbate and adsorbent (L/mol).

Besides, to evaluate the favorability of an adsorption process, a separation factor (R_{L}) parameter can be calculated, as given below (Atta et al. 2015; Naseem et al. 2019):

$$R_{\mathrm{L}} = \frac{1}{(1 + bC_e)} \tag{1.21}$$

Depending on R_{L} value, it is possible determining the feasibility of the adsorption process. If $R_{\mathrm{L}} > 1$, the process is unfavorable, for $R_{\mathrm{L}} = 0$, adsorption is irreversible, and if $0 < R_{\mathrm{L}} < 1$ adsorption process is favorable (Chaudhry et al. 2017).

1.4.2 Freundlich Model

This model is used when the adsorption sites cannot be hypothesized as uniform and an energetic distribution exists; usually, increasing their energy level results in a decrease in the number of active sites that can be associated with that level (i.e., a decreasing exponential distribution of energies of the active sites can be assumed) (Chaudhry et al. 2017). In fact, the Freundlich isotherm is applied when heterogeneous adsorbents are considered, and it is expressed by the following equation (Dada et al. 2012):

$$q_e = K_f C_e^{1/n} \tag{1.22}$$

which can be also linearized:

$$\mathrm{Log} q_e = \mathrm{Log} K_f + {}^1\!/\!_n \mathrm{Log} C_e \tag{1.23}$$

where q_e is the adsorption capacity at equilibrium (mg/g), K_f is the Freundlich constant and corresponds to the maximum adsorption capacity, approximately, C_e corresponds to the equilibrium concentration of pollutant (mg/L), and n is a Freundlich constant related to the binding energy (Mohan and Karthikeyan 1997). In particular, when $n > 1$ or $1/n < 1$ the adsorption condition is favorable (Aslan et al. 2016; Dada et al. 2012; El-Mallah and Hassouba 2014).

1.4.3 Temkin Model

This isotherm model assumes that in heterogeneous systems the adsorbent–adsorbate interaction generates a linearly and not logarithmic decreasing trend of the heat of adsorption of molecules on the surface with the coverage. Moreover, the process occurs by uniform distribution of binding energy up to some maximum binding energy (Varmazyar et al. 2017). The model is expressed as (Dada et al. 2012; Goel et al. 2015):

$$q_\mathrm{e} = B\ \ln\left(K_T C_e\right) \tag{1.24}$$

while the linearized expression is written as

$$q_e = B \ln K_T + B \ln C_e \tag{1.25}$$

where $B = \frac{\mathrm{RT}}{b}$ is a constant associated with the heat of adsorption, in which b (kJ/mol), R is the universal gas constant (8.314 J/K/mol), T is absolute temperature expressed in kelvin (K), K_T (L/atm or L/g) is the Temkin constant related to equilibrium binding constant, and C_e is the equilibrium concentration of adsorbate (mg/L).

1.4.4 Dubinin–Radushkevich Model

This model was developed for microporous solids and it hypothesizes the existence of a force field in the immediate proximity of the surface of the adsorbent, assuming an adsorption potential defined as the work done by the adsorption forces in binding a certain molecule and equals to the variation of free energy of a substance between its state in liquid bulk and the condition of adsorbed molecule. It can be applied when it is possible assuming a Gaussian energy distribution onto a heterogeneous surface, where the principal characteristic is its independence on the temperature (Sengupta et al. 2016). The mathematical equation is (Dada et al. 2012):

$$q_\mathrm{e} = (q_\mathrm{s}) \exp\left(-\beta\varepsilon^2\right) \tag{1.26}$$

The linearized form is:

$$\ln q_e = \ln q_s - \beta\varepsilon^2 \tag{1.27}$$

where q_e is the adsorption capacity at equilibrium (mg/g), q_s corresponds to the theoretical isotherm saturation capacity (mg/g), β is the Dubinin–Radushkevich isotherm constant (mol^2/kJ^2) related to the mean free energy of adsorption, and ε is the Polanyi potential associated with the equilibrium concentration (C_e) (Chaudhry et al. 2017):

$$\varepsilon = \mathrm{RTln}\left(1 + \frac{1}{C_e}\right) \tag{1.28}$$

where R (8.314 J/K/mol) is the gas universal constant and T is the absolute temperature (K). Finally, the energy (E) per molecule of adsorbate (adsorption energy) can be obtained by (Albadarin et al. 2012):

$$E = \frac{1}{\sqrt{-2\beta}} \tag{1.29}$$

1.4.5 *Sips Model*

The Sips isotherm is an extension of the Langmuir isotherm equation, which includes some of the properties and the mathematical form of the Freundlich isotherm. It overcomes the limits of Freundlich equation, by assuming a limit to the maximum value of the adsorption capacity for a stated adsorbent–adsorbate couple (Hamdaoui and Naffrechoux 2007; Febrianto et al. 2009). The model proposed by Sips (1948) can be expressed by the following equation:

$$q_e = \frac{q_S(K_S C_e)^{\mathrm{ms}}}{1 + (K_S C_e)^{\mathrm{ms}}} \tag{1.30}$$

where q_e is the adsorption capacity at equilibrium (mg/g), q_s is the maximum adsorption capacity of the Sips model (mg/g), C_e is the liquid concentration of the adsorbate at equilibrium (mg/L), K_S is the equilibrium constant of the Sips isotherm model (L/mg), and m_s is the exponent of the model of the Sips isotherm.

1.4.6 *Redlich–Peterson Model*

Redlich–Paterson is an empirical equation with three parameters that can accurately represent adsorption equilibria for a wide concentration range. This model incorporates the characteristics of the Langmuir and Freundlich isotherms into a single equation and hypothesizes a hybrid adsorption mechanism, not following the ideal monolayer adsorption as for Langmuir model. The isotherm of Redlich–Peterson is expressed by the following equation:

$$q_e = \frac{CeK_{rp}}{\left(1 + a_{rp}Ce^{b_{rp}}\right)} \tag{1.31}$$

where k_{rp} and a_{rp} are the R-P constant (L/g and L^{β}/mg^{β}, respectively) and β is the exponent, which can vary between 1 and 0.

As previously reported in kinetics models, the best fitting among the isotherm models is assessed by the coefficient of determination (R^2), chi-square test (χ^2), root mean square error (RMSE), and standard deviation Δq (%) (see Eqs. 1.15, 1.16, 1.17, and 1.18).

1.4.7 Batch Experimental Procedure

As for the kinetic study, also the assessment of the equilibrium condition is based on experimental studies, usually carried out in batch mode (for the sake of simplicity).

The batch experiments can be carried out in glass flasks, capped polyethylene bottles, or Falcon tubes containing an appropriated concentration of adsorbent material and a constant volume of solution. In the study, the concentration of adsorbate should vary so to cover the range of interest, which depends on the specific analyte. The suspension or reaction mixture obtained is shaken for a constant time and kept at constant temperature. Finally, the samples are centrifuged and filtered to quantify the residual concentration of the pollutant in the supernatant and the adsorption capacity, using Eq. (1.2).

It is necessary to mention that this analysis should be carried out with optimized but constant physicochemical parameters for each data set, such as pH, temperature, ionic strength, and background electrolyte. In addition, the isotherm study can be made at different temperatures in order to determine important thermodynamic parameters.

As examples retrieved in the literature, different researchers have used isotherm studies to determine the removal mechanism of different pollutants present in water. Harja and Ciobanu (2018) reported an isotherm analysis for the study of adsorption from aqueous solution using 10–300 mg/L of oxytetracycline, assuming a set of constant adsorption parameters (pH = 8, adsorbent concentration 2 g/L, contact time 1 h and temperature = 20 °C). Similarly, Ding et al. (2016) performed an isotherm study of Pb^{2+} and Cd^{2+} adsorption from polluted water, conducted at different concentrations of analytes (50–900 mg/L), with 0.1 g of adsorbent (biochar), pH 5.0, using 50 mL of solution, at 30 °C and shaking the samples for 24 h. Zhou et al. (2017) worked on a Fe_3O_4 magnetic polypyrrole–graphene oxide nanocomposite as adsorbent, at 0.05 g/L concentration, with an adsorption solution at pH 7.0 having a Hg concentration of 20–100 mg/L, in the temperature range 300–320 K.

In Table 1.2, some isotherm studies and the corresponding best-fitting models are reported together with the indications of the experimental conditions.

Table 1.2 Isotherm studies, best fitting isotherm model, and equilibrium constant for different adsorption studies

Pollutant	Adsorbent material	Temperatures (K)	Best fitting model	References
Diclofenac sodium	MgAl/layered double hydroxide supported on *Syagrus coronata* biochar	303.15 313.15 323.15 333.15	Sips q_{max} = 168.04 mg. g^{-1} (at 333.15 K)	de Souza dos Santos et al. (2020)
17β-estradiol removal	Graphene-like magnetic biochar	288.00 298.00 308.00	Langmuir q_{max} = 121.179 mg. g^{-1} (at 308 K)	Liu et al. (2020)
Rifampicin	Calcined *Mytella falcata* shells	303.15 313.15 323.15 333.15	Redlich–Peterson q_{max} = 10.00 mg. g^{-1} at 333.15 K)	Henrique et al. (2020)
Caffeine	*Elaeis guineensis*-activated carbon	303.15 313.15 323.15 333.15	Sips q_{max} = 13.5 mg.g^{-1} (at 333.15 K)	Melo et al. (2020)
Nitrate	Mg–Fe layered double hydroxide intercalated with chloride	303.15 323.15 333.15	Sips q_{max} = 18.17 mg. g^{-1} (at 303.15 K)	Santos et al. (2019)
Methylene blue	*Wodyetia bifurcata* biochar	298.15 313.15 323.15 333.15	Langmuir q_{max} = 216.66 mg. g^{-1} (at 333.15 K)	dos Santos et al. (2019a)
Methylene blue	*Syagrus oleracea*-activated carbon	298.15 313.15 323.15 333.15	Freundlich q_{max} = 30.0 mg.g^{-1} (at 333.15 K)	dos Santos et al. (2019b)
2-nitrophenol	Zn–Al and Mg–Fe layered double hydroxides	298.00 308.00 318.00 328.00	Freundlich q_{max} = 290 and 165 mg g^{-1}	Dalla Nora et al. (2020)

1.5 Thermodynamics of Adsorption

From thermodynamic (equilibrium) tests carried out at different temperatures, it is possible to retrieve indications about important thermodynamic parameters, such as Gibbs free energy ($\Delta G°$), enthalpy ($\Delta H°$), and entropy ($\Delta S°$), which characterize the adsorption of the investigated system.

From the values assumed by these parameters, it is possible to determine the adsorption characteristics of the investigated system, that is, if adsorption is endothermic or exothermic, favorable, spontaneous, if there is an increase or a reduction in system disorder, the process nature (i.e., physisorption or chemisorption), as well as if enthalpy or entropy control the operation (Do 1998; Piccin et al. 2017).

A thermodynamic analysis of the adsorption phenomena can include the measurement of the magnitude of the heat of adsorption and its variation with surface loading, which can provide useful indications about the nature of the adsorbent surface and the adsorbed molecules/ions interactions. The heat of adsorption determined at constant amounts of sorbate adsorbed is commonly known as the “isosteric heat of adsorption” or “differential heat of adsorption” (ΔH, kJ/mol). It is defined as the ratio of the infinitesimal change in the adsorbate enthalpy and the infinitesimal change in the amount adsorbed (Do 1998).

The isosteric heat of adsorption can be calculated from the following thermodynamic relationship, derived from the Van’t Hoff equation (Do 1998):

$$\frac{d \ln (C)}{dT} = -\frac{\Delta H}{RT^2} \text{ or } \frac{d \ln (C)}{d(1/T)} = \frac{\Delta H}{R} \tag{1.32}$$

The adsorbate equilibrium concentrations (C) at constant adsorption capacity (q) are taken from the adsorption isotherm data at different temperatures. Following this path, the ΔH is calculated from the slope of the plot of ln(C) versus ($1/T$) for different amounts of q (see Eq. 1.32).

In many applications, in order to evaluate the thermodynamic parameters of the process ($\Delta G°$, $\Delta H°$, and $\Delta S°$) and to investigate the adsorption mechanism, a basic assumption can be made, that is, the independence of the thermodynamic parameters on temperature. This simplified approach allows building a relationship between ln (K_e) and $1/T$, in which K_e is the adsorption constant determined starting from one of the models presented in the previous section (e.g., Langmuir constant):

$$\Delta G^0 = -RT \ln K_e \tag{1.33}$$

$$\Delta G^0 = \Delta H^0 - T\Delta S^0 \tag{1.34}$$

Starting from this relationship, it is possible calculating $\Delta H°$ and $\Delta S°$ as the slope and the intercept, respectively (Dotto et al. 2013; Milonjić 2007):

$$\ln (k_e) = \frac{-\Delta H^0}{RT} + \frac{\Delta S^0}{R} \tag{1.35}$$

where R (8.314 J/mol/K) is the universal constant of gases, T (K) is the absolute temperature, and K_e is the equilibrium constant obtained by the isotherm model that best fitted the equilibrium data. K_e must be dimensionless, while it is expressed by L/mol, L/g, or L/mg, then it should be multiplied by a correction factor. When the value is expressed in L/mol, it can be multiplied by 55.55 mol/L, equivalent to the number of moles of water per L of solution. However, when it is expressed in L/g, it must be multiplied by water density (Milonjić 2007).

As examples, in Table 1.3 some studies reported in the literature are listed, with the indication of the retrieved thermodynamic parameters.

Table 1.3 Thermodynamic parameters in different adsorption studies

Adsorbent	Pollutant	Temperature (K)	ΔG (kJ. mol^{-1})	ΔH° (kJ. mol^{-1})	ΔS° (J. mol^{-1})	References
Calcined *Mytella falcata* shells	Rifampicin	303.15 313.15 323.15 333.15	−29.28 −29.59 −31.32 −32.1	1.65	−10.1	Henrique et al. (2020)
Mussels shell	Safranin	298.15 308.15 318.15	−1.96 −2.16 −2.46	2.73	15.85	El Haddad et al. (2014)
MgAl/layered double hydroxide supported on *Syagrus coronata* biochar	Diclofenac sodium	303.15 313.15 323.15 333.15	−29.6 −28.57 −24.71 −20.09	−111.61	−0.271	de Souza dos Santos et al. (2020)
Elaeis guineensis-activated carbon	Caffeine	303.15 313.15 323.15 333.15	−17.190 −17.818 −16.615 −19.285	− 1.553	0.0508	Melo et al. (2020)
Powdered oyster shell	Cu^{2+}	298.15 308.15 333.15	−18.81 −20.42 −22.41	20.88	130	Hsu (2009)
Mg–Fe layered double hydroxide intercalated with chloride	Nitrate	303.15 323.15 333.15	−24.37 −18.64 −18.10	−90.78	−0.22	Santos et al. (2019)
Wodyetia bifurcata biochar	Methylene blue	298.15 313.15 323.15 333.15	−14.40 −15.41 −16.09 −16.77	5.79	0.068	dos Santos et al. (2019a)
Zn–Al–LDH and Mg–Fe–LDH	2-nitrophenol	298.00 308.00 318.00 328.00	−5.72 and −3.25 −9.45 and −5.04 −11.02 and −5.44 −12.91 and −7.30	63.45 and 4.52	0.234 and 0.004	Dalla Nora et al. (2020)

1.6 Conclusions

Adsorption is a fundamental technique that can be adopted for the removal of different compounds, both organic and inorganic, from polluted water. It is commonly known as an efficient, versatile, and cost-effective treatment that has a very wide applicability, due to a general unselectively and adaption to different polluting scenarios. Large applications can be recognized worldwide, mostly based on fixed-bed reactors in which many different adsorbents, depending on the adsorbate characteristics and on polluted stream properties, can be proficiently adopted.

Adsorption phenomena depend on many different parameters, among which are the physical and chemical characteristics of the adsorbent, the chemical characteristics of the pollutant, and the characteristics of the aqueous matrix. In general, adsorption is a surface phenomenon, and hence the adsorbents are porous solids with defined physical (e.g., porosity) and chemical properties (e.g., composition). The influence of all these properties is of fundamental importance in the practical application of this technique. The performance of an adsorption process mainly depends on the thermodynamic aspects of solute–solvent–sorbent interactions and on the diffusive–convective transport phenomena within the porous adsorbent. The characterization of an adsorption process can be done either from a thermodynamic point of view, in order to have information about the equilibrium (maximum) adsorption capacity and its dependence on the main process parameters, as those previously indicated, or by a kinetic standpoint, in order to determine the fluid-dynamic properties and the mass transfer from the fluid to adsorbent and, more in general, the rate of the overall process.

A meaningful application of this technique requires the knowledge of both the thermodynamic and kinetic properties of the investigated adsorbent/adsorbate couple, in the specific conditions of the polluted water. To this aim, dedicated experimental campaigns might be made in order to investigate the properties of the adsorption system and to retrieve important information about the adsorption mechanisms. The interpretation of the retrieved experimental data can be made by the fitting of a wide number of possible adsorption models, separately for the thermodynamic (equilibrium) and kinetic analysis. The availability of reliable theoretical models for the interpretation of adsorption data is an invaluable tool for process optimization (i.e., determination of the best operating parameters set) and for design purposes.

References

Albadarin AB, Mangwandi C, Al-muhtaseb AH, Walker GM, Allen SJ, Ahmad MNM (2012) Kinetic and thermodynamics of chromium ions adsorption onto low-cost dolomite adsorbent. Chem Eng J 179:193–202. https://doi.org/10.1016/j.cej.2011.10.080

AL-Othman Z, Naushad M (2012) Hexavalent chromium removal from aqueous medium by activated carbon prepared from peanut shell: adsorption kinetics, equilibrium and thermodynamic studies. Chem Eng J 184:238–247. https://doi.org/10.1016/j.cej.2012.01.048

Alver E, Metin AÜ, Brouers F (2020) Methylene blue adsorption on magnetic alginate/rice husk bio-composite. Int J Biol Macromol 154:104–113. https://doi.org/10.1016/j.ijbiomac.2020.02.330

Arancibia-Miranda N, Baltazar SE, García A, Romero AH, Rubio MA, Altbir D (2014) Lead removal by nano-scale zero valent iron: surface analysis and pH effect. Mater Res Bull 59:341–348. https://doi.org/10.1016/j.materresbull.2014.07.045

Aslan S, Yalçin K, Hanay Ö, Yildiz B (2016) Removal of tetracyclines from aqueous solution by nanoscale Cu/Fe bimetallic particle. Desalin Water Treat 3994:0. https://doi.org/10.1080/19443994.2015.1067870

Atta AM, Al-Lohedan HA, ALOthman ZA, Abdel-Khalek AA, Tawfeek AM (2015) Characterization of reactive amphiphilic montmorillonite nanogels and its application for removal of toxic cationic dye and heavy metals water pollutants. J Ind Eng Chem 31:374–384. https://doi.org/10.1016/j.jiec.2015.07.012

Azizian S (2004) Kinetic models of sorption: a theoretical analysis. J Colloid Interface Sci 276:47–52. https://doi.org/10.1016/j.jcis.2004.03.048

Babel S, Kurniawan TA (2003) Low-cost adsorbents for heavy metal uptake from contaminated water: a review. J Hazard Mater B97:219e243

Benjamin MM (2002) Aquatic chemistry: chemical Equilibria and rates in natural waters. Stumm W, Morgan JJ, 3rd edn. Wiley-Interscience. Water Chemistry McGraw Hill, New York

Bertocchi AF, Ghiani M, Peretti R, Zucca A (2006) Red mud and fly ash for remediation of mine sites contaminated with as, cd, cu, Pb and Zn. J Hazard Mater 134(1–3):112–119. https://doi.org/10.1016/j.jhazmat.2005.10.043

Boparai HK, Joseph M, Carroll DMO (2011) Kinetics and thermodynamics of cadmium ion removal by adsorption onto nano zerovalent iron particles. J Hazard Mater 186:458–465. https://doi.org/10.1016/j.jhazmat.2010.11.029

Chaudhry SA, Zaidi Z, Siddiqui SI (2017) Isotherm, kinetic and thermodynamics of arsenic adsorption onto Iron-Zirconium Binary Oxide-Coated Sand (IZBOCS): modelling and process optimization. J Mol Liq 229:230–240. https://doi.org/10.1016/j.molliq.2016.12.048

Chung HK, Kim WH, Park J, Cho J, Jeong TY, Park PK (2015) Application of Langmuir and Freundlich isotherms to predict adsorbate removal efficiency or required amount of adsorbent. J Ind Eng Chem 28:241–246. https://doi.org/10.1016/j.jiec.2015.02.021

Dada A, Olalekan AP, Olatunya A, DADA O (2012) Langmuir, Freundlich, Temkin and Dubinin–Radushkevich isotherms studies of equilibrium sorption of Zn 2 + unto phosphoric acid modified Rice husk. J Appl Chem 3:38–45. https://doi.org/10.9790/5736-0313845

Dalla Nora FB, Lima VVC, Oliveira MLS, Hosseini-Bandegharaei A, de Lima Burgo TA, Meili L, Dotto GL (2020) Adsorptive potential of Zn–Al and Mg–Fe layered double hydroxides for the removal of 2–nitrophenol from aqueous solutions. J Environ Chem Eng. https://doi.org/10.1016/j.jece.2020.103913

de Souza dos Santos GE, Ide AH, Duarte JLS, McKay G, Silva AOS, Meili L (2020) Adsorption of anti-inflammatory drug diclofenac by MgAl/layered double hydroxide supported on Syagrus coronata biochar. Powder Technol 364:229–240. https://doi.org/10.1016/j.powtec.2020.01.083

Di Natale F, Erto A, Lancia A (2013) Desorption of arsenic from exhaust activated carbons used for water purification. J Hazard Mater 260:451–458

Ding Y, Liu Y, Liu S, Li Z, Tan X, Huang X, Zeng G, Zhou Y, Zheng B, Cai X (2016) Competitive removal of Cd(II) and Pb(II) by biochars produced from water hyacinths: performance and mechanism. RSC Adv 6:5223–5232. https://doi.org/10.1039/c5ra26248h

Do DD (1998) Adsorption analysis: equilibria and kinetics. Imperial College Press, London

dos Santos KJL, de Souza dos Santos GE, de Sá ÍMGL, Ide AH, da Silva Duarte JL, de Carvalho SHV, Soletti JI, Meili L (2019a) Wodyetia bifurcata biochar for methylene blue removal from aqueous matrix. Bioresour Technol 293:122093. https://doi.org/10.1016/j.biortech.2019.122093

dos Santos KJL, de Souza dos Santos GE, de Sá ÍMGL, de Carvalho SHV, Soletti JI, Meili L, da Silva Duarte JL, Bispo MD, Dotto GL (2019b) Syagrus oleracea–activated carbon prepared by vacuum pyrolysis for methylene blue adsorption. Environ Sci Pollut Res. https://doi.org/10.1007/s11356-019-05083-4

Dotto GL, Vieira MLG, Esquerdo VM, Pinto LAA (2013) Equilibrium and thermodynamics of azo dyes biosorption onto Spirulina platensis. Brazilian J Chem Eng 30:13–21. https://doi.org/10.1590/S0104-66322013000100003

El Haddad M, Regti A, Slimani R, Lazar S (2014) Assessment of the biosorption kinetic and thermodynamic for the removal of safranin dye from aqueous solutions using calcined mussel shells. J Ind Eng Chem. https://doi.org/10.1016/j.jiec.2013.05.038

El-Mallah NM, Hassouba HM (2014) Kinetic and thermodynamic studies for the removal of nickel ions from an aqueous solution by adsorption technique. J Dispers Sci Technol 35:130–142. https://doi.org/10.1080/01932691.2013.769173

Febrianto J, Kosasih AN, Sunarso J, Ju YH, Indraswati N, Ismadji S (2009) Equilibrium and kinetic studies in adsorption of heavy metals using biosorbent: a summary of recent studies. J Hazard Mater. https://doi.org/10.1016/j.jhazmat.2008.06.042

Ghosal PS, Gupta AK (2017) Determination of thermodynamic parameters from Langmuir isotherm constant-revisited. J Mol Liq 225:137–146. https://doi.org/10.1016/j.molliq.2016.11.058

Goel C, Bhunia H, Bajpai PK (2015) Resorcinol-formaldehyde based nanostructured carbons for CO2 adsorption: kinetics, isotherm and thermodynamic studies. RSC Adv 5:93563–93578. https://doi.org/10.1039/c5ra16255f

Hamdaoui O, Naffrechoux E (2007) Modeling of adsorption isotherms of phenol and chlorophenols onto granular activated carbon part II. Models with more than two parameters. J Hazard Mater 147:401–411

Han Y, Cao X, Ouyang X, Sohi SP, Chen J (2016) Adsorption kinetics of magnetic biochar derived from peanut hull on removal of Cr (VI) from aqueous solution: effects of production conditions and particle size. Chemosphere 145:336–341. https://doi.org/10.1016/j.chemosphere.2015.11.050

Harja M, Ciobanu G (2018) Studies on adsorption of oxytetracycline from aqueous solutions onto hydroxyapatite. Sci Total Environ 628–629:36–43. https://doi.org/10.1016/j.scitotenv.2018.02.027

Henrique DC, Uchoa Quintela D, Honjo Ide A, Erto A, da Silva Duarte JL, Meili L (2020) Calcined Mytella falcata shells as alternative adsorbent for efficient removal of rifampicin antibiotic from aqueous solutions. J Environ Chem Eng:103782. https://doi.org/10.1016/j.jece.2020.103782

Ho YS, McKay G (1998) A comparison of chemisorption kinetic models applied to pollutant removal on various sorbents. Process Saf Environ Prot 76:332–340. https://doi.org/10.1205/095758298529696

Ho YS, McKay G (1999) Pseudo-second order model for sorption processes. Process Biochem 34:451–465. https://doi.org/10.1016/S0032-9592(98)00112-5

Hsu TC (2009) Experimental assessment of adsorption of Cu2+ and Ni2+ from aqueous solution by oyster shell powder. J Hazard Mater. https://doi.org/10.1016/j.jhazmat.2009.06.105

Hu X j, Liu Y g, Zeng G m, You S h, Wang H, Hu X, Guo Y m, Tan X f, Guo F y (2014) Effects of background electrolytes and ionic strength on enrichment of Cd(II) ions with magnetic graphene oxide-supported sulfanilic acid. J Colloid Interface Sci 435:138–144. https://doi.org/10.1016/j.jcis.2014.08.054

Huang Z, Wang X, Yang D (2015) Adsorption of Cr(VI) in wastewater using magnetic multi-wall carbon nanotubes. Water Sci Eng 8:226–232. https://doi.org/10.1016/j.wse.2015.01.009

Iovino P, Canzano S, Capasso S, Erto A, Musmarra D (2015) A modeling analysis for the assessment of ibuprofen adsorption mechanism onto activated carbons. Chem Eng J 277:360–367. https://doi.org/10.1016/j.cej.2015.04.097

Karimi S, Tavakkoli Yaraki M, Karri RR (2019) A comprehensive review of the adsorption mechanisms and factors influencing the adsorption process from the perspective of bioethanol dehydration. Renew Sust Energ Rev 107:535–553. https://doi.org/10.1016/j.rser.2019.03.025

Koilraj P, Sasaki K (2016) Fe3O4/MgAl-NO3 layered double hydroxide as a magnetically separable sorbent for the remediation of aqueous phosphate. J Environ Chem Eng 4:984–991. https://doi.org/10.1016/j.jece.2016.01.005

Largitte L, Pasquier R (2016) A review of the kinetics adsorption models and their application to the adsorption of lead by an activated carbon. Chem Eng Res Des 109:495–504. https://doi.org/10.1016/j.cherd.2016.02.006

Li Z, Sellaoui L, Dotto GL, Ben Lamine A, Erto A (2019) Interpretation of the adsorption mechanism of reactive black 5 and Ponceau 4R dyes on chitosan/polyamide nanofibers via advanced statistical physics model. J Mol Liq 285:165–170

Liu N, Liu Y, Tan X, Li M, Liu S, Hu X, Zhang P, Dai M, Xu W, Wen J (2020) Synthesis a graphene-like magnetic biochar by potassium ferrate for 17β-estradiol removal: effects of Al2O3 nanoparticles and microplastics. Sci Total Environ 715:136723. https://doi.org/10.1016/j.scitotenv.2020.136723

Marques SCR, Marcuzzo JM, Baldan MR, Mestre AS, Carvalho AP (2017) Pharmaceuticals removal by activated carbons: role of morphology on cyclic thermal regeneration. Chem Eng J 321:233–244

Ma F, Zhao B, Diao J (2019) Synthesis of magnetic biochar derived from cotton stalks for the removal of Cr(VI) from aqueous solution. Water Sci Technol 79:2106–2115. https://doi.org/10.2166/wst.2019.208

Melo LLA, Ide AH, Duarte JLS, Zanta CLPS, Oliveira LMTM, Pimentel WRO, Meili L (2020) Caffeine removal using Elaeis guineensis activated carbon: adsorption and RSM studies. Environ Sci Pollut Res. https://doi.org/10.1007/s11356-020-09053-z

Milonjić SK (2007) A consideration of the correct calculation of thermodynamic parameters of adsorption. J Serbian Chem Soc 72:1363–1367. https://doi.org/10.2298/JSC0712363M

Mohan SV, Karthikeyan J (1997) Removal of lignin and tannin colour from aqueous solution by adsorption onto activated charcoal. Environ Pollut 97:183–187

Nascimento RF, Lima ACA, Vidal CB, Melo DQ, Raulino GSC (2014) Adsorção: aspectos teóricos e aplicações ambientais. Imprensa Universitária, Fortaleza

Naseem K, Begum R, Wu W, Usman M, Irfan A, Al-Sehemi AG, Farooqi ZH (2019) Adsorptive removal of heavy metal ions using polystyrene-poly(N-isopropylmethacrylamide-acrylic acid) core/shell gel particles: adsorption isotherms and kinetic study. J Mol Liq 277:522–531. https://doi.org/10.1016/j.molliq.2018.12.054

Noh JS, Schwarz JA (1990) Effect of HNO_3 treatment on the surface acidity of activated carbons. Carbon 28(5):675–682

Oladejo J, Shi K, Chen Y, Luo X, Gang Y, Wu T (2020) Closing the active carbon cycle: regeneration of spent activated carbon from a wastewater treatment facility for resource optimization. Chemical Engineering and Processing - Process Intensification 150:107878

Piccin JS Jr, Sant'Anna Cadaval TR, Almedia de Pinto LA, Dotto GL (2017) Adsorption isotherms in liquid phase: experimental, modeling, and interpretations. In: Bonilla-Petriciolet A, Mendoza-Castillo DI, Reynel-Avila HE (eds) Adsorption processes for water treatment. Springer International Publishing, Cham, pp 19–51

Plazinski W, Rudzinski W, Plazinska A (2009) Theoretical models of sorption kinetics including a surface reaction mechanism: a review. Adv Colloid Interf Sci 152:2–13. https://doi.org/10.1016/j.cis.2009.07.009

Ruthven DM (1984) Principles of adsorption and adsorption processes, 7th edn. John Wiley and Sons, New York

Salari M, Dehghani MH, Azari A, Motevalli MD, Shabanloo A, Ali I (2019) High performance removal of phenol from aqueous solution by magnetic chitosan based on response surface methodology and genetic algorithm. J Mol Liq 285:146–157. https://doi.org/10.1016/j.molliq.2019.04.065

Sanghi R, Bhattacharya B (2002) Review on decolorisation of aqueous dye solutions by low cost adsorbents. Color Technol 118:256–269. https://doi.org/10.1111/j.1478-4408.2002.tb00109.x

Santos LC, da Silva AF, dos Santos Lins PV, da Silva Duarte JL, Ide AH, Meili L (2019) Mg-Fe layered double hydroxide with chloride intercalated: synthesis, characterization and application for efficient nitrate removal. Environ Sci Pollut Res. https://doi.org/10.1007/s11356-019-07364-4

Sellaoui L, Soetaredjo FE, Ismadji S, Bonilla-Petriciolet A, Belver C, Bedia J, Ben Lamine A, Erto A (2018) Insights on the statistical physics modeling of the adsorption of Cd^{2+} and Pb^{2+} ions on bentonite-chitosan composite in single and binary systems. Chem Eng J 354:569–576

Sen Gupta S, Bhattacharyya KG (2011) Kinetics of adsorption of metal ions on inorganic materials: a review. Adv Colloid Interf Sci 162:39–58. https://doi.org/10.1016/j.cis.2010.12.004

Sengupta A, Jayabun S, Boda A, Musharaf Ali S (2016) An amide functionalized task specific carbon nanotube for the sorption of tetra and hexa valent actinides: experimental and theoretical insight †. RSC Adv 6:39553–39562. https://doi.org/10.1039/c6ra07986e

Sepúlveda P, Rubio MA, Baltazar SE, Rojas-Nunez J, Sánchez Llamazares JL, Garcia AG, Arancibia-Miranda N (2018) As(V) removal capacity of FeCu bimetallic nanoparticles in aqueous solutions: the influence of Cu content and morphologic changes in bimetallic nanoparticles. J Colloid Interface Sci 524:177–187. https://doi.org/10.1016/j.jcis.2018.03.113

Toumi K-H, Benguerba Y, Erto A, Dotto GL, Khalfaoui M, Tiar C, Nacef S, Amrane A (2018) Molecular modeling of cationic dyes adsorption on agricultural Algerian olive cake waste. J Mol Liq 264:127–133

Uslu H, Datta D, Azizian S (2016) Separation of chromium (VI) from its liquid solution using new montmorillonite supported with amine based solvent. J Mol Liq 215:449–453. https://doi.org/10.1016/j.molliq.2016.01.023

Varmazyar A, Sedaghat S, Khalaj M (2017) Highly efficient removal of methylene blue by a synthesized TiO2/montmorillonite-albumin nanocomposite: kinetic and isothermal analysis in water. RSC Adv 7:37214–37219. https://doi.org/10.1039/c7ra07096a

Vocciante M, Trofa M, Rodríguez Estupiñán P, Giraldo L, D'Auria T, Moreno-Piraján JC, Erto A (2014) A rigorous procedure for the design of adsorption units for the removal of cadmium and nickel from process wastewaters. J Clean Prod 77:35–46

Wan Z, Chen W, Liu C, Liu Y, Dong C (2015) Preparation and characterization of ??-AlOOH @CS magnetic nanoparticle as a novel adsorbent for removing fluoride from drinking water. J Colloid Interface Sci 443:115–124. https://doi.org/10.1016/j.jcis.2014.12.012

Wang L, Zhang J, Zhao R, Li Y, Li C, Zhang C (2010) Adsorption of Pb(II) on activated carbon prepared from Polygonum orientale Linn.: kinetics, isotherms, pH, and ionic strength studies. Bioresour Technol 101:5808–5814. https://doi.org/10.1016/j.biortech.2010.02.099

Westwater JW, Drickamer HG (1957) The mathematics of diffusion. J Am Chem Soc 79:1267–1268. https://doi.org/10.1021/ja01562a072

Yousef RI, El-Eswed B, Al-Muhtaseb AH (2011) Adsorption characteristics of natural zeolites as solid adsorbents for phenol removal from aqueous solutions: kinetics, mechanism, and thermodynamics studies. Chem Eng J 171:1143–1149. https://doi.org/10.1016/j.cej.2011.05.012

Zhang M, Gao B, Varnoosfaderani S, Hebard A, Yao Y, Inyang M (2013) Preparation and characterization of a novel magnetic biochar for arsenic removal. Bioresour Technol 130:457–462. https://doi.org/10.1016/j.biortech.2012.11.132

Zhou C, Zhu H, Wang Q, Wang J, Cheng J, Guo Y, Zhou X, Bai R (2017) Adsorption of mercury (ii) with an Fe3O4 magnetic polypyrrole-graphene oxide nanocomposite. RSC Adv 7:18466–18479. https://doi.org/10.1039/c7ra01147d

Chapter 2
Methods of Synthesis of Magnetic Adsorbents

Thiago Lopes da Silva, Talles Barcelos da Costa, Henrique Santana de Carvalho Neves, Meuris Gurgel Carlos da Silva, and Melissa Gurgel Adeodato Vieira

Contents

Abstract The development of magnetic adsorbents is a relevant topic in research studies aimed at improving water and wastewater treatments due to their potentiality in spreading a number of contaminants. These adsorbents are synthesized to acquire magnetic properties, which enable them to be easily recovered by magnetic field after adsorption. The main metallic oxides are those formed from iron (Fe), cobalt (Co), nickel (Ni), and copper (Cu). Magnetic adsorbents are generally micro- and nanosized, which provides a large surface area. The effectiveness of the adsorptive process can be improved by changes in the surface of the solid phase, such as the binding of molecules and functional groups to the adsorbent. The use of

T. L. da Silva · T. B. da Costa · H. S. de C. Neves · M. G. C. da Silva · M. G. A. Vieira (✉)
School of Chemical Engineering, Department of Processes and Products Design, University of Campinas, Campinas, Sao Paulo, Brazil
e-mail: meuris@unicamp.br; melissagav@feq.unicamp.br

L. Meili, G. L. Dotto (eds.), *Advanced Magnetic Adsorbents for Water Treatment*, Environmental Chemistry for a Sustainable World 61,
https://doi.org/10.1007/978-3-030-64092-7_2

nonmagnetic adsorbents in nano- and microsize range implies high separation costs and operating difficulties in batch and continuous flow adsorption system, and the synthesis and use of magnetic adsorbents functionalized with chemical groups minimizes these problems. In this chapter, we present the main aspects about magnetic properties and the coprecipitation method to synthesize magnetic adsorbents based on metallic oxides, carbon matrix, graphene oxide, zeolites, polymer beads, and bioadsorbents, and theirs applications.

Keywords Bioadsorbents · Carbon matrix · Graphene oxide · Magnetic adsorbents · Magnetic oxides · Magnetism · Polymer beads · Silica · Synthesis methods · Zeolites

2.1 Introduction

The application of magnetic adsorbents to minimize environmental problems has received significant attention in recent years. Magnetic adsorbents have been applied to adsorb emerging pollutants and toxic metals from water and wastewater. These contaminants are present in the effluents of various processes such as petrochemical, paper, wood, metallurgy, and agribusiness (olive oil, tomato, wine, and coffee) (Barreto et al. 2012; Murthy and Naidu 2012). Studies have been developed to explore the use of magnetism as a means of separating adsorbent materials from aqueous medium.

Many studies focus on the use of magnetic particles for contaminants uptake, such as magnetic chitosan-grafted graphene oxide composite for ciprofloxacin (Wang et al. 2016a, b), magnetite nanoparticles coated with shells of a quaternary chitosan siliceous for diclofenac (Soares et al. 2019), carbon nanotube-based magnetic for diquat dibromide herbicide (Duman et al. 2019), Fe_3O_4-coated polymer clay composite for atenolol and gemfibrozil (Arya and Philip 2016), magnetic nanoparticles altered with Fe–Mn binary oxide for As^{3+} (Shan and Tong 2013), magnetic chitosan nanoparticles for Cr^{6+} (Thinh et al. 2013), shellac-coated iron oxide nanoparticles for Cd^{2+} (Gong et al. 2012), magnetite nanoparticles dithiocarbamate for Hg^{2+} (Figueira et al. 2011), goethite and hematite nanoparticles for Cu^{2+} (Chen and Li 2010), and magnetic ferrites for Cd^{2+} uptake from wastewater (Liu et al. 2018), among others.

Activated carbon is a classic adsorbent that can be used to capture numerous species present in liquid and gaseous media due to its high adsorption capacity. However, the reuse of this material in adsorption cycles is limited by the difficulty of regeneration of the adsorbate. The application of pulverized coals, as well as very small size adsorbents, entails high costs for separating the adsorbent from the fluid phase after the adsorptive process (Reddy and Lee 2013; Gupta and Ali 2006; Dallago et al. 2005). Other adsorbents with smaller size and thereby larger surface area also present the challenge of costly and complicated recovery processes. The removal of very small adsorbents from suspension in batch systems is difficult or

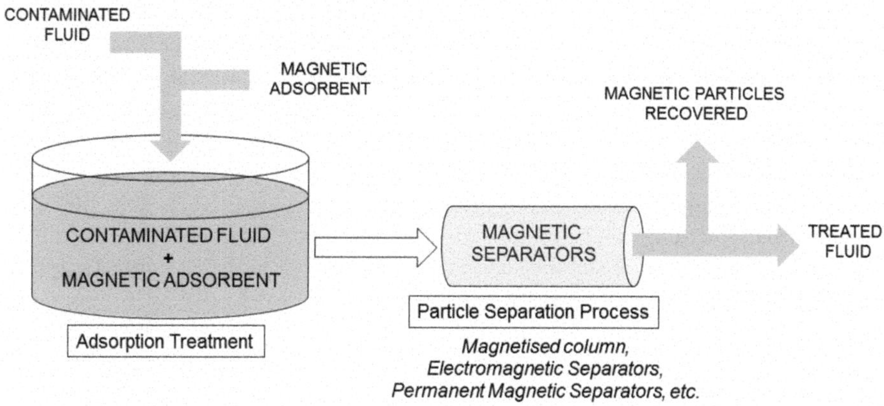

Fig. 2.1 Simples scheme of industrial adsorption process using magnetic adsorbents

impractical in many cases, and small sizes make the use of these adsorbent materials prohibitive in continuous systems, such as fixed beds. The application of magnetic adsorbents minimizes this problem because it provides a practical way for adsorbent separation by a magnetic field. Moreover, since magnetic adsorbent are micro- and nanosized, they present a large surface area, which implies rapid adsorption kinetics (Dallago et al. 2005; Wu et al. 2005).

Clay minerals (Hashem et al. 2015; Da Rosa et al. 2015), zeolites (Ostroski et al. 2009), biomass (Bilal et al. 2013; Simate and Ndlovu 2015), and industrial wastes (Nassar 2010) have been tested as alternative adsorbents to activated carbons. Although these materials generally have relatively lower costs, they may have small surface area, limited reuse, and low adsorption capacity when contaminants are at low concentrations (Nassar 2010). Therefore, new adsorbents with high removal efficiency and low cost should be studied and explored.

Currently, nanotechnology has been widely applied in several scientific and technological areas, such as biomedicine, biotechnology, magnetic resonance imaging, catalysis, and wastewater treatment (Kaur et al. 2014). Nanoparticles are materials with small average sizes between 1 and 100 nm, which influence their physical/chemical properties, such as immense specific surface area and reactivity (Xu et al. 2012). In particular, iron oxide nanomaterials, beside others characteristics, have low toxicity, biocompatibility, low cost, and high adsorption capacity (Hua et al. 2012). The iron oxides most commonly used in adsorption process are magnetite (Fe_3O_4) (Nassar 2010; Singh et al. 2011), maghemite (γ-Fe_2O_3) (Tuutijärvi et al. 2009; Jiang et al. 2013), and hematite (α-Fe_2O_3) (Chen and Li 2010; Shan et al. 2014). The industrial treatment using magnetic adsorbents is based on two steps: first, the actual adsorption process, and second, the recovery of the particles by magnetic attraction. Figure 2.1 shows a simple scheme of this process.

Although information around cost-effectiveness of the magnetic separation, high gradient magnetic separation (HGMS), and other separation processes are rare in the literature, the little information available indicates that HGMS presents advantages.

Comparing HGMS and sand filter (ratio of HGMS:Sand Filter) for steel mill wastewater, the relative installation cost is 1.1:1 (almost the same), the relative operation cost is 0.7:1 (less costs), the relative space requirement is 1:1.8 (smallest area), and the relative feed time is 1:24 (less consuming time). These metrics show the benefits of magnetic separation (Ambashta and Sillanpää 2010) and motivate further research studies and developments of magnetic adsorbent particles.

The use of magnetic adsorbents is promising and their synthesis is versatile. The surface of nanoparticles can be modified by the addition of functional groups, which can improve the colloidal stability and adsorption capacity of the solid (Zargoosh et al. 2013; Soares et al. 2014; Kirillov et al. 2014).

Magnetic adsorbents combine favorable chemical and magnetic properties in a single material, which allows easy separation from the fluid phase using an external magnetic field (Ambashta and Sillanpää 2010), making the process simpler when compared to conventional filtration, membrane separation, or centrifugation (Yamaura and Fungaro 2013).

In this chapter, we summarize the main aspects about magnetic adsorbents, their magnetic and chemical properties, as well as some of the synthesis methods. The following aspects are prioritized: the characteristics of the magnetic species, which provides magnetic features to the adsorbents; the method of coprecipitation synthesis that runs through almost all synthesis methods, from metallic adsorbents to those supported in organic and inorganic matrices; and the use of organic agents and polymers for surface functionalization.

2.2 Magnetism and Magnetic Adsorbents

Magnetic adsorbents belong to a class of adsorbents that have magnetic properties and are influenced by the magnetic field. The magnetic properties of the adsorbents are conferred by magnetic species embedded on its base. Usually, the magnetic species are oxides of iron, cobalt, nickel, and copper. On the influence of a magnetic field, the magnetic adsorbents are easily and efficiently attracted and separated from the fluid (liquid or gas) due to the presence of metal species in the adsorbent (Mehta et al. 2015).

2.2.1 Magnetism

Substances can generally be classified, according susceptibility to magnetic fields, as diamagnetic, paramagnetic, and ferromagnetic. When a substance is placed in a magnetic field of intensity H (oersteds, or Maxwell/cm^2), the intensity of the magnetic field within the substance can be higher or lower than H (Lee 1996).

H can be multiplied by the value of the section area (cm^2) where the current *i* is established to create the magnetic field to obtain $HA = \Phi$ as the magnetic flux. In an empty magnetic field, the Φ generated by the current (Φcurrent) is equal to the Φ observed (Φobserved). However, when a substance is placed in this field, the value of Φobserved differs from Φcurrent. The kind of magnetism that a substance exhibits is based on the relative magnitude of Φobserved and Φcurrent (Cullity and Graham 2010):

- Φobserved < Φcurrent: diamagnetic (i.e., copper, helium);
- Φobserved > Φcurrent: paramagnetic (i.e., sodium aluminum) or antiferromagnetic (i.e., MnO and FeO);
- Φobserved >> Φcurrent: ferromagnetic (i.e., iron, cobalt, nickel) or ferrimagnetic (i.e., Fe_3O_4).

The magnetism moments of substances are related to the electron motion, changes in forces because of external field and the spins of the electrons. Magnetic fields (or magnets) repel diamagnetic molecules because they have all electron spins paired and spin in opposite directions cancels the magnetic fields of each other. Therefore, no net magnetic field exists (Burrows et al. 2017). In this case, the field in the diamagnetic substance is less than *H*, so it tends to repel the lines of force (Lee 1996). Paramagnetism occurs because of unpaired electron spins in the atoms. The molecules have one or more unpaired electrons, thus the field in the paramagnetic substances is greater than H, and so they can be attracted by magnetic fields (Lee 1996). In paramagnetism, the electrons behave as tiny magnets, having the ability to align with the magnetic field in which the material is inserted (Burrows et al. 2017). Depending on the substance, temperature may have a strong influence on paramagnetism due to the orbital angular momentum that contributes positively or negatively to the magnetic moment (Lee 1996). Although magnetic fields influence both paramagnetic and diamagnetic materials, the paramagnetic effect is much larger than the diamagnetic effect. The diamagnetic effect is only observed in the absence of other types of magnetism.

Ferromagnetism can be considered as a special case of paramagnetism. In ferromagnetic materials the electrons of the atoms are unpaired, but the effect is more intense because all the moments on individual atoms become aligned at the same direction. Because individual spins work cooperatively, the magnetic effect is greatly enhanced. Fe, Co, and Ni are known as ferromagnetic elements (Lee 1996; Russel 1994). Ferromagnetic properties have been observed in several transition metals and their compounds (Lee 1996).

Paramagnetic and antiferromagnetic, as well as ferromagnetic and ferromagnetic substances, can be distinguished from each other only if the magnetic measurements extend over a range of temperature (Cullity and Graham 2010).

Another term that often appears in studies on magnetic adsorbents is superparamagnetism. The superparamagnetism occurs in small ferrimagnetic or ferromagnetic nanoparticles (Benz 2012). This is related to the nanosize that implies different magnetization behaviors (Ambashta and Sillanpää 2010). Similar to paramagnetism, the magnetization is observed solely in the presence of external magnetic field (Motornov et al. 2009), and because of the single domain of the superparamagnetic particles, all their magnetic moments are aligned in the same direction under the influence of magnetic fields (Benz 2012; Dinali et al. 2017).

Magnetic adsorbents mainly present superparamagnetic behaviors. The application of external magnetic fields instantaneously alters the magnetization of these particles, which allows precise and rapid drive and positioning of nanoparticles and manipulation by field gradients after adsorption processes. Superparamagnetism in nanoparticles is more related to their alignment and thermal fluctuation, rather than to individual dipoles in molecules or atoms (Motornov et al. 2009).

2.2.2 *Magnetic Adsorbents*

The use of magnetic adsorbents expands the possibilities and efficiency of adsorbents. The possibility of recovering the solid from the fluid phase using external magnets, considered as efficient separation method, allows the use of nanosized adsorbents. The advantages of using nanoparticles as adsorbent include ease and low production costs, lower amounts of adsorbent required, and large adsorptive capacities due to large surface areas (Chen and Li 2010).

Once magnetic adsorbents are mainly synthesized in nanometer size range, their shape and size have key role on the final properties of the materials. In addition, the colloidal stability of the magnetic adsorbent suspension is of vital importance for adsorption process (Safdarian and Ramezani 2018). The colloidal stability prevents the adsorbent aggregation and solvent dispersion. Colloidal solutions are severely affected by interactions between the fluid molecules and the dispersed solid (Motornov et al. 2009), which can be manipulated, when necessary, controlling the surface chemistry by surface modification of the adsorbent (Safdarian and Ramezani 2018).

The superparamagnetism phenomenon is a prerequisite for applying materials as magnetic adsorbents. Superparamagnetic nanoparticles are single-domain, usually designed from constituents that are ferromagnetic in bulk (Motornov et al. 2009). Usually, the magnetic particles have been synthesized with oxides of metals such as iron (Fe), cobalt (Co), nickel (Ni), and copper (Cu) (Mehta et al. 2015) and the most used are those with iron in composition.

There are many methods of synthesis of magnetic nanoparticles, including calcination, hydrothermal synthesis, microwave, microemulsion, ultrasound, sol-gel reactions, pyrolysis, and coprecipitation (Rane et al. 2018; Cui et al. 2013; Gnanaprakash et al. 2007). This chapter mainly focuses on the production of magnetic adsorbents by the coprecipitation method.

In the precipitation synthesis methods, also called bottom-up methods, nanoparticles are formed from a supersaturated homogeneous solution. There are also the top-down methods in which the bulk material is disintegrated to nanosize. Top-down methods include milling processes and lithographical approaches (Motornov et al. 2009).

2.2.3 Synthesis by Coprecipitation

The precipitation, or coprecipitation technique, is the most employed procedure to form magnetic adsorbents. In this approach, the solute concentration must be above the saturation concentration (C_0), that is, the concentration must be supersaturated (C_S); otherwise, there is no precipitation. At supersaturated concentration, nucleation and particle formation occur rapidly, and particle growth occurs until the solute concentration decreases to maximum solubility under those conditions, that is, the concentration decreases to C_0 (Motornov et al. 2009). From the initial nucleation to final particles, the growth of the particles is associated with diffusion mechanisms that allow, by molecular addition mechanism, the deposition of new molecules or combination among particles (Privman et al. 1999).

During precipitation or coprecipitation, a new phase (solid nanoparticles) is formed from the supersaturated homogeneous mother phase, where the nucleation process takes place for particle growth (Cui et al. 2013). The homogeneous nucleation process from supersaturated solution is initiated by physical changes, such as pressure and temperature shifts, or chemical changes that affect the pH, the solubility of dissolved material, or lead to chemical reactions. For example, the solute can be dissolved at a higher temperature (higher solubility) and, as the temperature decreases, there is a decrease in solubility and the nucleation and precipitation processes occur. In the heterogeneous nucleation, small seeds are added into supersaturated solution.

The basic theory about precipitation is applied to monodispersed solution, but in the case of magnetic adsorbents, the precipitation process occurs with more than one component in solution. In this case, the process is called coprecipitation. In the coprecipitation process, soluble substances are incorporated into the precipitates during their formation. It can occur by superficial adsorption and occlusion. In surface adsorption, the precipitate tends to drag substances present in the solution as a result of surface adsorption, while in coprecipitation by occlusion, substances may be retained in the structure or imperfections in the formed solid.

In the nanosize range, the physical/chemical properties are size-dependent. In the coprecipitation technique, the nucleated particle size mostly depends on parameters related to solubility and interfacial energy. Frequently, the physical properties are connected to the shape and size of particles due to the interplay between the energy of particles' interaction with the environment and the energy of thermal motion (Motornov et al. 2009). During the particle production, the control of parameter such as temperature, pH, and ionic strength (imposed by noncomplexing salts) is very important (Gnanaprakash et al. 2007).

The formation of nuclei and the growth of solid are the two main steps that are related to the nanoparticles development. Although, compared to other methods, the precipitation generates larger quantities of nanoparticles, it is difficult to control the size distribution and the incorporation of impurities (Dinali et al. 2017). Uniform nanosize distribution of particles is accomplished through a short nucleation time,

Table 2.1 General characteristics of the coprecipitation process for magnetic nanoparticles production (Rane et al. 2018)

Main characteristics	**Typical coprecipitation synthetic methods**
Generally, insoluble products are formed under high saturation concentration	Decomposition and electrochemical reduction of metal-organic precursors
Nucleation is related to the number of particles formed	Oxides synthesized from aqueous and non-aqueous solutions
Secondary processes, such as aggregation and Ostwald ripening, strongly affect morphology, size, and product properties	Molecular precursors react to produce metal chalcogenides
Usually, precipitation is a result of chemical reaction on supersaturation conditions	Coprecipitation with assisted sonication or microwave
Coprecipitation advantages	**Coprecipitation disadvantages**
Easy and rapid synthesis	Uncharged species of not precipitate
Control of composition and particle size dependent on synthesis conditions	Impurities may coprecipitate with the solid
	Time consuming
Possibilities to modify the overall homogeneity and particle morphology	Problems in experimental reproducibility
Mild temperature	Species with different precipitation rates produce nonuniform and inhomogeneous particles
Energy efficiency	
Organic solvents are not required	Broad size distribution

which produces all the particles at the end of the reaction and by coprecipitation procedure conditions (Cui et al. 2013).

Some characteristics of the coprecipitation process for the production of magnetic nanoparticles are presented in Table 2.1.

The use of iron oxides has advantages over other oxides, such as ease of synthesis, magnetism, low toxicity, biocompatibility, high adsorption capacity, and low cost (Xu et al. 2012; Hua et al. 2012). Several studies report the coprecipitation studies using iron oxides, for example, Fe_3O_4 (magnetite), Fe_2O_3, γ-Fe_2O_3 (maghemite, the ferromagnetic cubic form of Fe^{3+} oxide, or the oxidized form of magnetic Fe_2O_3), α-Fe_2O_3 (hematite), $Fe_3O_4.\alpha$-Fe_2O_3, α-FeO(OH) (goethite), and $CoFe_2O_4$, among others. Among those, magnetite and maghemite are preferable due to the required magnetic properties (Ramzannezhad and Bahari 2017; Kharissova et al. 2015; Kakavandi et al. 2015; Vayssieres et al. 1998).

Magnetite (Fe_3O_4) is an iron oxide formed by Fe^{2+} and Fe^{3+} ions, in a ratio of 1:2 and its molecular formula can be written as $FeO.Fe_2O_3$. In this inverted spinel cubic crystal structure, the Fe^{3+} cations occupy the tetrahedral sites and the octahedral sites are occupied by Fe^{2+} and Fe^{3+} cations in equivalent quantities (Teja and Koh 2009; Schwertmann and Cornell 2000).

Maghemite (γ-Fe_2O_3) has physical properties like those of magnetite; both are ferrimagnetic and have an inverted spinel-like cubic structure. However, the inverted spinel-like structure in maghemite is deficient in Fe^{3+}, as there are not enough cations to fill all octahedral coordination sites. Maghemite can be obtained by oxidizing magnetite as shown in Eq. 2.1 (Cornell and Schwertmann 2003):

$$4Fe_3O_{4(s)} + O_{2(g)} \rightarrow 6\gamma Fe_2O_{3(s)} \quad (2.1)$$

In the synthesis of superparamagnetic particles (e.g., Fe_3O_4 or γ-Fe_2O_3), the wet chemical routes are preferable due to their efficacy and simplicity, and because of the appreciable control over shape, size and composition of the nanoparticles (Motornov et al. 2009; Gupta and Gupta 2005). The most commonly used method to synthesize magnetic nanoparticles is the coprecipitation of Fe^{2+} and Fe^{3+} under different settings, or the decomposition of the organic iron (Cui et al. 2013). The coprecipitation of Fe^{2+} and Fe^{3+} by alkali addition, such as NaOH, NH_4OH or ammonia, is the most common procedure (Dinali et al. 2017). Eq. 2.2 shows the basic reaction of the coprecipitation of Fe^{2+}/Fe^{3+} salt solutions by alkali:

$$2Fe^{3+} + Fe^{2+} + 8OH^- \rightarrow FeO.Fe_2O_3 + 4H_2O \quad (2.2)$$

The characteristics of magnetic nanoparticles and properties of the material precipitated are intensely dependent on the synthesis conditions. The reaction performed in alkali solutions allows the control of oversized material by suitable selection of iron salts, pH, Fe^{2+}/Fe^{3+} stoichiometric ratio, and ionic strength of the solution (Motornov et al. 2009).

At pH 3, Fe^{3+} ions are precipitated as soluble ferrihydrite that can react with Fe^{2+} present in solution to form magnetite. The transfer of electrons from Fe^{2+} to Fe^{3+} is an important factor in the crystallization process and a low amount of Fe^{2+} ions (less than 10 mol%) promotes the formation of all metal species into spinel. The superficial charges in acidic and alkaline media occur due to high electron mobility and specific ions adsorption (Gnanaprakash et al. 2007).

The Fe^{2+}/Fe^{3+}concentration ratio and the nature of salts are factors that influence the shape and size of nanoparticles during the coprecipitation synthesis. For spinel iron oxides (magnetite nanoparticles), the synthesis of homogenous particles in size and chemical composition is attained at stoichiometry, or ratio concentration, of Fe^{2+}:Fe^{3+} equal to 1:2 (Fe^{2+}/Fe^{3+} equal 0.5) and with pH adjustments (Gnanaprakash et al. 2007; Vayssières et al. 1998). The impurities in magnetite may be related to partial oxidation of Fe^{2+} to Fe^{3+}, which leads to a Fe^{2+}/Fe^{3+} ratio less than 0.5. If the solution is incubated for a long period, at high pH, goethite can be fashioned, and the use of strong alkaline media (NaOH, KOH, and LiOH, among others) as hydrolyzing agent can cause the development of nonmagnetic iron species (Gnanaprakash et al. 2007).

The solution containing Fe^{2+} and Fe^{3+} can be prepared with iron salts in acid medium and then, with the increase of pH with alkali addition, the magnetic particles can be formed by coprecipitation. The apparatus employed in particle preparation must avoid the presence of oxygen due to the possibility of Fe^{2+} oxidation. To prevent this, usually the water used in the process is purged with nitrogen, and/or the procedure is done under nitrogen atmosphere.

Magnetic ferrites can be formed by solid-state reactions, or directly from aqueous media. Conventional formation of magnetic ferrite in aqueous solution by coprecipitation methods starts with dissolving iron molecules (salts like ferrous sulfate), mixing and heating the mixture up to 90 °C or autoclave temperatures. The particle products are crystallized at temperatures between 500 °C and 1100 °C (Liu et al. 2018). For heavy metal removal, magnetic ferrites can be prepared at temperature above 60 °C and the heavy metals are captured into the lattice points of the ferrite (Liu et al. 2018; Tamaura et al. 1991).

Magnetite nanoparticles (Fe_3O_4) can be prepared by coprecipitation, dissolving iron salts like $FeCl_3$ (Fe^{3+} source) and $FeCl_2$ (Fe^{2+} source) in acid solution (HCl, H_2SO_4, etc.), followed by adding an alkali solution (NaOH, NH_4OH, etc.) (Safdarian and Ramezani 2018; Gnanaprakash et al. 2007). Low temperature can be applied (~ 60 °C) and the stirring process must be vigorous, about 800 rpm (Safdarian and Ramezani 2018) or a sonicator can be used.

Iron oxides like goethite (α-FeO(OH)) and hematite (α-Fe_2O_3) have been applied to adsorb heavy metals. Nano hematites can be created dissolving $FeCl_3$ and HCl in preheated water (90 °C) and storing the solution at 100 °C for long periods of time (2 days) to obtain hematite nanoparticles. As for the synthesis of nano-goethite, $Fe_2(SO_4)_3$ can be added in NaOH solution and stored at 40 °C for 2 days. After coprecipitation, the solid phase (particles) can be separated and dried at 100 °C to develop the goethite nanoparticles (Chen and Li 2010).

In Shan et al. (2014), hematite-coated magnetic nanoparticle was made up through heterogeneous nucleation for antimony (III) removal. At room temperature, NaOH solution was dripped on iron solution ($FeCl_3$: Fe^{3+} source + $FeSO_4$: Fe^{2+} source) until pH reached 10; under vigorous agitation, it can produce hematite nanoparticle. Via a heterogeneous nucleation, the produced hematite nanoparticles can be coated adding polyethylene glycol at nanoparticles, redispersed in water, and then adding simultaneously the precursor solution of $FeCl_3$ and precipitant solution of NaOH (Shan et al. 2014).

The iron nanoparticles can be coated with organic compounds during the coprecipitation, causing changes and functionalization on the solid surface. Superparamagnetic core-shell nanoparticles can be synthesized by adding organic compounds to the alkali solution used in the coprecipitation. Safdarian and Ramezani (2018) synthesized 4 different magnetic nanoadsorbents using Fe^{2+}/Fe^{3+} solution, alkali solution, methacrylic acid (MAA), and 2,2′-azobisisobutyronitrile (AIBN). The Fe_3O_4 nanoparticles were produced by coprecipitation changing the pH of Fe^{2+}/Fe^{3+} solution from acid to basic, with NaOH alkali solution. For the production of "methacrylic acid coated magnetic nanoparticle" (Fe_3O_4@MAA), MAA was added in the alkali solution used to initiate the coprecipitation. For the production of "poly methacrylic acid coated magnetic nanoparticles" (Fe_3O_4@PMAA), before collecting Fe_3O_4@MAA, the AIBN was poured to the mixture and stirred for 1 h. Finally, for "amphiphilic shell of methacrylic acid polymer on magnetic nanoparticles core" (Fe_3O_4@AMPH) formation, the MAA and AIBN are added together in alkali solution and dropped into Fe^{2+}/Fe^{3+} solution. The coating process promotes changes in the surface of the nanoparticles, giving to Fe_3O_4@MAA abundant double bonds, to Fe_3O_4@PMAA numerous surface carboxylic groups, and to Fe_3O_4@AMPH carboxyl and alkyl groups. Polymer-coated

Fe_3O_4 nanoparticles with the presence of carboxylic groups can adsorb metal ions from solution through complex formation. The amphiphilic nanoparticles are easily dispersed in hydrophobic and hydrophilic media, which allows the application in separation of non-polar and polar solutes.

The difficulty in obtaining ultrafine magnetic particles is to fit the mean size around values within this range. Various experimental procedures are applied to limit the space for particle growth and thereby to control its size. Precipitation can be performed on micro emulsions, gels, vesicles, and polymer solution, among others. Complexing agents and surfactants can be used, which usually promote changes on the surface of the material, which can also modify the magnetic properties of nanoparticles (Vayssières et al. 1998). The synthesis of magnetic nanoparticles in large scale by conventional methods is a challenge. In large-scale production, the control of nucleation and growth of particles is complex because the processes often involve reaction solutions with nonhomogeneous compositions, secondary nucleation, polydispersivity of particles, and nonuniformity in size distribution. Such common problems can be minimized by alternative production methods, such as sol-gel (Cui et al. 2013).

The synthesis of multicomponent magnetic nanoparticles has been proposed, since they are expected to present new functions and improved performance in comparison to single-constituent materials. Core-shell nanoparticles and the incorporation of magnetic species into various adsorbent matrices, such as carbon matrix and zeolite polymers, among others, can enhance the magnetic properties of the as-prepared magnetic materials, allowing their separation by magnetic fields, as well as improving their role in adsorption process.

2.3 Synthesis Procedures of Magnetic Adsorbents Used for Contaminants Adsorption Process

In recent years, a wide range of studies have been reported involving the synthesis of magnetic adsorbents in different adsorbent matrices. The most well-known synthesis routes include calcination, hydrothermal synthesis, microwave, microemulsion, ultrasound, sol-gel reactions, pyrolysis, and mainly coprecipitation methods. The most commonly used magnetic adsorbents for adsorption of emerging contaminants and toxic metals from aqueous media are generally iron-based embedded in various supports. These magnetic adsorbents can be categorized into the following groups: magnetic adsorbents based on carbon matrix, silica, graphene oxide, zeolites, polymer beads, and bioadsorbents.

2.3.1 Adsorbents Based on the Carbon Matrix

Magnetic adsorbents with carbon matrices have been extensively studied owing to their high surface area, low cost, lack of internal diffusion resistance, and low toxicity (Kakavandi et al. 2015). Biochars, as well as other carbon matrices, are produced by pyrolysis in inert atmosphere, in total or partial absence of oxygen, condition considered fundamental for the formation of adsorbents with high specific surface area. Biomass from innumerous sources can be used for carbon production, such as wood, agroindustrial waste, wood sawdust (Wang et al. 2018; Yang et al. 2016; Shah et al. 2016), and vegetable peel (Gupta and Nayak 2012).

Most of the magnetic adsorbents based on carbon matrices have been synthesized by coprecipitation and pyrolysis methods with magnetic oxide precursors.

Carbon-based magnetic particles can be produced by coprecipitation of carbon in iron solution, by addition of alkali in Fe^{2+}/Fe^{3+} solution, and by coprecipitation with the presence of carbon in the magnetic nanoparticle suspension to precipitate together. Similarly, this methodology can be applied to synthesize activated charcoal with magnetic characteristics (Kakavandi et al. 2015). Coprecipitation should be performed in poor oxygen atmosphere (nitrogen atmosphere) to avoid iron oxidation, with vigorous stirring and in mild temperature (Shah et al. 2016). These processes promote the adhesion of iron oxides and magnetic particles on carbon matrix surface, thus producing magnetic adsorbents based on carbon. Table 2.2 presents studies in which carbon matrix-based magnetic particles were applied. The functional groups and reagents used in the preparation of each adsorbent are also included.

In general, the synthesis procedures of magnetic adsorbent based on carbons are simpler than those for adsorbents based on other matrices, and the magnetic nanoparticles are loaded only in carbon surface. The traditional method of coprecipitation has the disadvantage of reducing the porosity of magnetic biochar due to the formation of iron oxides and magnetic nanoparticles inside carbon pores (Oliveira et al. 2002). An alternative to minimize this problem is simultaneous activation and magnetization during pyrolysis.

The surface functionalization can be performed during the coprecipitation and pyrolysis methods. In pyrolysis, with the addition of ferric chloride ($FeCl_3$) to the biomass during heat treatment, it is possible to add functional groups to the sorbent surface due to the catalytic effect of $FeCl_3$ (Zhu et al. 2014). Furthermore, it is possible to reduce the time and cost of magnetic biochar synthesis because it is a one-step process (Yang et al. 2016).

The synthesis of carbon adsorbent can be done by pyrolyzing the biomass with iron charged on its surface. The immersion of biomass in $FeCl_3$ solution carries the material surface with iron. After drying of the biomass, it can be directly pyrolyzed at elevated temperatures (>400 °C) in an inert atmosphere for the production of carbonaceous magnetic adsorbent. The reactions involved in the pyrolysis process that lead to the creation of iron oxides on the material surface are represented by chemical Eqs. 2.3, 2.4, 2.5, 2.6, 2.7, and 2.8 (Yang et al. 2016; Liu et al. 2013):

Table 2.2 Synthesis data and surface functional groups of magnetic adsorbents based on carbon matrix

Magnetic adsorbents	Synthesis method	Materials	Surface functional group	References
Magnetic activated carbon (Fe_3O_4@C)	Fe_3O_4 Coprecipitation	$Fe(NO_3)_3{\cdot}9H_2O$, nitric acid, powder activated carbon	–	Kakavandi et al. (2015)
Sawdust modified with magnetic nanoparticles (SD-MNP)	Fe_3O_4 Coprecipitation	Mulberry wood sawdust (SD), H_2O, $FeCl_3{\cdot}7H_2O$, $FeSO_4{\cdot}7H_2O$, $NH_3{\cdot}H_2O$	–	Shah et al. (2016)
Magnetic biochar (MBC)	Fe_3O_4 Coprecipitation	Sawdust, $FeCl_3{\cdot}6H_2O$	-OH, -C=O, -C=C, -FeO	Yang et al. (2016)
Carbon Nanotubes (Fe_3O_4/CNT)	Fe_3O_4 Chemical deposition	$FeCl_3{\cdot}6H_2O$, $FeCl_2{\cdot}4H_2O$, carbon nanotubes (CNT), H_2O, H_2SO_4, NH_4OH	-OH, -C=O, -FeO	Elmi et al. (2017)
Magnetic carbon composite (MCC)	γ-Fe_2O_3 Hydrothermal	Pinewood sawdust, $FeSO_4{\cdot}7H_2O$, $FeCl_3{\cdot}6H_2O$, NaOH, H_2O	-OH, -NH, -C=C, -NO, -CO, -CH, -FeO	Wang et al. (2018)
Magnetic biochar composite (MBC)	γ-Fe_2O_3 Hydrothermal	Pinewood sawdust, $FeSO_4{\cdot}7H_2O$, $FeCl_3{\cdot}6H_2O$, NaOH, H_2O	-C=C, -NO, -CO, -CH	Wang et al. (2018)

$$FeCl_3 + 3H_2O \rightarrow Fe(OH)_3 + 3HCl \uparrow \tag{2.3}$$

$$Fe(OH)_3 \rightarrow FeO(OH) + H_2 \tag{2.4}$$

$$6FeO(OH) + 4H_2 \rightarrow 2Fe_2O_3 + 4H_2O \tag{2.5}$$

$$6FeO(OH) + 4CO \rightarrow 2Fe_2O_3 + 4CO_2 \uparrow \tag{2.6}$$

$$6FeO(OH) + 4C \rightarrow 2Fe_2O_3 + 2CO \uparrow \tag{2.7}$$

$$2FeCL_3 + H_2 \rightarrow 2FeCl_2 + 2HCl \tag{2.8}$$

Initially, $FeCl_3$ in solution is hydrolyzed and precipitated on the surface of biomass as goethite (FeO(OH)) during the drying process (Eqs. 2.3 and 2.4). Goethite is reduced to magnetic Fe_3O_4, under mesothermal conditions (i.e., 400–600 °C), by reaction with H_2, CO, and amorphous carbon-reducing compounds, which were created during the pyrolysis process (Eqs. 2.5, 2.6, and 2.7). Also, $FeCl_3$ residual (not turned in goethite during the drying process) can be reduced to $FeCl_2$ at 500 °C (Eq. 2.8).

Yang et al. (2016) synthesized magnetic carbon adsorbent for Hg^0 uptake from gaseous steam by one-step pyrolysis of sawdust loaded with iron on surface. Sawdust was immersed in $FeCl_3$ solution, and after the drying process, the biomass was pyrolyzed at 600 °C. The particles with magnetic Fe_3O_4 dispersed on its surface removed Hg^0 in Fe_3O_4 structure, due to the occurrence of Fe^{3+}, lattice oxygen, and oxygen-rich functional groups (C=O) that act as electrons acceptors.

Despite the benefits of inert atmospheres in charcoal production, in the synthesis of magnetic adsorbents, the pyrolysis process can be executed in the existence of oxygen, which may lead to surface functionalization of the adsorbent.

Wang et al. (2018) synthesized magnetic adsorbent for Hg^{2+} capture from pinewood sawdust by coprecipitation, hydrothermal process, and heat treatment in the absence and presence of oxygen. The raw sawdust was first coprecipitated in iron solution ($FeSO_4/FeCl_3$ solution) and then the solution was transferred to autoclave (200 °C, 24 h) for hydrothermal treatment. The particles produced were dried, calcined (with oxygen at atmosphere), and pyrolyzed (without oxygen at atmosphere), both at 400 °C, for 1 h, to custom 2 types of magnetic adsorbents: MCC (calcined) and MBC (pyrolized). The MCC surface presented crystalline γ-Fe_2O_3 that can be related to the satisfactory magnetization (15.58 emu/g), and functionalization with carboxyl, lactone and hydroxyl groups, that led to an adsorption capacity 5 times higher than adsorbent MBC (5.68 emu/g with predominant amorphous structure).

2.3.2 Adsorbents Based on Silica

Magnetic nanoparticles are subject to the negative effects of oxidation that can reduce their surface area and reusability. Silica coating is ideal to minimize the oxidative effects and to enhance the dispersion of magnetic nanoparticles. Other advantages are good thermal and mechanical stability, insolubility in organic and aqueous solvents, ecofriendly character, and easy surface functionalization (Zhang et al. 2013; Sanati et al. 2019). Although functionalization is usually beneficial, it should be well evaluated due to the possibility of changes that also can cause reduction in the magnetic effect of the particles (Adibmehr and Faghihian 2018). Particles based on systems of Ni/SiO_2, Fe/SiO_2, Fe_3O_4/SiO_2, $NiFe_2O_4/SiO_2$, and $CoFe_2O_4/SiO_2$ have been intensively studied in this field (Kharissova et al. 2015). The synthesis of magnetic adsorbent based on silica usually performs with more than one step, and more than one method is applied: coprecipitation, core-shell, sol-gel, functionalization by solvents, etc. Table 2.3 shows some synthesis methods, reagents, and functional groups formed on silica-based magnetic adsorbents.

The magnetic core coating can be synthesized through various silicon-based compounds. The hydrolysis of these compounds in the magnetic particle dispersion makes that an amorphous silica shell adheres to the magnetic core. Some of the main reagents applied are tetraethylorthosilicate (TEOS) (Adibmehr and Faghihian 2018; Zhang et al. 2013), (3-Chloropropyl), trimethoxysilane (Shahabuddin et al. 2018), and sodium silicate (Na_2SiO_3) (Zhang et al. 2013). Silicon can also be sourced from biological materials, for example, from rice husk by applying the acid leaching technique (Sanati et al. 2019).

Table 2.3 Synthesis data and surface functional groups of magnetic adsorbents based on silica

Magnetic adsorbents	Synthesis method	Materials	Surface functional group	References
$Fe_3O_4@SiO_2$–SH Core-Shell	Fe_3O_4 Coprecipitation	$FeCl_3{\cdot}6H_2O$, $FeCl_2{\cdot}4H_2O$, $Na_2SiO_3{\cdot}5H_2O$, 3-mercaptopropyltrimethoxysilane (3-MPTS), HCl, H_2O, NaOH, toluene	-FeO, -OH, -Si–O–Si, -SH, -CH	Zhang et al. (2013)
$Fe_3O_4@SiO_2$-SH Core-Shell	Fe_3O_4 Coprecipitation	$FeCl_3{\cdot}6H_2O$, $FeSO_4{\cdot}7H_2O$, $NH_3{\cdot}H_2O$, H_2O, poly-ethylene glycol (PEG), tetraethoxysilane (TEOS), MetOH, trimethoxysilylpropanethiol (TMMPS)	-FeO, -SiO, -SH	Wang et al. (2016a, b)
Magnetic mesoporous silica nanospheres (MSCMNPs) Core-Shell	Fe_3O_4 Chemical precipitation	$FeCl_3{\cdot}6H_2O$, $FeCl_2{\cdot}4H_2O$, H_2O, NH_4OH, cetyltrimethylammonium bromide (CTAB), tetraethylorthosilicate (TEOS), NH_4NO_3, EtOH	-FeO, -SiO-Si, -Si-O–H, -CH_2, -CS, -CH	Vojoudi et al. (2017)
Fe_3O_4@SBA-15-NH_2 Core-Shell	Fe_3O_4 Coprecipitation method	$FeCl_3{\cdot}6H_2O$, $FeCl_2{\cdot}4H_2O$, HCl, NaOH, EtOH, H_2O, NH_4F, tetraorthosilicate, toluene, 3-amino-propyltriethoxysilane (APTES)	-OH, -FeO^-, -NH, -CH, -C=O, -NN, -CO, -CN	Adibmehr and Faghihian (2018)
Mesoporous silica (SBA-15) nanocomposite	Fe_3O_4 Simple on-pot coprecipitation	(3-chloropropyl)trimethoxysilane, metformin, tolu-ene, HCl, $FeCl_3{\cdot}6H_2O$, $FeCl_2{\cdot}4H_2O$, NH_4OH	-NH, -CN/-C=N, -SiO, -FeO^-	Shahabuddin et al. (2018)
$Fe_3O_4@SiO_2$@APTMS Core-Shell	Mix with Fe_3O_4 from sol-gel method	$FeCl_3{\cdot}6H_2O$, $FeCl_2{\cdot}4H_2O$, NH_4OH, 3-aminopropyltrimethoxysilane (APTMS), *n*–hex-ane, rice husk silica extraction (RHS), HCl, H_2O	-FeO^-, -OH, -Si-O-Si	Sanati et al. (2019)

Other functional groups or polymer chains that have high affinity for adsorbate can easily alter hydroxyl groups presented on the surface of these types of adsorbent. These modifications, or functionalization, aim to increase or modify functional groups, improving the capacity and selectivity of the adsorbent. Functionalizing groups with organic chains that have electronegative elements (oxygen, nitrogen and sulfur) increase the affinity for various cations (Shahabuddin et al. 2018). Some nitrogen-functionalizing agents have been applied in the synthesis of magnetic adsorbents, such as 3-aminopropyltrimethoxysilane, that was used to selectively remove Cd^{2+} from aqueous environments (Sanati et al. 2019); 3-amino-propyltriethoxysilane that can bind to larger molecules such as complexed metals (Adibmehr and Faghihian 2018); melamine, aminosilane, and metformin that have high affinity for toxic metals (Alizadeh et al. 2012; Shahabuddin et al. 2018); and cetyltrimethylammonium bromide (CTAB) (Vojoudi et al. 2017). Sulfonated agents are also an alternative to the surface functionalization. Adsorbents with bis (3-triethoxysilylpropyl) tetrasulfide compound were used to remove heavy metals (Vojoudi et al. 2017), and 3-mercaptopropyltrimethoxysilane was employed to functionalize magnetic nanoparticles for Hg^{2+} removal (Zhang et al. 2013).

It is possible to develop selective magnetic silica-based adsorbents. These adsorbents can remove very specific components of complex fluid phases. For this purpose, adsorbents with ions and molecule-imprinted polymers can be synthesized with structures capable of recognizing and making bonds with ions and molecules, respectively (Ramakrishnan and Rao 2006; Wang et al. 2017). As an example, the synthesis of a Pb^{2+} selective ion-imprinted magnetic nanoadsorbent could be performed with silicate-coated magnetite (Fe_3O_4) nuclei functionalized with a complex of 1,5-diphenylcarbazide and Pb^{2+}. After lead elution, the ion-printed adsorbent was ready for application (Adibmehr and Faghihian 2018).

Functionalized magnetic adsorbents can also be manufactured from commercial mesoporous silicas. SBA-15 is an effective silica in removing heavy metals and has uniform hexagonal pore shapes with high pore volume. Alizadeh et al. (2012) synthesized magnetic based-silica adsorbent with silica SBA-15 with (3-chloroproprryl) trimethoxysilane and metformin to form a highly functionalized surface. The method consisted of treating commercial silica SBA-15 with (3-chloroproprryl) trimethoxysilane and metformin to form a highly functionalized surface. Finally, the functionalized particles were magnetized by the conventional coprecipitation procedure and separated by means of a magnetic field. Although the applied methodology did not completely protect the magnetic particles from oxidation because the nanocomposite is a heterogeneous material with differing properties in its different phases, the magnetic nanoparticle doping of SBA-15 improves the dispersion of material in the effluent, as Fe_3O_4 increases particle density and electrostatic attraction with water molecules generating a synergistic effect (Shahabuddin et al. 2018).

The application of the sol-gel synthesis method can help to develop ingenious nanoadsorbents with better located positions on the surface of the adsorbent. One of these methodologies is to coat magnetic nanoparticles with TEOS (tetraethylorthosilicate) and CTAB (cetyltrimethylammonium bromide) surfactant, and then replace the surfactant with the bis (3-triethoxysilylpropyl) tetrasulfide functionalizing agent. In the research of Vojoudi et al. (2017), the magnetic nanoparticles were coated with TEOS by acid hydrolysis and then the particles were added to a CTAB solution under constant stirring and TEOS was dripped into the solution at 80 °C. CTAB was removed by ethanolic ammonium nitrate solution forming mesoporous magnetic nanoparticles. The final adsorbent was composed of 0.5% Fe_3O_4 core, 13.9% amorphous silica, and 65.6% functionalized mesoporous silica.

2.3.3 Adsorbents Based on Graphene Oxide

Graphene oxide is an ultra-thin carbon-based material that has attracted attention from many researches due to its properties of very high specific surface area, hydrophilicity, good thermal and mechanical resistance, high electron conduction, and possibility of modification by functional groups (Zhao et al. 2012). The development of clusters that reduce the effective surface area can occur when graphene oxide is applied as an adsorbent (Zhang et al. 2014). To overcome these problems, the combination with magnetic composites such as magnetite (Fe_3O_4) and cobalt ferrite (Fe_2CoO_4) can reduce agglomeration effectively, increase affinity with metallic pollutants and dyes, and facilitate the separation of nanoadsorbent from solution (Bai et al. 2012; Jiang et al. 2015). Table 2.4 shows some synthesis methods, reagents, and functional groups formed on graphene oxide magnetic adsorbents.

Adsorbents based on graphene oxide are produced from commercial graphene oxide or they are synthesized from commercial graphite. The application of this high efficient adsorbent is attractive due to its possibility of large-scale production (Yang et al. 2009). The Hummers method of graphene oxide production is one of the best-known and simple method, where graphite is oxidized successive times until it forms carbon sheets bound to hydroxyl, carboxyl, epoxide, and carbonyl functional groups. This method consists of dispersing commercial graphite in an acid solution (H_2SO_4) at low temperature, and then adding potassium permanganate ($KMnO_4$) to initiate oxidation. Subsequently, the solution is diluted and hydrogen peroxide (H_2O_2, 30% v/v) is added to form more oxidized groups. The solid part (graphene oxide) obtained is separated and washed with HCl to remove oxidizing agents. Acids and oxidizing agents can be substituted for others that have the same effect (Hummers and Offeman 1958; Kovtyukhova et al. 1999; Kumar et al. 2014).

The occurrence of functional groups on the graphene oxide surface is an essential factor for a relevant affinity with pollutant cations. These groups may be substituted or linked to other organic groups with larger amount of electronegative species. These groups help to prevent oxidation of magnetic particles (Le et al. 2017). Some of the functionalizing agents studied to improve the adsorptive characteristics of these materials are “Reversible Addition–Fragmentation Chain Transfer” (RAFT) reaction agents, acrylic acid, acrylamide (Hosseinzadeh et al. 2019), sulfanilic acid

Table 2.4 Synthesis data and surface functional groups of magnetic adsorbents based on graphene oxide

Magnetic adsorbents	Synthesis method	Materials	Surface functional group	References
Sulfanilic acid-grafted magnetic graphene oxide sheets (MGOs/SA)	Fe_3O_4 Coprecipitation	Graphite oxide, $FeCl_3{\cdot}6H_2O$, $FeCl_2{\cdot}4H_2O$, $NH_3{\cdot}H_2O$, aryl diazonium, $NaNO_2$, NaOH, HCl	–	Hu et al. (2014)
GO−$MnFe_2O_4$ Nanohybrids	$MnFe_2O_4$ Mix Mn-sulfate and Fe-chloride	Graphite oxide, H_2O, $FeCl_3{\cdot}6H_2O$, $MnSO_4{\cdot}H_2O$, NaOH, acetone	-C=O, -C=C, -OH, -CO, -COOH, -FeO	Kumar et al. (2014)
Magnetic graphene oxide composite (MGO)	Fe_3O_4 Coprecipitation	Graphite oxide, H_2O, $FeCl_2{\cdot}4H_2O$, $FeCl_3{\cdot}6H_2O$, $NH_3{\cdot}H_2O$	-OH, -C=C, -O=C–O, -C–O–C, -FeO	Guo et al. (2016)
Magnetic polypyrrole-graphene oxide nanocomposite (PPy–GO)	Fe_3O_4 Chemical precipitation	Graphite oxide, H_2O, Py monomer, $FeCl_3{\cdot}6H_2O$, $FeSO_4{\cdot}7H_2O$, $NH_3{\cdot}H_2O$	-C=O, -C=C, -CH, -CO, -CN, -C–O–C, -NH, -FeO	Zhou et al. (2017)
Functionalized graphene oxide (GO-RAFT NPs)	Fe_3O_4 Precipitation	Graphite oxide, H_2O, $FeCl_3{\cdot}6H_2O$, $FeCl_2{\cdot}4H_2O$, NH_4OH, S,S′-bis(α,α′-dimethyl-α′′-acetic acid)-trithiocarbonate (BDATC), chloroform, acetone, carbon disulfide, tetrabutylammonium hydrogen sulfate, toluene, NaOH, HCl, N,N′-dicyclohexylcarbodiimide (DCC), 4-dimethylaminopyrdine (DMAP), acrylic acid, acrylamide, azoisobutyronitrile (AIBN), diethyl ether, β-cyclodextrin	-OH, -COOH, -C–O–C–, -CO, -CH, -C=O, -C=S, -CS, -NH, -CH_2, -FeO	Hosseinzadeh et al. (2019)

(Hu et al. 2014), ethylenediaminetetraacetic acid (EDTA) (Madadrang et al. 2012), and polymer polypyrrol (Zhou et al. 2017).

Guo et al. (2016) presented another simple method that consists of synthesis of magnetic adsorbent coprecipitating commercial graphene with magnetic iron particles. In detail, the magnetic particles were prepared coprecipitating iron from $FeCl_3$/$FeCl_2$ solution by addition of ammonium hydroxide. A suspension of graphene oxide was added to the mixture, heated to 80 °C to make the mixture viscous, and then autoclaved at 180 °C. The magnetic adsorbent produced presented a specific surface area of 1260 m^2/g, acidic groups with negative charges on the exterior surface, and desirable magnetic properties.

The functionalization of surface plays an important role in the adsorption efficiency. Hu et al. (2014) produced superparamagnetic adsorbent with 15.24 emu/g of magnetization by coprecipitation of graphene oxide (synthesized by the modified Hummers method applying other oxidizing agents) in solution of $FeCl_3$ and $FeCl_2$, by ammonia solution adding. Aryl diazonium salt, prepared with sulfonic acid and sodium nitrite ($NaNO_2$) mixture, was used for cold grafting (0 °C) and to functionalize magnetic graphene oxide. This adsorbent material was effective in removing inorganic (Cd^{2+}) and organic (aniline) pollutants.

One way to produce a multifunctionalized graphene oxide-based magnetic adsorbent is by using the "Reversible Addition-Fragmentation Chain Transfer" (RAFT) polymerization process to form long structures that can hold several functional groups. Graphene oxide was prepared by the Hummers method and the magnetic adsorbent synthesized by coprecipitation. The selected RAFT agent "S,S′-bis(α,α-′-dimethyl-α″-acetic acid)-trithiocarbonate" (BDATC) was added to the graphene magnetic oxide, binding to oxygen and magnetic composites. The acrylic acid and acrylamide monomers were added to the polymer chain by the initiator Azoisobutyronitrile (AIBN), forming an adsorbent with -N, -O, and -S functions. The magnetic adsorbent presented high selectivity for Hg^{2+} ions, a sufficient saturation magnetization (34 emu/g) for magnetic separation and high chemical and mechanical stability, allowing its reuse (Hosseinzadeh et al. 2019).

Manganese ferrite magnetic nanoparticle graphene oxides ($MnFe_2O_4$) have excellent properties for the adsorption of toxic metals Pb^{2+}, As^{3+}, As^{5+}, Cr^{6+} and Azo dye acid Red B (ARB) (Wu and Qu 2005). Manganese oxide has the potential to oxidize As^{3+} and As^{5+} to a less toxic state. $MnFe_2O_4$ may also be synthesized by coprecipitation. For the preparation of magnetic graphene oxide, $FeCl_3$ and $MnSO_4$ are added at a molar ratio of 1:2 (Mn:Fe) and precipitated with the addition of NaOH at pH 10.5. The magnetic adsorbent presented superparamagnetic behavior due $MnFe_2O_4$ nanoparticles, and high adsorption capacity to Pb^{2+}, As^{3+}, and As^{5+} (Kumar et al. 2014).

2.3.4 Adsorbents Based on Zeolites

Synthetic and natural zeolites are also important inorganic matrices for immobilization of magnetic micro and nanoparticles. These magnetic adsorbent materials can be obtained by different routes (Wang and Peng 2010; Attia et al. 2013). The main method is coprecipitation, in which zeolite is added to an iron ion solution (Fe^{2+}/Fe^{3+}), and the precipitation occurs through the addition of alkaline solution. The adsorbents based on zeolites may also be prepared by coprecipitating metallic nanoadsorbent previously prepared with zeolite in iron solution with the increase of pH. Table 2.5 presents a summary of studies on the synthesis data and surface functional groups of magnetic adsorbents based on zeolites.

Oliveira et al. (2004) and Pergher et al. (2005) evaluated different magnetic materials based on commercial zeolites (Mordenite, Beta, NaY and ZSM-5). These magnetic composites were prepared by zeolites suspension in $FeCl_3$ and $FeSO_4$

Table 2.5 Synthesis data and surface functional groups of magnetic adsorbents based on zeolites

Magnetic adsorbents	Synthesis method	Materials	Surface functional group	References
NaY Zeolite: iron oxide composites	Fe_3O_4 Precipitation	Commercial zeolite (NaY), $FeCl_3{\cdot}6H_2O$, $FeSO_4{\cdot}7H_2O$, NaOH	–	Oliveira et al. (2004)
Magnetic zeolites (Mordenita, Beta, NaY, and ZSM-5)	Fe_3O_4 Precipitation	Commercial zeolites (Mordenita, Beta, NaY, and ZSM-5), $FeCl_3{\cdot}6H_2O$, $FeSO_4{\cdot}7H_2O$, H_2O, NaOH,	–	Pergher et al. (2005)
FeO(OH)/Fe_3O_4 nanoparticles encapsulated in zeolite matrix	Fe_3O_4 Coprecipitation	Natural zeolite clinoptilolite, H_2O, NH_4OH, $FeCl_3{\cdot}9H_2O$, $Fe(NO_3)_2{\cdot}6H_2O$	–	Bosînceanu and Sulitanu (2008)
magMCM-41	Fe_3O_4 Coprecipitation method under sonication	$FeCl_3{\cdot}6H_2O$, $FeCl_2{\cdot}4H_2O$, H_2O, NH_4OH, cetyltrimethylammonium bromide (CTABr), tetraethyl orthosilicate (TEOS), EtOH, NH_4NO_3	-Si–O–Si, -SiOH, $-NH_2$, -C=O, -OH	Chen et al. (2009)
Zeolite from fly ash-iron oxide magnetic composite (ZFAM)	Fe_3O_4 Precipitation	Zeolite synthesized from fly ash (ZFA), NaOH, H_2O, $FeSO_4{\cdot}7H_2O$	–	Fungaro et al. (2012)
Magnetite nanoparticles and zeolite from coal fly ash	Fe_3O_4 Precipitation	Zeolite synthesized from coal fly ash, NaOH, H_2O, $FeSO_4{\cdot}7H_2O$	-T–O–T (T = Si, Al), -Si–OH, -FeO	Yamaura and Fungaro (2013)
Zeolite clinoptilolite (Fe_3O_4@Z)	Fe_3O_4 Coprecipitation method under sonication	Clinoptilolite zeolite ($AlCaH_6KNaO_3Si$), H_2O, $FeCl_3{\cdot}6H_2O$, $FeCl_2{\cdot}4H_2O$, NaOH	-OH, -CO, -CH, -C=O, -FeO	Jorfi et al. (2019)

solution. NaOH solution was added to precipitate iron oxide. The amount of zeolite used was measured to give a mass ratio of (zeolite): (iron oxide) of 3:1. The authors verified that 4 types of iron oxides were formed under the reaction conditions employed in this work: Fe_3O_4, α-Fe_2O_3, γ-Fe_2O_3, and α-FeO(OH). Among these oxides, only Fe_3O_4 and γ-Fe_2O_3 are magnetic with saturation magnetization of 100 and 60 J/T.kg, respectively. The characterization analyzes of magnetization, XRD, and Mössbauer spectroscopy showed that the major magnetic phase molded was γ-Fe_2O_3, possibly with small amounts of α-FeO(OH).

Bosînceanu and Sulitanu (2008) observed that FeO(OH)/Fe_3O_4 nanoparticles and nanoclusters, when incorporated by chemical coprecipitation method in the natural zeolite clinoptilolite (CLI), presented an intriguing aspect due to the cohabitation of paramagnetic and superparamagnetic Fe^{3+} in the clinoptilolite. The binding of Fe^{3+} to clinoptilolite channels resulted in the formation of Fe_3O_4 and FeO (OH) nanoclusters and the stabilization of tetrahedral AlO_4 in the aluminosilicate matrix. The authors concluded that this "non-conventional" structure and the bound Fe^{3+} presented paramagnetic and superparamagnetic properties, respectively.

Chen et al. (2009) produced a magnetic silicate and aluminosilicate solid material (magMCM-41) with magnetization (8.3 emu/g) and high surface area (800 m^2/g). The colloidal suspension of Fe_3O_4 nanoparticles was prepared by coprecipitation method under sonication in $FeCl_3$, $FeCl_2$, cetyltrimethylammonium bromide (CTABr), and NH_4OH at room temperature. The suspension of Fe_3O_4 nanoparticles (10 wt%) was incorporated in the MCM-41, and there were no observable effects on magnetic nanoparticle morphology and pore symmetry. However, small changes (<20%) were detected in textural properties and surface chemistry. After application of the produced magnetic nanoadsorbent, it was well dispersed in solution and could be quickly magnetically removed (1550 G, 1 cm/min).

Zeolites synthesized from industrial coal fly ash waste, by hydrothermal treatment with alkaline solutions, can be used to produce magnetite nanoparticles (Fungaro et al. 2012; Yamaura and Fungaro 2013). The magnetic nanocomposite was prepared by the precipitation method and partial oxidation of Fe^{2+}, through the addition of zeolite and Fe_3O_4 nanoparticles in solution. The application of magnetic nanoadsorbent showed promising results for the removal of uranium ions (U^{6+}) and reactive orange 16 dye by means of adsorptive process. Thus, the combination of magnetite with the zeolite obtained from coal ash had a great adsorption efficiency, proposed a noble finality to an abundant residue discarded by the coal industries, and was easily separated from aqueous solution by magnetization.

Jorfi et al. (2019) synthesized a new nanoadsorbent coated with natural clinoptilolite zeolite ($AlCaH_6KNaO_3Si$) combined with Fe_3O_4 nanoparticles (Fe_3O_4@Z). Clinoptilolite is the most abundant natural zeolite, which has three-dimensional structures formed by silica and tetrahedral alumina connected by their oxygen vertices. In this work, Fe_3O_4@Z (Eq. 2.9) was produced by clinoptilolite zeolite suspension mixture and $FeCl_3$ and $FeCl_2$ solution with a molar ratio of 2:1. The coprecipitation method under sonication was used in this study for synthesis of magnetic nanocomposite:

$$Z(g) + Fe^{2+} + 2Fe^{3+} + 8OH^- \rightarrow Fe_3O_4@Z + 4H_2O \quad (2.9)$$

The magnetic characterization of the nanoadsorbent showed no difference in the magnetic value of the nanocomposites and the value of the saturation magnetization (MS), which was 15.856 emu/g. The functional groups present on the solid surface were hydroxyl, alkyl, carbonyl, and -FeO groups.

2.3.5 Adsorbents Based on Polymer Beads

Polymers are generally the most common supports for magnetic particle immobilization for the production of micro or nano magnetic adsorbents. The use of polymers offers different magnetic adsorbent functional groups, depending on the polymer characteristics and procedure conditions.

Different studies have been conducted using polymeric matrices for magnetic adsorbents based on cellulose (Luo et al. 2016), chitosan (Fan et al. 2013; Kyzas and Deliyanni 2013; Azari et al. 2017), alginate (Alves et al. 2019), thiol-functionalized polymer (Jainae et al. 2015), polydopamine (Davodi et al. 2017), plant polyphenol (Jiang et al. 2018), polyacrylic acid (Liao and Chen 2002), and polyvinyl acetate polymer (Tseng et al. 2007). Table 2.6 summarizes the synthesis data and surface functional groups of magnetic adsorbents based on natural and synthetic polymeric beads.

The production of magnetic adsorbent can be performed from the combination of chemically modified magnetic nanoparticles, activated carbon, and α-cellulose (MCB). MCB beads were synthesized by extrusion technology using Fe_3O_4 nanoparticles with activated carbon (chemically modified with citric acid). This technology promoted to the magnetic beads a porous surface with a surface area of about 90.05 m^2/g (Luo et al. 2016).

Kyzas and Deliyanni (2013) and Azari et al. (2017) synthesized chitosan in magnetic matrices. This polymeric material is one of the most abundant, low-cost, nontoxic, hydrolytic, and biocompatible natural polymers. Kyzas and Deliyanni (2013) evaluated the synthesis of the commercially modified chitosan-based magnetic nanoadsorbent covalently crosslinked by glutaraldehyde and functionalized with Fe_3O_4 nanoparticles. Chitosan-supported Fe_3O_4 nanoparticles were molded by a coprecipitation method in pH around 10. Azari et al. (2017) synthesized chitin and chitosan magnetic nanoadsorbent beads extracted from shrimp shells using demineralization and deproteinization, respectively. Chitosan was also crosslinked with glutaraldehyde from the deacetylation of chitin extracted from shrimp shells. According to the characterization analyses, the modified chitosan-based magnetic particles presented rough structure and good porosity. From the analysis of the functional groups, the main bands observed on the surface of the magnetic particles were the amine (-NHCO) and hydroxyl (-OH) groups, besides -FeO.

Table 2.6 Synthesis data and surface functional groups of magnetic adsorbents based on natural and synthetic polymer beads

Magnetic adsorbents	Synthesis method	Materials	Surface functional group	References
PAA-bound Fe_3O_4 nanoparticles	Fe_3O_4 Coprecipitation	$FeCl_3·6H_2O$, $FeCl_2·4H_2O$, H_2O, $NH_3·H_2O$, H_2O, NH_4OH, EtOH, polyacrylic acid (PAA), phosphate, NaCl, carbodiimide	-OH, $-NH_2$	Liao and Chen (2002)
Magnetite-polyvinyl acetate polymer and iminodiacetic acid (M-PVAC-IDA)	Fe_3O_4 Chemical coprecipitation	$FeCl_3·6H_2O$, $FeCl_2·4H_2O$, H_2O, $NH_3·H_2O$, polyvinyl alcohol (PVAL), vinyl acetate (VAC), divinylbenzene, MetOH, epichlorohydrin, acetone, iminodiacetic acid (IDA)	-C=O, $-COO^-$	Tseng et al. (2007)
Modified Chitosan	Fe_3O_4 Coprecipitation	$FeCl_3·6H_2O$, $Co(NO_3)_2·6H_2O$, pure chitosan, glutaraldehyde	-OH, -NH, -CH, -C=O, -CN, -C – O – C, -CH–OH, -FeO	Kyzas and Deliyanni (2013)
Polystyrene coated $CoFe_2O_4$ particles (AEPE-PS-MPs)	$CoFe_2O_4$ Coprecipitation	$FeCl_3·6H_2O$, $Co(NO_3)_2·6H_2O$, NaOH, polystyrene, EtOH, 2-(3-(2-aminoethylthio)propylthio) ethanamine (AEPE)	–	Jainae et al. (2015)
Magnetic cellulose Based Beads (MCB)	Fe_3O_4 Modifying by citric acid	Fe_3O_4 nanoparticle powder, citric acid, acetone, H_2O, activated carbon, nitric acid, α-cellulose, NaOH, urea, sodium chloride	-OH, -CO, $-CH_3$, -CH	Luo et al. (2016)
Modified Chitosan	Fe_3O_4 Coprecipitation	Chitin and chitosan extracted from shrimp shells, NaOH, H_2O, acetone, EtOH, HCl, $FeCl_3·6H_2O$, $FeCl_2·4H_2O$, glutaraldehyde	-OH, -NH, -C=O, -CH, -CN, -C–O–C, -CH–OH, -FeO	Azari et al. (2017)
Polydopamine@F_eO_4 nanocomposite	Fe_3O_4 Chemical precipitation	$FeCl_3·6H_2O$, $FeCl_2·4H_2O$, $NH_3·H_2O$, H_2O, EtOH, 4-(2-aminoethyl) benzene-1,2-diol (dopamine, DA)	-OH, -NH, -C=C, -FeO	Davodi et al. (2017)
Plant polyphenol-coated Fe_3O_4 material (Fe_3O_4/PP)	Covered by Fe_3O_4 particles	Raw plant polyphenol powder, H_2O, naked Fe_3O_4 powder	-OH, -CO, -C=C, -FeO	Jiang et al. (2018)
Alginate activated carbon beads	Fe_3O_4 Reverse coprecipitation	$FeCl_3·6H_2O$, $FeSO_4·7H_2O$, NH_4OH, $CaCl_2$, sodium alginate, activated carbon	-OH, -CH, -C–O–C, -FeO	Alves et al. (2019)

Alginate is an unbranched linear polysaccharide extracted from brown algae and made up of two types of monomers: β-D-manuronic acid and α-L-guluronic acid. This polymer has promising and attractive characteristics as adsorbent due to the large number of hydroxyl and carboxylic functional groups, which provide a high affinity and binding capacity with metal ions (Alnaief et al. 2011; Cuadros et al. 2015). Hybrid nanostructured material as a magnetic adsorbent can be obtained by encapsulation of Fe_3O_4 nanoparticles synthesized by reverse coprecipitation method with activated carbon in alginate-based beads (MAAC). The use of alginate allows the crosslinking by ionic crosslinking gelation method due to the exchange of Na^+ ions present in the alginate structure by Ca^{2+} ions in solution. The MAAC beads present superparamagnetic behavior (saturation magnetization of 48 emu/g) with good distribution of Fe_3O_4 nanoparticles and activated carbon in its structure besides the strong bond between alginate and activated carbon (Alves et al. 2019).

Jainae et al. (2015) prepared magnetic adsorbent using $CoFe_2O_4$ particles coated with polystyrene and modified with "2-(3-(2-aminoethylthio)propylthio) ethanamine" (AEPE-PS-MPs) and noncoated particles. $CoFe_2O_4$ magnetic particles were synthesized by the coprecipitation method using $FeCl_3$ and $Co(NO_3)$ and addition of NaOH. These magnetic particles were coated with polystyrene by polymerization reaction and chemically modified with "2-(3-(2-aminoethylthio) propylthio)ethanamine." The AEPE-PS-MPs particles coated with polystyrene presented the same value saturation magnetization of $CoFe_2O_4$ particles. The coating did not alter the saturation magnetization, but improved the resistance to acid conditions, and increased uptake affinity of Hg^{2+} in the adsorptive processes.

Another polymer investigated as a matrix for synthesis of magnetic adsorbents is polydopamine. Davodi et al. (2017) evaluated the polidopamine coating method for surface functionalization of Fe_3O_4 nanoparticles (PDA@Fe_3O_4). Fe_3O_4 nanoparticles were synthesized by the coprecipitation from $FeCl_3$ and $FeCl_2$ salts at pH around 7. PDA@Fe_3O_4 nanocomposite was obtained by the addition of polydopamine to the Fe_3O_4 nanoparticles synthesized by one-step self-polymerization reaction in a sonicated ice water bath. Polydopamine was formed in surface of superparamagnetic Fe_3O_4 nanoparticles (13–15 nm) with magnetic saturation of 44.7 emu/g.

Plant polyphenols (PP), or tannin, are low-cost natural polymers obtained from all parts of higher plants, such as leaves, fruits, bark, wood, and roots (Hemingway 1989; Wang et al. 2013). Jiang et al. (2018) developed a magnetic adsorbent coating Fe_3O_4 particles with PP extracted from *Larix gmelinii* bark. Fe_3O_4/PP composite presented high roughness but lower magnetization saturation values, compared with Fe_3O_4 particles, 61.3 and 59.1 emu/g, respectively.

Magnetic nanoadsorbent can be developed using superparamagnetic Fe_3O_4 nanoparticles as cores and polyacrylic acid (PAA) as ionic exchange groups (Liao and Chen 2002). The Fe_3O_4 nanoparticles were synthesized by coprecipitation method in ammonia solution under hydrothermal conditions. The covalent binding of PAA with magnetic nanoparticles was performed by carbodiimide activation. The magnetic adsorbent prepared by the procedure presented ion exchange capacity higher than commercial ion exchange resins.

The adsorbent based on magnetic polymer attached with metal chelating ligands of iminodiacetic acid (IDA) was synthesized based on Fe_3O_4 nanoparticles by Tseng et al. 2007. The magnetic nanoparticles prepared by coprecipitation procedure were coated by polyvinyl acetate (PVAC) and vinyl acetate (PVAC), producing magnetic PVAC nanoparticles (M-PVAC). Three sequential procedures (alcoholysis, epoxide activation, and coupling of IDA) were employed to introduce functional groups on the surfaces of magnetic nanoparticles of M-PVAC, producing magnetite-polyvinyl alcohol (MPVAL), magnetite-polyvinylpropene oxide (M-PVEP), and magnetite-acetate. Polyvinyl-IDA (M-PVAC-IDA).

2.3.6 Bioadsorbents

Although the matrices of the previously mentioned magnetic adsorbent materials present satisfactory results in the applications in adsorptive processes, their use has some restrictions, such as the high cost of obtaining, the difficulty of regeneration and the significant reduction of adsorptive capacity after continuous cycles of regeneration. As alternative, bioadsorbents present a variety of low-cost and abundant source of matrices for magnetic adsorbent. These materials can provide various functional groups in its surface (Fomina and Gadd 2014).

Recent studies involving the synthesis of innovative magnetic bioadsorbents are presented in Table 2.7. *Ananas comosus* (pineapple) peel pulp (Venkateswarlu and Yoon 2015), anaerobic granule sludge/chitosan (Liu et al. 2017), fungal biomass *Actinomucor* sp. (Masoudi et al. 2018), activated *Cyclosorus interruptus* (Zhou et al. 2018), and CTS/*Serratia marcescens* (He et al. 2019) are some examples of promising magnetic bioadsorbent materials that have been synthesized for contaminants removal from wastewater.

Venkateswarlu and Yoon (2015) produced magnetic nanoparticles from *Ananas comosus* peel pulp extract by dissolving $FeCl_3.6H_2O$ and sodium acetate, used as reducing agent. The magnetic surface was further functionalized with "diethyl-4-(4-amino-5-mercapto-4H-1,2,4-triazol-3-yl)phenyl phosphonate" (DEAMTPP). The green magnetic nanoparticles synthesized present Fe_3O_4 cubic inverse spinel structure, 12 nm particle size, and functionalization with phosphonate groups. The surface area was 11.25 m^2/g, and magnetization of saturation was 16.9 emu/g, having ferromagnetic character.

Liu et al. (2017) proposed magnetic bioadsorbent of anaerobic granule sludge and chitosan composite (M-CS-AnGS). This composite was synthesized through crosslinking method with glutaraldehyde and anaerobic granular sludge (AnGS). Bioadsorbents based on AnGS have advantages as higher specific surface area, rich binding sites, and richer in functional groups, such as carbonyl, carboxyl, hydroxyl, and amino groups. The magnetic nanoparticles were synthesized by solvothermal method. The preparation of M-CS-AnGS composite was started by the addition of Fe_3O_4 into the chitosan solution forming magnetic chitosan (M-CS) for heavy metal removal (Pb^{2+} and Cu^{2+}).

Table 2.7 Synthesis data and surface functional groups of different magnetic bioadsorbents

Magnetic bioadsorbents	Synthesis method	Materials	Surface functional group	References
DEAMTPP@Fe_3O_4 MNPs	Fe_3O_4 Precipitation	*Ananas comosus* (pineapple) peel pulp, H_2O, $FeCl_3{\cdot}6H_2O$, sodium acetate, EtOH, diethyl-4-(4-amino-5-mercapto-4H-1,2,4-triazol-3-yl)phenyl phosphonate (DEAMTPP), NaOH	-OH, -CH, -SH, -P=O, -PO, -FeO	Venkateswarlu and Yoon (2015)
Anaerobic granule sludge/chitosan (M-CS-AnGS) composite	Fe_3O_4 Solvothermal	$FeCl_3{\cdot}6H_2O$, ethylene glycol, NaAc, ethylenediamine, H_2O, chitosan, acetic acid, glutaraldehyde, anaerobic granular sludge, NaOH	-NH, -OH, -CH_2, -C=O, -CN, -C – O – C, -C – O – H	Liu et al. (2017)
Fe_3O_4 – Fungal biomass *Actinomucor* sp.	Immobilization of Fe_3O_4 particles on fungal surface	*Actinomucor* sp., iron oxide magnetic nanoparticles (Fe_3O_4), H_2O	-OH, -NH, -CO, -FeO	Masoudi et al. (2018)
Fe_3O_4 – Activated *Cyclosorus interruptus* (AS-ACI)	Fe_3O_4 Coprecipitation	Raw *Cyclosorus interruptus*, NaOH, H_2O, $FeSO_4{\cdot}7H_2O$, $FeCl_3{\cdot}6H_2O$, $NH_3{\cdot}H_2O$, 3-aminopropyltriethoxysilane (AS), EtOH, epichlorohydrin (EC), ethylenediamine (EA)	–	Zhou et al. (2018)
IMB (CTS/*Serratia marcescens*)	Fe_3O_4 Hydrothermal	Chitosan (CTS), acetic acid, $Pb(NO_3)_2$, H_2O, EtOH, ether, magnetic nano-Fe_3O_4, epichlorohydrin, *Serratia marcescens*, tripolyphosphate, EDTA, NaOH	-OH, -CH, -C=O, -FeO	He et al. (2019)

Masoudi et al. (2018) proposed a novel magnetic bio-nanocomposite prepared with fungal biomass *Actinomucor* sp. and Fe_3O_4 for Cd^{2+} adsorption. Magnetic bioadsorbent was synthesized by immobilization of Fe_3O_4 nanoparticles on fungal biomass surface. Fe_3O_4 nanoparticles were added into wet biomass dispersed in water with a ratio of 1:10 (dry weight). The nanocomposite was dried for 72 h, at 60 °C, and Fe_3O_4 nanoparticles stabilized on the *Actinomucor* sp. biomass. The nanoparticles were attached on the fungal cell wall with a uniform distribution.

Zhou et al. (2018) developed a magnetic biomaterial based on NaOH-activated *Cyclosorus interruptus* (ACI), a fern plant from Asia, which has a multiporous structure and rugged morphology. Magnetic bioadsorbent was produced coating 3-aminopropyltriethoxysilane on ACI, and then Fe_3O_4 particles (synthesized by chemical coprecipitation technique at pH around 10) were incorporated on the rough surface of the ACI. The crosslinking process provides hydroxyl and amine groups on material surface improving the Pb^{2+} adsorption capacity.

He et al. (2019) proposed another approach to the synthesis of magnetic biomaterials *Serratia marcescens* bacterium as base adsorbent. $Pb(NO_3)_2$ was added in chitosan solution, and then Fe_3O_4 magnetic particles, epichlorohydrin and bacteria powder were mixed together. Pb^{2+} was eluted with EDTA and the particles were immersed in NaOH. The particles with Pb^{2+} imprinted in its matrix formed a chelated structure with nitrogen and oxygen from amino and hydroxyl groups. The particles present good separation under magnetic field and the functionalized surface allows the applications in adsorptive processes.

2.4 Conclusion

While in conventional adsorbents the large surface area is provided by the porous in the adsorbent, in magnetic adsorbents, on nanosize range, the large surface area is obtained by the very small size of the particles. The use of adsorbent particles with magnetic properties provides a simple and effective way for adsorbent separation from water after adsorption by a simple magnet. The superparamagnetic behavior is due to the metal oxide species present in solid phase. The magnetic adsorbent synthesis is performed under mild conditions and the coprecipitation method is, so far, the most widely used method for incorporating magnetic species into the adsorbents, which may be associated with other methods of magnetic adsorbents synthesis. Multicomponent magnetic nanoparticles can be synthesized in different organic and inorganic matrices, which can provide improved performance. The surface can be functionalized with chemical groups, which enables the use of adsorbent for specific purposes. Surface changes can be easily achieved by coating process with polymers, controlling the synthesis conditions, and with the use of different solvents and organic chemicals. Magnetic adsorbents have shown to be promising candidates for sustainable adsorption processes.

Acknowledgments The authors thank the Coordination for the Improvement of Higher Education Personnel (CAPES, n° 88887.200617/2018), National Council for Scientific and Technological Development (CNPq, n° 28/2018 - 406193/2018-5), and São Paulo Research Foundation (FAPESP, n° 2017/18236-1), for the financial support.

References

Adibmehr Z, Faghihian H (2018) Novel ion-imprinted adsorbent for lead removal from aqueous solutions, selectivity and adsorption capacity improvement, and evaluation of adsorption isotherms and kinetic. Sep Sci Technol 53:2388–2400. https://doi.org/10.1080/01496395.2018.1459703

Alizadeh A, Khodaei MM, Kordestani D, Fallah AH, Beygzadeh M (2012) The successful synthesis of biguanide-functionalized mesoporous silica catalysts: excellent reactivity combined with facile catalyst recyclability. Microporous Mesoporous Mater 159:9–16. https://doi.org/10.1016/j.micromeso.2012.03.019

Alnaief M, Alzaitoun MA, Garcia-González CA, Smirnova I (2011) Preparation of biodegradable nanoporous microspherical aerogel based on alginate. Carbohydr Polym 84:1011–1018. https://doi.org/10.1016/j.carbpol.2010.12.060

Alves LC, Yáñez-Vilar S, Piñeiro-Redondo Y, Rivas J (2019) Novel magnetic nanostructured beads for cadmium(II) removal. Nano 9:356. https://doi.org/10.3390/nano9030356

Ambashta RD, Sillanpää M (2010) Water purification using magnetic assistance: a review. J Hazard Mater 180:38–49. https://doi.org/10.1016/j.jhazmat.2010.04.105

Arya V, Philip L (2016) Adsorption of pharmaceuticals in water using Fe_3O_4 coated polymer clay composite. Microporous Mesoporous Mater 232:273–280. https://doi.org/10.1016/j.micromeso.2016.06.033

Attia TMS, Hu XL, Qiang YD (2013) Synthesized magnetic nanoparticles coated zeolite for the adsorption of pharmaceutical compounds from aqueous solution using batch and and column studies. Chemosphere 93:2076–2085. https://doi.org/10.1016/j.chemosphere.2013.07.046

Azari A, Gharibi H, Kakavandi B, Ghanizadeh G, Javid A, Mahvi AH, Sharafi K, Khosravia T (2017) Magnetic adsorption separation process: an alternative method of mercury extracting from aqueous solution using modified chitosan coated Fe_3O_4 nanocomposites. J Chem Technol Biotechnol 92:188–200. https://doi.org/10.1002/jctb.4990

Bai S, Shen XP, Zhong X, Liu Y, Zhu GX, Xu X, Chen KM (2012) One-pot solvothermal preparation of magnetic reduced graphene oxide-ferrite hybrids for organic dye removal. Carbon 50:2337–2346. https://doi.org/10.1016/j.carbon.2012.01.057

Barreto WJ, Ando RA, Estevão BM, Zanoni KPS (2012) Adsorption of caffeic acid on titanium dioxide: a spectroscopic study. Spectrochim Acta A Mol Biomol Spectrosc 2:16–20. https://doi.org/10.1016/j.saa.2012.02.046

Benz M (2012) Superparamagnetism: theory and application. Discussion of two papers on magnetic nanoparticles

Bilal M, Shah JA, Ashfaq T, Gardazi SMH, Tahir AA, Pervez A, Haroon H, Mahmood Q (2013) Waste biomass adsorbents for copper removal from industrial wastewater – a review. J Hazard Mater 263:322–333. https://doi.org/10.1016/j.jhazmat.2013.07.071

Bosînceanu R, Sulitanu N (2008) Synthesis and characterization of FeO(OH)/Fe_3O_4 nanoparticles encapsulated in zeolite matrix. J Optoelectron Adv Mater 10:3482–3486

Burrows A, Holman J, Parsons A, Pilling G, Price G (2017) Chemistry 3: introducing inorganic, organic and physical chemistry. 3rd edn. Oxford University Press, Oxford. ISBN: 9780198733805

Chen YH, Li FA (2010) Kinetic study on removal of copper (II) using goethite and hematite nanophotocatalysts. J Colloid Interface Sci 347:277–281. https://doi.org/10.1016/j.jcis.2010.03.050

Chen X, Lam KF, Zhang Q, Pan B, Arruebo M, Yeung KL (2009) Synthesis of highly selective magnetic mesoporous adsorbent. J Phys Chem C 113:9804–9813. https://doi.org/10.1021/jp9018052

Cornell RM, Schwertmann U (2003) The iron oxides: structure, properties, reactions, occurrences and uses, 2nd edn. Wiley-VCH, Weinheim. https://doi.org/10.1002/3527602097

Cuadros TR, Erices AA, Aguilera JM (2015) Porous matrix of calcium alginate/gelatin with enhanced properties as scaffold for cel culture. J Mech Behav Biomed Mater 46:331–342. https://doi.org/10.1016/j.jmbbm.2014.08.026

Cui H, Liu Y, Ren W (2013) Structure switch between α-Fe_2O_3, γ-Fe_2O_3 and Fe_3O_4 during the large scale and low temperature sol–gel synthesis of nearly monodispersed iron oxide nanoparticles. Adv Powder Technol 24:93–97. https://doi.org/10.1016/j.jwpe.2015.07.001

Cullity BD, Graham CD (2010) Introduction to magnetic materials, 2nd edn. IEEE Press, Wiley. ISBN 978-0-471-47741-9

Da Rosa CA, Ostroski IC, Meneguin JG, Gimenes ML, Barros MASD (2015) Study of Pb^{2+} adsorption in a packed bed column of bentonite using CFD. Appl Clay Sci 104:48–58. https://doi.org/10.1016/j.clay.2014.11.02

Dallago R, Smaniotto AE, Oliveira LCA (2005) Resíduos sólidos de curtumes como adsorventes para a remoção de corantes em meio aquoso. Quim Nova 28:433–437. https://doi.org/10.1590/s0100-40422005000300013

Davodi B, Ghorbani M, Jahangiri M (2017) Adsorption of mercury from aqueous solution on synthetic polydopamine nanocomposite based on magnetic nanoparticles using Box–Behnken design. J Taiwan Inst Chem Eng 80:363–378. https://doi.org/10.1016/j.jtice.2017.07.024

Dinali R, Ebrahiminezhad A, Manley-Harris M, Ghasemi Y, Berenjian A (2017) Iron oxide nanoparticles in modern microbiology and biotechnology. Crit Rev Microbiol. https://doi.org/10.1080/1040841X.2016.1267708

Duman O, Ozcan C, Polat TG, Tunç S (2019) Carbon nanotube-based magnetic and non-magnetic adsorbents for the high-efficiency removal of diquat dibromide herbicide fromwater: OMWCNT, OMWCNT-Fe_3O_4 and OMWCNT-k-carrageenan-Fe_3O_4 nanocomposites. Environ Pollut 244:723–732. https://doi.org/10.1016/j.envpol.2018.10.071

Elmi F, Hosseini T, Taleshi MS, Taleshi F (2017) Kinetic and thermodynamic investigation into the lead adsorption process from wastewater through magnetic nanocomposite Fe_3O_4/CNT. Nanotechnol Environ Eng 2:1–13. https://doi.org/10.1007/s41204-017-0023-x

Fan L, Luo C, Sun M, Li X, Qiu H (2013) Highly selective adsorption of lead ions by water-dispersible magnetic chitosan/graphene oxide composites. Colloids Surf B Biointerf 103:523–529. https://doi.org/10.1016/j.colsurfb.2012.11.006

Figueira P, Lopes C, Daniel-da-Silva A, Pereira E, Duarte A, Trindade T (2011) Removal of mercury (II) by dithiocarbamate surface functionalized magnetite particles: application to synthetic and natural spiked waters. Water Res 45:5773–5784. https://doi.org/10.1016/j.watres.2011.08.057

Fomina M, Gadd GM (2014) Biosorption: current perspectives on concept, definition and application. Bioresour Technol 160:3–14. https://doi.org/10.1016/j.biortech.2013.12.102

Fungaro DA, Yamaura M, Craesmeyer GR (2012) Uranium removal from aqueous solution by zeolite from fly ash-iron oxide magnetic nanocomposite. Int Rev Chem Eng 4:353–358

Gnanaprakash G, Mahadevan S, Jayakumar T, Kalyanasundaram P, Philip J, Raj B (2007) Effect of initial pH and temperature of iron salt solutions on formation of magnetite nanoparticles. Mater Chem Phys 103:168–175. https://doi.org/10.1016/j.matchemphys.2007.02.011

Gong J, Chen L, Zeng G, Long F, Deng J, Niu Q, He X (2012) Shellac-coated iron oxide nanoparticles for removal of cadmium(II) ions from aqueous solution. J Environ Sci 24:1165–1173. https://doi.org/10.1016/S1001-0742(11)60934-0

Guo Y, Deng J, Zhu J, Zhou X, Bai R (2016) Removal of mercury(II) and methylene blue from a wastewater environment with magnetic graphene oxide: adsorption kinetics, isotherms and mechanism. RSC Adv 6:82523–82536. https://doi.org/10.1039/c6ra14651a

Gupta VK, Ali I (2006) Adsorbents for water treatment: development of low-cost alternatives to carbon. In: Hubbard AT (ed) Encyclopedia of surface and colloid science. Taylor & Francis US, New York, pp 149–184

Gupta AK, Gupta M (2005) Synthesis and surface engineering of iron oxide nanoparticles for biomedical applications. Biomaterials 26:3995–4021. https://doi.org/10.1016/j.biomaterials.2004.10.012

Gupta VK, Nayak A (2012) Cadmium removal and recovery from aqueous solutions by novel adsorbents prepared from orange peel and Fe_2O_3 nanoparticles. Chem Eng J 180:81–90. https://doi.org/10.1016/j.cej.2011.11.006

Hashem FS, Amin MS, El-Gamal SMA (2015) Chemical activation of vermiculite to produce highly efficient material for Pb^{2+} and Cd^{2+} removal. Appl Clay Sci 115:189–200. https://doi.org/10.1016/j.clay.2015.07.042

He Y, Wu P, Xiao W, Li G, Yi J, He Y, Chen C, Ding P, Duan Y (2019) Efficient removal of Pb (II) from aqueous solution by a novel ion imprinted magnetic biosorbent: adsorption kinetics and mechanisms. PLoS One 14:1–17. https://doi.org/10.1371/journal.pone.0213377

Hemingway RW (1989) Structural variations in proanthocyanidins and their derivatives. In: Hemingway RW, Karchesy JJ, Branham SJ (eds) Chemistry and significance of condensed tannins. Springer US, Boston, pp 83–107

Hosseinzadeh H, Hosseinzadeh S, Pashaei S (2019) Fabrication of novel magnetic graphene oxide nanocomposites for selective adsorption of mercury from aqueous solutions. Environ Sci Pollut Res 26:26807–26821. https://doi.org/10.1007/s11356-019-05918-0

Hu X-j, Y-guo L, G-ming Z, Wang H, Hu X, A-wei C, Wang Y-q, Guo Y-m, T-ting L, Zhou L, S-heng L, X-xia Z (2014) Effect of aniline on cadmium adsorption by sulfanilic acid-grafted magnetic graphene oxide sheets. J Colloid Interface Sci 426:213–220. https://doi.org/10.1016/j.jcis.2014.04.016

Hua M, Zhang S, Pan B, Zhang W, Lv L, Zhang Q (2012) Heavy metal removal from water/wastewater by nanosized metal oxides: a review. J Hazard Mater 211–212:317–331. https://doi.org/10.1016/j.jhazmat.2011.10.016

Hummers WS, Offeman RE (1958) Preparation of graphitic oxide. J Am Chem Soc 80:1339. https://doi.org/10.1021/ja01539a017

Jainae K, Sukpirom N, Fuangswasdi S, Unob F (2015) Adsorption of Hg(II) from aqueous solutions by thiol-functionalized polymer-coated magnetic particles. J Ind Eng Chem 23:273–278. https://doi.org/10.1016/j.jiec.2014.08.028

Jiang W, Pelaez M, Dionysiou DD, Entezari MH, Tsoutsou D, O'shea K (2013) Chromium (VI) removal by maghemite nanoparticles. Chem Eng J 222:527–533. https://doi.org/10.1016/j.cej.2013.02.049

Jiang GD, Chang Q, Yang FF, Hu XY, Tang HQ (2015) Sono-assisted preparation of magnetic ferroferric oxide/graphene oxide nanoparticles and application on dye removal. Chin J Chem Eng 23:510–515. https://doi.org/10.1016/j.cjche.2014.06.037

Jiang X, Zhao Y, Wang X, Liu L, Wang Y, Zhang W, Jiao L, Liang W (2018) Adsorption of aqueous Cd(II) over a Fe_3O_4/plant polyphenol magnetic material. J Water Supply Res Technol AQUA 67:738–753. https://doi.org/10.2166/aqua.2018.051

Jorfi S, Shooshtarian MR, Pourfadakari S (2019) Decontamination of cadmium from aqueous solutions using zeolite decorated by Fe_3O_4 nanoparticles: adsorption modeling and thermodynamic studies. Int J Environ Sci Technol:1–14. https://doi.org/10.1007/s13762-019-02350-2

Kakavandi B, Kalantary RR, Jafari AJ, Nasseri S, Ameri A, Esrafili A, Azari A (2015) Pb (II) adsorption onto a magnetic composite of activated carbon and superparamagnetic Fe_3O_4 nanoparticles: experimental and modeling study. Clean (Weinh) 43:1157–1166. https://doi.org/10.1002/clen.201400568

Kaur R, Hasan A, Iqbal N, Alam S, Saini MK, Raza SK (2014) Synthesis and surface engineering of magnetic nanoparticles for environmental cleanup and pesticide residue analysis: a review. J Sep Sci 37:1805–1825. https://doi.org/10.1002/jssc.201400256

Kharissova OV, Dias HVR, Kharisov BI (2015) Magnetic adsorbents based on micro- and nanostructured materials. RSC Adv 5:6695–6719. https://doi.org/10.1039/c4ra11423j
Kirillov VL, Balaev DA, Semenov SV, Shaikhutdinov KA, Martyanov ON (2014) Size control in the formation of magnetite nanoparticles in the presence of citrate ions. Mater Chem Phys 145:75–81. https://doi.org/10.1016/j.matchemphys.2014.01.036
Kovtyukhova NI, Ollivier PJ, Martin BR, Mallouk TE, Chizhik SA, Buzaneva EV, Gorchinskiy AD (1999) Layer-by-layer assembly of ultrathin composite films from micron-sized graphite oxide sheets and Polycations. Chem Mater 11:771–778. https://doi.org/10.1021/cm981085u
Kumar S, Nair RR, Pillai PB, Gupta SN, Iyengar MAR, Sood AK (2014) Graphene oxide-$MnFe_2O_4$ magnetic nanohybrids for efficient removal of lead and arsenic from water. ACS Appl Mater Interfaces 6:17426–17436. https://doi.org/10.1021/am504826q
Kyzas GZ, Deliyanni EA (2013) Mercury(II) removal with modified magnetic chitosan adsorbents. Molecules 18:6193–6214. https://doi.org/10.3390/molecules18066193
Le MQC, Cao XT, Lee WK, Hong SS, Lim KT (2017) Fabrication and adsorption properties of novel magnetic graphene oxide composites for removal of methylene blue. Mol Cryst Liq Cryst 644:160–167. https://doi.org/10.1080/15421406.2016.1277467
Lee JD (1996) Concise inorganic chemistry, 5th edn. Chapmann and Hall Publication, London. ISBN: 978-0-632-05293-6
Liao MH, Chen DH (2002) Preparation and characterization of a novel magnetic nano-adsorbent. J Mater Chem 12:3654–3659. https://doi.org/10.1039/B207158D
Liu WJ, Tian K, Jiang H, Yu HQ (2013) Facile synthesis of highly efficient and recyclable magnetic solid acid from biomass waste. Sci Rep 3:2419. https://doi.org/10.1038/srep02419
Liu T, Han X, Wang Y, Yan L, Du B, Wei Q, Wei D (2017) Magnetic chitosan/anaerobic granular sludge composite: synthesis, characterization and application in heavy metal ions removal. J Colloid Interface Sci 508:405–414. https://doi.org/10.1016/j.jcis.2017.08.067
Liu F, Zhou K, Chen Q, Wang A, Chen W (2018) Comparative study on the synthesis of magnetic ferrite adsorbent for the removal of Cd(II) from wastewater. Adsorpt Sci Technol 36:1456–1469. https://doi.org/10.1177/0263617418779729
Luo X, Lei X, Cai N, Xie X, Xue Y, Yu F (2016) Removal of heavy metal ions from water by magnetic cellulose-based beads with embedded chemically modified magnetite nanoparticles and activated carbon. ACS Sustain Chem Eng 4:3960–3969. https://doi.org/10.1021/acssuschemeng.6b00790
Madadrang CJ, Kim HY, Gao GH, Wang N, Zhu J, Feng H, Gorring M, Kasner ML, Hou SF (2012) Adsorption behavior of EDTA-graphene oxide for Pb (II) removal. ACS Appl Mater Interfaces 4:1186–1193. https://doi.org/10.1021/am201645g
Masoudi R, Moghimi H, Azin E, Taheri RA (2018) Adsorption of cadmium from aqueous solutions by novel Fe_3O_4 – newly isolated *Actinomucor* sp. bio-nanoadsorbent: functional group study. Artif Cells Nanomed Biotechnol 46:1092–1101. https://doi.org/10.1080/21691401.2018.1533841
Mehta D, Mazumdar S, Singh SK (2015) Magnetic adsorbents for the treatment of water/wastewater – a review. J Water Process Eng 7:244–265. https://doi.org/10.1016/j.jwpe.2015.07.001
Motornov M, Roiter Y, Tokarev I, Minko S (2009) Colloidal systems on the nanometer length scale 5. https://doi.org/10.1201/9781420007206.ch5
Murthy PS, Naidu MM (2012) Conservation and recycling sustainable management of coffee industry by-products and value addition—a review. Resour Conserv Recycl 66:45–58. https://doi.org/10.1016/j.resconrec.2012.06.005
Nassar NN (2010) Rapid removal and recovery of Pb(II) from wastewater by magnetic nanoadsorbents. J Hazard Mater 184:538–546. https://doi.org/10.1016/j.jhazmat.2010.08.069
Oliveira LCA, Rios RVRA, Fabris JD, Garg V, Sapag K, Lago RM (2002) Activated carbon/iron oxide magnetic composites for the adsorption of contaminants in water. Carbon 40:2177–2183. https://doi.org/10.1016/S0008-6223(02)00076-3

Oliveira LCA, Petkowicz DI, Smaniotto A, Pergher SBC (2004) Magnetic zeolites: a new adsorbent for removal of metallic contaminants from water. Water Res 38:3699–3704. https://doi.org/10.1016/j.watres.2004.06.008

Ostroski IC, Barros MASD, Silva EA, Dantas JH, Arroy PA, LIMA OCMA (2009) A comparative study fot the ion exchange of Fe(III) and Zn(II) on zeolite NaY. J Hazard Mater 161:1404–1412. https://doi.org/10.1016/j.jhazmat.2008.04.111

Pergher SBC, Oliveira LCA, Smaniotto A, Petkowicz DI (2005) Magnetic zeolites for removal of metals in water. Quim Nova 28:751–755. https://doi.org/10.1590/S0100-40422005000500003

Privman V, Goia DV, Park J, Egon M (1999) Mechanism of formation of monodispersed colloids by aggregation of nanosize precursors. J Colloid Interface Sci 213:36–45. https://doi.org/10.1006/jcis.1999.6106

Ramakrishnan K, Rao TP (2006) Ion imprinted polymer solid phase extraction (IIP-SPE) for preconcentrative separation of Erbium(III) from adjacent lanthanides and yttrium. Sep Sci Technol 41:233–246. https://doi.org/10.1080/01496390500446327

Ramzannezhad A, Bahari A (2017) Characteristics of Fe_3O_4, α-Fe_2O_3, and γ-Fe_2O_3 nanoparticles as suitable candidates in the field of nanomedicine. J Supercond Nov Magn 30:2165–2174. https://doi.org/10.1007/s10948-017-4014-8

Rane AV, Kanny K, Abitha VK, Thomas S (2018) Methods for synthesis of nanoparticles and fabrication of nanocomposites. Synthesis of inorganic nanomaterials, advances and key technologies, micro and nano technologies, pp 121–139. https://doi.org/10.1016/B978-0-08-101975-7.00005-1

Reddy DHK, Lee S (2013) Application of magnetic chitosan composites for the removal of toxic metal and dyes from aqueous solutions. Adv Colloid Interf Sci 201–202:68–93. https://doi.org/10.1016/j.cis.2013.10.002

Russel, JB (1994) Química Geral 2, 2nd ed. Makron Books, São Paulo. ISBN: 8534601925

Safdarian M, Ramezani Z (2018) Sequential synthesis of a magnetic nano-adsorbent: how the first step identifies the final product. Colloids Surf A 541:97–107. https://doi.org/10.1016/j.colsurfa.2018.01.004

Sanati AM, Kamari S, Ghorbani F (2019) Application of response surface methodology for optimization of cadmium adsorption from aqueous solutions by Fe_3O_4@SiO_2@APTMS core–shell magnetic nanohybrid. Surf Interfaces 17:100374. https://doi.org/10.1016/j.surfin.2019.100374

Schwertmann U, Cornell RM (2000) Iron oxides in the laboratory: preparation and characterization, 2nd edn. Wiley-VCH, Weinheim, 204 f. ISBN: 978-3-527-29669-9

Shah J, Jan MR, Khan M, Amir S (2016) Removal and recovery of cadmium from aqueous solutions using magnetic nanoparticle-modified sawdust: kinetics and adsorption isotherm studies. Desalin Water Treat 57:9736–9744. https://doi.org/10.1080/19443994.2015.1030777

Shahabuddin S, Tashakori C, Kamboh MA, Korrani ZS, Saidur R, Nodeh HR, Bidhendi ME (2018) Kinetic and equilibrium adsorption of lead from water using magnetic metformin-substituted SBA-15. Environ Sci Water Res Technol 4:549–558. https://doi.org/10.1039/c7ew00552k

Shan C, Tong M (2013) Efficient removal of trace arsenite through oxidation and adsorption by magnetic nanoparticles modified with Fe–Mn binary oxide. Water Res 15:3411–3421. https://doi.org/10.1016/j.watres.2013.03.035

Shan C, Ma Z, Tong M (2014) Efficient removal of trace antimony(III) through adsorption by hematite modified magnetic nanoparticles. J Hazard Mater 268:229–236. https://doi.org/10.1016/j.jhazmat.2014.01.020

Simate GS, Ndlovu S (2015) The removal of heavy metals in a packed bed column using immobilized cassava peel waste biomass. J Ind Eng Chem 21:635–643. https://doi.org/10.1016/j.jiec.2014.03.031

Singh S, Barick KC, Bahadur D (2011) Surface engineered magnetic nanoparticles for removal of toxic metal ions and bacterial pathogens. J Hazard Mater 192:1539–1547. https://doi.org/10.1016/j.jhazmat.2011.06.074

Soares PIP, Alves AMR, Pereira LCJ, Coutinho JT, Ferreira IMM, Novo CMM, Borges JPMR (2014) Effects of surfactants on the magnetic properties of iron oxide colloids. J Colloid Interface Sci 419:46–51. https://doi.org/10.1016/j.jcis.2013.12.045

Soares SF, Fernandes T, Sacramento M, Trindade T, Daniel-da-Silva AL (2019) Magnetic quaternary chitosan hybrid nanoparticles for the efficient uptake of diclofenac from water. Carbohydr Polym 203:35–44. https://doi.org/10.1016/j.carbpol.2018.09.030

Tamaura Y, Katsura T, Rojarayanont S, Yoshida T, Abe H (1991) Ferrite process; heavy metal ions treatment system. Water Sci Tech 23:1893–1900. https://doi.org/10.2166/wst.1991.0645

Teja AS, Koh PY (2009) Synthesis, properties, and applications of magnetic iron oxide nanoparticles. Prog Cryst Growth Charact Mater 55:22–45. https://doi.org/10.1016/j.pcrysgrow.2008.08.003

Thinh NN, Hanh PT, Ha LT, Anh LN, Hoang TV, Hoang VD, Dang LH, Khoi NV, Lam TD (2013) Magnetic chitosan nanoparticles for removal of Cr(VI) from aqueous solution. Mater Sci Eng C 33:1214–1218. https://doi.org/10.1016/j.msec.2012.12.013

Tseng JY, Chang CY, Chen YH, Chang CF, Chiang PC (2007) Synthesis of micro-size magnetic polymer adsorbent and its application for the removal of Cu(II) ion. Colloids Surf A Physicochem Eng Asp 295:209–216. https://doi.org/10.1016/j.colsurfa.2006.09.001

Tuutijärvi T, Lu J, Sillanpää M, Chen G (2009) As(V) adsorption on maghemite nanoparticles. J Hazard Mater 166:1415–1420. https://doi.org/10.1016/j.jhazmat.2008.12.069

Vayssières L, Chanéac C, Tronc E, Pierre JJ (1998) Size tailoring of magnetite particles formed by aqueous precipitation: an example of thermodynamic stability of nanometric oxide particles. J Colloid Interface Sci 205:205–212. https://doi.org/10.1006/jcis.1998.5614

Venkateswarlu S, Yoon M (2015) Rapid removal of cadmium ions using green-synthesized Fe_3O_4 nanoparticles capped with diethyl-4-(4 amino-5-mercapto-4H-1,2,4-triazol-3-yl)phenyl phosphonate. RSC Adv 5:65444–65453. https://doi.org/10.1039/c5ra10628a

Vojoudi H, Badiei A, Bahar S, Ziarani GM, Faridbod F, Ganjali MR (2017) A new nano-sorbent for fast and efficient removal of heavy metals from aqueous solutions based on modification of magnetic mesoporous silica nanospheres. J Magn Magn Mater 441:193–203. https://doi.org/10.1016/j.jmmm.2017.05.065

Wang S, Peng Y (2010) Natural zeolites as effective adsorbents in water and wastewater treatment. Chem Eng J 156:11–24. https://doi.org/10.1016/j.cej.2009.10.029

Wang L, Liang WY, Yu J, Liang ZX, Ruan LL, Zhang YC (2013) Flocculation of *Microcystis aeruginosa* using modified larch tannin. Environ Sci Technol 47:5771–5777. https://doi.org/10.1021/es400793x

Wang F, Yang B, Wang H, Song Q, Tan F, Cao Y (2016a) Removal of ciprofloxacin from aqueous solution by a magnetic chitosan grafted graphene oxide composite. J Mol Liq 222:188–194. https://doi.org/10.1016/j.molliq.2016.07.03

Wang Z, Xu J, Hu Y, Zhao H, Zhou J, Liu Y, Lou Z, Xu X (2016b) Functional nanomaterials: study on aqueous Hg(II) adsorption by magnetic Fe_3O_4@SiO_2-SH nanoparticles. J Taiwan Inst Chem Eng 60:394–402. https://doi.org/10.1016/j.jtice.2015.10.041

Wang J, Han Y, Li J, Wei J (2017) Selective adsorption of thiocyanate anions using straw supported ion imprinted polymer prepared by surface imprinting technique combined with RAFT polymerization. Sep Purif Technol 177:62–70. https://doi.org/10.1016/j.seppur.2016.12.038

Wang H, Liu Y, Ifthikar J, Shi L, Khan A, Chen Z, Chen Z (2018) Towards a better understanding on mercury adsorption by magnetic bio-adsorbents with γ-Fe_2O_3 from pinewood sawdust derived hydrochar: influence of atmosphere in heat treatment. Bioresour Technol 256:269–276. https://doi.org/10.1016/j.biortech.2018.02.019

Wu R, Qu J (2005) Removal of water-soluble azo dye by the magnetic material MnFe2O4. J Chem Technol Biotechnol 80:20–27. https://doi.org/10.1002/jctb.1142

Wu R, Qu J, Chen Y (2005) Magnetic powder MnO-Fe_2O_3 composite – a novel material for the removal of azo-dye from water. Water Res 39:630–638. https://doi.org/10.1016/j.watres.2004.11.00

Xu P, Zeng GM, Huang DL, Feng CL, Hu S, Zhao MH, Lai C, Wei Z, Huang C, Xie GX, Liu ZF (2012) Use of iron oxide nanomaterials in wastewater treatment: a review. Sci Total Environ 424:1–10. https://doi.org/10.1016/j.scitotenv.2012.02.023

Yamaura M, Fungaro DA (2013) Synthesis and characterization of magnetic adsorbent prepared by magnetite nanoparticles and zeolite from coal fly ash. J Mater Sci 48:5093–5101. https://doi.org/10.1007/s10853-013-7297-6

Yang X, Zhang X, Ma Y, Huang Y, Wang Y, Chen Y (2009) Superparamagnetic graphene oxide–Fe_3O_4 nanoparticles hybrid for controlled targeted drug carriers. J Mater Chem 19:2710–2714. https://doi.org/10.1039/b821416f

Yang J, Zhao Y, Ma S, Zhu B, Zhang J, Zheng C (2016) Mercury removal by magnetic biochar derived from simultaneous activation and magnetization of sawdust. Environ Sci Technol 50:12040–12047. https://doi.org/10.1021/acs.est.6b03743

Zargoosh K, Abedini H, Abdolmaleki A, Molavian MR (2013) Effective removal of heavy metal ions from industrial wastes using thiosalicylhydrazide-modified magnetic nanoparticles. Ind Eng Chem Res 52:14944–14954. https://doi.org/10.1021/ie401971w

Zhang S, Zhang Y, Liu J, Xu Q, Xiao H, Wang X, Xu H, Zhou J (2013) Thiol modified Fe_3O_4@SiO_2 as a robust, high effective, and recycling magnetic sorbent for mercury removal. Chem Eng J 226:30–38. https://doi.org/10.1016/j.cej.2013.04.060

Zhang YK, Yan T, Yan LG, Guo XY, Cui LM, Wei Q, Du B (2014) Preparation of novel cobalt ferrite/chitosan grafted with graphene composite as effective adsorbents for mercury ions. J Mol Liq 198:381. https://doi.org/10.1016/j.molliq.2014.07.043

Zhao J, Ren W, Cheng HM (2012) Graphene sponge for efficient and repeatable adsorption and desorption of water contaminations. J Mater Chem 22:20197–20202. https://doi.org/10.1039/c2jm34128j

Zhou C, Zhu H, Wang Q, Wang J, Cheng J, Guo Y, Zhou X, Bai R (2017) Adsorption of mercury (ii) with an Fe_3O_4 magnetic polypyrrole-graphene oxide nanocomposite. RSC Adv 7:18466–18479. https://doi.org/10.1039/c7ra01147d

Zhou J, Liu Y, Zhou X, Ren J, Zhong C (2018) Magnetic multi-porous bio-adsorbent modified with amino siloxane for fast removal of Pb(II) from aqueous solution. Appl Surf Sci 427:976–985. https://doi.org/10.1016/j.apsusc.2017.08.110

Zhu X, Liu Y, Qian F, Zhou C, Zhang S, Chen J (2014) Preparation of magnetic porous carbon from waste hydrochar by simultaneous activation and magnetization for tetracycline removal. Bioresour Technol 154:209–214. https://doi.org/10.1016/j.biortech.2013.12.019

Chapter 3
Magnetic Biosorbents and Their Applications in Water Remediation

Pamela Y. Quintas, Emiliano F. Fiorentini, María N. Oviedo, and Leticia B. Escudero

Contents

Abstract Water pollution is caused by the presence of toxic inorganic and organic compounds that come from both natural and anthropogenic sources. The development of the industrial sector has caused the discharge of poor or nontreated residual liquids containing high concentrations of pollutants, such as arsenic, mercury, lead, cadmium, hydrocarbons, and hormones. Although different remediation treatments have been proposed to remove pollutants from contaminated matrices, it is necessary to focus on additional efficient and green technologies. Adsorption has shown promising results for the treatment of pollutants from natural water. In this sense,

P. Y. Quintas · E. F. Fiorentini · M. N. Oviedo
Laboratory of Analytical Chemistry for Research and Development (QUIANID), Interdisciplinary Institute of Basic Sciences (ICB–UNCUYO–CONICET), Faculty of Natural and Exact Sciences, National University of Cuyo, Mendoza, Argentina

L. B. Escudero (✉)
Laboratory of Environmental Biotechnology (BioTA), Interdisciplinary Institute of Basic Sciences (ICB), UNCUYO – CONICET, Faculty of Natural and Exact Sciences, National University of Cuyo, Mendoza, Argentina

L. Meili, G. L. Dotto (eds.), *Advanced Magnetic Adsorbents for Water Treatment*, Environmental Chemistry for a Sustainable World 61,
https://doi.org/10.1007/978-3-030-64092-7_3

various types of adsorbents have been synthesized and applied for the removal of different contaminants. Within adsorption, the use of biodegradable materials goes along for the care of the environment and it is equally efficient as or even more efficient than classical chemical adsorbents. However, the phase separation of the biosorbent containing the pollutant from the aqueous phase after the biosorption process is usually made by time-consuming steps like filtration or centrifugation. Moreover, in the case of using some bacteria, it is necessary to make the phase separation with an ultracentrifuge, which is not always available in all laboratories. In this way, magnetic separation has arisen as an alternative to the separation step, offering besides other advantages on the biosorbent response to remove pollutants.

In this chapter, we present the most recent advances of using magnetic biosorbents for decontamination of water. A wide variety of contaminants, including metals, metalloids, dyes, and other organic pollutants, are considered. A brief explanation from the creation to the present day of magnetic biosorbents is presented. Different techniques used to characterize magnetic biosorbents are also detailed. The reviewed literature evidences the great potential of magnetic biosorbents for the elimination of different contaminants from aqueous matrices. The contribution of some magnetic nanomaterials to form magnetic biosorbents with biological substrates causes an improvement of the biosorption efficiency due to the high specific surface area, pore volume, and versatility of surface active sites coming from the magnetic nanomaterials. Thus, removal percentages from 70 to 100% are obtained in the majority of the reported works. Other considerations related to biosorption are also covered in this chapter.

Keywords Adsorption · Aqueous solutions · Biosorption · Dyes · Isotherms · Kinetics · Magnetic biosorbents · Metalloids · Metals · Organic contaminants · Remediation · Removal · Thermodynamic · Wastewater · Water

3.1 Introduction

Water is fundamental to guarantee the existence of any living species. As water is a nonrenewable resource, many awareness campaigns are carried out by governments to avoid water wasting (Addo et al. 2019; Benedict and Hussein 2019; Ouda et al. 2013). On the other hand, the maintenance of water quality that is available in the planet is also a great challenge for today's society. Chemical pollutants are one of the great threats for natural waters, contributing greatly to the creation of contaminated waters that turn into sources of direct exposure to the environment, plants, and animals (Moore and Ramamoorthy 1984; Stremilova et al. 2001). Human living can also be affected by contaminated natural waters because contaminants could be moved to seafood or drinking water (Jezierska and Witeska 2006; Sharma and Bhattacharya 2016). Taking these aspects into account, it is necessary to work on the development of remediation technologies that collaborate with water quality through pollutant removal processes. Clearly, it is a priority to develop decontamination processes that are environmentally sustainable, in order to achieve two objectives: i) the elimination of pollutants from contaminated aqueous matrices and ii) the no generation of adverse environmental effects as a result of remediation treatments.

3.1.1 Polluted Water: The Cause of the Problem

The pollutants present in water can originate from natural sources such as volcanic activities or natural fires (Mottana et al. 2016). However, anthropogenic sources are the main contribution of contaminants in water, through the industrial activity, the dumping of untreated or poorly treated industrial effluents to water bodies, oil spills, and the use of pesticides in agriculture (Madhav et al. 2019). The industrial development of the last decades has generated large amounts of wastes. For instance, Europe generates more than 850 million tons of industrial wastes per year (Stanners and Bourdeau 2008). Moreover, electronic wastes are increasing in the world and highly contribute to the environment pollution if no suitable treatments or disposals are applied (Kiddee et al. 2013).

The pollutants present in waters include inorganic compounds, such as mercury, lead, cadmium, antimony, thallium, tin, cobalt, arsenic, chromium, selenium, cyanide, bromate, nitrite, nitrate, fluoride, and chloramines (Xagoraraki and Kuo 2008). Organic compounds are also present in waters such as hydrocarbons, benzene, polychlorinated biphenyls, toluene, pesticides, fertilizers, hormones, and antibiotics, among others (Xagoraraki and Kuo 2008). Both inorganic and organic compounds causes adverse effects on human health, affecting the nervous, endocrine, reproductive, pulmonary, circulatory, and/or gastrointestinal systems (Jaishankar et al. 2014; Kim et al. 2013; Kimbrough 1995). Most of these compounds show mutagenic, teratogenic, and carcinogenic effects (Gebhart 2008; Safe 1989; Vainio et al. 1985).

The physicochemical properties of water influence the toxicity of pollutants, besides the concentration of contaminants and the chemical species naturally present in the samples. Moreover, some pollutants such as metals can persist in the environment for a long time, which affects the ecosystem, drinking water, and food. For this reason, effective remediation treatments for the removal of contaminants from waters have been developed. To date, ion exchange, precipitation, membrane filtration, biodegradation, electrochemical oxidation, and adsorption have been performed for the removal of pollutants (Abdullah et al. 2019; Hubicki and Kołodyńska 2012; Mahmood et al. 2011; Ontanon et al. 2017; Patel 2019; Weng and Yu 2019). The majority of these technologies are of high cost and require the use of non-environmentally friendly reagents. As adsorption shows advantages of simplicity and low cost, various adsorbents have been explored in the last years with decontamination purposes, including polymers, activated carbon, nanomaterials, chelating materials, industrial and agricultural wastes (Bulgariu et al. 2019; Mansour et al. 2018; Sajid et al. 2018; Yang et al. 2019).

3.1.2 Improved Biosorbents: The Origin of the Solution

Biosorbents are a group of adsorbents that offer great advantages to eliminate water pollutants because of their ease to be obtained, economy, and biodegradability, which are in good agreement with the environment care (Escudero et al. 2018). This group includes microorganisms, plant derivatives, biomolecules, and industrial and domestic wastes (Fig. 3.1) (Escudero et al. 2019).

Biosorbents can also exhibit some disadvantages like poor mechanical strength, which generate a limited reutilization capacity. On the other hand, phase separation involved after the biosorption process is restricted to filtration or centrifugation processes only, which are time-consuming steps. In order to solve these drawbacks, biosorbents with magnetic properties have arisen as novel materials with superlative adsorbent potential.

The synthesis of magnetic biosorbents offers materials with a high specific surface area, pore volume, versatility of surface active sites capable of interacting with contaminants, easy magnetic separation of the solid phase from the aqueous media, and the possibility of reuse of the solid material during several biosorption–desorption cycles (Hassan et al. 2020). Considering that the final properties of the magnetic biosorbents depend on the biosorbent, the magnetic material, and the synthesis methods, great efforts need to be carried out in order to obtain efficient materials (Hassan et al. 2020). Nowadays, a limited variety of magnetic biosorbents have been investigated for the removal of pollutants from waters and effluents (Hassan et al. 2020; Soares et al. 2019b; Tural et al. 2017; Zheng et al. 2020). Thus, it is necessary the creation of additional magnetic biosorbents to

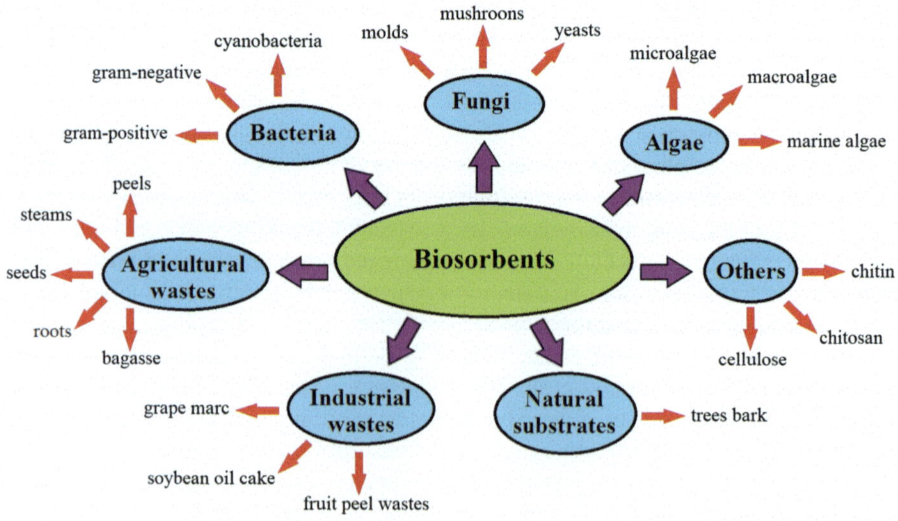

Fig. 3.1 Biosorbents often used for the removal of pollutants from aqueous environmental matrices. A brief classification and examples of some biosorbents are exhibited

obtain new materials with specific properties, such as solubility and polarity (Haas and Franz 2009), that can be efficient to retain inorganic and organic pollutants.

This book chapter is focused on the most recent applications of magnetic biosorbents for water remediation. The removal of metals, metalloids, dyes, and other organic compounds from water is considered. Biosorption capacities and removal efficiencies are compared and detailed. Kinetic, equilibrium, and thermodynamic parameters are also commented to better understand and characterize the biosorption processes. The reutilization capacity of the magnetic biosorbents is also discussed. Special attention is given to the preparation and characterization techniques used to synthesize the biosorbents with magnetic properties. The magnetic susceptibility is analyzed according to different preparation methods.

3.2 Magnetic Biosorbents

3.2.1 From the Creation to the Present

The use of biological substrates for the elimination of toxic compounds has been studied since the eighteenth century (Escudero et al. 2018). In the last years, the field of biosorption has grown and become more complex, where it is sought to understand the mechanisms by which the biosorption process occurs, methods that allow the characterization of the biomass, and the quantification of analytes, besides equilibrium, kinetic, and thermodynamic studies.

The addition of magnetic properties to a sorbent is a relatively new concept, which has drawn the attention of the scientific community and has improved preexisting methodologies. Magnetic adsorbents are basically a new generation of adsorbents where a conventional adsorbent is embedded with magnetic particles (Mehta et al. 2015). The most commonly used magnetic particles are oxides of metals such as Fe, Co, Ni, and Cu (Philippova et al. 2011). With the simple application of an external magnetic field provided by a magnet, the magnetic adsorbent can be quickly and easily separated from an aqueous phase.

Magnetic ion exchange resins were first used in 1995 with the aim of eliminating organic matter (Ambashta and Sillanpää 2010). Subsequently, polymers (Philippova et al. 2011) and carbon nanotubes (Hu et al. 2010) modified with magnetic particles have been used commercially. Currently, among its various applications, magnetic nanomaterials have been used in bioremediation processes of contaminated water as part of hybrid materials with biosorbents in order to increase the biosorption efficiency for the removal of pollutants and improve the phase separation stage (Simeonidis et al. 2016). As it was previously mentioned, magnetic adsorbents are responsible to avoid the use of centrifugation or filtration steps, which are normally required with conventional adsorbents, representing an advantage by simplifying and decreasing the time of the decontamination process (Reddy and Yun 2016). The most used nanoparticles in polluted water treatment studies are iron oxides, such as magnetite, on the basis of their low cost and low environmental impact (Tang and Lo

2013). Moreover, their chemical functionalization has allowed a significant improvement in the adsorption capacity and in turn has improved their selectivity toward different pollutants (Su 2017).

Recently, magnetic nanomaterials based on biopolymers have also begun to be used for water treatment (Carpenter et al. 2015; Dehabadi and Wilson 2014). It is well known that, during growth cycles of living organisms, various biopolymers occur naturally. In this way, several studies have emerged obtaining biopolymers through the use of microorganisms, such as algae, yeasts, fungi, and bacteria. These biopolymers have striking advantages, including high environmental availability of biomass, low cost, biodegradability and biocompatibility, reduced toxicity, and easy cultivation of microorganisms (Rebelo et al. 2017; Resch-Fauster et al. 2017). In this context, the coating of magnetic nanoparticles with biopolymers allows to improve the colloidal stability of magnetic nanoparticles in aqueous media, thus avoiding the formation of magnetic aggregates, which would decrease the available surface area and thus the biosorption capacity (Mehta et al. 2015). The improvement of colloidal stability is due to steric shielding of the biopolymer chains or due to electrostatic repulsions that are produced, thanks to charges present in the biopolymer. Polysaccharides are the most widely used biopolymers in magnetic biosorbents, such as starch, cellulose, dextran, and chitosan, among others (Bohara et al. 2016).

3.2.2 Preparation Techniques of Magnetic Nanoparticles and Biosorbents

Based on the fact that the properties of magnetic nanoparticles depend largely on the size and shape of the particles, the development of synthetic strategies to achieve the synthesis of magnetic nanoparticles with controlled morphology and uniform size distribution has been explored in the last decade (Ling et al. 2015; Wu et al. 2015). The different methods for the synthesis of magnetic nanoparticles are detailed below.

Sol-Gel Method

It is based on a wet chemical technique to synthesize monodispersed pure nanoparticles (Gupta et al. 2018). Solvent mixing, subsequent hydrolysis, and polycondensation of metal alkoxides or inorganic salts are carried out surrounded by ligand reagents so as to form a colloidal system called "sol," added to the elimination of solvents by a series of chemical reactions that form a crosslinked system known as "gel." The desired crystalline structure of the nanoparticles is obtained by heating (Fig. 3.2a). A narrow size distribution is obtained and it is possible to control the size of the nanoparticles based on the established reaction conditions and the stirring speed. Ethanol is the most widely used solvent because it improves the hydrophilic character of nanoparticles, which can reach a minimum

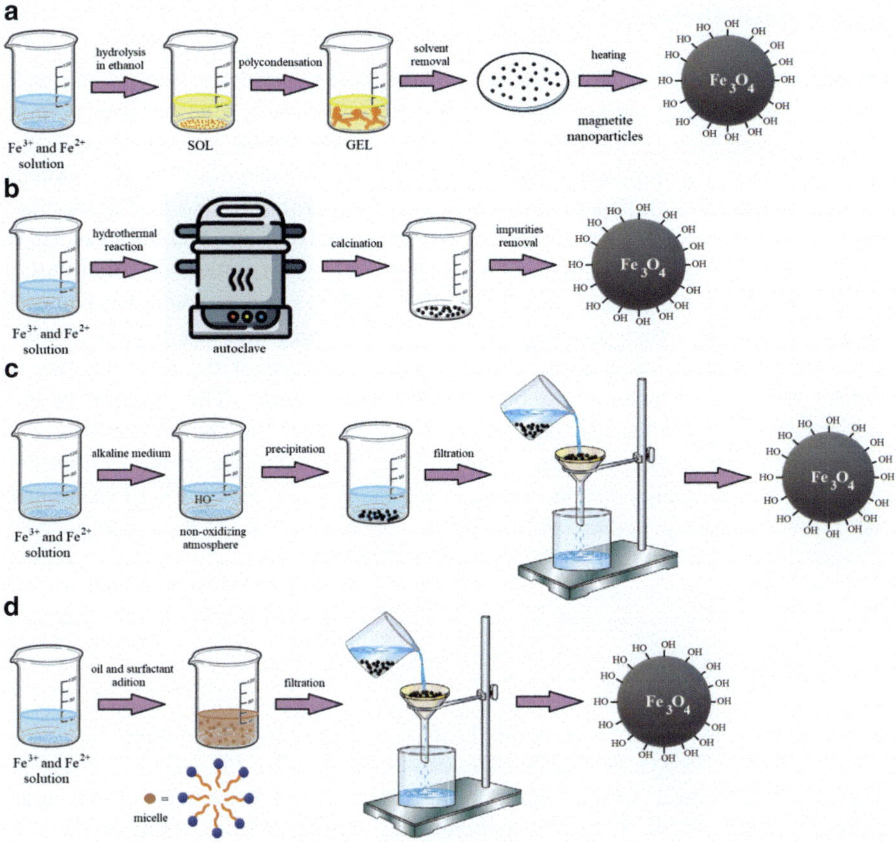

Fig. 3.2 Methods of synthesis of magnetite nanoparticles (Fe_3O_4). (**a**) Sol-gel method: dissolution of ferrous and ferric salts, solvent mixing (ethanol), hydrolysis, and polycondensation process, conformation of colloidal system "sol," solvent removal, formation of crosslinked system "gel," obtaining nanoparticles by heating. (**b**) Hydrothermal method: dissolution of ferrous and ferric salts, autoclave transfer, separation of solid–liquid phases in the solution, calcination, cooling, removal of impurities, obtaining nanoparticles. (**c**) Coprecipitation method: dissolution of ferrous and ferric salts in alkaline medium in a nonoxidizing atmosphere, precipitation and filtration, obtaining nanoparticles. (**d**) Microemulsion method: dissolution of ferrous and ferric salts, addition of oil and surfactant, micelle formation, filtration, obtaining nanoparticles. (Modified after Gupta et al. 2017)

size of up to 10 nm. Despite certain limitations, as the need to extract the precursors of the reaction and the impossibility of characterizing the nanoparticles formed by magnetic resonance due to instability in aqueous medium, it is a widely used method (Valverde-Aguilar et al. 2015).

Hydrothermal Method

An aqueous medium is used where most of the inorganic substances are dissolved and then transferred to an autoclave where the synthesis is carried out over several hours at temperatures above 200 °C and vapor pressure above 200 psi (Gupta et al. 2018). There is a separation of solid–liquid phases in the solution. Then, a calcination step is performed, followed by the removal of impurities using different solvents (Fig. 3.2b). The formation of ferrites can be carried out by hydrolysis and oxidation of ferrous salts or by neutralization of mixed metal hydroxides (Velinov et al. 2016). The properties of the nanoparticles can be modified using a certain mixture of solvents, controlling the reaction conditions so as to control the nucleation and growth rates, as well as coating with polymers. The size of the nanoparticles is reduced when the concentration of the solvents decreases. The stability of the synthesized nanoparticles will depend on the polarity of the chosen solvents, the use of stabilizing agents, and the steric factors. This synthesis method offers the formation of high-quality nanoparticles; however, it is not widely used because the high pressures used generate safety problems, even more so as the reaction scale increases (Qiao and Swihart 2017; Song et al. 2015a).

Coprecipitation Method

It is one of the most widely used methods of synthesis of magnetic nanoparticles due to its simplicity. The synthesis procedure can be observed in Fig. 3.2c. The magnetite preparation consists of the mixture of ferric (Fe^{3+}) and ferrous (Fe^{2+}) ions in an alkaline medium of pH 8–14 and nonoxidizing atmosphere, according to the following reaction:

$$2\ Fe^{3+} + Fe^{2+} + 8\ OH^- \rightarrow Fe_3O_4 + 4\ H_2O$$

The morphology and composition of the synthesized magnetite depend on the type of used salts, including chlorides, sulfates, and nitrates, on the relationship between ferric and ferrous cations, and on the reaction conditions, such as temperature, pH, and ionic strength (Su 2017). The advantages of this methodology are that a large number of nanoparticles can be synthesized, use of less dangerous reagents, its implementation is simple, and it is highly profitable and compatible with the presence of biopolymers based on mild synthesis conditions (Gautam et al. 2015). Particle sizes between 2 and 17 nm can be obtained, but particle size control is not possible because nucleation can occur during the growth stage, which results in polydispersivity. However, uniform particles can be obtained by homogeneous precipitation reactions, where nucleation and growth processes occur individually, by using molecules or polymers that have chelating groups, which allow the control of the particle size and its colloidal stabilization (Chiaradia et al. 2015).

Microemulsion Method

A microemulsion consists of an optically isotropic and thermodynamically stable immiscible system, consisting of three components: i) aqueous phase of polar character, ii) oil phase of nonpolar character and surfactant. Surfactant molecules stabilize microemulsions, which act as reactors for the formation of magnetic nanoparticles. This type of synthesis allows spherical, oval, or tubular nanoparticles to be obtained by normal micellar or inverse micellar formation (Gupta et al. 2018). Subsequently, filtration is carried out by using porous materials. In Fig. 3.2d, the synthesis process is observed. This method allows to easily controlling the morphology of the particles by altering the composition of the polar phase; however, the amount of nanoparticles generated is low compared to other synthesis methodologies. Other disadvantages involved in this procedure are its high cost and difficulty in separating the surfactant from the product obtained (Hrubovčák et al. 2015).

Other Synthesis Methods

- Sonochemical Method

Sonochemical reactions can be used in the synthesis of magnetic nanoparticles in order to obtain a controlled size distribution in a much more efficient way than other proposed methods. This method is based on acoustic cavitation to arrive at the synthesis of nanoparticles. Basically, by placing the solution in an ultrasonic reactor at high temperatures, pressures and cooling rates, the formation of nanoparticles is achieved by the collapse of bubbles with intense shock waves, where the growth of nuclei is avoided. This method is more environmentally sustainable and profitable than others reported in the literature. Moreover, it improves the magnetic response of the synthesized nanoparticles and allows to obtain finer particles with a narrow size distribution (Wang et al. 2015a).

- Oxidative Hydrolysis

It consists of preparing magnetite by suspensions of ferrous hydroxide in aqueous solution, which is oxidized with mild oxidizing agents, such as nitrate ions. It allows obtaining particles of nanometric sizes up to 1 μm, based on the relationship between match products and pH. However, it is possible to achieve strict control over the nanoparticle size distribution (Carvalho et al. 2016).

- Thermal Decomposition

It is based on the thermal decomposition of Fe organic precursors, together with organic solvents and surfactants. Basically, the mixture is heated generating monomers with subsequent burst nucleation (Ling et al. 2015). Alternatively, the organometallic precursor is injected in a hot surfactant solution and the simultaneous formation of many nuclei is induced (Mendoza-Garcia and Sun 2016). The nuclei grow from the monomers and monodispersed nanoparticles are formed. The morphology of the nanoparticles is controlled with time, temperature, and proportion of reactants.

- Green Synthesis

The efforts of the scientific community are today focused on replacing preexisting synthesis methods by methods that are consistent with the principles of green chemistry—for example, using nontoxic substances as starting materials, such as plant tissues, microorganisms, and fruits (Nayak et al. 2015). Plant extracts have proven to be the most appropriate for the production of large quantities of nanoparticles, which have the function of reducing and stabilizing them (Wang et al. 2015b). The biomolecules present in the extracts are responsible for reducing metal ions (Anuradha et al. 2015). As advantages, it shows high efficiency, smooth reaction conditions, cost efficiency, stability, and environmental sustainability. Although the biosynthesis is considered complex, it is more efficient than syntheses by classical processes since it allows the morphology of nanoparticles to be more effectively regulated (Dhand et al. 2016).

Preparation of Magnetic Biosorbents

Once the magnetic nanoparticles were synthetized by one of the aforementioned routes, the magnetic biosorbent can be prepared. The magnetization is produced by the addition of magnetic particles inside the pores or on the surface of the bisorbents under study (Safarik et al. 2018). A general procedure of synthesis of magnetic biosorbents can be seen in Fig. 3.3. Different strategies of preparation of magnetic biosorbents are discussed below:

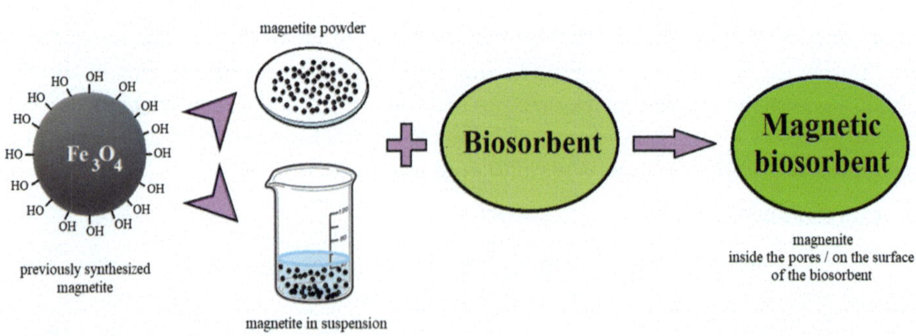

Fig. 3.3 General synthesis of magnetic biosorbents. Previously synthesized magnetite is mixed with the biosorbent, in powder form or in suspension in some organic solvent. Magnetic particles are added inside the pores or on the surface of the biomass

(a) The simplest and fastest strategy is the mixing of magnetite with the biosorbent, which has allowed the synthesis of magnetic biosorbents with different percentages of magnetite on its surface (Khandanlou et al. 2013a). Often, slightly modified procedures have been developed, where an inert gas has been used during the magnetization process (Khandanlou et al. 2013b), or the use of ferrites, that is to say magnetic substances where the ferrous ion has been replaced by another divalent cation (Hashemian and Salimi 2012).
(b) Mixing of suspension of magnetic particles with the biosorbent and subsequent drying by temperature. The suspension has been achieved by adding water-soluble organic solvent, such as methanol, ethanol, and acetone (Safarik et al. 2016). Those biosorbents sensitive to high temperatures have been modified analogously, but using temperatures below zero to dry the magnetized biosorbent (Pospiskova and Safarik 2015).
(c) Use of mechanochemical procedures: Hydrated salts of $FeCl_2$ and $FeCl_3$ have been mixed in excess of NaCl at room temperature; subsequently, the biosorbent has been added as a powder and the mixture has been ground for an appropriate time. Finally, NaOH powder was added to the mixture and mixing was continued (Safarik et al. 2014).

3.2.3 Characterization Techniques

It is well known that the physical–chemical characteristics that magnetic biosorbents present could be useful to understand the mechanism of union between the pollutant and the surface of the biomass. In this way, the characterization is of considerable interest on the development of biosorption processes (Arief et al. 2008). According to the biosorbent, it is possible to use different techniques for characterization, including Fourier transform infrared spectroscopy, scanning electron microscopy, X-ray photoelectron spectroscopy, energy dispersive X-ray spectroscopy, X-ray diffraction spectroscopy, and thermogravimetric analysis. The measurement of magnetic susceptibility is carried out using vibrating sample magnetometry.

Fourier Transform Infrared Spectroscopy

It is a vibrational spectroscopic technique used to characterize the functional groups present on the surface of a magnetic biosorbent. It allows providing a qualitative description, where the availability of certain functional groups on the surface of the magnetic biosorbent is determined. Among its benefits, it provides high-quality spectra, direct analysis without destruction of the sample, and largely avoids the use of solvent and reagents (Escudero et al. 2018).

Scanning Electron Microscopy

It provides topographical information of the magnetic biosorbents. It is a technique with a high resolution that allows obtaining a clear and real image. However, the particle size that it can detect is limited (Goldstein et al. 2017).

Energy-Dispersive X-ray Spectroscopy

This technique is useful for elemental analysis and chemical characterization of a biosorbent. In general, it is combined with scanning electron microscopy in order to obtain more complete results. Basically, this technique is based on the bombardment of the biosorbent with an electron beam, which allows the emission of X-rays of specific energy from the lower zone of the volume of interaction of the bisorbent (Bonnelle and Mandé 2013).

X-ray Photoelectron Spectroscopy

It is a technique that allows you to examine the surface and determine the chemical structure of the magnetic biosorbents because it offers information on the composition and electronic state of the elements found in a given surface region of the magnetic biomass. Each element present is associated with a certain region of union, which is characteristic of each atomic orbital. The short range of photoelectrons that are excited from the biosorbent allows this technique to be specific to its surface (Woicik and Woicik 2016).

X-ray Diffraction Spectroscopy

It is a nondestructive technique of the sample that allows obtaining information about the crystallographic structure of a biosorbent. In addition, it offers high precision for different crystalline systems. On the other hand, this technique contributes to understand the biosorption mechanism (Chen and Yip 2017).

Thermogravimetric Analysis

This technique allows the measurement of the mass of a biosorbent while it is heated or cooled in a defined atmosphere. It is mainly used for the characterization of physicochemical properties of materials with regard to their composition. It is possible to measure even thermal events that do not cause a mass change, such as melting, glass transition, or other solid–solid phase transitions (Prime et al. 2009).

Vibrating Sample Magnetometry

This technique allows measuring the magnetic moment of a magnetic biosorbent when it vibrates in the presence of a static magnetic field, from which it is possible to determine the magnetic susceptibility of the adsorbent. As advantages, this technique is versatile, simple, and allows obtaining precise and reliable results with low cost. The operation of a vibrating sample magnetometer is based on Lenz's law of induction. The movement of the sample can be carried out by means of a Mössbauer transductor that allows the control of the mechanical vibrations of the sample, its amplitude, and its frequency. The sensitivity of a magnetometer is sufficient to detect a minimal variation in the magnetic moment of 5×10^{-4} emu (El-Alaily et al. 2015).

3.3 Magnetic Biosorbents Used for Water Remediation

3.3.1 Removal of Metals and Metalloids

Toxic metals ions are considered pollutants associated with environmental and human diseases (He et al. 2019b). Some metals have beneficial biochemical effects at low concentrations, but high levels of these elements produce different degrees of toxicity (Jaishankar et al. 2014). Nevertheless, metals such as lead, cadmium, silver, and mercury generate toxic effects even at low levels (Bagal-Kestwal et al. 2008). Contrary to organic compounds, metals and metalloids present in the environment are more difficult to remove from contaminated matrices due to the fact that they cannot biodegrade into nontoxic compounds, although sometimes they may be converted into less toxic chemical forms (Ayangbenro and Babalola 2017).

In order to remove metal and metalloids through simple, economical, and effective processes, magnetically modified biosorbents have been applied. Daneshfozoun et al. synthesized a novel agro-based magnetic biosorbent from *Ceiba pentandra*, oil palm empty fruit bunches and celluloses (Daneshfozoun et al. 2017). The agro-fibers were magnetized with Fe_3O_4 nanoparticles for the biosorption of Pb(II), Cu(II), Zn (II), Mn(II), and Ni(II) ions from aqueous solutions. Several parameters were optimized including effects of pH, contact time, initial ion concentration, temperature, and the sorbent reusability (Table 3.1). The authors demonstrated that the equilibrium was achieved in less time with the use of the magnetic biosorbent with respect to the nonmagnetically modified biosorbent. This behavior could be attributed to an increased surface area and therefore, to more sites available to interact with the metal. The result showed that reusability efficiency remained above 98% and 95% after 5 cycles of reuse, which is in good agreement with other magnetic biosorbents applied for the removal of pollutants from water (He et al. 2019a; Fan et al. 2011).

The removal of Pb(II) was also developed by He et al. through Pb(II) ion imprinting technology and crosslinking reactions among chitosan, *Serratia*

Table 3.1 Recently reported magnetic biosorbents for the removal of metals and metalloids from aqueous solutions and water samples

Biosorbent	Metal	Biosorption conditions				Biosorption capacity ($mg\ g^{-1}$)	Saturation magnetization ($emu\ g^{-1}$)	References
		pH	T (°C)	Time[a] (min)	C_0[b] ($mg\ L^{-1}$)			
Agro-based magnetic biosorbents with Fe_3O_4 nanoparticles	Pb(II) Cu(II) Zn(II) Mn(II) Ni(II)	5–7	n.r.	45–60	500	49	n.r.	Daneshfozoun et al. (2017)
Chitosan, *Serratia marcescens* and Fe_3O_4	Pb(II)	5	n.r.	480	200	116.28[e]	21.5	He et al. (2019a)
Fe_3O_4 nanoparticles with *Chlorella vulgaris*	Cd(II) Pb(II)	4.5	28 ± 1	720	10	13.16[e] 20.00[e]	n.r.	Gupta et al. (2017)
Magnetic oak wood and oak Bark fast pyrolysis biochars	Cd(II) Pb(II)	5	25	2880	50	8.33[e] 55.91[e]	8.87	Mohan et al. (2014)
Cyclosorus interruptus as support to load Fe_3O_4 nanoparticles	Hg(II)	5	30	720	100	385.3[e]	9.5	Zhou et al. (2017)
Chitosan cross/glutaraldehyde magnetically modified	Hg(II)	5	25	1440	100	152[e]	5.79	Kyzas and Deliyanni (2013)
Magnetically modified *Saccharomyces cerevisiae* subsp. uvarum	Cu(II)	4	25	n.r.	30	1.2[d]	n.r.	Uzun et al. (2011)
Saccharomyces cerevisiae and chitosan-coated magnetic nanoparticles as support	Cu(II)	4.5	28	120	60	144.9[e]	n.r.	Peng et al. (2010)
Magnetic chitosan nanoparticles	Cu(II)	5	25.15	480	100	35.5[e]	36	Yuwei and Jianlong (2011)
Magnetic chitosan beads	Cu(II)	5	25	720	30	129.6[e]	39.5	Jiang et al. (2014)
Dendrimer-like magnetic biosorbent based on modified orange peel waste	As(V)	6	25	300	25	81.3	n.r.	Meng et al. (2017)
Magnetic wheat straw with Fe_3O_4 nanoparticles	As(III) As(V)	7–9	30	720	10	3.90[e] 8.06[e]	11.87	Tian et al. (2011)

Saccharomyces cerevisiae cells modified with iron nanoparticles	Ni(II)	5	n.r.	180	50	54.23[e]	n.r.	Guler and Ersan (2016)
Magnetic nanoparticle impregnated onto tea waste	Ni(II)	4	30	180	100	38.3[e]	n.r.	Panneerselvam et al. (2011)
Crosslinked magnetic chitosan-2-aminopyridine glyoxal Schiff's base	Cu(II) Cd(II) Ni(II)	5	30	180	100	124 ± 1[e] 84 ± 2[e] 67 ± 2[e]	24.3	Monier et al. (2012)
Rhizopus cohnii and Fe_3O_4 nanoparticles coated with alginate and polyvinyl alcohol	Cr(VI)	1	28	720	40	6.73	n.r.	Li et al. (2008)
Eucalyptus leaf residue and Fe_3O_4 nanoparticles	Cr(VI)	3.00	25 ± 0.5	720	200	n.r.	16.00	Wang et al. (2014
Magnetic nanosorbent functionalized with N-methyl-D-glucamine	B	8.2	30	30	8	13.44[e]	37.00	Tural et al. (2018)
Chitosan extracted from shrimp shells with magnetic nanoparticles	V(V) Pd(II)	5.0 6.0	20 ± 1	10	10	186.6[e] 192.3[e]	n.r.	Omidinasab et al. (2018)
Magnetic thiourea–chitosan	Ag(I)	5	30	50	20[c]	5.29[d]	10.16	Fan et al. (2011)
Magnetically modified *Kluyveromyces fragilis*	Sr(II)	7	20	60	80	140.8[e]	n.r.	(Ji et al. 2010)
Sawdust magnetically modified with Fe_3O_4 nanoparticles	Sr(II)	6.74	20	30	20	12.59 [e]	n.r.	Cheng et al. (2012)
Paecilomyces catenlannulatus with iron oxide nanoparticles.	Eu(III)	3.5	25	1440	60	69.45	24.51	Li et al. (2018a)
Saccharomyces Cerevisiae onto crosslinked chitosan coated with magnetic nanoparticles	U(VI)	4.0	25	180	50	72.4[e]	n.r.	Saifuddin and Dinara (2012)
Magnetically modified *Rhodotorula glutinis*	U	6	20	30	100	187 ± 26[e]	n.r.	Bai et al. (2012)

NR Not reported
[a]Equilibrium time
[b]C_0: Initial concentration
[c]Expressed as mmoles L^{-1}
[d]Expressed as mmoles g^{-1}
[e]Theoretical Q_{max}

marcescens, and Fe_3O_4 nanoparticles (He et al. 2019a). X-ray diffraction analysis, Fourier transform infrared spectroscopy, scanning electron microscope, X-ray photoelectron spectroscopy, Energy-dispersive X-ray spectroscopy, and magnetism analysis techniques were applied to characterize the magnetic biosorbent. X-ray diffraction analysis elucidated a crystalline structure primarily attributed by Fe_3O_4 nanoparticles. Moreover, it was shown by comparison of the Fourier transform infrared spectroscopy spectra between the biomass and biomass with Pb(II), some changes in the region of the peak at 3434 cm^{-1} due to complex formation between hydroxyl or amino groups and the metal. Crosslinking reaction was confirmed because the peaks at 2805 cm^{-1} (C-H stretching of $-CH_3$ and $-CH_2$) and 1641 cm^{-1} (C = O stretching) shifted. In addition, the peak at 580 cm^{-1} (Fe-O stretching) demonstrated that the Fe_3O_4 nanoparticles successfully coated the biosorbent.

On the other hand, Gupta et al. investigated the biosorption process of *Chlorella vulgaris* coated with Fe_3O_4 nanoparticles to remove Cd(II) and Pb(II) from aqueous solutions (Gupta et al. 2017). Magnetic biosorbent was prepared by mixing Fe_3O_4 nanoparticles with dry *Chlorella vulgaris* in distilled water. Then, the mixture was stirred at room temperature for 24 h allowing a suitable incorporation of the magnetic nanoparticles onto the surface of biosorbent. Different parameters were optimized to obtain the highest removal efficiency and biosorption capacity. For instance, the effect of ionic strength was evaluated in a concentration range of 0.001–0.01 mol L^{-1} of $NaNO_3$ and the results showed that high levels of salt decrease the removal of Cd(II) from 88.54 to 47.54%. On the other hand, the biosorption of Pb(II) does not depend on the saline effect; this may be because Pb (II) ions are specifically bound to the active sites.

Mohan et al. also studied the removal of Cd(II) and Pb(II) from water (Mohan et al. 2014). In this work, the researchers used low-cost oak wood and oak bark fast pyrolysis biochars magnetized with iron oxide nanoparticles. Biosorbents were synthesized from rapid pyrolysis at 400 and 450 °C in an auger-fed reactor. After this process, biochar suspensions were mixed with Fe(II) and Fe(III) salts, and finally treated with NaOH. The saturation magnetization of biosorbents was 4.47 and 8.87 emu g^{-1} for magnetic oak bark char and magnetic oak wood biochar, respectively. This result can be explained because the magnetic oak wood biochar has a greater surface with magnetic nanoparticles than the magnetic oak bark char.

Mercury is a metal with high toxicity, especially under the methyl mercury species that bioaccumulates primarily in the tissue of fishes (Carrasco et al. 2011). It has been shown that this element causes neurotoxicity in humans; therefore, the removal of mercury in the environment is very important (Rice et al. 2014). Zhou et al. prepared a novel thiourea-functionalized magnetic biosorbent for adsorption of Hg(II) based on the ability of this metal to strongly bind to thiol and amine groups (Zhou et al. 2017). In this work, *Cyclosorus interruptus* was used as support to load Fe_3O_4 nanoparticles, due to this biomass having several functional groups and multipores on its surface. The pH played an important role in the biosorption. It was observed that pH_{pzc} of the magnetic biosorbent was about 4.2; therefore, mostly functional groups were protonated at pH less than 4.2. Thus, electrostatic interaction between Hg(II) ions and positively charged active sites was not favored and predominates repulsive forces. The results showed a high bioadsorption capacity at pH 5.

Kyzas et al. developed a work where they compared the biosorption capacity of two modified chitosan to remove Hg(II) ions from aqueous solutions (Kyzas and Deliyanni 2013). Both biosorbents were crosslinked with glutaraldehyde; however, only one of them was magnetically modified with magnetic nanoparticles. The biosorption process was performed at pH 5, while desorption was achieved at pH2. After 4 cycles, the reuse proportion decreased by 26% with the use of the biosorbent without magnetic properties, in contrast to the 10% loss achieved by the magnetic derivative. Therefore, the magnetic biosorbent has a more complex structure and is more resistant than the nonmagnetic one, being able to be reused in more cycles, and more suitable for the removal of Hg(II) from wastewater.

Copper is an essential micronutrient; however, high concentrations of this element can cause hematological disorders and kidney disease (Ashish et al. 2012). In addition, cooper produces a negative impact on the environment. In the last years, the use of copper-based nanoparticles in paints and pesticides has been increasing, favoring copper accumulation in the environment (Keller et al. 2017). Moreover, it is well known that the bioaccumulation of copper in aquatic systems can produce alterations in fish olfaction, preventing them from avoiding predators, among others (Kiaune and Singhasemanon 2011). Therefore, the removal of this metal from contaminated environmental samples has been extensively studied (Al-Homaidan et al. 2014). Uzun et al. prepared a biosorbent based on noncultivated yeast cells of *Saccharomyces cerevisiae* subsp. uvarum magnetically modified with iron oxide nanoparticles for the study of Cu(II) removal (Uzun et al. 2011). In addition, this magnetic biosorbent was used for the removal of Cu(II), Ni(II), and Hg(II) from artificial wastewater containing Zn(II), Fe(III), Co(II), Cd(II), Cr(III), and Al(III). It was observed that the biosorption capacity is affected by the coexistence of other metal ions in the sample. This phenomenon is due to complex interactions that occur between metals and the surface of the biomass, which depend on the biomass nature, type, and concentration of metals and the experimental parameters, among others.

Peng et al. also studied the removal of Cu(II) from aqueous solution using *Saccharomyces cerevisiae* yeast (Peng et al. 2010). However, in this work chitosan coated with Fe_3O_4 was used as support. The magnetic nanoparticles were synthesized by coprecipitating Fe(III) and Fe(II) with molar ratio 2:1 by ammonia solution at pH 10 in hydrothermal conditions. Concentrations of Cu(II) were measured by atomic absorption spectrometry. The time to reach the equilibrium was 1 h; however, more than 90% of the metal was removed at 10 min. It was observed that Langmuir model was most adequate to fit the experimental data. The maximum biosorption capacity was 144.9 mg g^{-1}, being greater than other results previously reported using chitosan as a biosorbent.

Yuwei et al. investigated the removal of Cu(II) using magnetic chitosan nanoparticles in aqueous solutions (Yuwei and Jianlong 2011). The magnetic hysteresis curve showed the ferromagnetic property of the biosorbent and indicated a saturation magnetization of 36 emu g^{-1}. The saturation magnetization was higher compared to others reported on chitosan-based Fe_3O_4 beads. Thus, the magnetic biosorbent could be easily separated with the use of an external magnetic field. Desorption process was carried out using an ethylenediaminetetraacetic acid solution by adjusting the pH to 5.0 and with a contact time of 4 h. The researchers highlighted that the biosorption removal remained greater than 90% after 4 cycles of reuse. In addition, the results showed a maximum biosorption capacity of 35.5 mg g^{-1}, based on Langmuir model properly.

Jiang et al. also employed magnetic chitosan for the removal of Cu(II) from aqueous solutions (Jiang et al. 2014). According to the authors, for Fe_3O_4 nanoparticles synthesis, Fe(III) salt was mixed with ethylene glycol as a reducing agent and sodium acetate. Then, the mixture was agitated and gradually heated in an Ar atmosphere at 185 °C. Then, it was kept at the same temperature for 48 h. After separation with an external magnetic field, the nanoparticles were washed and dried properly. Scanning electron microscopy, X-ray diffraction, Fourier transform infrared spectrometry, and X-ray photoelectron spectroscopy, among other techniques, were used for biosorbent characterization. X-ray photoelectron spectroscopy spectra showed that the interaction between amino group and Fe_3O_4 nanoparticles allows the formation of crosslinking bridges, favoring the stability of the magnetic biosorbent. On the other hand, the biosorption process was fast, achieving the equilibrium in the first 10 min. In addition, the biosorption isotherms of Cu(II) were in good agreement with the Langmuir model, indicating a monolayer biosorption process. Moreover, the evaluation of the kinetic models showed that pseudo first-order kinetic model was the most suitable. In this work, ethylenediaminetetraacetic acid solution was also used to evaluate the reuse efficiency of the magnetic sorbent. It was shown that the adsorption capacity was highly effective after six cycles of reuse.

Arsenic is a highly toxic pollutant widely found in nature. Its concentration in the environment can be increased by both natural and anthropogenic sources producing toxic and carcinogenic effects in humans (Chung et al. 2014). Meng et al. investigated the biosorption–reduction of this metalloid with the use of dendrimer-like magnetic porous sorbent based on modified orange peel waste (Meng et al. 2017). In this paper, it was demonstrated that the magnetic biosorbent can be useful to remove both As(V) and As(III) species from aqueous solutions. The magnetic biosorbent was obtained by mixing previously washed and dried orange peel pieces with Fe(III) and Fe(II) solutions in stainless steel autoclave. Then, the mixture was stirred for 24 h and heated at 180 °C in oven for 10 h. Finally, the biosorbent obtained was adequately filtered and washed with deionized water. In the biosorption process, it was observed that in the presence of As(V), Fe(II) ions were significantly oxidized to Fe(III) while As(V) were reduced to As(III) (Fig. 3.4). Therefore, the reduction mechanism explains the biosorption of both As species on the surface of the dendrimer-like through an oxide-reduction reaction between As and Fe. The removal percentage of 89% was obtained after 3 cycles of reuse. Thus, this biosorbent can be applied several times and regenerated with NaOH.

On the other hand, Tian et al. also studied the removal of As from water samples using a wheat straw with different concentrations of magnetic nanoparticles (Tian et al. 2011). It is well known that the chemistry of As depends on pH. Moreover, the

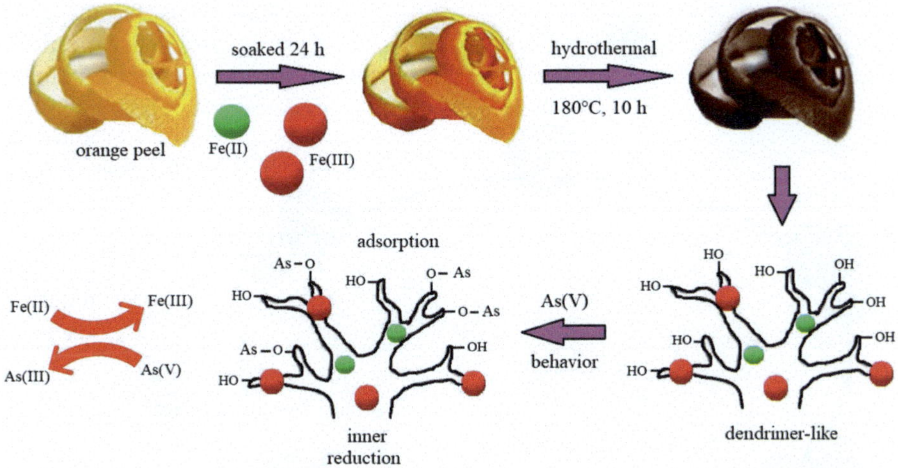

Fig. 3.4 Schematic process for the removal of As using dendrimer-like coated with Fe_3O_4 nanoparticles. The magnetic biosorbent was prepared by adding dried orange peel pieces to a solution of Fe(III) and Fe(II) in stainless steel autoclave. The mixture was stirred for 24 h and then was heated at 180 °C in oven for 10 h. Finally, the magnetic biosorbent obtained was adequately filtered and washed with deionized water. It was shown that the biosorption of As(V) was carried out through a biosorption–reduction procedure on the surface of the biosorbent. As(V) was partially reduced to As(III) while Fe(II) was significantly oxidized to Fe(III), facilitating arsenic adsorption (Modified after Meng et al. 2017)

protonation or deprotonation on the surface of the iron oxide nanoparticles at the level of the hydroxyl molecules also depends on the pH. At low pH, the surface of the magnetite is protonated, acquiring a positive charge; therefore, the interaction with As(V) present under an anionic form is favored. However, this phenomenon is not observed for As(III) species since at pH less than 9.2, the uncharged form in aqueous solution predominates.

Nickel is a toxic metal frequently present in wastewater from various industries, especially related to metallurgy and manufacturing batteries (Coman et al. 2013). Guler et al. studied the removal of Ni(II) from water samples using *Saccharomyces cerevisiae* cells modified with iron nanoparticles as a magnetic biosorbent (Guler and Ersan 2016). The biosorption process involves five-step mechanisms including oxidation and reduction reactions, complexation, biosorption, translocation, and diffusion (Fig. 3.5). The concentration of Ni(II) was measured using a flame atomic absorption spectrometer at 351.5 nm wavelength. The results showed that the maximum biosorption capacity was 54.23 mg/g at pH 5 and 180 min of contact time (Table 3.1). In addition, the biomass covered with magnetic nanoparticles was 2.5 times more efficient than the nonderived biosorbent.

Panneerselvam et al. used tea waste coated with Fe_3O_4 nanoparticles to remove Ni(II) from aqueous samples (Panneerselvam et al. 2011). The magnetic nanoparticles were prepared through the conventional coprecipitation method. Biomass is mainly composed of cellulose, presenting a large area with active sites. In

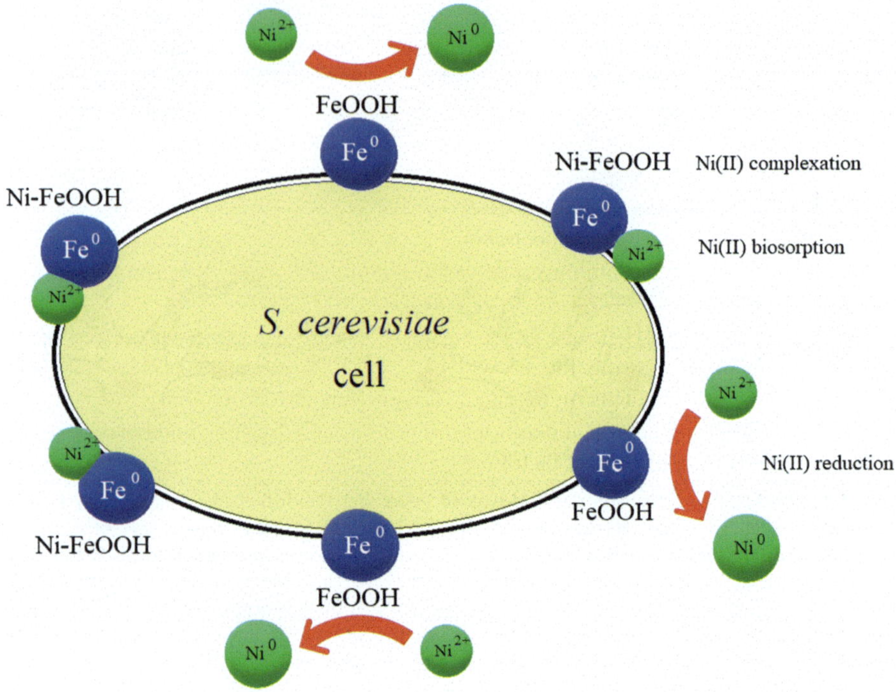

Fig. 3.5 Schematic process for the removal of Ni(II) using *S. cerevisiae* cells modified with iron nanoparticles through five steps including reduction of Ni(II) with the subsequent oxidation of Fe^0 to Fe^{3+} on the surface of the biosorbent, followed by the formation of complexes with hydroxyl groups, biosorption of metal via active groups, translocation of Ni via the oxide shell and diffusion into the iron core and *S. cerevisiae* cell wall. (Modified after Guler and Ersan 2016)

addition, the structure of the tea waste is completely covered with magnetic nanoparticles that have a negative charge, which allows greater electrostatic attraction to Ni(II) cations. Under optimal experimental conditions, the removal percentages of tea waste, Fe_3O_4 nanoparticles, and tea waste with Fe_3O_4 were 29.8%, 46.5%, and 87%, respectively. It is hence evidenced the great potential of using the magnetic biosorbent to remove Ni(II) from aqueous samples.

Monier et al. studied the removal of Cu(II), Cd(II), and Ni(II) ions by crosslinked magnetic chitosan-2-aminopyridine glyoxal Schiff's base resin (Monier et al. 2012). It was observed that among other parameters, pH affects the biosorption process being negligible at pH less than 2. This behavior can be explained because at low pH values, H^+ ions compete with metal ions for active sites of the biosorbent. On the other hand, the researchers confirm that crosslinked magnetic chitosan resin is insoluble in both acidic and basic solutions and is more resistant than unmodified chitosan. The maximum biosorption capacities at pH 5 according to Langmuir isotherm were 12, 84, and 67 mg g^{-1} for Cu(II), Cd(II), and Ni(II), respectively.

One of the most common pollutants in industrial effluents is hexavalent chromium. This chemical form is considered highly harmful because it is a carcinogenic and teratogenic agent for animals and humans (Patra et al. 2010). The removal of Cr (VI) was developed by Li et al. through the application of *Rhizopus cohnii* and magnetic particles coated with alginate and polyvinyl alcohol (Li et al. 2008). According to the results, Cr(VI) was completely retained in the magnetic biosorbent after 12 h of contact with Cr(VI) species. However, as time went on, the concentration of Cr(III) increased on the biosorbent. This phenomenon can be explained due to a partial transformation of Cr(VI) to Cr(III) on the surface of the magnetic biosorbent. The maximum biosorption capacity was 6.73 mg g^{-1} at pH 1.0 (Table 3.1). After 5 cycles of biosorption–desorption, there was no significant decrease in the biosorption efficiency. Therefore, the magnetic biosorbent used in this work provides a low cost and eco-friendly treatment for Cr(V) in wastewater.

Wang et al. investigated the biosorption potential of magnetic biochar from eucalyptus leaf residue obtained after the essential oil extraction process and using magnetic nanoparticles of iron for the removal of Cr(VI) (Wang et al. 2014). The magnetic hysteresis curve indicated a saturation magnetization of 16.00 emu g^{-1}, and the high magnetic susceptibility allowed that the magnetically modified biosorbent can be separated easily by applying a magnetic field. Under optimal experimental conditions, the removal proportions for Cr(VI) and total Cr were 97.11% and 97.63%, respectively.

Boron is a pollutant frequently present in contaminated water; high concentrations of this element are toxic to human health (Shaaban 2010). Tural et al. studied boron removal through the application of magnetic nanosorbent based on N-methyl-D-glucamine coated with iron oxide nanoparticles (Tural et al. 2018). The magnetic biosorbent was synthesized through two different processes, direct coupling and click chemistry. Direct coupling was performed in two steps. First, propylbromide was bound to magnetic nanoparticles. Then, the magnetic product obtained reacted with N-methyl-D-glucamine to produce the tertiary amine functionalized nanoparticles. On the other hand, click chemistry procedure implies a covalent reaction between brominated magnetic nanoparticles and NaN_3. After this process, the compound obtained was reacted with methyl propiolate, using Cu(I) to catalyze azide–alkyne cycloaddition, and then with N-methyl-D-glucamine in the presence of pure methanol and argon atmosphere to finally obtain the magnetic biosorbent. High magnetic susceptibility of 38.2 emu g^{-1} and 37.0 emu g^{-1} was obtained for the magnetic biosorbent prepared by both direct coupling and click chemistry, respectively. This property allows that the magnetically modified biosorbent can be easily separated from the solution, avoiding more time-consuming steps such as centrifugation or filtration. The results showed that the biosorbent obtained via click coupling provided a greater biosorption capacity and a higher removal percentage than that obtained by the direct coupling technique.

Omidinasab et al. used chitosan extracted from shrimp shells with magnetic nanoparticles for V(V) and Pd(II) from wastewaters (Omidinasab et al. 2018). In the first step, an amount of chitosan flakes was diluted in an acetic glacial solution. Thereafter, this solution was added dropwise to a suspension of magnetic nanoparticles, previously synthesized by coprecipitation of Fe(III) and Fe(II) in NH_4OH medium. The mixture was agitated under controlled conditions and then, the magnetic biosorbent was separated by applying an external magnetic field. The equilibrium time was achieved at 10 min for both metals. According to the results, the biosorption process conforms appropriately to the Freundlich model, ensuring that the surface of the sorbent is surrounded by multilayers of metal ions. On the other hand, kinetic data showed that pseudo second-order kinetic model was the most suitable, demonstrating that the biosorption is a combination of physical and chemical biosorption. In addition, thermodynamic studies showed that the biosorption was exothermic and spontaneous, with values of ΔG^0 between 0 and $-$ 20 kJ mol^{-1}, being in good agreement with isothermal model.

Fan et al. developed an interesting study for the removal of Ag(I) with magnetic thiourea–chitosan synthesized using Ag(I) as imprinted ions (Fan et al. 2011). The biosorption process was developed in batch at 30 °C, an optimum pH of five and a contact time of 50 min. After achieving equilibrium, the magnetic biosorbent was extracted with the use of a magnet. After this stage, Ag(I) concentrations were determined by an atomic absorption spectrophotometer, equipped with a hydride vapor generator at 200 nm. The biosorption capacity of the magnetized thiourea–chitosan impregnated with Ag(I) compared to the magnetized biosorbent without the cover of Ag(I) was studied. It was observed that the magnetic biosorbent impregnated with metal ions was 2.5 times better than the noncovered biosorbent. In addition, it was shown that the biosorbent maintains a removal percentage of 90% after five cycles of biosorption–desorption.

Magnetically modified *Kluyveromyces fragilis* cells was synthesized to remove Sr(II) from aqueous solutions (Ji et al. 2010). The mixture was stirred at a controlled temperature for 1 h. After that, the magnetic biomass was separated by applying a magnet. Finally, Sr(II) concentration was determined using a inductively coupled plasma mass spectrometer. In this work, the study of different experimental parameters was performed, including the effect of pH. It was observed that the biosorption capacity increased with increasing pH; this is due to an increase of anionic groups on the surface of the magnetically modified yeast, and subsequently it allows a greater interaction with Sr(II) ions positively charged. The maximum adsorption capacity was 140.8 mg g^{-1}. However, this parameter decreased at 13 mg g^{-1} after 3 cycles of reuse.

Cheng et al. also investigated Sr(II) removal using sawdust magnetically modified with Fe_3O_4 nanoparticles and chitosan as the bridging reagent (Cheng et al. 2012). The magnetic biosorbent prepared in this work was properly characterized by Fourier transform infrared spectroscopy and scanning electron microscopy. The effect of ionic strength was evaluated, and the results showed that an increase in $NaNO_3$ concentration significantly affected the biosorption of Sr(II) ions. This result may be due to the fact that the biosorption process is governed by electrostatic interactions. Contact time is another parameter that affects the biosorption capacity. It was observed that the biosorption equilibrium was achieved at 30 min; however, after 20 min the amount of Sr(II) ions biosorbed slightly decreased. The authors highlighted that this phenomenon may be due to the dissolution of metal ions adsorbed onto the magnetic biosorbent.

The emission of radioactive substances present in wastewater from nuclear industries has been increasing in recent years. Therefore, environmental protection controls and human health care are highly necessary (Sharma and Tomar 2008). In this sense, Li et al. investigated the removal of Eu(III) using *Paecilomyces catenlannulatus* with iron oxide nanoparticles (Li et al. 2018a). The magnetic biosorbent was synthesized by coprecipitation of Fe(II) and Fe(III) salts on the surface of the biomass in an atmosphere of N_2 and an alkaline medium. The point of zero charge was 3.68; therefore, at pH greater than 4.0, the magnetic biosorbent acquires a negative charge on its surface, which would allow a greater interaction with the Eu(III) cation. However, a high-level biosorption could be related to the formation of metal complexes, for example, $Eu(OH)_3(s)$ or $EuOHCO_3(s)$. The biosorption procedure was performed under optimal experimental conditions, obtaining a maximum biosorption capacity of 69.45 mg g^{-1} at pH 3.5 and 25 °C (Table 3.1). Moreover, it was shown that the biosorbent can be regenerated for 5 cycles without significantly decreasing its biosorption capacity.

Uranium is another radioactive pollutant that generates a great environmental impact. Saifuddin et al. investigated its removal through *Saccharomyces Cerevisiae* biomass onto crosslinked chitosan coated with Fe_3O_4 nanoparticles (Saifuddin and Dinara 2012). The biosorption process was performed by mixing an amount of magnetic biosorbent with a uranium solution, and after the biosorption, the suspension was easily separated by using an external magnet. A removal proportion of 98% was achieved at 110 min. Bai et al. applied yeast cells of *Rhodotorula glutinis* magnetically modified with Fe_3O_4 nanoparticles for the removal of uranium in wastewater (Bai et al. 2012). This sorbent was prepared by mixing a yeast cells suspension, previously washed several times with acetate buffer (pH 4.6), and Fe_3O_4 nanoparticles, keeping the mixture at a fixed temperature for 2 h without stirring. The equilibrium was reached within 20 min at pH 6 and with initial concentration of uranium of 100 mg L^{-1}. According to the authors, the biosorption conforms appropriately to the Langmuir model, demonstrating that uranium biosorption was performed through monolayer coverage on the surface of the magnetic biosorbent. Moreover, it was shown that the biosorption was a spontaneous and endothermic physicochemical procedure.

3.3.2 *Removal of Dyes*

Dyes are widely used to confer color to products in many industries such as textile, paper, and plastic (Tarhan et al. 2019). In this sense, a large quantity of colored wastewater that endangers human health and aquatic organisms is produced. These pollutants have proven to be toxic, carcinogenic, mutagenic, and teratogenic (Sivashankar et al. 2014). Therefore, from the environment point of view, their removal is of utmost importance. Among the different dyes, methylene blue is a cationic dye widely used in textiles and cosmetics industries, paper coatings and hygienic and medical applications (Ahsaine et al. 2018). However, it can cause adverse effects on humans, such as cyanosis, vomiting, nausea, diarrhea, tachycardia, and mental confusion (Clifton and Jerrold 2003). For these reasons, this dye has been particularly studied by several researchers.

Li et al. synthesized magnetic biosorbents derived from renewable resources (Li et al. 2018c). The researchers proposed the fabrication of a magnetic peach gum bead biosorbent in a simple one-step reaction. This one was based on the simultaneous formation of iron oxide nanoparticles and crosslinking of peach gum polysaccharide. Different parameters that influence the biosorption process of methylene blue on the magnetic biosorbent were studied, including pH, ionic strength, initial dye concentration, contact time, and temperature. The removal proportion remains greater than 85% after six cycles of reuse. The kinetics data were successfully described by the pseudo second-order kinetics model, while the biosorption isotherms are well fitted by Langmuir model with maximum biosorption capacity of 231.5 mg g^{-1}. Furthermore, the thermodynamic study suggested that biosorption was spontaneous and endothermic.

Magnetic iron oxide nanoparticles with cress seed mucilage were synthesized for removing methylene blue from aqueous solutions (Allafchian et al. 2019). First, the seeds were carefully cleaned and cress seed mucilage was extracted at pH 10, 35 °C, and a water-to-seed ratio of 30:1. Thereafter, the material was separated from the mixture and dried at room temperature for 48 h. The magnetic nanoparticles were synthesized in situ when powdered cress seed mucilage had been put in contact with Fe(III)/Fe(II) molar ratio of 2:1. As per the conclusion of the authors, the magnetic biosorbent with anionic functional groups would electrostatically attract positively charged dye molecules. Magnetic properties were studied by vibrating sample magnetometry at ambient temperature. The result demonstrated that the biosorbent has a superparamagnetic behavior with a saturation magnetization value of 48.96 emu g^{-1} (Table 3.2).

Saber–Samandari et al. developed the gelatin beads, high-molecular weight polypeptide, with carbon nanotubes that were functionalized together with iron-based magnetic nanoparticles (Saber-Samandari et al. 2017). This biosorbent was applied for methylene blue and red 80 dyes removal. The magnetic properties of the synthesized magnetic biosorbent were evaluated. The results showed a material with high saturation magnetization properties, which demonstrated its superparamagnetic behavior with no coercivity and hysteresis. The maximum biosorption efficiency was 73% of methylene blue and 96% of red 80 dyes.

Yang et al. showed a simple synthesis of monodispersed activated hierarchical porous carbon spheres from corn starch and its application for the removal of methylene blue from wastewater (Yang et al. 2016). The synthetized biosorbent had good magnetic properties and morphology. The work does not inform the conditions in which the biosorption study was performed, but it showed that about 97% of dye was biosorbed quickly and within 5 min.

Tural et al. also studied the biosorption of methylene blue from contaminated water using *Bacillus Subtilis* bacteria immobilized with nanosized magnetic silica (Tural et al. 2017). Techniques such as Fourier transform infrared spectroscopy, thermogravimetric analysis, vibrating sample magnetometry, and scanning electron

Table 3.2 Recently reported magnetic biosorbents for the removal of dyes from aqueous solutions and water samples

Biosorbent	Dye	Biosorption conditions				Biosorption capacity (mg g^{-1})	Saturation magnetization (emu g^{-1})	References
		pH	T (°C)	Time[a] (min)	C_0[b] (mg L^{-1})			
Magnetic peach gum bead	Methylene blue	7	20	60	1.25[c]	231.5[f]	n.r.	Li et al. (2018c)
Magnetic iron oxide with cress seed mucilage	Methylene blue	3	25	40	20[d]	44.6[f]	48.96	Allafchian et al. (2019)
Gelatin–carbon nanotubes–iron nanoparticles	Methylene blue Red 80	n.r.	21	≈140	1000	465.5 380.7	31	Saber-Samandari et al. (2017)
Monodispersed activated hierarchical porous carbon spheres	Methylene blue	n.r.	n.r.	5	n.r.	n.r.	11.91	Yang et al. (2016)
Bacillus subtilis with magnetic silica	Methylene blue	6.8	30	3	50	59 ± 0.6[f]	37.7	Tural et al. (2017)
Penicillium janthinellum strain fungus with Fe_3O_4	Congo red	n.r.	28 ± 1	360	50	102.4	n.r.	Zhang et al. (2016)
Magnetic chitosan nanoparticles	Metanil yel-low Reactive black 5	3	30	1020	0.67[c]	620 2549	n.r.	Tarhan et al. (2019)
Magnetically modified macroalgae *Cymopolia barbata*	Safranin O	6	n.r.	90	100	192.2[f]	n.r	Mullerova et al. (2019)
Graphene oxide functionalized magnetic chitosan composite	Acid red-17 Bromophenol blue	6 2	n.r.	60	10[e]	6.6 9.9	n.r.	Sohni et al. (2018)
κ-carrageenan-coated Fe_3O_4 nanoparticles.	Methylene blue	9	23	5	120	185.3	32.4	Salgueiro et al. (2013)
Magnetic sorghum husks	Methylene blue Crystal violet	6.14 10.41	30.18 30.84	≈40	48.92 47.59	30.04[f] 18.78[f]	n.r.	Adeogun et al. (2019)

(continued)

Table 3.2 (continued)

Biosorbent	Dye	Biosorption conditions				Biosorption capacity (mg g^{-1})	Saturation magnetization (emu g^{-1})	References
		pH	T (°C)	Time[a] (min)	C_0[b] (mg L^{-1})			
Magnetic biochar	Crystal violet	6	40	5	400	349.40	61.48	Sun et al. (2015)
Magnetic chicken bone biochar	Rhodamine B	10	26	≈500	100	113.31[f]	64.7	Oladipo and Ifebajo (2018)
Magnetic polydopamine–chitosan nanoparticles	Methylene blue Malachite green	5 6.5	30	≈30	30	109.25 56.79	n.r.	Wang et al. (2016)
Magnetic poly(vinyl alcohol)/modified gum tragacanth/graphene oxide hydrogel beads	Crystal violet Congo red	8 5	n.r.	900	20	94.0 101.74	n.r.	Sahraei et al. (2017)
Magnetic rye straw native and modified with citric acid–NaOH	Acridine Orange Methyl green	n.r.	n.r.	120	5	Native: 36.9[f] 78.1[f] Modified: 208.3[f] 357.1[f]	n.r.	Baldikova et al. (2015)
Carbon dots with magnetic $ZnFe_2O_4$	Methyl orange	5	20	About 40	20	109.7	n.r.	Shi et al. (2018)
Rice husk ash with magnetic iron oxide nanoparticles	Methyl orange	2	n.r.	120	150	254	78.8	Hosseinzadeh and Mohammadi (2016)
Magnetite reduced graphene oxide/chitosan nanocomposite	Remazol black Acid red 2	3	n.r.	150	20	n.r.	n.r.	Sheshmani and Mashhadi (2018)

Mn0.4Zn0.6Fe_2O_4 supported on *Yarrowia lipolytica*	Tartrazine Ponceau	2	25	5	12 16	90.827[f] 101.461[f]	n.r.	Asfaram et al. (2018)
Raspberry-like Fe_3O_4@yeast	Methylene blue	11	n.r.	n.r.	100	1221	n.r.	Song et al. (2015b)
Glutaraldehyde crosslinked magnetic chitosan nanoparticles	Metanil yellow	4	25	1020	12.5[c]	625[f]	n.r.	Tural et al. (2016b)
Magnetic polypyrrole nanocomposite	Methylene blue	6	30	60	150	270.27[f]	n.r.	Bai et al. (2015)

NR Not reported
[a]Equilibrium time
[b]C_0: Initial concentration
[c]Expressed as mmoles L^{-1}
[d]Expressed as mg g^{-1}
[e]Expressed as μg mL^{-1}
[f]Theoretical Q_{max}

microscope were used to characterize the magnetic biosorbent. Magnetic property analysis showed that the saturation magnetization of magnetic nanoparticles decreased intensely with the increase of 3-bromopropyl-trimethoxysilane and *B. Subtilis* organic groups. Desorption experiments were carried out in acid medium. The results showed that the reusability of biosorbent was decreased from 90 to 60% in five biosorption–desorption cycles.

Zhang et al. also used microorganism-based magnetic biosorbents to remove Congo red from aqueous solutions (Zhang et al. 2016). The fungal pellets were obtained from P*enicillium Janthinellum* strains. To provide magnetic properties to the biosorbent, Fe_3O_4 nanoparticles were incorporated during the culturing process. In this work, the authors researched the magnetic induction heating. The results reflected that, under the same biosorption temperature, magnetic induction heating had faster biosorption rates than that based on a traditional heating process. Also, the researchers highlighted that the use of magnetic induction heating can help to save energy for biosorption processes that require higher temperatures.

Tarhan et al. demonstrated the competitive behavior between two azo dyes—metanil yellow and reactive black 5—for a magnetic biosorbent composed of chitosan nanoparticles and glutaraldehyde crosslinked (Tarhan et al. 2019). Deionized water was used as eluent for regeneration of the magnetic biosorbent. Intraparticle diffusion model was used to analyze the competition biosorption of the two dyes under study. The results showed that the biosorption rates of the metanil yellow onto the biosorbent were slower than that of the reactive black 5 dye. Finally, the researchers explained that the high biosorption capacity of the biosorbent is attributed to the sharp electrostatic interaction between the $–NH_3^+$ of chitosan and dye anions.

Mullerova et al. showed that magnetically modified marine macroalgae *Cymopolia Barbata* is an excellent biosorbent to cationic dye safranin O, also known as basic red 2 (Mullerova et al. 2019). The biomass was magnetically modified using microwave synthesized magnetic iron oxide particles. The authors highlighted that modification did not require the drying step. However, the results suggested that *Cymopolia Barbata* biopolymers were principally responsible for dye biosorption. The kinetic data show greater adjustment through a pseudo second-order model. A time of 90 min was required to reach equilibrium. Adsorption isotherms were better correlated with both Freundlich and Langmur models. A maximum adsorption capacity of 192.2 mg g^{-1} was obtained, where the biosorption process was exothermic and spontaneous.

A magnetic chitosan functionalized graphene oxide composite has been also synthesized and applied for the removal of bromophenol blue and acid red-17 dyes from industrial wastewater (Sohni et al. 2018). The synthesis of the magnetic biosorbent was made in some simple steps (Fig. 3.6). The effect of the ionic strength was studied and the results showed that the ions did not interfere on the biosorption capacity of the biosorbent. Thus, the researchers emphasize its use to remove dyes from textile effluents that are rich in salts.

κ-carrageenan comprises a water-soluble sulfated polysaccharides family extracted from red sea weeds. The sulfonate anion groups of carrageenan are mainly

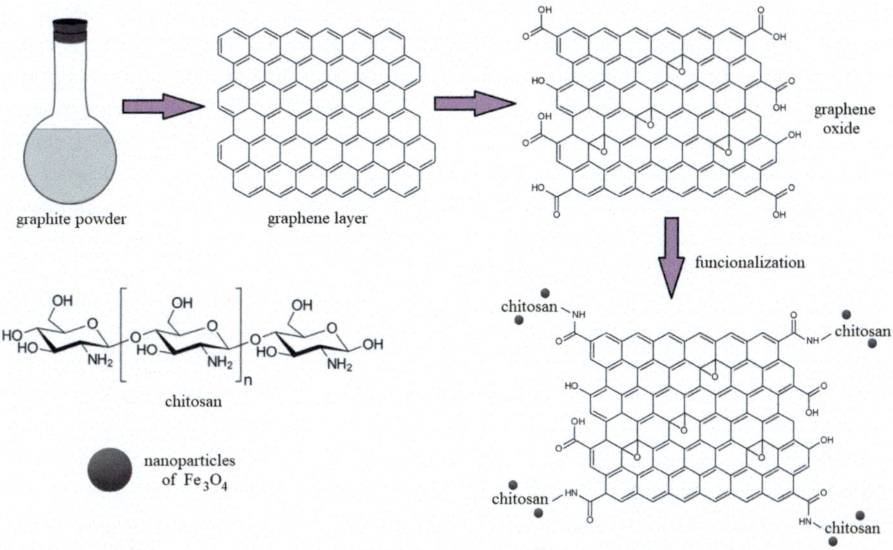

Fig. 3.6 Synthesis of graphene oxide functionalized magnetic chitosan composite biosorbent. Initially, graphene oxide was synthesized using modified Hummer's method. The, magnetic nanoparticles of Fe_3O_4 were impregnated with chitosan. Finally, the magnetic biosorbent was obtained by mixing graphene oxide and magnetic chitosan in continuously stirring at 50 °C for 90 min (Modified after Sohni et al. 2018)

responsible for the interaction between these polysaccharides and cationic dyes. Also, these compounds form hydrogels stabilized by alkali−metal cations. Therefore, thanks to these properties, the κ-carrageenans are interesting biosorbents for dyes. In this sense, Salgueiro et al. used this compound together with magnetic iron oxide nanoparticles for methylene blue removal (Salgueiro et al. 2013). The magnetic nanoparticles were synthetized through the typical coprecipitation method. The stability of magnetic nanoparticles was studied by determining the iron leached from these particles. The results showed that at an optimal pH (Table 3.2), Fe concentration in water was 8.6 mg L^{-1}. In concordance with the authors, under these experimental conditions, the stability of the magnetic nanoparticles is acceptable.

Adeogun et al. investigated the use of magnetic sorghum husk for the effective removal of methylene blue and crystal violet dyes (Adeogun et al. 2019). For the synthesis of the magnetic biosorbent, a suspension of sorghum husk was put in contact with ferric and ferrous chloride solution in ratio 2:1. The absorbance was determined at 665 nm for methylene blue and 586 nm for crystal violet. The optimization of the factors studied for the biosorption of dyes was made by the Box–Behnken method. Under optimum conditions, removal proportions of 86.1% and 99.0% were obtained for crystal violet and methylene blue, respectively (Table 3.2).

Sun et al. studied the biosorption potential of magnetic biochar, using biomass of plants processed by pyrolysis method, coated with magnetic nanoparticles of iron for the removal of crystal violet (Sun et al. 2015). The magnetic hysteresis curve showed the ferromagnetic property of the biosorbent and indicated a saturation magnetization of 61.48 emu g^{-1}. Thermodynamic evaluation established that the biosorption was a spontaneous and endothermic physical process. The magnetic biochar also was used by Oladipo et al. for rhodamine B dye biosorption (Oladipo and Ifebajo 2018). In this work, waste chicken bones were selected for the synthesis of biochar. The magnetism was achieved via coprecipitations using iron salts. The authors assure that this biosorbent is adequate to minimize investment costs in a small-to-medium scale industrial treatment plants.

Wang et al. developed an attractive hybrid nanobiosorbent by assembling a biomimetic polymer and chitosan onto magnetic nanoparticles based on Schiff base reaction (Wang et al. 2016). The removal of methylene blue and malachite green was studied. Under optimal conditions (Table 3.2), 96.9% and 92.5% of removal was achieved for methylene blue and malachite green, respectively. Due to the high surface area of nanoparticles and the many active sites of polydopamine and chitosan, the magnetic biosorbent presented a great capacity of biosorption.

Sahraei et al. created magnetic nanocomposite hydrogel beads by using the gelation method in a solution of boric acid to remove crystal violet and Congo red dyes (Sahraei et al. 2017). This biosorbent was synthetized by gum tragacanth, polyvinyl alcohol, and graphene oxide, Fe_3O_4 nanoparticles, and finally was crosslinking with glutaraldehyde. The authors underscore the importance of using gum tragacanth as low-cost natural polysaccharide because it is nontoxic, biocompatible, and stable over a wide pH range. The results demonstrated the great capacity of this new magnetic biosorbent to simultaneously remove several pollutants.

Baldikova et al. used native and chemically modified rye straw as magnetic biosorbent to remove acridine orange and methyl green from aqueous solutions (Baldikova et al. 2015). Microwave was used for magnetic biosorbents synthesis. In accordance with the authors, the magnetic biosorbents could be kept in water suspension at 4 °C for more than 6 months without changes. It was observed that chemical modification with citric acid–NaOH increased more than four times the maximum biosorption capacity. This behavior could be due to the presence of additional functional groups on the biosorbent surface that could retain the pollutants.

On the other hand, many hydroxyl and carboxyl groups can be found on the surface of carbon dots, turning these biosorbents as good candidates to remove dyes. Shi et al. combined carbon dots with magnetic $ZnFe_2O_4$ nanoparticles for the removal of methyl orange from water samples (Shi et al. 2018). The biosorption capacity of $ZnFe_2O_4$ nanoparticles increases from 75.5 to 109.7 mg g^{-1} when increasing the contents of carbon dots from 1 to 5%. This behavior demonstrated that the use of hybrid materials like magnetic biosorbents is useful for decontamination of water, in this case through the introduction of carbon dots in the nanomaterial, providing a new hybrid material with more vacant surface sites.

Hosseinzadeh et al. investigated the potential use of an agricultural waste of rice husk ash with magnetic iron oxide nanoparticles for the removal methyl orange (Hosseinzadeh and Mohammadi 2016). The reuse ability of the magnetic biosorbent was studied with ethanol during some biosorption–desorption cycles. After the fifth cycle, the removal efficiency of methyl orange was decreased to 61%. The isotherm models demonstrated that the Langmuir model fits well the biosorption equilibrium data, reaching a maximum biosorption capacity of 254 mg g^{-1}.

Sheshmani et al. studied the efficiency of magnetite reduced graphene oxide/chitosan nanocomposite as biosorbent for biosorption of remazol black and acid red 22 from aqueous solutions (Sheshmani and Mashhadi 2018). The authors studied different parameters that influence the biosorption of dyes, such as pH, sorbent mass, contact time, and dye concentrations. The optimal conditions of these parameters are shown in Table 3.2. In accordance with the authors, the biosorption equilibrium data of remazol black and acid red 22 could be fitted with Langmuir and Freundlich models, respectively.

Asfaram et al. developed a new ferromagnetic biosorbent formed by $Mn0.4Zn0.6Fe_2O_4$ nanoparticles supported on dead *Yarrowia lipolytica* to simultaneously remove tartrazine and ponceau 4R (Asfaram et al. 2018). Ultrasound, magnetic stirrer, and vortex-assisted biosorption were studied. Under optimal conditions, more than 99.0% of removal efficiencies were obtained for both dyes. Finally, the result showed that electrostatic interactions were one of the main mechanisms responsible for the dye's biosorption.

An interesting work was developed by Song et al. where raspberry-like Fe_3O_4@yeast magnetic microspheres were synthesized by a simple electrostatic interaction-driven self-assembly heterocoagulation (Song et al. 2015b). This new material was applied in a continuous, upflow, fixed-bed column system to remove methylene blue from wastewater. The flow rate and bed height were kept constant at 5 mL min^{-1} and 1.2 cm, respectively. The retention process was controlled by electrostatic interactions between the magnetic biosorbent and the dye.

Tural et al. researched the use of glutaraldehyde crosslinked magnetic chitosan nanoparticles synthetized by coprecipitation method for the removal of metanil yellow (Tural et al. 2016b). The data showed that the biosorption of the dye onto the biosorbent was performed by a physical process. The magnetic biosorbent can be regenerated and reused efficiently using alkaline solution at pH 10.0 for three repeated cycles. Bai et al. showed the use of a magnetic polypyrrole nanocomposite as biosorbent for the removal of methylene blue from water (Bai et al. 2015). The simple synthesis of the biosorbent is shown in Fig. 3.7. The results of characterization demonstrated that the specific surface area of biosorbent increased more than 5 times with the addition of graphene oxide (162.6 m^2 g^{-1}). Within 5 min, it achieved 90% of the biosorption removal through strong electrostatic interactions between the negatively charged surface of biosorbent and the cationic dye.

3.3.3 Removal of Other Contaminants

The use of pesticides to control pests is an important factor for agriculture, which plays a critical role in worldwide economy. However, the pesticide residues used in

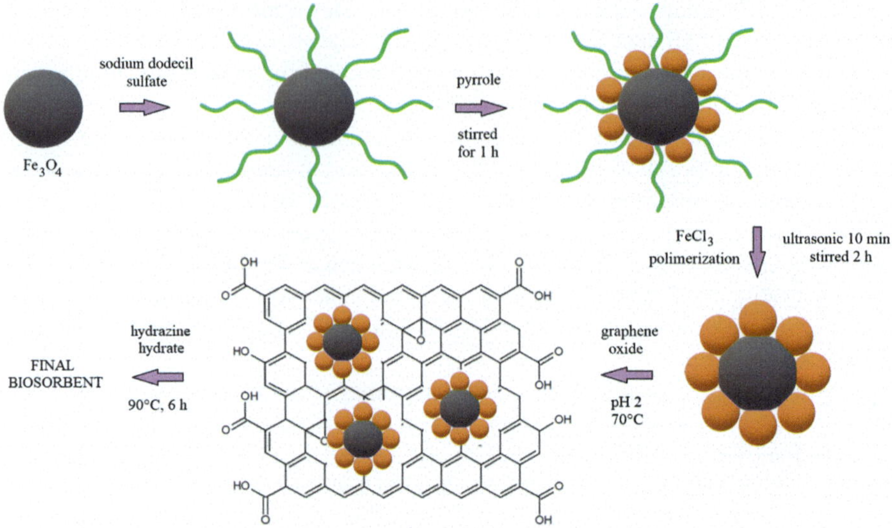

Fig. 3.7 Schematic illustration of the magnetic polypyrrole nanocomposite synthesis. Initially, Fe_3O_4 nanoparticles were prepared through a modified solvothermal reaction. Then, these microspheres together with sodium dodecyl sulfate were stirred in a flask. Pyrrole was added and the mixture was stirred for 1 h and by adding $FeCl_3$, the polymerization was started thereafter. Finally, graphene oxide solution prepared according to the Hummers method was added dropwise. The resulting composite was then reduced with hydrazine hydrate to synthesize polypyrrole with reduced graphene oxide–Fe_3O_4 magnetic biosorbent (Modified after Bai et al. 2015)

agriculture such as herbicide, insecticides, and fungicides have been incorporated into the waterbodies impacting on the ecosystem (Hussein et al. 2017). Furthermore, these compounds could cause adverse effects on the human health such as totipotency, absorption blocks of nutrients, and disruption for endocrine system (ul Haq et al. 2020).

In this regard, paraquat is an herbicide extensively used due to its high efficiency, for being cheap, nonselective, and fast action. However, this compound is highly toxic to both humans and animals. Moreover, its high water solubility appreciably contributes to a greater risk of water contamination (Tsai 2013). Fernandez et al. studied the removal of paraquat from water using magnetic nanosorbents constituted by iron oxide nanoparticles coupled with the *k*-carrageenan biopolymer into the siliceous shells (Fernandes et al. 2017). According to the data, under optimal conditions, 95% of paraquat was removed by using this magnetic biosorbent (Table 3.3). The authors highlight that this behavior is due to the anionic sulfonate groups of the biopolymer that interacted with the cationic form of paraquat at pH 7.3. A good regeneration capacity could be attained and four biosorption–desorption cycles were successfully applied.

Phenol and its compounds are widely used as pesticides. Tural et al. researched the removal of phenol, 2-chlorophenol, and 4-chlorophenol from wastewater with chitosan-coated magnetite nanoparticles (Tural et al. 2016a). Variables such as pH,

Table 3.3 Recently reported magnetic biosorbents for the removal of other pollutants from aqueous solutions and water samples

		Biosorption conditions						
Biosorbent	Pollutants	pH	T (°C)	Time (min)	Initial concentration (mg L^{-1})[a]	Biosorption capacity (mg g^{-1})[b]	Saturation magnetization (emu g^{-1})	References
Iron oxide nanoparticles coupled with the k-carrageenan biopolymer into the siliceous shells	Paraquat	7.3	25	180	90	257	n.r.	Fernandes et al. (2017)
Chitosan-coated magnetite nanoparticles	Phenol 2-chlorophenol 4-chlorophenol	About 6.5	30	30	50	196[c] 244[c] 323[c]	66.0	Tural et al. (2016a)
Magnetite nanoparticles functionalized by k-carrageenan with siliceous shells	Metoprolol	7	25	480	n.r.	78.5	n.r.	Soares et al. (2016)
Sulfhydryl-terminated, silica-coated iron oxide beads	Patulin	7.2	25	240	5	n.r.	n.r.	Bayraç and Camızcı (2019)
Nanoparticles and extracellular polymeric substances from *Klebsiella sp.* J1	Tetracycline	6	30	n.r.	10	56.04	18.45	Pi et al. (2017)
Magnetic nano-ferromanganese oxide-modified biochar	Tetracycline	4	45	900	100	171.71	n.r.	Liang et al. (2019)
Genipin crosslinked chitosan and this, coupled with graphene oxide–SO_3H	Tetracycline Ibuprofen	6	40 35	120	200 10	492.7 158.77	n.r.	Liu et al. (2019)
Cysteine-modified silane-coated magnetic nanomaterial	Ibuprofen	6	n.r.	30	50	138.[c]	8.86	Kollarahithlu and Balakrishnan (2018)
Magnetite cores within a siliceous network enriched with quaternary chitosan	Diclofenac	6	25	300	100	240.[c]	n.r.	Soares et al. (2019a)

(continued)

Table 3.3 (continued)

Biosorbent	Pollutants	Biosorption conditions				Biosorption capacity (mg g^{-1})[b]	Saturation magnetization (emu g^{-1})	References
		pH	T (°C)	Time (min)	Initial concentration (mg L^{-1})[a]			
Magnetic carbon from *Salix psammophila* hydrochar	Triclosan	n.r.	30	n.r.	50	892.9[c]	6.40–14.98	Zhu et al. (2014)
Magnetic activated carbon from waste rice straw hydrochar	Triclosan	n.r.	n.r.	n.r.	40	292	n.r.	Liu et al. (2014)
ZnO nanoparticle modified magnetic biochar derived from camphor leaves	Ciprofloxacin	4	40	240	n.r.	168.4	3.18	Hu et al. (2019)

NR Not reported
[a]Equilibrium time
[b]C_0: Initial concentration
[c]Theoretical Q_{max}

contact time, biosorbent dosage, and initial pollutants concentration were studied by batch tests. Molecular interactions between phenols and the magnetic biosorbent included hydrogen bonding, hydrophobic interaction, and van der Waals forces. In addition, the magnetic biosorbent could be regenerated and reused for five repeated cycles using a basic solution at pH 11.0.

Metoprolol is a selective beta-blocker used in the treatment of severe diseases of the cardiovascular system, especially hypertension and infarction. Although ecotoxicity data are not fully available, beta-blockers can adversely impact on environment and human health even at low concentrations (Assimon et al. 2018). In this regard, Soares et al. developed a hybrid nanoadsorbent composed of magnetite nanoparticles functionalized by *k*-carrageenan with siliceous shells to remove metoprolol from aqueous solutions (Soares et al. 2016). It seems that the metoprolol biosorption very quickly reached the highest response after 15 min. However, the biosorption process was stabilized after 8 h of contact time. The FTIR analysis suggested that the biosorption mechanism is mostly based on electrostatic interactions between protonated amine groups of pollutant and sulfonate groups of polymers *k*-carrageenan.

Patulin, 4-hydroxy-4H-furo[3,2c]pyran-2(6H)-one, is a naturally produced toxic secondary metabolite during the growth of certain fungi on fruits, especially on apples (Luo et al. 2016). This mycotoxins cause different health problems for human and animals including edema, ulceration, inflammation, vomiting, and hemorrhages. Thus, its control is an important factor for providing food safety (WHO 1995). In this context, Bayraç et al. studied the removal of patulin using sulfhydryl-terminated, silica-coated iron oxide beads from both aqueous solution and apple juice (Bayraç and Camızcı 2019). Patulin levels were measured by high-performance liquid chromatography–diode array detector. The results showed that the removal efficiency was 71.25%.

Antibiotics have been worldwide used for the treatment of human and animal diseases. In spite of this, antibiotics are one of the most frequently detected organic contaminants in medical and aquacultural wastewater (Li et al. 2018b). Hence, pharmaceutical pollution has become an important problem for human, animals, and environment over the world. Among all kinds of antibiotics, tetracycline is one the most typical used in human and veterinary medicine. Its residues have been detected in various waterbodies that results in a huge threat to human health and aquatic ecosystems. So, tetracycline has been widely studied by the scientific communities. Pi et al. studied the use of a magnetic biosorbent formed by Fe_3O_4 nanoparticles and extracellular polymeric substances from *Klebsiella sp.* J1 to remove tetracycline from aqueous solutions (Pi et al. 2017). In concordance with the results, both physical and chemical adsorption took place in the removal process. The first process is occasioned from Van der Waals forces, while ion exchange caused chemical biosorption. Furthermore, Fourier transform infrared spectroscopy analysis demonstrated that carboxyl groups intervened in the biosorption of tetracycline on the new magnetic biosorbent.

Liang et al. also studied the removal of tetracycline but by applying magnetic nanoferromanganese oxide-modified biochar derived from pine sawdust (Liang et al.

2019). The pine sawdust wastes were taken from the furniture manufacturing factory and used as raw material for biochar synthesis. The biochar was then mixed with iron nanoparticles and the tube furnace pyrolysis method was made to synthesize the magnetic biosorbent. The researchers evaluated the environmental toxicity of magnetic biosorbent with seeds of Chinese cabbage, and the results demonstrated that it had no damage on the environment. The data showed that electrostatic biosorption, hydroxyl binding, and π–π interactions were responsible for the removal process of tetracycline.

Liu et al. developed a new biosorbent which is composed of genipin crosslinked chitosan and this coupled with graphene oxide–SO_3H (Liu et al. 2019). According to previous studies, the genipin is usually used to improve the mechanical stability of chitosan in an extensive pH range. As the presence of sulfo groups allows stable complexes with many pharmaceutical compounds, this work studied the use of this magnetic biosorbent for the removal of tetracycline and ibuprofen from aqueous solutions through batch experiments. Good biosorption capacities were obtained for both pollutants (Table 3.3). Kollarahithlu et al. also studied ibuprofen biosorption (Kollarahithlu and Balakrishnan 2018). In this work, nickel ferrite nanoparticles by coprecipitation method were synthetized and then these nanoparticles were functionalized with 3 glycidyloxypropyltrimethoxysilane. Finally, the surface of the biosorbent was modified with L-cysteine. Under optimal experimental conditions, 83% of removal efficiency was obtained.

Soares et al. synthetized a magnetic hybrid biosorbent formed of magnetite cores within a siliceous network enriched with quaternary chitosan by biosorption of sodium diclofenac from water (Soares et al. 2019a). Pollutant concentration was determined spectrophotometrically at 276 nm. The zeta potential analysis demonstrated that there was a positive surface charge on biosorbent between pH 2 and 7. This behavior was highlighted for the researchers due to its application being possible in real samples from acidic to neutral pH values to remove anionic contaminants.

Triclosan, 5-chloro-2-(2,4-dichlorophenoxy)phenol is another antibiotic widely evaluated in removal studies. Zue et al. analyzed the removal of triclosan from aqueous solutions (Zhu et al. 2014). In this work, the fabrication of magnetic carbon composites via the thermal pyrolysis of hydrochar that has been pretreated with mixtures $FeCl_3$ and $ZnCl_2$ was proposed. The hydrochar used was a solid residual of the hydrothermal carbonization of *Salix psammophila* biomass. The antibiotic level was determined by HPLC with ultraviolet detector and the wavelength was set at 280 nm. In concordance with the researchers, the biosorbent synthetized was highly effective and its fabrication was very simple. Liu et al. also applied a magnetic activated carbon prepared from hydrochar for triclosan biosorption (Liu et al. 2014). In this work, waste rice straw was used for the fabrication of hydrochar via chemical activation by K_2CO_3. Hydrothermal liquefaction of residual rice straw allows the production of biofuels and hydrocarbons, which cannot be used as biosorbents due to their limited porosity and surface area. The addition of activated magnetic carbon allowed improving the porosity and adsorption capacity, reaching a surface area of $674\ m^2\ g^{-1}$.

Hu et al. developed an efficient biosorbent ZnO nanoparticle modified magnetic biochar derived from camphor leaves for ciprofloxacin removal (Hu et al. 2019). Ciprofloxacin is an antibiotic from the fluoroquinolones group with bactericidal effects. The data showed that a biosorption capacity is reduced 76% after 3 cycles. Finally, the researchers demonstrated that both physisorption and chemisorption have taken place in biosorption process.

3.4 Conclusions, Trends, and Future Perspectives

Magnetic biosorbents have emerged in recent decades as promising materials in the field of environmental remediation. Magnetic biosorbents have been used in numerous processes for the removal of inorganic and organic pollutants from different environmental matrices, such as natural and industrial waters. The presence of a magnetic component in this type of biomaterials has surpassed the conventional biosorbents, punctually in the phase separation phase that occurs between the biosorbent that has the retained contaminant and the aqueous phase. Magnetic biosorbents have very good magnetic susceptibility values, which allow performing the separation step with magnet only, avoiding time-consuming stages such as filtration or centrifugation. On the other hand, magnetic biosorbents have a large specific surface area and the presence of functional groups on their surface is varied, which improves their biosorption potential. Moreover, the versatility of magnetic biosorbents toward both organic (e.g., dyes, antibiotics, and hormones) and inorganic (e.g., lead, cadmium, antimony, and thallium) pollutants is demonstrated in this chapter. Thus, all the works mentioned in the chapter report high removal percentages of contaminants, attaining values between 70% and 100%.

However, so far there are numerous studies based on the use of magnetic biosorbents for the removal of pollutants from aqueous solutions, so it is mandatory that researchers continue working on the application of these materials in real samples. On the other hand, most of the biosorption processes are carried out in batches, which is a limitation when it is necessary to clean large volumes of water. In this sense, it is desirable that additional studies using continuous processes will be made, evaluating the different varieties of the flow systems.

Finally, there is a tendency toward the synthesis and application of new hybrid materials for the removal of pollutants from environmental matrices, which exceed the properties of basic binary magnetic biosorbents. In this way, new complex materials that are composed of three or more components are trends for the adsorption studies. For instance, hybrid materials that have a biodegradable component, such as bacteria; a magnetic component, such as magnetic nanoparticles; and a component that provides greater variety and functional groups, such as a chelating molecule as ethylenediaminetetraacetic acid are required in the near future for environmental care.

Acknowledgments The authors thank Consejo Nacional de Investigaciones Científicas y Técnicas (CONICET), Agencia Nacional de Promoción Científica y Tecnológica (FONCYT) (Project PICT−2015–1338), and Universidad Nacional de Cuyo (Project M073) for the financial support.

References

Abdullah N, Yusof N, Lau WJ, Jaafar J, Ismail AF (2019) Recent trends of heavy metal removal from water/wastewater by membrane technologies. J Ind Eng Chem 76:17–38. https://doi.org/10.1016/j.jiec.2019.03.029

Addo IB, Thoms MC, Parsons M (2019) The influence of water-conservation messages on reducing household water use. Appl Water Sci 9:126. https://doi.org/10.1007/s13201-019-1002-0

Adeogun AI, Akande JA, Idowu MA, Kareem SO (2019) Magnetic tuned sorghum husk biosorbent for effective removal of cationic dyes from aqueous solution: isotherm, kinetics, thermodynamics and optimization studies. Appl Water Sci 9:160. https://doi.org/10.1007/s13201-019-1037-2

Ahsaine HA, Zbair M, Anfar Z, Naciri Y, El Alem N, Ezahri M (2018) Cationic dyes adsorption onto high surface area 'almond shell' activated carbon: kinetics, equilibrium isotherms and surface statistical modeling. Mater Today Chem 8:121–132. https://doi.org/10.1016/j.mtchem.2018.03.004

Al-Homaidan AA, Al-Houri HJ, Al-Hazzani AA, Elgaaly G, Moubayed NM (2014) Biosorption of copper ions from aqueous solutions by *Spirulina platensis* biomass. Arab J Chem 7:57–62. https://doi.org/10.1016/j.arabjc.2013.05.022

Allafchian A, Mousavi ZS, Hosseini SS (2019) Application of cress seed musilage magnetic nanocomposites for removal of methylene blue dye from water. Int J Biol Macromol 136:199–208. https://doi.org/10.1016/j.ijbiomac.2019.06.083

Ambashta RD, Sillanpää M (2010) Water purification using magnetic assistance: a review. J Hazard Mater 180:38–49. https://doi.org/10.1016/j.jhazmat.2010.04.105

Anuradha J, Abbasi T, Abbasi S (2015) An eco-friendly method of synthesizing gold nanoparticles using an otherwise worthless weed pistia (*Pistia stratiotes L.*). J Adv Res 6:711–720. https://doi.org/10.1016/j.jare.2014.03.006

Arief VO, Trilestari K, Sunarso J, Indraswati N, Ismadji S (2008) Recent progress on biosorption of heavy metals from liquids using low cost biosorbents: characterization, biosorption parameters and mechanism studies. Clean Soil Air Water 36:937–962. https://doi.org/10.1002/clen.200800167

Asfaram A, Ghaedi M, Dashtian K, Ghezelbash GR (2018) Preparation and characterization of Mn0. 4Zn0. $6Fe_2O_4$ nanoparticles supported on dead cells of Yarrowia lipolytica as a novel and efficient adsorbent/biosorbent composite for the removal of azo food dyes: central composite design optimization study. ACS Sustain Chem Eng 6:4549–4563. https://doi.org/10.1021/acssuschemeng.7b03205

Ashish B, Neeti K, Himanshu K (2012) Copper toxicity: a comprehensive study. J Recent Sci 2:58–67. Not available doi

Assimon MM, Brookhart MA, Fine JP, Heiss G, Layton JB, Flythe JE (2018) A comparative study of carvedilol versus metoprolol initiation and 1-year mortality among individuals receiving maintenance hemodialysis. Am J Kidney Dis 72:337–348. https://doi.org/10.1053/j.ajkd.2018.02.350

Ayangbenro AS, Babalola OO (2017) A new strategy for heavy metal polluted environments: a review of microbial biosorbents. Int J Environ Res Public Health 14:94. https://doi.org/10.3390/ijerph14010094

Bagal-Kestwal D, Karve MS, Kakade B, Pillai VK (2008) Invertase inhibition based electrochemical sensor for the detection of heavy metal ions in aqueous system: application of ultra-microelectrode to enhance sucrose biosensor's sensitivity. Biosens Bioelectron 24:657–664. https://doi.org/10.1016/j.bios.2008.06.027

Bai J, Wu X, Fan F, Tian W, Yin X, Zhao L, Fan F, Li Z, Tian L, Qin Z (2012) Biosorption of uranium by magnetically modified *Rhodotorula glutinis*. Enzym Microb Technol 51:382–387. https://doi.org/10.1016/j.enzmictec.2012.08.007

Bai L, Li Z, Zhang Y, Wang T, Lu R, Zhou W, Gao H, Zhang S (2015) Synthesis of water-dispersible graphene-modified magnetic polypyrrole nanocomposite and its ability to efficiently adsorb methylene blue from aqueous solution. Chem Eng J 279:757–766. https://doi.org/10.1016/j.cej.2015.05.068

Baldikova E, Safarikova M, Safarik I (2015) Organic dyes removal using magnetically modified rye straw. J Magn Magn Mater 380:181–185. https://doi.org/10.1016/j.jmmm.2014.09.003

Bayraç C, Camızcı G (2019) Adsorptive removal of patulin from apple juice via sulfhydryl-terminated magnetic bead-based separation. J Hazard Mater 366:413–422. https://doi.org/10.1016/j.jhazmat.2018.12.001

Benedict S, Hussein H (2019) An analysis of water awareness campaign messaging in the case of Jordan: water conservation for state security. Water 11:1156. https://doi.org/10.3390/w11061156

Bohara RA, Thorat ND, Pawar SH (2016) Role of functionalization: strategies to explore potential nano-bio applications of magnetic nanoparticles. RSC Adv 6:43989–44012. https://doi.org/10.1039/C6RA02129H

Bonnelle C, Mandé C (2013) Advances in X-ray spectroscopy: contributions in honour of Professor Y. Cauchois. Elsevier, Burlington

Bulgariu L, Escudero LB, Bello OS, Iqbal M, Nisar J, Adegoke KA, Alakhras F, Kornaros M, Anastopoulos I (2019) The utilization of leaf-based adsorbents for dyes removal: a review. J Mol Liq 276:728–747. https://doi.org/10.1016/j.molliq.2018.12.001

Carpenter AW, de Lannoy CF, Wiesner MR (2015) Cellulose nanomaterials in water treatment technologies. Environ Sci Technol 49: 5277–5287. https://doi.org/10.1021/es506351r

Carrasco L, Benejam L, Benito J, Bayona JM, Díez S (2011) Methylmercury levels and bioaccumulation in the aquatic food web of a highly mercury-contaminated reservoir. Environ Int 37:1213–1218. https://doi.org/10.1016/j.envint.2011.05.004

Carvalho RS, da Silva AL, Trindade T (2016) Uptake of europium (III) from water using magnetite nanoparticles. Part Part Syst Charact 33:150–157. https://doi.org/10.1002/ppsc.201500170

Chen S-H, Yip S (2017) Spectroscopy in biology and chemistry: neutron, X-ray, laser. Academic, New York

Cheng Z, Gao Z, Ma W, Sun Q, Wang B, Wang X (2012) Preparation of magnetic Fe_3O_4 particles modified sawdust as the adsorbent to remove strontium ions. Chem Eng J 209:451–457. https://doi.org/10.1016/j.cej.2012.07.078

Chiaradia V, Valério A, Feuser PE, de Oleiveira D, Araújo PH, Sayer C (2015) Incorporation of superparamagnetic nanoparticles into poly (urea-urethane) nanoparticles by step growth interfacial polymerization in miniemulsion. Colloids Surf A Physicochem Eng Asp 482:596–603. https://doi.org/10.1016/j.colsurfa.2015.06.035

Chung J-Y, Yu S-D, Hong Y-S (2014) Environmental source of arsenic exposure. J Prev Med Public Health 47:253. https://doi.org/10.3961/jpmph.14.036

Clifton J, Jerrold L (2003) Methylene blue. Am J Ther 10:289–291

Coman V, Robotin B, Ilea P (2013) Nickel recovery/removal from industrial wastes: a review. Resour Conserv Recycl 73:229–238. https://doi.org/10.1016/j.resconrec.2013.01.019

Daneshfozoun S, Abdullah M, Abdullah B (2017) Preparation and characterization of magnetic biosorbent based on oil palm empty fruit bunch fibers, cellulose and *Ceiba pentandra* for heavy metal ions removal. Ind Crop Prod 105:93–103. https://doi.org/10.1016/j.indcrop.2017.05.011

Dehabadi L, Wilson LD (2014) Polysaccharide-based materials and their adsorption properties in aqueous solution. Carbohydr Polym 113:471–479. https://doi.org/10.1016/j.carbpol.2014.06.083

Dhand V, Soumya L, Bharadwaj S, Chakra S, Bhatt D, Sreedhar B (2016) Green synthesis of silver nanoparticles using *coffea arabica* seed extract and its antibacterial activity. Mater Sci Eng C 58:36–43. https://doi.org/10.1016/j.msec.2015.08.018

El-Alaily T, El-Nimr M, Saafan S, Kamel M, Meaz T, Assar S (2015) Construction and calibration of a low cost and fully automated vibrating sample magnetometer. J Magn Magn Mater 386:25–30. https://doi.org/10.1016/j.jmmm.2015.03.051

Escudero LB, Quintas PY, Wuilloud RG, Dotto GL (2018) Biosorption of metals and metalloids. In: Crini G, Lichtfouse E (eds) Green adsorbents for pollutant removal: innovative materials. Springer International Publishing, Cham, pp 35–86

Escudero LB, Quintas PY, Wuilloud RG, Dotto GL (2019) Recent advances on elemental biosorption. Environ Chem Lett 17:409–427. https://doi.org/10.1007/s10311-018-0816-6

Fan L, Luo C, Lv Z, Lu F, Qiu H (2011) Removal of Ag(I) from water environment using a novel magnetic thiourea-chitosan imprinted Ag(I). J Hazard Mater 194:193–201. https://doi.org/10.1016/j.jhazmat.2011.07.080

Fernandes T, Soares S, Trindade T, Daniel-da-Silva A (2017) Magnetic hybrid nanosorbents for the uptake of paraquat from water. Nano 7:68. https://doi.org/10.3390/nano7030068

Gautam RK, Rawat V, Banerjee S, Sanroman MA, Soni S, Singh SK, Chattopadhyaya MC (2015) Synthesis of bimetallic Fe–Zn nanoparticles and its application towards adsorptive removal of carcinogenic dye malachite green and Congo red in water. J Mol Liq 212:227–236. https://doi.org/10.1016/j.molliq.2015.09.006

Gebhart E (2008) Mutagenicity, carcinogenicity, and teratogenicity, Elements and their compounds in the environment: occurrence, analysis and biological relevance, pp 433–457

Goldstein JI, Newbury DE, Michael JR, Ritchie NW, Scott JHJ, Joy DC (2017) Scanning electron microscopy and X-ray microanalysis. Springer, New York

Guler UA, Ersan M (2016) *S. cerevisiae* cells modified with nZVI: a novel magnetic biosorbent for nickel removal from aqueous solutions. Desalin. Desalin Water Treat 57:7196–7208. https://doi.org/10.1080/19443994.2015.1013992

Gupta PL, Jung H, Tiwari D, Kong S-H, Lee S-M (2017) Insight into the mechanism of Cd(II) and Pb(II) removal by sustainable magnetic biosorbent precursor to *Chlorella vulgaris*. J Taiwan Inst Chem Eng 71:206–213. https://doi.org/10.1016/j.jtice.2016.12.007

Gupta N, Pant P, Gupta C, Goel P, Jain A, Anand S, Pundir A (2018) Engineered magnetic nanoparticles as efficient sorbents for wastewater treatment: a review. Mater Res Innov 22:434–450. https://doi.org/10.1080/14328917.2017.1334846

Haas KL, Franz KJ (2009) Application of metal coordination chemistry to explore and manipulate cell biology. Chem Rev 109:4921–4960. https://doi.org/10.1021/cr900134a

Hashemian S, Salimi M (2012) Nano composite a potential low cost adsorbent for removal of cyanine acid. Chem Eng J 188:57–63. https://doi.org/10.1016/j.cej.2012.02.008

Hassan M, Naidu R, Du J, Liu Y, Qi F (2020) Critical review of magnetic biosorbents: their preparation, application, and regeneration for wastewater treatment. Sci Total Environ 702:134893. https://doi.org/10.1016/j.scitotenv.2019.134893

He Y, Wu P, Xiao W, Li G, Yi J, He Y, Chen C, Ding P, Duan Y (2019a) Efficient removal of Pb (II) from aqueous solution by a novel ion imprinted magnetic biosorbent: adsorption kinetics and mechanisms. PLoS One 14:e0213377. https://doi.org/10.1371/journal.pone.0213377

He Y, Xiao W, Li G, Yang F, Wu P, Yang T, Chen C, Ding P (2019b) A novel lead-ion-imprinted magnetic biosorbent: preparation, optimization and characterization. Environ Technol 40:499–507. https://doi.org/10.1080/09593330.2017.1397762

Hosseinzadeh H, Mohammadi S (2016) Biosorption of anionic dyes from aqueous solutions using a novel magnetic nanocomposite adsorbent based on rice husk ash. Sep Sci Technol 51:939–953. https://doi.org/10.1080/01496395.2016.1142564

Hrubovčák P, Zeleňáková A, Zeleňák V, Kováč J (2015) The study of magnetic properties and relaxation processes in Co/Au bimetallic nanoparticles. J Alloys Compd 649:104–111. https://doi.org/10.1016/j.jallcom.2015.07.044

Hu J, Shao D, Chen C, Sheng G, Li J, Wang X, Nagatsu M (2010) Plasma-induced grafting of cyclodextrin onto multiwall carbon nanotube/iron oxides for adsorbent application. J Phys Chem B 114:6779–6785. https://doi.org/10.1021/jp911424k

Hu Y, Zhu Y, Zhang Y, Lin T, Zeng G, Zhang S, Wang Y, He W, Zhang M, Long H (2019) An efficient adsorbent: simultaneous activated and magnetic ZnO doped biochar derived from camphor leaves for ciprofloxacin adsorption. Bioresour Technol 288:121511. https://doi.org/10.1016/j.biortech.2019.121511

Hubicki Z, Kołodyńska D (2012) Selective removal of heavy metal ions from waters and waste waters using ion exchange methods, ion exchange technologies. Ayben Kilislioğlu, IntechOpen

Hussein M, Abdullah A, Badr El-Din N, Mishaqa E (2017) Biosorption potential of the microchlorophyte *chlorella vulgaris* for some pesticides. J Fertil Pestic 8:1000177. https://doi.org/10.4172/2471-2728.1000177

Jaishankar M, Tseten T, Anbalagan N, Mathew B, Beeregowda K (2014) Toxicity, mechanism and health effects of some heavy metals. Interdiscip Toxicol 7:60–72. https://doi.org/10.2478/intox-2014-0009

Jezierska B, Witeska M (2006) The metal uptake and accumulation in fish living in polluted waters. Soil and water pollution monitoring, protection and remediation. Springer Netherlands, Dordrecht, pp 107–114

Ji Y-Q, Hu Y-T, Tian Q, Shao X-Z, Li J, Safarikova M, Safarik I (2010) Biosorption of strontium ions by magnetically modified yeast cells. Sep Sci Technol 45:1499–1504. https://doi.org/10.1080/01496391003705664

Jiang W, Wang W, Pan B, Zhang Q, Zhang W, Lv L (2014) Facile fabrication of magnetic chitosan beads of fast kinetics and high capacity for copper removal. ACS Appl Mater Interfaces 6:3421–3426. https://doi.org/10.1021/am405562c

Keller AA et al (2017) Comparative environmental fate and toxicity of copper nanomaterials. NanoImpact 7:28–40. https://doi.org/10.1016/j.impact.2017.05.003

Khandanlou R, Ahmad M, Kamyar S, Kalantari K (2013a) Investigation of the role of reductant on the size control of Fe3O4 nanoparticles on rice straw. BioResources 9:642–655. Not available doi

Khandanlou R, Ahmad M, Kamyar S, Kalantari K (2013b) Synthesis and characterization of rice straw/Fe3O4 nanocomposites by a quick precipitation method. Molecules (Basel, Switzerland) 18:6597–6607. https://doi.org/10.3390/molecules18066597

Kiaune L, Singhasemanon N (2011) Pesticidal copper (I) oxide: environmental fate and aquatic toxicity. Rev Environ Contam Toxicol 213:1–26. https://doi.org/10.1007/978-1-4419-9860-6_1

Kiddee P, Naidu R, Wong MH (2013) Electronic waste management approaches: an overview. J Waste Manag 33:1237–1250. https://doi.org/10.1016/j.wasman.2013.01.006

Kim K, Jahan S, Kabir E, Brown R (2013) A review of airborne polycyclic aromatic hydrocarbons (PAHs) and their human health effects. Environ Int 60:71–80. https://doi.org/10.1016/j.envint.2013.07.019

Kimbrough R (1995) Polychlorinated biphenyls (PCBs) and human health: an update. Crit Rev Toxicol 25:133–163

Kollarahithlu SC, Balakrishnan RM (2018) Adsorption of ibuprofen using cysteine-modified silane-coated magnetic nanomaterial. Environ Sci Pollut Res 26:1–10. https://doi.org/10.1007/s11356-018-3272-8

Kyzas G, Deliyanni E (2013) Mercury (II) removal with modified magnetic chitosan adsorbents. Molecules 18:6193–6214. https://doi.org/10.3390/molecules18066193

Li H, Li Z, Liu T, Xiao X, Peng Z, Deng L (2008) A novel technology for biosorption and recovery hexavalent chromium in wastewater by bio-functional magnetic beads. Bioresour Technol 99:6271–6279. https://doi.org/10.1016/j.biortech.2007.12.002

Li F, Li X, Cui P (2018a) Effect of water chemistry on Eu(III) biosorption by magnetic bioadsorbent. Radiochim Acta 106:593–599. https://doi.org/10.1515/ract-2018-2927

Li M-F, Liu Y-G, Liu S-B, Zeng G-M, Hu X-J, Tan X-F, Jiang L-H, Liu N, Wen J, Liu X-H (2018b) Performance of magnetic graphene oxide/diethylenetriaminepentaacetic acid nanocomposite for the tetracycline and ciprofloxacin adsorption in single and binary systems. J Colloid Interface Sci 521:150–159. https://doi.org/10.1016/j.jcis.2018.03.003

Li C, Wang X, Meng D, Zhou L (2018c) Facile synthesis of low-cost magnetic biosorbent from peach gum polysaccharide for selective and efficient removal of cationic dyes. Int J Biol Macromol 107:1871–1878. https://doi.org/10.1016/j.ijbiomac.2017.10.058

Liang J, Fang Y, Luo Y, Zeng G, Deng J, Tan X, Tang N, Li X, He X, Feng C (2019) Magnetic nanoferromanganese oxides modified biochar derived from pine sawdust for adsorption of tetracycline hydrochloride. Environ Sci Pollut Res 26:5892–5903. https://doi.org/10.1007/s11356-018-4033-4

Ling D, Lee N, Hyeon T (2015) Chemical synthesis and assembly of uniformly sized iron oxide nanoparticles for medical applications. Acc Chem Res 48:1276–1285. https://doi.org/10.1021/acs.accounts.5b00038

Liu Y, Zhu X, Qian F, Zhang S, Chen J (2014) Magnetic activated carbon prepared from rice straw-derived hydrochar for triclosan removal. RSC Adv 4:63620–63626. https://doi.org/10.1039/c4ra11815d

Liu Y, Liu R, Li M, Yu F, He C (2019) Removal of pharmaceuticals by novel magnetic genipin-crosslinked chitosan/graphene oxide-SO_3H composite. Carbohydr Polym 220:141–148. https://doi.org/10.1016/j.carbpol.2019.05.060

Luo Y, Li Z, Yuan Y, Yue T (2016) Bioadsorption of patulin from kiwi fruit juice onto a superior magnetic chitosan. J Alloys Compd 667:101–108. https://doi.org/10.1016/j.jallcom.2016.01.143

Madhav S, Ahamad A, Singh AK, Kushawaha J, Chauhan JS, Sharma S, Singh P (2019) Water pollutants: sources and impact on the environment and human health. In: Pooja D, Kumar P, Singh P, Patil S (eds) Sensors in water pollutants monitoring: role of material. Springer Singapore, Singapore, pp 43–62

Mahmood M, Barbooti M, Balasim A, Altameemi A, Najah M, Al-Shuwaiki N (2011) Removal of heavy metals using chemicals precipitation. Eng Technol J 29:595–612. Not available doi

Mansour F, Al-Hindi M, Yahfoufi R, Ayoub GM, Ahmad MN (2018) The use of activated carbon for the removal of pharmaceuticals from aqueous solutions: a review. Rev Environ Sci Biotechnol 17:109–145. https://doi.org/10.1007/s11157-017-9456-8

Mehta D, Mazumdar S, Singh S (2015) Magnetic adsorbents for the treatment of water/wastewater: a review. J Water Process Eng 7:244–265. https://doi.org/10.1016/j.jwpe.2015.07.001

Mendoza-Garcia A, Sun S (2016) Recent advances in the high temperature chemical synthesis of magnetic nanoparticles. Adv Funct Mater 26:3809–3817. https://doi.org/10.1002/adfm.201504172

Meng F, Yang B, Wang B, Duan S, Chen Z, Ma W (2017) Novel dendrimerlike magnetic biosorbent based on modified orange peel waste: adsorption–reduction behavior of arsenic. ACS Sustain Chem Eng 5:9692–9700. https://doi.org/10.1021/acssuschemeng.7b01273

Mohan D, Kumar H, Sarswat A, Alexandre-Franco M, Pittman CU Jr (2014) Cadmium and lead remediation using magnetic oak wood and oak bark fast pyrolysis bio-chars. Chem Eng J 236:513–528. https://doi.org/10.1016/j.cej.2013.09.057

Monier M, Ayad D, Abdel-Latif D (2012) Adsorption of Cu(II), Cd(II) and Ni(II) ions by cross-linked magnetic chitosan-2-aminopyridine glyoxal Schiff's base. Colloids Surf B Biointerfaces 94:250–258. https://doi.org/10.1016/j.colsurfb.2012.01.051

Moore JW, Ramamoorthy S (1984) Impact of heavy metals in natural waters, heavy metals in natural waters: applied monitoring and impact assessment. Springer New York, New York, pp 205–233

Mottana A, Carrà S, Doglioni C (2016) Levels of water and soil natural pollutions in Italy. Rendiconti Lincei 27:3–6. https://doi.org/10.1007/s12210-015-0496-0

Mullerova S, Baldikova E, Prochazkova J, Pospiskova K, Safarik I (2019) Magnetically modified macroalgae *Cymopolia barbata* biomass as an adsorbent for safranin O removal. Mater Chem Phys 225:174–180. https://doi.org/10.1016/j.matchemphys.2018.12.074

Nayak D, Pradhan S, Ashe S, Rauta PR, Nayak B (2015) Biologically synthesised silver nanoparticles from three diverse family of plant extracts and their anticancer activity against epidermoid A431 carcinoma. J Colloid Interface Sci 457:329–338. https://doi.org/10.1016/j.jcis.2015.07.012

Oladipo AA, Ifebajo AO (2018) Highly efficient magnetic chicken bone biochar for removal of tetracycline and fluorescent dye from wastewater: two-stage adsorber analysis. J Environ Manag 209:9–16. https://doi.org/10.1016/j.jenvman.2017.12.030

Omidinasab M, Rahbar N, Ahmadi M, Kakavandi B, Ghanbari F, Kyzas GZ, Martinez SS, Jaafarzadeh N (2018) Removal of vanadium and palladium ions by adsorption onto magnetic chitosan nanoparticles. Environ Sci Pollut Res 25:34262–34276. https://doi.org/10.1007/s11356-018-3137-1

Ontanon OM, Gonzalez PS, Barros GG, Agostini E (2017) Improvement of simultaneous Cr (VI) and phenol removal by an immobilised bacterial consortium and characterisation of biodegradation products. New Biotechnol 37:172–179. https://doi.org/10.1016/j.nbt.2017.02.003

Ouda OKM, Shawesh A, Al-Olabi T, Younes F, Al-Waked R (2013) Review of domestic water conservation practices in Saudi Arabia. Appl Water Sci 3:689–699. https://doi.org/10.1007/s13201-013-0106-1

Panneerselvam P, Morad N, Tan KA (2011) Magnetic nanoparticle (Fe_3O_4) impregnated onto tea waste for the removal of nickel(II) from aqueous solution. J Hazard Mater 186:160–168. https://doi.org/10.1016/j.jhazmat.2010.10.102

Patel H (2019) Fixed-bed column adsorption study: a comprehensive review. Appl Water Sci 9:45. https://doi.org/10.1007/s13201-019-0927-7

Patra RC, Malik S, Beer M, Megharaj M, Naidu R (2010) Molecular characterization of chromium (VI) reducing potential in Gram positive bacteria isolated from contaminated sites. Soil Biol Biochem 42:1857–1863. https://doi.org/10.1016/j.soilbio.2010.07.005

Peng Q, Liu Y, Zeng G, Xu W, Yang C, Zhang J (2010) Biosorption of copper (II) by immobilizing *Saccharomyces cerevisiae* on the surface of chitosan-coated magnetic nanoparticles from aqueous solution. J Hazard Mater 177:676–682. https://doi.org/10.1016/j.jhazmat.2009.12.084

Philippova O, Barabanova A, Molchanov V, Khokhlov A (2011) Magnetic polymer beads: recent trends and developments in synthetic design and applications. Eur Polym J 47:542–559

Pi S, Li A, Wei W, Feng L, Zhang G, Chen T, Zhou X, Sun H, Ma F (2017) Synthesis of a novel magnetic nano-scale biosorbent using extracellular polymeric substances from *Klebsiella sp. J1* for tetracycline adsorption. Bioresour Technol 245:471–476. https://doi.org/10.1016/j.biortech.2017.08.190

Pospiskova K, Safarik I (2015) Low-temperature magnetic modification of sensitive biological materials. Mater Lett 142:184–188. https://doi.org/10.1016/j.matlet.2014.11.163

Prime RB, Bair HE, Vyazovkin S, Gallagher PK, Riga A (2009) Thermogravimetric analysis (TGA), Thermal analysis of polymers: fundamentals and applications, pp 241–317

Qiao L, Swihart MT (2017) Solution-phase synthesis of transition metal oxide nanocrystals: morphologies, formulae and mechanisms. Adv Colloid Interf Sci 244:199–266. https://doi.org/10.1016/j.cis.2016.01.005

Rebelo R, Fernandes M, Fangueiro R (2017) Biopolymers in medical implants: a brief review. Procedia Eng 200:236–243. https://doi.org/10.1016/j.proeng.2017.07.034

Reddy DHK, Yun YS (2016) Spinel ferrite magnetic adsorbents: alternative future materials for water purification? Coord Chem Rev 315:90–111. https://doi.org/10.1016/j.ccr.2016.01.012

Resch-Fauster K, Klein A, Blees E, Feuchter M (2017) Mechanical recyclability of technical biopolymers: potential and limits. Polym Test 64:287–295. https://doi.org/10.1016/j.polymertesting.2017.10.017

Rice KM, Walker EM Jr, Wu M, Gillette C, Blough ER (2014) Environmental mercury and its toxic effects. J Prev Med Public Health 47:74. https://doi.org/10.3961/jpmph.2014.47.2.74

Saber-Samandari S, Saber-Samandari S, Joneidi-Yekta H, Mohseni M (2017) Adsorption of anionic and cationic dyes from aqueous solution using gelatin-based magnetic nanocomposite beads comprising carboxylic acid functionalized carbon nanotube. Chem Eng J 308:1133–1144. https://doi.org/10.1016/j.cej.2016.10.017

Safarik I, Horska K, Pospiskova K, Filip J, Safarikova M (2014) Mechanochemical synthesis of magnetically responsive materials from non-magnetic precursors. Mater Lett 126:202–206. https://doi.org/10.1016/j.matlet.2014.04.045

Safarik I, Baldikova E, Pospiskova K, Safarikova M (2016) Magnetic modification of diamagnetic agglomerate forming powder materials. Particuology 29:169–171. https://doi.org/10.1016/j.partic.2016.05.002

Safarik I, Baldikova E, Prochazkova J, Safarikova M, Pospiskova K (2018) Magnetically modified agricultural and food waste: preparation and application. J Agric Food Chem 66:2538–2552. https://doi.org/10.1021/acs.jafc.7b06105

Safe S (1989) Polychlorinated biphenyls (PCBs): mutagenicity and carcinogenicity. Mutat Res 220:31–47. https://doi.org/10.1016/0165-1110(89)90007-9

Sahraei R, Pour ZS, Ghaemy M (2017) Novel magnetic bio-sorbent hydrogel beads based on modified gum tragacanth/graphene oxide: removal of heavy metals and dyes from water. J Clean Prod 142:2973–2984. https://doi.org/10.1016/j.jclepro.2016.10.170

Saifuddin N, Dinara S (2012) Immobilization of *Saccharomyces cerevisiae* onto cross-linked chitosan coated with magnetic nanoparticles for adsorption of uranium (VI) ions. Adv Nat Appl Sci 6:249–267. Not available doi

Sajid M, Nazal MK, Ihsanullah BN, Osman AM (2018) Removal of heavy metals and organic pollutants from water using dendritic polymers based adsorbents: a critical review. Sep Purif Technol 191:400–423. https://doi.org/10.1016/j.seppur.2017.09.011

Salgueiro AM, Daniel-da-Silva AL, Girão AV, Pinheiro PC, Trindade T (2013) Unusual dye adsorption behavior of κ-carrageenan coated superparamagnetic nanoparticles. Chem Eng J 229:276–284. https://doi.org/10.1016/j.cej.2013.06.015

Shaaban M (2010) Role of boron in plant nutrition and human health. Am J Plant Physiol 5:224–240. https://doi.org/10.3923/ajpp.2010.224.240

Sharma S, Bhattacharya A (2016) Drinking water contamination and treatment techniques. Appl Water Sci 7:1043–1067. https://doi.org/10.1007/s13201-016-0455-7

Sharma P, Tomar R (2008) Synthesis and application of an analogue of mesolite for the removal of uranium (VI), thorium (IV), and europium (III) from aqueous waste. Microporous Mesoporous Mater 116:641–652. https://doi.org/10.1016/j.micromeso.2008.05.036

Sheshmani S, Mashhadi S (2018) Potential of magnetite reduced graphene oxide/chitosan nanocomposite as biosorbent for the removal of dyes from aqueous solutions. Polym Compos 39:E457–E462. https://doi.org/10.1002/pc.24608

Shi W, Guo F, Wang H, Liu C, Fu Y, Yuan S, Huang H, Liu Y, Kang Z (2018) Carbon dots decorated magnetic $ZnFe_2O_4$ nanoparticles with enhanced adsorption capacity for the removal of dye from aqueous solution. Appl Surf Sci 433:790–797. https://doi.org/10.1016/j.apsusc.2017.10.099

Simeonidis K, Mourdikoudis S, Kaprara E, Mitrakas M, Polavarapu L (2016) Inorganic engineered nanoparticles in drinking water treatment: a critical review. Environ Sci-Wat Res 2:43–70. https://doi.org/10.1039/C5EW00152H

Sivashankar R, Sathya A, Vasantharaj K, Sivasubramanian V (2014) Magnetic composite an environmental super adsorbent for dye sequestration–a review. Environ Nanotechnol Monit Managet 1:36–49. https://doi.org/10.1016/j.enmm.2014.06.001

Soares SF, Simões TR, António M, Trindade T, Daniel-da-Silva AL (2016) Hybrid nanoadsorbents for the magnetically assisted removal of metoprolol from water. Chem Eng J 302:560–569. https://doi.org/10.1016/j.cej.2016.05.079

Soares SF, Fernandes T, Sacramento M, Trindade T, Daniel-da-Silva AL (2019a) Magnetic quaternary chitosan hybrid nanoparticles for the efficient uptake of diclofenac from water. Carbohydr Polym 203:35–44. https://doi.org/10.1016/j.carbpol.2018.09.030

Soares SF, Fernandes T, Trindade T, Daniel-da-Silva AL (2019b) Recent advances on magnetic biosorbents and their applications for water treatment. Environ Chem Lett. https://doi.org/10.1007/s10311-019-00931-8

Sohni S, Gul K, Ahmad F, Ahmad I, Khan A, Khan N, Bahadar Khan S (2018) Highly efficient removal of acid red-17 and bromophenol blue dyes from industrial wastewater using graphene oxide functionalized magnetic chitosan composite. Polym Compos 39:3317–3328. https://doi.org/10.1002/pc.24349

Song HJ, You S, Jia XH, Yang J (2015a) MoS_2 nanosheets decorated with magnetic Fe_3O_4 nanoparticles and their ultrafast adsorption for wastewater treatment. Ceram Int 41:13896–13902. https://doi.org/10.1016/j.ceramint.2015.08.023

Song R, Bai B, Puma GL, Wang H, Suo Y (2015b) Biosorption of azo dyes by raspberry-like Fe_3O_4@ yeast magnetic microspheres and their efficient regeneration using heterogeneous Fenton-like catalytic processes over an up-flow packed reactor. ReacT Kinet Mech Cat 115:547–562. https://doi.org/10.1007/s11144-015-0854-z

Stanners D, Bourdeau P (2008) Waste production and management, Europe's environment – the dobris assessment. European Environment Agency

Stremilova NN, Viktorovskii IV, Zigel VV (2001) Concentration of contaminants in analysis of natural waters. Russ J Gen Chem 71:19–22. https://doi.org/10.1023/a:1012312902629

Su C (2017) Environmental implications and applications of engineered nanoscale magnetite and its hybrid nanocomposites: a review of recent literature. J Hazard Mater 322:48–84. https://doi.org/10.1016/j.jhazmat.2016.06.060

Sun P, Hui C, Khan RA, Du J, Zhang Q, Zhao Y-H (2015) Efficient removal of crystal violet using Fe_3O_4-coated biochar: the role of the Fe_3O_4 nanoparticles and modeling study their adsorption behavior. Sci Rep 5:12638. https://doi.org/10.1038/srep12638

Tang SC, Lo IM (2013) Magnetic nanoparticles: essential factors for sustainable environmental applications. Water Res 47:2613–2632. https://doi.org/10.1016/j.watres.2013.02.039

Tarhan T, Tural B, Boga K, Tural S (2019) Adsorptive performance of magnetic nano-biosorbent for binary dyes and investigation of comparative biosorption. SN Appl Sci 1:8. https://doi.org/10.1007/s42452-018-0011-1

Tian Y, Wu M, Lin X, Huang P, Huang Y (2011) Synthesis of magnetic wheat straw for arsenic adsorption. J Hazard Mater 193:10–16. https://doi.org/10.1016/j.jhazmat.2011.04.093

Tsai W-T (2013) A review on environmental exposure and health risks of herbicide paraquat. Toxicol Environ Chem 95:197–206. https://doi.org/10.1080/02772248.2012.761999

Tural B, Ertaş E, Tural S (2016a) Removal of phenolic pollutants from aqueous solutions by a simple magnetic separation. Desal Water Treat 57:26153–26164. https://doi.org/10.1080/19443994.2016.1162202

Tural S, Tarhan T, Tural B (2016b) Removal of hazardous azo dye Metanil yellow from aqueous solution by cross-linked magnetic biosorbent; equilibrium and kinetic studies. Desalin Water Treat 57:13347–13356. https://doi.org/10.1080/19443994.2015.1056842

Tural B, Ertaş E, Enez B, Fincan SA, Tural S (2017) Preparation and characterization of a novel magnetic biosorbent functionalized with biomass of *Bacillus Subtilis*: kinetic and isotherm studies of biosorption processes in the removal of methylene blue. J Environ Chem Eng 5:4795–4802. https://doi.org/10.1016/j.jece.2017.09.019

Tural S, Ece MŞ, Tural B (2018) Synthesis of novel magnetic nano-sorbent functionalized with N-methyl-D-glucamine by click chemistry and removal of boron with magnetic separation method. Ecotox Environ Safe 162:245–252. https://doi.org/10.1016/j.ecoenv.2018.06.066

ul Haq A, Saeed M, Usman M, SAR N, Bokhari TH, Maqbool T, Ghaus H, Tahir T, Khalid H (2020) Sorption of chlorpyrifos onto zinc oxide nanoparticles impregnated Pea peels (*Pisum sativum L*): equilibrium, kinetic and thermodynamic studies. Environ Technol Innov 17:100516. https://doi.org/10.1016/j.eti.2019.100516

Uzun L, Sağlam N, Safarikova M, Safarik I, Denizli A (2011) Copper biosorption on magnetically modified yeast cells under magnetic field. Sep Sci Technol 46:1045–1051. https://doi.org/10.1080/01496395.2010.541400

Vainio H, Waters MD, Norppa H (1985) Mutagenicity of selected organic solvents. Scand J Work Environ Health 1:75–82. Not available doi

Valverde-Aguilar G, Pérez-Mazariego JL, Marquina V, Gómez R, Aguilar-Franco M, Garcia-Macedo J (2015) Effect of PEO–PPO–PEO triblock copolymers in the synthesis of magnetic nanoparticles embedded in SiO_2 and TiO_2 matrices by sol–gel method. J Mater Sci 50:704–716. https://doi.org/10.1007/s10853-014-8630-4

Velinov N, Petrova T, Tsoncheva T, Genova I, Koleva K, Kovacheva D, Mitov I (2016) Auto-combustion synthesis, Mössbauer study and catalytic properties of copper-manganese ferrites. Hyperfine Interact 237:24. https://doi.org/10.1007/s10751-016-1222-8

Wang S-y, Y-k T, Li K, Y-y M, H-f L, Z-q G (2014) Combined performance of biochar sorption and magnetic separation processes for treatment of chromium-contained electroplating wastewater. Bioresour Technol 174:67–73. https://doi.org/10.1016/j.biortech.2014.10.007

Wang Y, Nkurikiyimfura I, Pan Z (2015a) Sonochemical synthesis of magnetic nanoparticles. Chem Eng Commun 202:616–621. https://doi.org/10.1080/00986445.2013.858039

Wang Z, Xu C, Li X, Liu Z (2015b) In situ green synthesis of Ag nanoparticles on tea polyphenols-modified graphene and their catalytic reduction activity of 4-nitrophenol. Colloids Surf A Physicochem Eng Asp 485:102–110. https://doi.org/10.1016/j.colsurfa.2015.09.015

Wang Y, Zhang Y, Hou C, Liu M (2016) Mussel-inspired synthesis of magnetic polydopamine–chitosan nanoparticles as biosorbent for dyes and metals removal. J Taiwan Inst Chem Eng 61:292–298. https://doi.org/10.1016/j.jtice.2016.01.008

Weng M, Yu X (2019) Electrochemical oxidation of para-aminophenol with rare earth doped lead dioxide electrodes: kinetics modeling and mechanism. Front Chem 7. https://doi.org/10.3389/fchem.2019.00382

WHO (1995) Evaluation of certain food additives and contaminants: forty-fourth report of the joint FAO/WHO expert committee on food additives. World Health Organization, Geneva

Woicik JC, Woicik J (2016) Hard X-ray photoelectron spectroscopy (HAXPES), 59. Springer, Cham

Wu W, Wu Z, Yu T, Jiang C, Kim WS (2015) Recent progress on magnetic iron oxide nanoparticles: synthesis, surface functional strategies and biomedical applications. Sci Technol Adv Mater 16:023501. https://doi.org/10.1088/1468-6996/16/2/023501

Xagoraraki I, Kuo D (2008): Water pollution: emerging contaminants associated with drinking water. In: Quah KHaS (Hrsg.) International encyclopedia of public health. Academic, San Diego, pp 539–550

Yang X, Jin D, Zhang M, Wu P, Jin H, Li J, Wang X, Ge H, Wang Z, Lou H (2016) Fabrication and application of magnetic starch-based activated hierarchical porous carbon spheres for the efficient removal of dyes from water. Mater Chem Phys 174:179–186. https://doi.org/10.1016/j.matchemphys.2016.02.073

Yang J, Hou B, Wang J, Tian B, Bi J, Wang N, Li X, Huang X (2019) Nanomaterials for the removal of heavy metals from wastewater. Nano 9:424. https://doi.org/10.3390/nano9030424

Yuwei C, Jianlong W (2011) Preparation and characterization of magnetic chitosan nanoparticles and its application for Cu(II) removal. Chem Eng J 168:286–292. https://doi.org/10.1016/j.cej.2011.01.006

Zhang Q, Lu T, Bai D-M, Lin D-Q, Yao S-J (2016) Self-immobilization of a magnetic biosorbent and magnetic induction heated dye adsorption processes. Chem Eng J 284:972–978. https://doi.org/10.1016/j.cej.2015.09.047

Zheng C, Zheng H, Hu C, Wang Y, Wang Y, Zhao C, Ding W, Sun Q (2020) Structural design of magnetic biosorbents for the removal of ciprofloxacin from water. Bioresour Technol 296:122288. https://doi.org/10.1016/j.biortech.2019.122288

Zhou J, Liu Y, Zhou X, Ren J, Zhong C (2017) Removal of mercury ions from aqueous solution by thiourea-functionalized magnetic biosorbent: preparation and mechanism study. J Colloid Interface Sci 507:107–118. https://doi.org/10.1016/j.jcis.2017.07.110

Zhu X, Liu Y, Luo G, Qian F, Zhang S, Chen J (2014) Facile fabrication of magnetic carbon composites from hydrochar via simultaneous activation and magnetization for triclosan adsorption. Environ Sci Technol 48:5840–5848. https://doi.org/10.1021/es500531c

Chapter 4
Lignocellulosic Wastes as Precursor of Carbonaceous Magnetic Adsorbents by Organic and Inorganic Pollutants Adsorption

Paola Rodríguez-Estupiñán, Yaned Milena Correa-Navarro, Liliana Giraldo, and Juan Carlos Moreno-Piraján

Contents

P. Rodríguez-Estupiñán · J. C. Moreno-Piraján (✉)
Facultad de Ciencias, Departamento de Química, Grupo de Investigación en Sólidos Porosos y Calorimetría, Universidad de los Andes, Bogotá, Colombia
e-mail: jp.rodrigueze@uniandes.edu.co; jumoreno@uniandes.edu.co

Y. M. Correa-Navarro
Facultad de Ciencias, Departamento de Química, Grupo de Investigación en Sólidos Porosos y Calorimetría, Universidad de los Andes, Bogotá, Colombia

Facultad de Ciencias Exactas y Naturales, Departamento de Química, Universidad de Caldas, Manizales, Colombia
e-mail: yaned.correa@ucaldas.edu.co

L. Giraldo
Facultad de Ciencias, Departamento de Química, Universidad Nacional de Colombia, Bogotá, Colombia
e-mail: lgiraldogu@unal.edu.co

L. Meili, G. L. Dotto (eds.), *Advanced Magnetic Adsorbents for Water Treatment*, Environmental Chemistry for a Sustainable World 61,
https://doi.org/10.1007/978-3-030-64092-7_4

Abstract It is evident that high consumption of several kind of products around the world increases the presence of pollutants in different water sources. For example, caffeine is an alkaloid widely used in foods, beverages, and commercial analgesic medicines as a central nervous system stimulant. Caffeine is an anthropogenic pollutant, and it is part of emergent contaminants. In addition, nickel (II) has been found in industrial effluents such as mining, oil refining, mineral processing, electroplating, silver refining, paint formulation, battery manufacturing, and steam electric power plants. This metal is toxic and it bio-accumulates, thus it has been classified as a priority pollutant. Thus, it is necessary to use innovative techniques to remove pollutants of water. In this sense, adsorption is an option to deal with this problem; therefore, carbonaceous materials could be a good alternative. Moreover, magnetic carbon composites obtained from carbonaceous materials have been widely studied for environmental applications, as it makes easy the process of separation from aqueous solutions.

In the present work, commercial granular activated carbon (GAC) modified with different oxidant agents—nitric acid, hydrogen peroxide, sodium hypochlorite, and sulfuric acid—and fique bagasse biochar (BC) were magnetized and used for removal of caffeine and nickel (II) from aqueous solution. The magnetization of GAC and BC were carried out by coprecipitation method. Characterization of carbonaceous materials were performed through different techniques: Fourier transform infrared spectroscopy, scanning electron microscopy, surface area measurement, and zero point charge measurement.

Adsorption capacities of caffeine and nickel were significantly different between magnetic carbonaceous materials and nonmagnetic carbonaceous materials, being better the starting activated carbons. Additionally, the results for GAC, BC, and GAC-M fit well with Redlich-Peterson isotherm. Finally, pseudo-first-order model described better kinetic data for magnetic carbonaceous materials, while pseudo-second-order model fits better for biochar, activated carbon, and their derivatives.

Keywords Activated carbon · Biochar · Surface chemistry · Caffeine and nickel simple adsorption · Magnetic nanoparticles · Water contaminants

4.1 Introduction

Water is an indispensable resource for life, due to its role in different biological and metabolic mechanisms, and also due to its use in different manufacturing processes. According to the World Health Organization, in 2015, only 91% of the world's population had access to an improved source of drinking water; it means that close to 663 million people did not have access to drinking water. Additionally, the constant increase of diverse sources of contamination and the accelerated growth population do not show a promising scenario (UNICEF and WHO 2015). Therefore, it is the duty of society to protect and reduce any harmful effect on this resource by establishing laws and environmentally friendly practices (Madrid 2012; World Health Organization and UN-Water 2014).

Contamination with heavy metals comes from different industries such as metallurgic, chemical, paint, textile, paper, crude refining, petrochemical, leather, fertilizers, and pesticides, among others, and it has become a high-impact environmental problem due to their toxicity even at a low concentration (Soto-Jimenez 2011). Heavy metals incorporated into the environment become potential pollutants of air, soil, surface water, and groundwater due to the hydrogeological cycle, which constitutes an important mechanism for the transport, diffusion, and biomagnification of contaminants presented in the water (Li et al. 2012b).

On the other hand, there are compounds which have been called emerging contaminants (EC), they have been detected in superficial water, groundwater, and drinking water around the world (Pal et al. 2014; Basheer 2018; Mansour et al. 2018). These molecules have been found in low concentrations and are not regulated despite the proven adverse effect they have on different organisms (Tran et al. 2018). Caffeine, a nervous central system stimulant of high human consumption, mainly in energy drinks and as adjuvant medicines, is an EC. This molecule when ingested by humans is metabolized or not into smaller molecules that are frequently detected in aqueous effluents (Álvarez-Torrellas et al. 2016).

To remove organic and inorganic pollutants of water bodies, different processes have been employed. However, conventional treatments are not effective enough, so harmful molecules remain and cause deterioration in ecosystems (de Andrade et al. 2018). For this reason, it is necessary to combine efforts in finding techniques that allow the total removal of the EC and heavy metals from aquifers. One of the options that is being used is the adsorption with different types of adsorbents such as active carbon and biochar. The development of new porous solids from lignocellulosic materials is an added value to this type of waste and an opportunity to solve two environmental problems. Recently, the use of magnetic nanoparticles is being combined to the most of the conventional adsorbents for the removal of organic and inorganic pollutants from aqueous systems.

In this chapter, we summarize the data obtained in the treatment of simulated waters contaminated with nickel and caffeine in simple adsorption. The solids used correspond to five activated carbons from coconut shell (GAC) and one biochar obtained from fique bagasse (FB). GAC was submitted to four oxidation processes with nitric acid (GACoxN), hydrogen peroxide (GACoxP), sodium hypochlorite (GACoxCl), and sulfuric acid (GACoxS). Subsequently, the solids were subjected to a coprecipitation of the magnetite nanoparticles on their surfaces. The synthesis of these kinds of nanomaterial and adsorption process has also been described here; we have focused on the effect of surface chemistry and the precursor on the functionalization process with magnetic nanoparticles, as well as the effect on nickel and caffeine adsorption capacity.

4.1.1 Synthesis of Adsorbents with Magnetic Nanoparticles

The use of magnetic nanoparticles in the preparation of adsorbents is related to the enhancement of the adsorbent performance, for example, by modifying adsorption mechanisms through the change of its chemistry nature surface. Also by the concentration increase of active adsorption sites and by facilitating the method of separation of the supernatant solution and the solid, due to the noteworthy advantages of using magnetic nanoparticles by applying an external magnetic field, it is a faster and efficient method in comparison to filtration methods. Another advantage of preparing hybrid materials consists of providing a support for nanoparticles to avoid aggregation and deposition in order to favor their reactivity and mobility (Baruah et al. 2019).

One of the most used porous solids is definitely activated carbon; its versatility has positioned it worldwide as an excellent adsorbent. Nevertheless, in the seeking of environmentally and economically accessible solids, materials with lower energy and chemical demands as biochar have been developed. Thus, a large adsorbent development field from a wide variety of lignocellulosic precursors has been opened. The synthesis of carbon-based nanoparticle-modified solids is discussed here, as well as the effect of the chemical nature surface of carbonaceous materials (Baruah et al. 2019).

The oxygenated surface functional groups of carbonaceous materials play an important role in the interaction with particles (atoms, ions, molecules, nanoparticles) dissolved in aqueous media through the establishment of specific interactions, such as covalent bonding, hydrogen bonding, electrostatic interactions, π–π interactions, and nonspecific or dispersive interactions. The change of surface chemistry by oxidation processes is a successful method to modify the selectivity against a target particle. Among the different grafting techniques for introducing functional groups on the surface, the use of the oxidant agents is the most popular one (Baruah et al. 2019).

4.1.2 What Is Magnetite?

The use of metal-based nanoparticles as modifying agents of conventional adsorbents is a method that has gained attention due to the multiple advantages that it can offer to the adsorption and degradation process of organic and inorganic pollutants. Among the wide variety of metal-based nanoparticles, iron oxides (Fe_xO_y) are one of the most used because of their low cost, toxicity and environmental effects, high pH, and hydrothermal stability. Besides, they have an outstanding property that they can be separated from a mixture by applying an external magnetic field (Santhosh et al. 2019).

The stoichiometric relationship between iron and oxygen ions and the oxidation state of iron allow to obtain different crystalline structures, among these are cubic hexoctahedral crystal system like wüstite (FeO), trigonal crystal system like hematite (α-Fe_2O_3), cubic gyroidal crystal system like maghemite (γ-Fe_2O_3), and isometric octahedral crystal system like magnetite ($FeO{\cdot}Fe_2O_3$ or Fe_3O_4). The latter configuration is what we have worked on in this research. Magnetite has an inverse spinel structure, in which the oxygen atoms organize in a face-centered cubic (FCC) system and iron atoms are located in interstitial tetrahedral sites, and remaining ferrous and ferric ions are randomly distributed at an octahedrally coordinated position (Fleet 1981).

As it was summarized by Mahmoudiab et al. 2011, the iron oxide nanoparticles can be synthesized by physical, chemical, and biological methods. Some of the most representative techniques for each method are gas-phase deposition, aerosol, pulsed laser ablation, and powder ball milling for the physical methods. Biological methods involve the use of fungi, bacteria, and proteins, are used for biological methods, and for chemical methods, coprecipitation and hydrothermal techniques are the most commonly used, followed by the techniques like microemulsion, sonochemical, thermal deposition, and finally electrochemical deposition (Mahmoudiab et al. 2011).

The chemical method by coprecipitation has demonstrated great versatility, simplicity, and accessibility for the production of iron oxide particles at the laboratory level and its possible escalation. In this technique, the iron oxide nanoparticles can be prepared in an oxygen-free aqueous solution (to prevent transformation into γ-Fe_2O_3) through the mixing ferrous (Fe^{2+}) and ferric ions (Fe^{3+}) in a stoichiometric ratio of 1:2 in a basic pH, for example, with the addition of NaOH or NH_4OH solutions (Boruah et al. 2019). The process of formation of the iron oxide nanoparticle under the conditions just described can be represented by the following reaction:

$$\begin{aligned} &FeSO_{4(aq)} + 2FeCl_{3(aq)} + 8NH_{3(aq)} + 8H_2O_{(l)} \\ &\quad \rightarrow Fe_3O_{4(aq)} + 8NH_4{}^+ + 6Cl_{(aq)} + SO^{2+}_{4\,(aq)} + 8OH^-_{(ac)} \end{aligned} \tag{4.1}$$

As mentioned before in the reaction, the proportion between Fe^{2+} and Fe^{3+} is 1:2, and the salts used can be chlorides, sulfates, nitrates, among others.

The surface chemistry of iron oxide magnetic nanoparticles is determined by electrically unsaturated sites by the presence of ferrous and ferric ions, also by the oxygen atoms that can form hydroxyl groups, according to simulated computed properties the structural units of magnetite would have the ability to form hydrogen bridge interactions, two hydrogen bond donor and four hydrogen bond acceptor (Fig. 4.1).

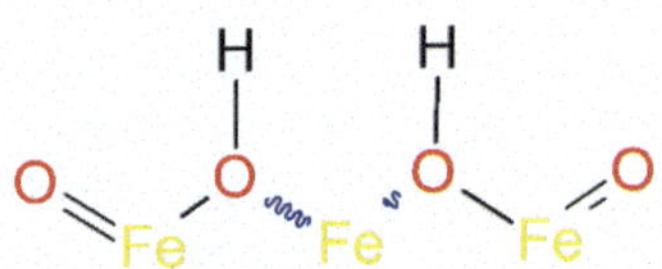

Fig. 4.1 Magnetite chemical structure depiction

Because of the composition and structure of the magnetite, it is possible to modify the net charge in a solid–liquid interface with the pH of the fluid phase. Thus at low pH values, the high proton concentration favors the presence of $-OH_2^+$ groups, therefore it shows a positive net charge. At neutral pH values, -OH is the predominant group, and the net charge is close to zero. And at high pH values, the deprotonation reaction allows the surface to be negatively charged and the O^{-2} group is present in a greater concentration. Clearly, the net surface charge will condition the type of mechanism by which the magnetite interacts with a cationic, anionic, or neutral particle. According to the nature of the target particle and net charge of the magnetite, different mechanisms can be carried out, for example, for organic molecules, π-stacking, hydrogen bond, and columbic interactions can occur to form inner or outer-sphere complexes (de Morais 2019). Their efficiency is the result of a variety of possible mechanisms occurring on their surface and their selectivity for nonreversible adsorption of pollutants through Fe-O bridges. Magnetite nanoparticles show a relative good affinity to a variety of heavy metals enabling direct adsorption of their oxy-ionic forms, whereas they preserve some reducing potential due to the presence of a Fe^{2+} ions in their crystal structure (Martinez-Boubeta, Carlos Simeonidis 2019).

4.2 Methodology

4.2.1 *Materials and Reagents*

Analytical grade ferric chloride hexahydrate ($FeCl_3 \cdot 6H_2O$), ferrous sulfate heptahydrate ($FeSO_4 \cdot 7H_2O$), ammonium hydroxide (28% w/v), sodium chloride (NaCl), potassium bromide (KBr) were purchased from Merck (Darmstadt, Germany). The biochar is prepared from locally available lignocellulosic waste known as fique bagasse, and the activated carbon is a commercial sample obtained from coconutshell and physical activation with CO_2. For organic and inorganic removal studies, two kinds of simulated pollutants were used, nickel chloride ($NiCl_2 \cdot 6H_2O$) and caffeine (CFN).

4.2.2 *Sample Preparation*

Oxidized Activated Carbon

Commercial activated carbon from coconut shell (GAC) was washed until constant pH, then it was dried for 24 h in an oven at 120 °C. For liquid-phase acid oxidation, 30 g of the GAC was refluxed with 250 mL of 6 M HNO_3 or 6 M H_2SO_4 solution, then the samples were washed several times with hot water until a constant pH of the wash water is reached, and then they were dried in an oven (GACoxN and

GACoxS). For liquid-phase peroxide oxidation, 30 g of the GAC was mixed with 250 mL of 12% H_2O_2 solution by 12 h, then the sample was washed several times with hot water until a constant pH of the wash water is reached, and then it was dried in an oven (GACoxP). For liquid-phase hypochlorite oxidation, 30 g of the GAC was mixed with 250 mL of 12% NaClO solution by 12 h (GACoxCl) (Rodríguez-Estupiñán et al. 2018).

Fique Bagasse Biochar Preparation

Fique bagasse (FB) was picked from a farm in Aranzazu (Caldas, Colombia) after fiber extraction. The FB was dried at 100 °C until constant weight. Pyrolysis of FB was performed under nitrogen atmosphere with a heating rate of 1 °C/min until 850 °C and residence time of 180 min was controlled, and for this process a horizontal furnace thermolyne was used. The sample was labeled as BC850 (Correa-Navarro et al. 2019).

Coated Magnetic Carbonaceous Materials

Magnetic carbonaceous materials were obtained by modified Li method by in situ coprecipitation technique (Li et al. 2012a). First, an aqueous solution was prepared with $FeCl_3 \cdot 6H_2O$ and $FeSO_4 \cdot 7H_2O$ with a molar ratio of $Fe^{3+}:Fe^{2+} = 2:1$. Second, this solution was heated at 80 °C and 2.5 g of GAC and 50 mL of water were slowly added. Third, in order to adjust the pH and precipitate the iron oxide nanoparticle, the ammonia solution was added quickly and the temperature was raised to 85 °C. After that, the mixture was stirred for 45 min. Then, the slurry was cooled. Finally, the solution was filtered, and the remaining solid was washed with distilled water until constant pH and dried at 100 °C. The products were designed with an **-M** at the end of the name previously assigned according to the precursor and oxidation method. Magnetite (Fe_xO_y) was prepared by the same chemical coprecipitation method without carbonaceous material.

4.2.3 Physicochemical Characterization Techniques

All materials were characterized using semiautomatic sortometer Quantachrome Autosorb IQ^2 to study their specific surface area, pore size distribution, and isotherm type. The morphology characteristics were studied by SEM images, which were obtained using a TESCAN scanning electron microscope, the samples were prepared in an ethanol suspension and a drop of each suspension was placed on the foil support, after the drops were dried at room temperature, the sample holder was entered into the microscope chamber.

The chemical characteristics evaluated by infrared (IR) spectra were recorded on a Shimadzu IRTracer 100 spectrometer; the spectra were performed at 8 cm^{-1} resolution between 400 and 4000 cm^{-1} wavenumber range. The point of zero charge measurements (pH_{PZC}) were determined by reverse mass titration, in brief: slurries of magnetic carbonaceous materials and NaCl (0.1 M) at various mass percentages were prepared. The pHs of these slurries were measured after shaking at least 48 h. The pHpzc was determined by plotting the equilibrium pH as a function of solid weight. Finally, potentiometric titrations were also performed, where a suspension of the solid in $NaNO_3$ 0.01 M solution is titrated whit standard solution of NaOH 0.1 M once the pH of the suspension was set at 3.0. According to the results of this technique, the surface chemistry nature can be described in terms of the protonic affinity of the surface groups giving their acid force, according to that, the group concentration can be described by a continuous pKa distribution (Bandosz et al. 1993).

4.2.4 *Determination of Caffeine and Nickel Adsorption Capacity*

The adsorption equilibrium was carried out using CFN solutions at concentrations from 1 to 1000 mg/L at pH = 6.0, and shaking for 72 hours, meanwhile for Ni^{+2} was employing solutions from 10 to 150 mg L^{-1} and there were shaking for 120 h. Studies were performed using 5.0 mL of CFN or Ni^{+2} solutions and 10.0 mg of GAC and BC samples, and also the modified carbonaceous materials. In addition, kinetic assays were performed with CFN solutions at 50 mg/L. The CFN amounts adsorbed at equilibrium time were obtained using Eq. (4.2), by UV-Vis spectrometry, and the Ni^{+2} quantity adsorbed at equilibrium time was determined with Eq. (4.2) by atomic spectrometry:

$$Q_e = \frac{V(C_0 - C_e)}{W} \tag{4.2}$$

where C_o is the initial concentration, C_e is the equilibrium concentration of CFN or Ni^{+2} (mg L^{-1}), V (L) is the volume of CFN or Ni^{+2} solution, and W (g) is the dry mass of carbonaceous material employed.

4.3 Results and Discussion

Lignocellulose-based carbonaceous materials have been extensively studied, specifically as adsorbents, and catalytic supports have presented easy availability, cost effectiveness, great versatility, and good performance, because of their high surface area, open surface area, and an easily designed surface chemistry, using different modification methods. The microstructure of the carbonaceous material depends critically on the carbon source, and for activated carbons it also depends on the activation method (physical or chemical). The macro and microstructures of the composites were studied by scanning electron microscopy (SEM) and N_2 adsorption-desorption isotherms at −196 °C.

4.3.1 Physical Properties of Magnetic Carbonaceous Composites

In the micrographs, carbonaceous materials morphology of the external surface is observed on a micrometric scale, which has cavities with a high degree of roughness, with diameters of 1 μm. However, the BC850 sample shows a uniform structure without the presence of cavities; this is congruent with a surface from a lignocellulosic residue and that has not been subjected to a physical or chemical activation process (Fig. 4.2).

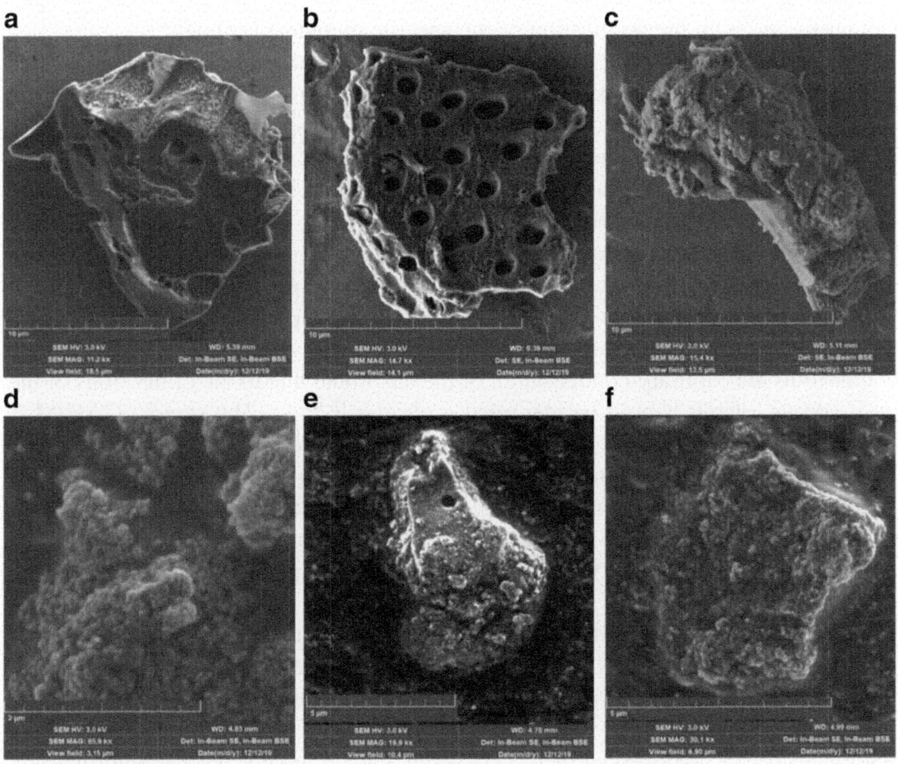

Fig. 4.2 SEM micrograph at different magnifications for (**a**) GAC, (**b**) GACoxN, (**c**) BC850, (**d**) Magnetite, (**e**) GAC-M, (**f**) GACoxN-M

For activated carbons, although micropores and mesopores are not visible, the micrographs present the shapes and location of the macropores on the surface of the solid. The macropores are formed during the activation by the effect of the oxidation gas (CO_2) and the loss of volatile matter during the pyrolysis process. On the other hand, oxidation with nitric acid has a visual impact on the morphology of the surface, due to the loss in uniformity, reaching an eroded appearance and the generation of cavities on the surface of solids. The SEM image in Fig. 4.2d shows the micrograph of magnetite; it shows a homogeneous grain size and sphere-like agglomerates. No significant morphological differences among the GAC-M series can be observed by comparing the micrographs, although the coverage of carbon granules by magnetite nanoparticles was observed (see Fig. 4.2e, f).

The textural characterization of porous solids is one of the most important stages to study adsorption process, in addition to defining the most specific use of an adsorbent material. The adsorption capacity of the material is related to the specific parameters from the gas adsorption measures (sortometry). The main parameters defined by this technique are the apparent surface area, pore volume and pore size distribution. The most used characterization technique to study the porous structure of micro and mesoporous materials is the adsorption of N_2 at -196 °C. The textural characteristics of the solids are calculated from the experimental data of the nitrogen isotherms and with the application of the specific theoretical models.

Figure 4.3a–f shows the N_2 adsorption/desorption isotherms, for GAC series have essentially a type I (a) shape, characteristic of microporous solids according to the IUPAC classification; the isotherms are concave with respect to the relative pressure axis and the maximum capacity approximates a limit value given by the micropores volume. About GAC magnetic series the isotherm can be classified as IV(a) type, characteristic of mesoporous solids according to the IUPAC classification, the isotherm has an initial stage related to the microporosity available in activated carbons, that was not obstructed by the anchoring of magnetite nanoparticles, then a mono-multilayer adsorption on the mesoporous structures takes place. A third stage was marked by the change in the nitrogen adsorption tendency, and this change is related to the capillary condensation of the adsorbate due to the dimensions of the mesoporous, also it should be identified a hysteresis loop. The sample GAC-M is clearly classified as H1 based on the IUPAC classification, this hysteresis is related to the presence of narrow range of uniform mesopores with minimal network effects. For the rest of the GAC-M series samples, a hybrid hysteresis loop (H1-H3) shows behaviors related to more complex pore structure and important network effects, like pore blocking and cavitation. They are characteristic phenomena of nonuniform diameter pore and nonrigid aggregates of particles, but also if the pore network is composed of macropores that are not completely filled with pore condensate (Sotomayor et al. 2018). All these phenomena are favored by the porosity blockage given by the presence of magnetite nanoparticles (Kwon et al. 2014).

In Figs. 4.4a–f, the pore size distributions evaluated by the QSDFT method are presented; it can be observed that GAC solids have a greater porosity contribution with structures with dimensions between 0.7 and 2 nm and a small contribution of larger pores between 2 and 15 nm, these types of structures decrease for oxidized

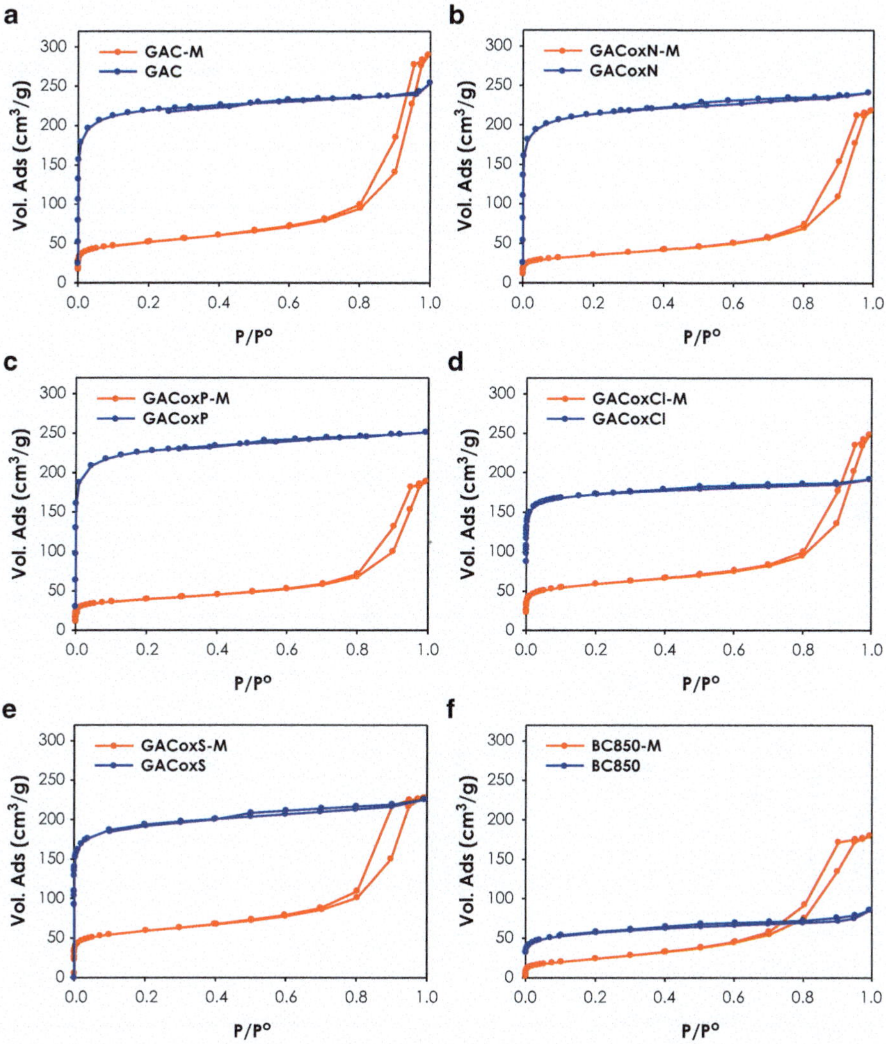

Fig. 4.3 N_2 adsorption-desorption isotherm by GAC series (**a–e**), BC850 (**f**) and magnetic modified solids

solids with HNO_3 and H_2O_2. Based on the histograms of the pore size distributions, it is observed that the effect of the oxidizing agents is mainly in the group of micropores with approximate dimensions at 1 nm, which decrease with respect to the starting solid.

The agents used for the modification of surface chemistry act, including oxygen atoms, on the surface of activated carbons. The attack of the oxidizing agents is mainly carried on carbon atoms that are found in the openings of the pores or on the surface of the solid, because these atoms do not have compensated cohesion forces,

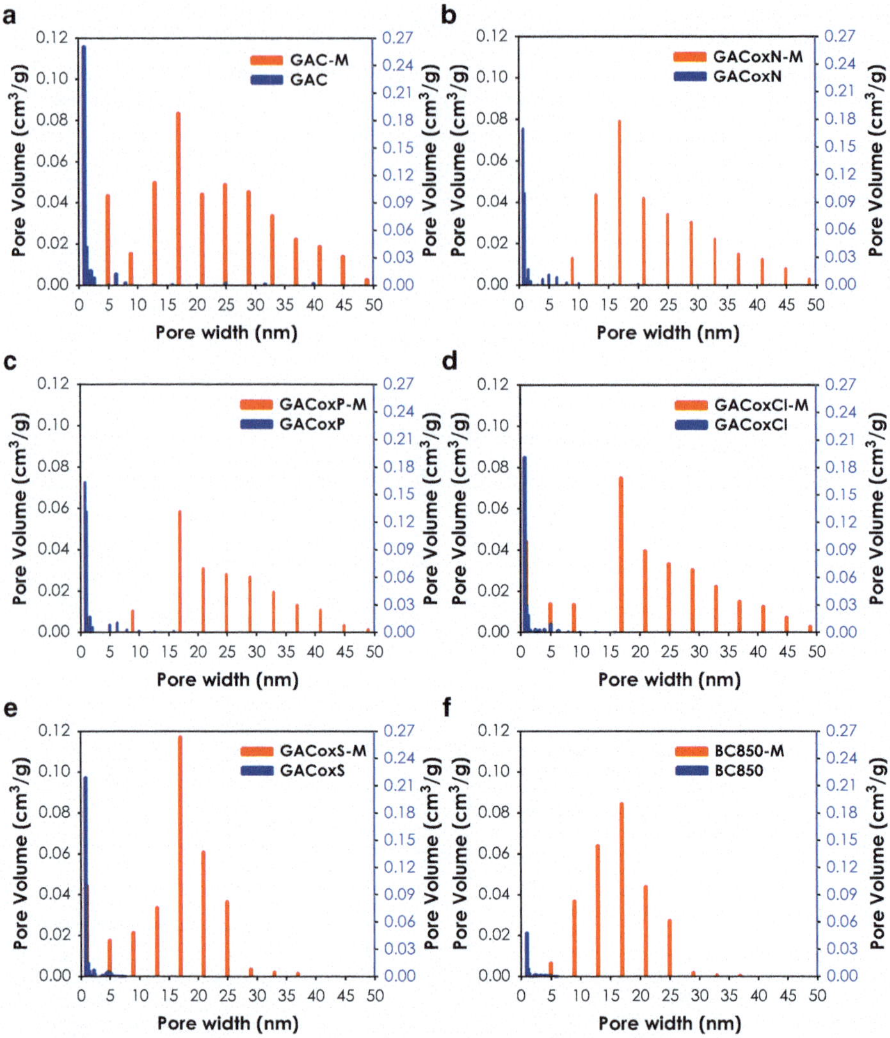

Fig. 4.4 Pore size distribution evaluated from DFT models by GAC series (**a**-**e**), BC850 (**f**) and magnetic modified solids

as it occurs in an atom located within the bulk solid, which makes them labile for interaction with oxidizing agents.

The porous system of the solids modified with the magnetite nanoparticles has a totally different PSD from the starting solids, where there is a microporous system remaining in the activated carbons and an intra-particle secondary system of the nanoparticles included in the porous system of activated carbon. The mesoporous nature could favor adsorption of big organic pollutants on its surface (Satheesh et al. 2014).

The experimental results were analyzed in different relative pressure ranges, using the BET method for the calculation of apparent surface areas (the P/P^o range was determined using the Rouquerol method). The microporosity analysis is performed by the Dubinin Astakhov model (at $P/P^o < 0.1$); the effect of different pore geometries and the effect of roughness/heterogeneity and homogeneity of the surface was studied by density functional theory models at P/P^o between 10^{-6} and 1, using the AsiQWin software. For mesoporous solids, the BJH model is used, for the analysis of mesoporosity and pore size distribution. Tables 4.1 and 4.2 summarized all the textural parameters evaluated.

The BET specific surface area for the magnetic modified samples were found to be among 129–220 m^2/g, which were much smaller than that for starting activated carbon, the S_{BET} parameter decreased between 68.0% and 84.2%, for the GACoxCl-

Table 4.1 Textural parameters of GAC and GAC-M series

	BET		DA ($P/P^0 < 0.1$)				QSDFT (P/P^0 10^{-5}-1)	
Sample	S_{BET} [$m^2.g^{-1}$]	C	V_{mic} [$cm^3.g^{-1}$]	Eo [$kJ.mol^{-1}$]	N	Pore diameter [nm]	V_P [$cm^3.g^{-1}$]	Pore width [nm]
GAC	849	117	0.351	7.638	1.80	1.420	0.344	0.785
GAC-M	192	388	0.068	6.845	2.20	1.480	0.428	1.182
GACoxN	815	121	0.349	8.447	1.40	1.340	0.337	0.753
GACoxN-M	129	389	0.047	6.650	2.00	1.480	0.327	13.13
GACoxP	871	113	0.357	7.659	1.80	1.420	0.352	0.785
GACoxP-M	148	535	0.054	7.719	1.90	1.420	0.285	0.723
GACoxCl	687	155	0.258	9.298	2.00	1.340	0.272	0.785
GACoxCl-M	220	569	0.080	7.424	2.00	1.440	0.363	0.723
GACoxS	751	1492	0.266	8.029	2.80	1.420	0.321	0.723
GACoxS-M	219	566	0.078	7.652	2.10	1.420	0.346	0.704
BC850	212	541	0.100	6.891	1.00	1.380	0.116	0.889
BC850-M	89.5	92.9	0.029	5.351	2.10	1.600	0.272	9.432

Table 4.2 Pore volumes and average pore width of the GAC-M series, evaluated by the B.J.H model, for the adsorption and desorption branch

	BJH adsorption branch		BJH desorption branch	
Sample	Pore volume [$cm^3.g^{-1}$]	Pore diameter [nm]	Pore volume [$cm^3.g^{-1}$]	Pore diameter [nm]
GAC-M	0.523	0.553	0.416	15.85
GACoxN-M	0.386	0.556	0.319	16.03
GACoxP-M	0.334	0.557	0.263	16.00
GACoxCl-M	0.457	0.552	0.334	16.01
GACoxS-M	0.582	0.516	0.307	15.91
BC850-M	0.324	0.523	0.285	8.922

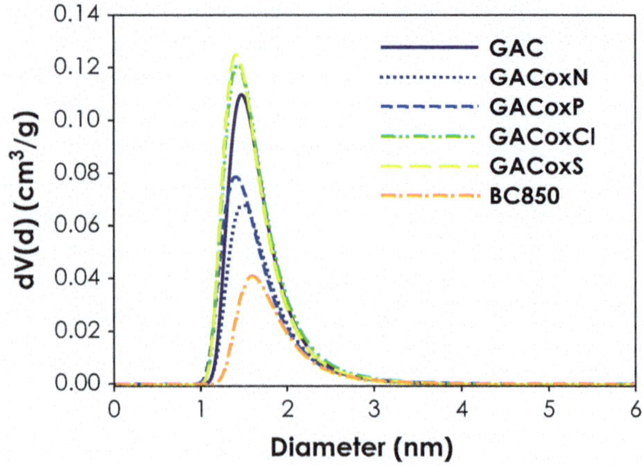

Fig. 4.5 Pore size distribution evaluated from Dubinin Astakhov model by magnetic modified GAC series and BC850

M and GACoxN-M samples, respectively. While the BC850 biochar sample decreases its specific area by only 58.0%. This suggests that the modification of activated carbon and biochar with magnetite by the coprecipitation method decreases the specific surface area of the materials; a more detailed analysis of the pore size distribution of the modified samples were evaluated from QSDFT model in the Fig. 4.4a–f, where it is shown for all samples with the exception of the GACoxCl and GACoxS, the porosity between 0.7 and 2 nm is mostly blocked by magnetite nanoparticles. The PSD obtained from the Dubinin Astakhov model also shows a change in the distribution of micropores; blocking this porosity is more evident in the GACoxN and GACox samples (Fig. 4.5), which also have the lowest textural parameters.

The analysis of the adsorption process in mesoporous solids is closely linked to the concept of capillary condensation, which is a physical process that occurs in the adsorption in pores that have a size between 2 and 50 nm. It is the final phase of the adsorption process and allows to determine the distribution of pore sizes in mesoporous solids, by relating the pressure in vapor balance with the meniscus curvature radius, surface tension, and molar volume. In Table 4.2, an increase of the pore diameter can be observed for all samples after the precipitation of the magnetite particles on the surface.

4.3.2 Chemical Properties of Magnetic Carbonaceous Composites

The FTIR spectra of carbonaceous materials, magnetite, and magnetic carbonaceous materials are shown in Fig. 4.6a, b. In the FTIR spectra of the GAC series, are show a broad adsorption bands centered at 3400 cm^{-1} related to stretching vibration of –OH groups, then at 1630 cm^{-1}, two bands related to stretching of C=O bonds and

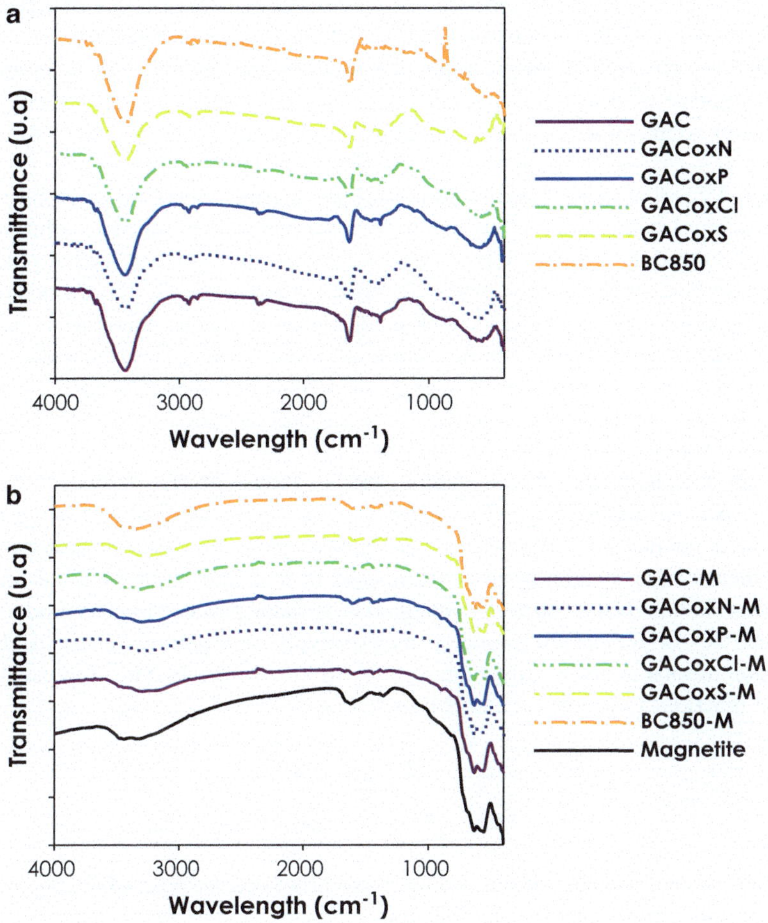

Fig. 4.6 Infrared spectra of (**a**) GAC series and BC850, (**b**) GAC-M series and BC850-M

stretching modes in aromatic ring structure were evidenced; at 2920 cm^{-1} bands are assigned to the vibrations of C-H of the alkanes, and these signals are typical bands for carbonaceous solids.In the FTIR spectra of the GAC-M series, it is observed a strong characteristic band at 640–540 cm^{-1} that was assigned to the stretching of Fe^{3+} ...O bonds located in tetrahedral and octahedral sites in magnetic materials; therefore, it is evident that all magnetic materials were coated with FexOy. In addition, broad band at 3500–3600 cm^{-1} was attributed to the hydroxyl (O–H) stretching (Kwon et al. 2014).

Another technique that allows to study the modifications on chemical nature of the surface of a porous solid is the potentiometric titration; by means of this technique it is possible to know the changes on surface charges due to the immersion of the particle in an aqueous solution with an acid initial pH. As the pH increases in the titration, the groups on the surface of the particle can exchange ions, because the groups are deprotonated according to their acid strength. The pKa is associated with the acid strength of the functional groups, which can be affected by the nearby groups by inductive effect. For example, the effect on the functional group of interest can be negative or positive if it is favoring its ionization or if the closest group enhances or weakens the acid strength (Figueiredo et al. 1999; Bandosz and Ania 2006).

Figure 4.7a shows the pKa distribution of the starting solid GAC and magnetite nanoparticles. In the case of the magnetite particles in an aqueous solution, its surface charges due to adsorption of ions like protons or hydroxyl (called potential determining ions) or due to dissociation of hydroxyl groups Fe---(OH) formed on the surface. Hydroxyl group can react with acidic or basic species when it is charged. The positive charge ($Fe\text{-}OH_2{}^{+}$) on the magnetite surface comes from the addition of a proton to the neutral surface hydroxyl group, while the negative charge ($Fe\text{-}O^{-}$) happens because of the acidic dissociation of the surface hydroxyl group and loss of the proton (Vidojkovic and Rakin 2017).

The surface chemistry of a magnetite particle is not uniform, in fact it is constituted by oxo and hydroxyl groups, which can be singly, doubly and triply metal coordinated, the unsaturated oxygen atoms have a high proton affinity, also the bond arrangement of the oxygen atoms with respect to iron atoms can be octahedral or tetrahedral. In aqueous solution, the magnetite particles produce an iron oxide/water interface, the surfaces of the particles are positively or negatively charged, due to adsorption of ions or due to dissociation of hydroxyl groups, like it is represented in the next reactions (Vidojkovic and Rakin 2017):

$$\mathrm{Fe - OH_2^+ \rightarrow M - OH + H^+ \; pK_{a1}} \tag{4.3}$$

$$\mathrm{Fe - OH \rightarrow M - O^- + H^+ \; pK_{a2}} \tag{4.4}$$

The groups represented in the reactions are produced because of the interaction between water molecules around the metal centers of the Fe_xO_y structure. The autoprotolysis of water molecules produced the hydroxyl surface groups, subsequently and according to the media pH, the protonation or deprotonation reactions of the hydroxyl groups can occur and determine the surface charge.

The potentiometric titration results summarized in Table 4.3 show three species on the magnetite surface with different pKa, 4.7 (0.701 μmol/g), 6.5 (3.256 μmol/g), and 11.1 (19.28), respectively. The lower concentration for the species with a pKa of 4.7 can be related with the proton located on oxo sites, which have a high proton affinity, while the species with a higher pKa, 6.5 and 11.1, can be related with -OH group dissociation (Zebardast et al. 2014).

On the other hand, carbonaceous materials are mainly constituted by unsaturated carbon atoms, and their valences are balanced with the interaction with heteroatoms like oxygen and hydrogen to form a variety of surface groups, some of acidic character and others of basic character. Also carbon atoms can form like-aromatic

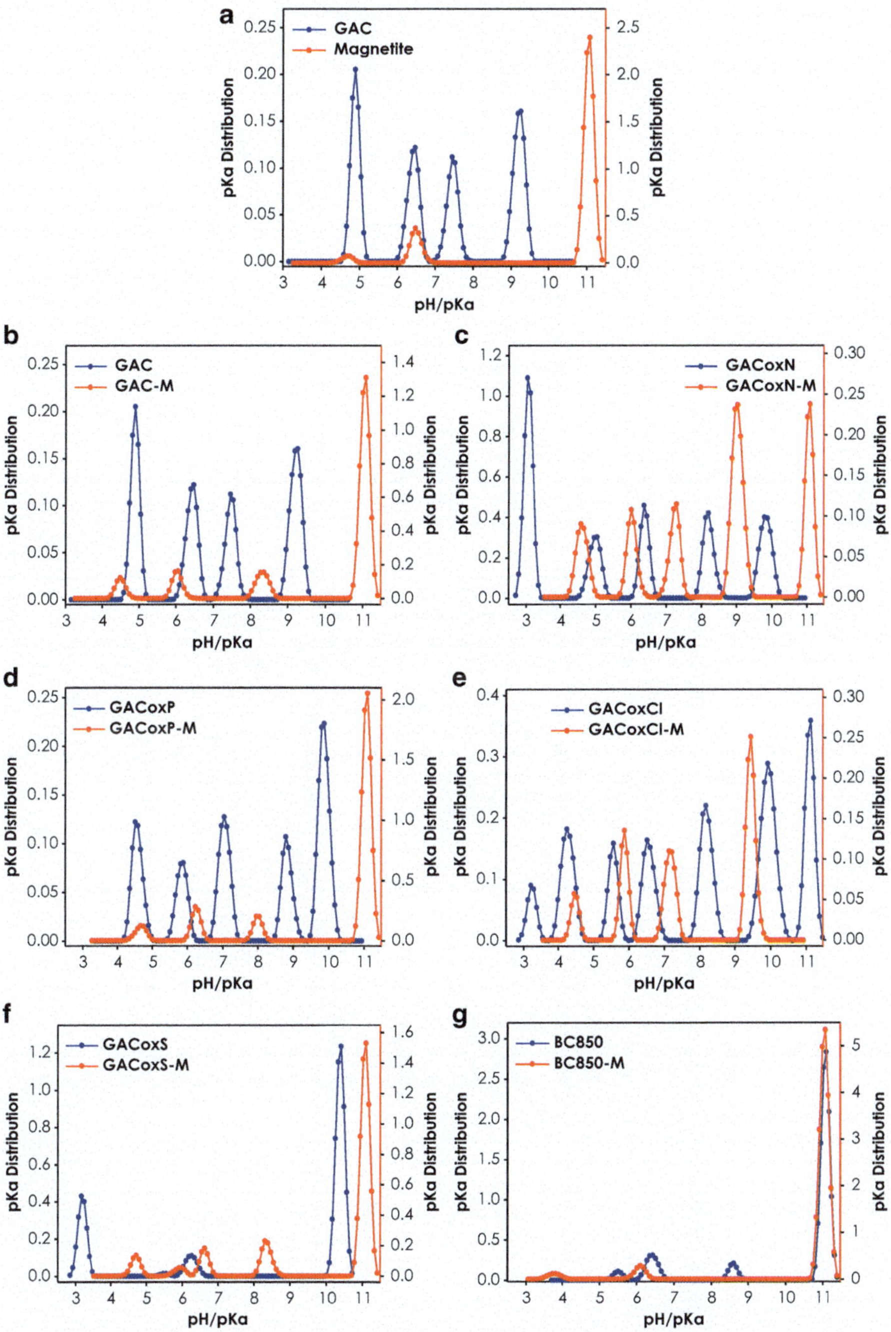

Fig. 4.7 pKa distribution and surface groups density by pKa value (μmol/g) determined by potentiometric titration. (**a**) GAC and Magnetite; (**b**) GAC and GAC-M; (**c**) GACoxN and GACoxN-M; (**d**) GACoxP and GACoxP-M; (**e**) GACoxCl and GACoxCl-M; (**f**) GACoxS and GACoxS-M; (**g**) BC850 and BC850-M

Table 4.3 pKa distribution and surface groups concentration by pKa value (molecules/nm^2) determined by potentiometric titration

pH									
Sample	3–4	4–5	5–6	6–7	7–8	8–9	9–10	10–11	Total
Magnetite	–	4.7 (0.701)	–	6.5 (3.256)	–	–	–	11.1 (19.28)	23.23
GAC	–	4.9 (0.045)	–	6.5 (0.039)	7.7 (0.027)	–	9.3 (0.046)	–	0.157
GAC-M	–	4.5 (0.133)	–	6.0 (0.180)	–	8.3 (0.225)	–	11.1 (1.373)	1.912
GACoxN	3.1 (0.273)	5.0 (0.101)	–	6.4 (0.113)	–	8.2 (0.123)	9.8 (0.138)	–	0.749
GACoxN-M	–	4.6 (0.182)	–	6.0 (0.175)	7.2 (0.196)	–	9.0 (0.461)	11.1 (0.359)	1.372
GACoxP	–	4.6 (0.033)	5.9 (0.024)	–	7.1 (0.037)	8.8 (0.028)	9.9 (0.062)	–	0.184
GACoxP-M	–	4.7 (0.202)	–	6.3 (0.376)	–	8.0 (0.269)	–	11.1 (3.646)	3.646
GACoxCl	3.3 (0.028)	4.3 (0.077)	5.6 (0.048)	6.6 (0.066)	–	8.2 (0.089)	9.9 (0.135)	11.1 (0.111)	0.555
GACoxCl-M	–	4.5 (0.047)	5.9 (0.110)	–	7.2 (0.125)	–	9.5 (0.224)	–	0.506
GACoxS	3.2 (0.111)	–	5.6 (0.004)	6.3 (0.039)	–	–	–	10.4 (0.320)	0.474
GACoxS-M	–	4.7 (0.125)	5.9 (0.062)	6.6 (0.160)	–	8.4 (0.234)	–	11.1 (1.364)	1.944
BC850	–	–	5.5 (0.103)	6.4 (0.380)	–	8.6 (0.186)	–	11.1 (2.548)	3.216
BC850-M	3.8 (0.370)	–	–	6.1 (0.718)	–	–	–	11.1 (12.33)	13.42

Table 4.4 pH_{PCZ} and magnetic carbonaceous materials and carbonaceous materials

Sample	$pH_{PZC} \pm 0{,}1$	Sample	$pH_{PZC} \pm 0{,}1$
GAC	5.40	GAC-M	6.80
GACoxP	6.22	GACoxP-M	6.86
GACoxCl	7.25	GACoxCl-M	6.98
GACoxN	3.45	GACoxN-M	6.72
GACoxS	3.62	GACoxS-M	7.02
BC850	7.82	BC850-M	7.82
		Magnetite	7.62

structures (graphenic layers), and their resonance can modify the acid strength of the nearby groups by inductive effect. The concentration of these functionalities can define the nature of the surface of carbonaceous materials; also they can maximize the affinity between the adsorbent and the adsorbates. It has been considered that the aromatic structures of carbon are more refined to interact with nonpolar and organic molecules, while oxygenated functional groups interact very well with polar molecules or metal ions (Jagiełło et al. 1995; Bandosz and Ania 2006).

About activated carbons, GAC pKa distribution shows four species, the first two associated with carboxylic acid groups (i.e., pKa < 7), with pKa values of 4.91 and 6.53, the next two species with pKa values at 7.69 and 9.32 can be associated with phenols and quinone groups, respectively. According to that, the starting sample has overall acidic character, and this is also evidenced by a $pH_{PZC} = 5.40$ (Table 4.4). As for the oxidized samples with nitric acid, hydrogen peroxide, sodium hypochlorite, and sulfuric acid, it is possible to observe the increase in concentration of the oxygenated surface groups, and a variety of different pKa values related to the change of surface chemistry for solids. The oxidation reaction with nitric acid favors rupture of bonds of the carbonaceous structure to produce mainly ketone and dicarboxylic groups; it can be evidenced in the shifting of peaks at different pKa values, and also it can be evidenced in the increase of their concentration. The new pKa values located at 3.1 (0.273 molecules/mn^2) and 5.0 (0.101 molecules/mn^2) were obtained, two different carboxylic acids. Additionally, the lactone and/or carboxylic anhydride groups are represented by the peaks at pKa values between 6.4 (0.113 molecules/mn^2) and 8.2 (0.123 molecules/mn^2) and the phenolic groups by pKa above 9.8 (0.138 molecules/mn^2) (Table 4.3). The increase in the concentration of the acid surface groups is also evidenced by the decrease in pH_{PZC} to 3.45 (Rodríguez-Estupiñán et al. 2018).

The effect of the hydrogen peroxide can be attributed mainly to the decomposition of the abduct between oxygen and water molecules when it contacts the activated carbon granules; the species formed interact with other molecules of the hydrogen peroxide to form new species and with the carbonaceous surface. The pKa distribution evidenced the formation of two new species with respect to GAC; the pKa values are centered at 5.9 (0.024 molecules/nm^2) and 8.8 (0.028 molecules/nm^2) and there is an increase of phenol groups 9.9(0.062 molecules/nm^2). The total amount of groups increased slightly compared with GAC, but through the oxidation with hydrogen peroxide, the formation of weak acid groups is favored; this leads to an increase in the pH_{PZC} to 6.22 for GACoxP.

About GACoxCl sample, sodium hypochlorite attacks carbon double bonds to form carboxylic acid groups and can also attack methyl or methylene groups to form phenolic, carbonyl, and quinones groups. The pKa distribution differs a lot from the starting activated carbon; in Fig. 4.7(e), it is possible to see seven species, and three of them can be classified as carboxylic acid group with pKa < 7, the two species at 6.2 and 8.8 can be attributed to lactone groups. Hypochlorite sodium also promoted the formation of species with pKa >10, and these species can be attributed to weak acids; basicity parameter also increased, and it was related to the increase of pH_{PZC} to 7.25.

Furthermore, the modification with sulfuric acid is related to the oxidation property of S^{6+} (Asasian Kolur et al. 2019). Figure 4.7f shows the pKa distribution of sulfuric-acid-treated sample, in which the presence of two new acid species with a pKa values at 3.2 and 5.6 can be observed, and the higher concentration of the species have a pKa value of 10.4(0.320), which can be related with weak acids. The increase in the concentration of the acid surface groups is also evidenced by the decrease in pH_{PZC} to 3.62 (Niu et al. 2015).

Finally, potentiometric titration for BC850 evidenced four species, including two strong acidic sites and weak acidic sites at pKa: 5.5 (0.103 μmol/g), 6.4 (0.308 μmol/g), 8.6 (0.186 μmol/g), and 11.1 (2.548), respectively. The carboxyl acid group (pKa 2–5) can be responsible for the first site detected and the second can be attributed for lactone groups (pKa 6–8.5). Also, phenols groups (pKa ~9) can be related with third site and last site can be assigned for pyrone group (pKa 9–13) (Contescu et al. 1997). In addition, the highest concentration for basic groups was can be related to presence of the alkaline minerals including K, Ca, and Mg, which were detected by SEM-EDS analysis. Meanwhile, in potentiometric titration for BC850-M was detected three sites, at pKa: 3.8 (0.370 μmol/g), 6.4 (0.718 μmol/g), and 11.1 (12.22), respectively. This can be related to the same groups presented in BC850: carboxylic, phenolic, and pyrone groups. In addition, the increased concentration of basic groups can be due to -OH group's dissociation after interaction of Fe–(OH) and water (Zebardast et al. 2014).

According to Table 4.4, all pH_{PZC} data of magnetic solids range from 6.72 to 7.82, due to dissociation of hydroxyl groups on the magnetite surface (Yoon et al. 2014).

Based on the above results of FTIR, potentiometric titration results, SEM images, and N_2 adsorption/desorption isotherm analyses, we can make a conclusion that carbonaceous materials were successfully modified by the iron oxide nanoparticles, now the effect of surface chemistry as of magnetite nanoparticles on the adsorption of caffeine and nickel ions will be analyzed, both from aqueous solution.

4.3.3 Caffeine and Nickel Removal from Aqueous Solution

Caffeine Equilibrium and Kinetic Studies

Four isotherm models have been evaluated in this study: Langmuir (Eq. (4.5)) (I. Langmuir 1916), Freundlich (Eq. (4.6)) (H. Freundlich 1906), Redlich-Peterson (Eq. (4.7)) (Redlich and Peterson 1958), and Toth (Eq. (4.8)) (Toth 1971). The experimental equilibrium data were analyzed with the mentioned models and the adjustment was evaluated through the determination coefficient (R^2):

$$Q_e = \frac{Q_0 * K_L * C_e}{1 + K_L * C_e} \tag{4.5}$$

$$Q_e = K_F * C_e^{1/n} \tag{4.6}$$

$$Q_e = \frac{K_{RP} * C_e}{1 + a_{RP} * C_e^{\beta}} \tag{4.7}$$

$$Q_e = \frac{K_T * C_e}{(a_T * C_e)^{1/t}} \tag{4.8}$$

where Q_e (mg g^{-1}) is the equilibrium of CFN concentration in the adsorbent, C_e (mg L^{-1}) is the equilibrium concentration of CFN in the aqueous phase, Q_o (mg g^{-1}) is the maximum adsorption, and K_L(L g^{-1}) is the Langmuir equilibrium adsorption constant. K_F(L g^{-1}) is the Freundlich constant or relative sorption capacity, and *n* is a constant indicating adsorption intensity. K_{RP}(L g^{-1}) and a_R (L mg^{-1}) are Redlich–Peterson isotherm constants, and *B* is the exponent. K_T (L g^{-1}) and a_T (mg L^{-1}) are Toth isotherm constants, and *t* is the exponent. Figures 4.8 and 4.9 presented graphs of the carbonaceous materials evaluated by different adsorption models employed.

Adsorption capacity of caffeine was significantly different between magnetic carbonaceous and nonmagnetic materials. However, for all isotherms, a different behavior was observed when the initial concentration of the caffeine solution is low or when it is high (between 1 and 1000 mg/L). For carbonaceous materials, it is observed that the adsorption capacity increases, due to the progressive occupation of the adsorption sites until saturation of the adsorbent, which is also evidenced by the formation of a plateau in the isotherm. While in the isotherms of the magnetized materials at low concentrations, a behavior similar to that of the starting materials is observed (Fig. 4.8b), but at high concentrations for materials such as GAC-M, GACoxS-M, and BC850-M, an increase of the adsorption capacity is observed with a second steep slope. In accordance with the above, it can be concluded that the dynamic equilibrium are influenced by the initial concentration. This behavior is explained by the increase in the driving force of the concentration gradient (Srivastava et al. 2009).

Comparing Q_{max} parameter of Langmuir model for caffeine adsorption, the best precursor material was CAG and the best derivate magnetic material was CAGoxS-M, reaching values of 232.9 mg g^{-1} and 42.21 mg g^{-1}, respectively (Table 4.5). The values of caffeine adsorption capacity achieved in this work were higher than that reported (Beltrame et al. 2018; dos Santos Lins et al. 2019), but they were lower compared to other researches (Chung et al. 2017; Portinho et al. 2017) (Table 4.6).

Linear correlation coefficient (R^2) values of four isotherm models evaluated were similar. However, Redlich-Peterson adsorption model was the best fit for all the experimental data. It suggested that the adsorption process of carbonaceous materials evaluated was a hybrid adsorption mechanism, since it is a combination between

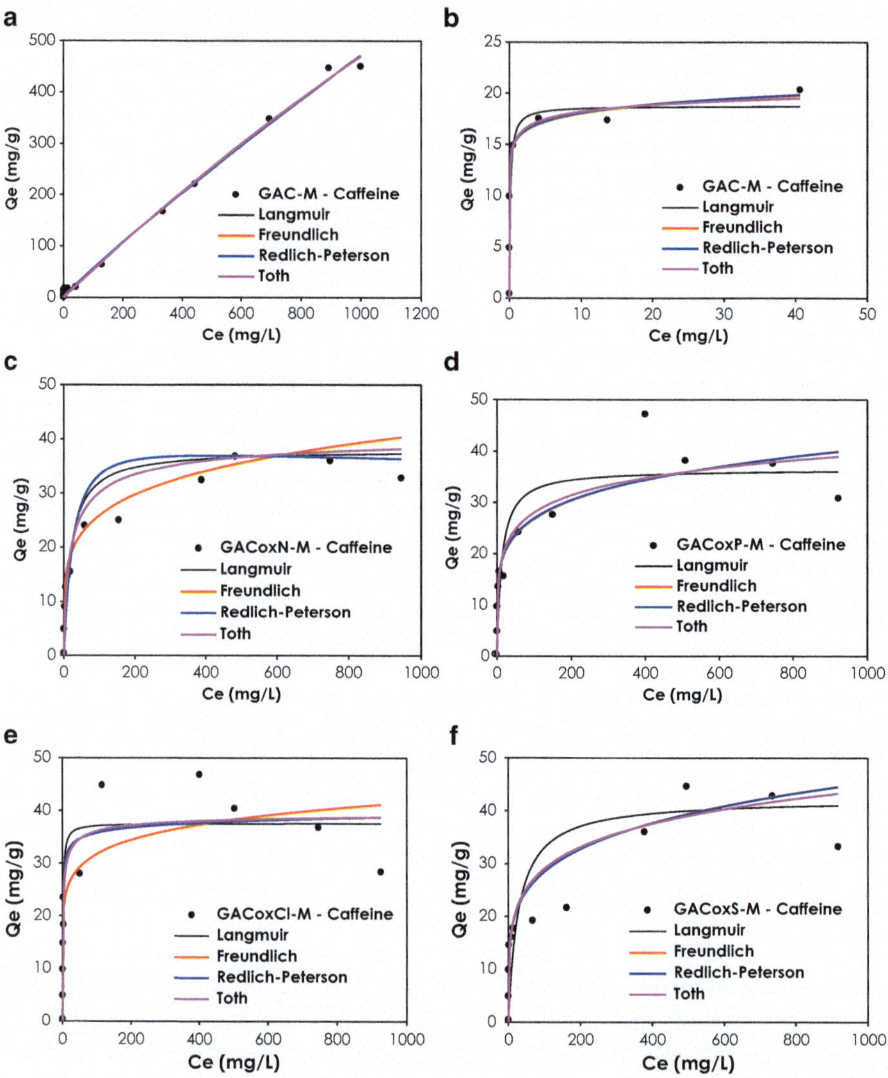

Fig. 4.8 Equilibrium data of caffeine adsorption on magnetic modified solids: (**a**) CAG-M; (**b**) GACoxN-M; (**c**) GACoxP-M; (**d**) GACoxCl-M; (**e**) GACoxS-M; and (**f**) BC 850-M, the colored solid lines correspond to Langmuir. Freundlich, Redlich-Peterson, and Toth models fitting

Langmuir and Freundlich models (Redlich-Peterson). Therefore, monolayer and multilayer adsorption processes were possible in caffeine adsorption onto carbonaceous materials evaluated. On the other hand, differences of adsorption capacity could be caused for porous obstruction of carbonaceous materials by magnetite; therefore, the surface area of magnetic materials decreased. In addition, superficial

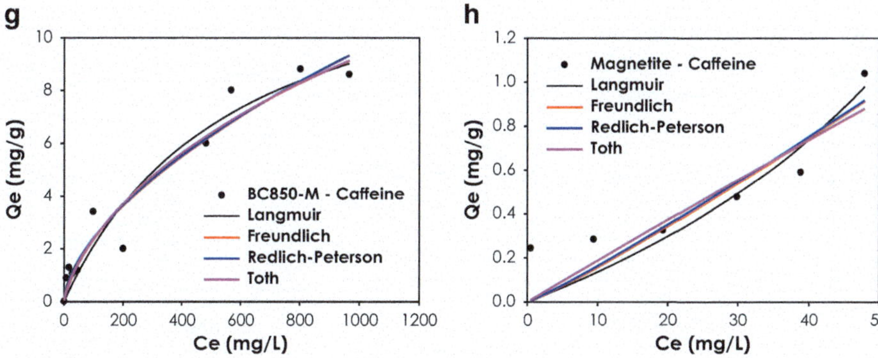

Fig. 4.8 (continued)

chemistry of carbon materials was modified by Fe_xO_y, and there were less oxygen superficial functional groups available (Reguyal et al. 2017).

To determine the mechanisms of adsorption on the carbonaceous materials with better adsorption capacity, GAC, GACoxCl, CAG-M, and GACoxCl-M, three kinetic models were evaluated: pseudo-first order (Eq. (4.9)), pseudo-second order (Eq. (4.10)), and Elovich (Eq. (4.11)) (Kajjumba et al. 2018). The equations were used with experimental data, and normalized standard deviation ($\Delta Q(\%)$), chi-square (χ^2), average relative error (ARE(%)), and hybrid fractional error function (HYBRID) were calculated using Eqs. (4.12), (4.13), (4.14), and (4.15); these tests allowed to obtain the errors in experimental data, and their values should be as close to "zero" as possible (Alahabadi et al. 2017):

$$Q_t = Q_e(1 - \exp(-kt) \tag{4.9}$$

$$Q_t = \frac{Q_e^2 k_2 t}{1 + Q_e k_2 t} \tag{4.10}$$

$$Q_t = \frac{1}{\beta} \ln(1 + \alpha\beta t) \tag{4.11}$$

where Q_t (mg g^{-1}) is the adsorption capacity at time t; t is the time (min), Q_e (mg g^{-1}) is the adsorption capacity at equilibrium, k is the rate constant of the pseudo-first-order model (1 min^{-1}), k_2 (g mg^{-1} min^{-1}) is the pseudo-second-order rate constant, α (mg g^{-1} min^{-1}), and β (g mg^{-1}) are the constants for Elovich model. Plots of Qe versus t are presented in Fig. 4.10. All curves showed a similar trend. Caffeine was retained onto carbonaceous materials and magnetic carbonaceous material rapidly at the start of the adsorption process, while slower adsorption rate was obtained at the end of the adsorption process:

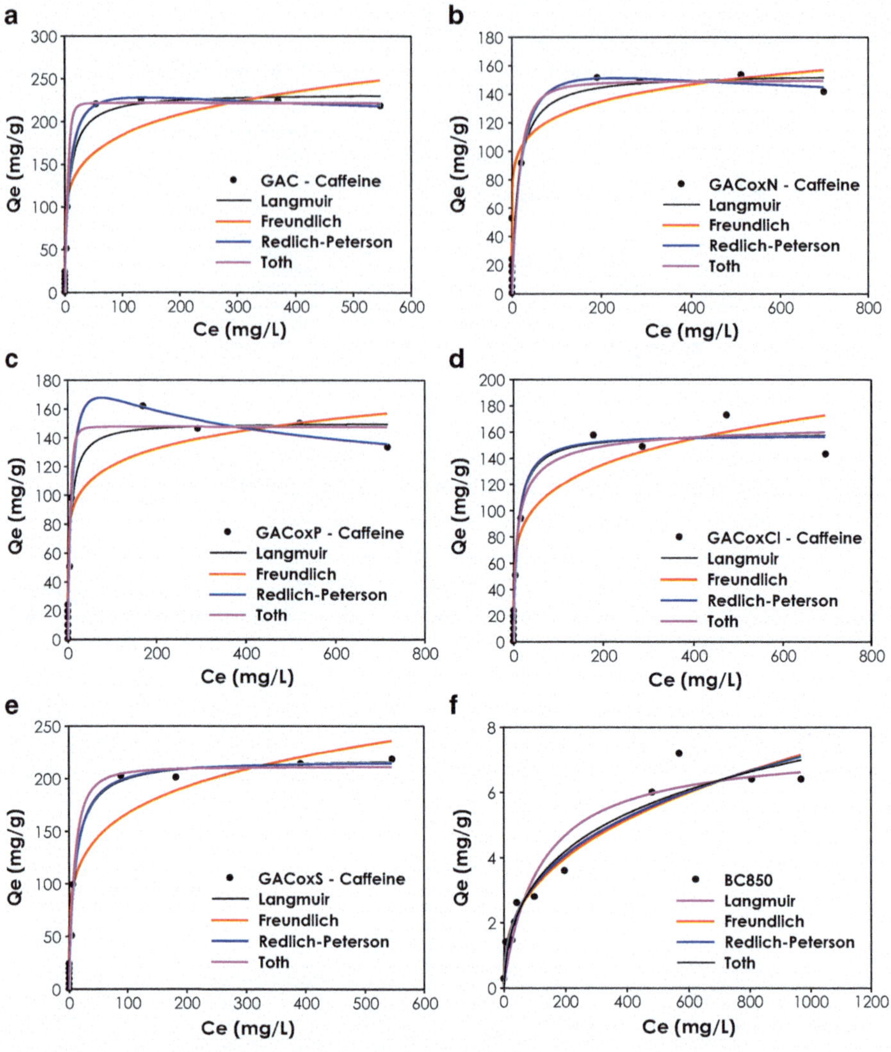

Fig. 4.9 Equilibrium data of caffeine adsorption on (**a**) CAG; (**b**) GACoxN; (**c**) GACoxP; (**d**) GACoxCl; (**e**) GACoxS; and (**f**) BC 850, the colored solid lines correspond to Langmuir. Freundlich, Redlich-Peterson, and Toth models fitting

$$\Delta Q(\%) = 100\sqrt{\frac{1}{N-1}\sum_{i=1}^{N}\left(\frac{(Q_{\text{exp}} - Q_{\text{cal}})}{Q_{\text{exp}}}\right)_i^2} \tag{4.12}$$

$$\text{ARE}(\%) = \frac{100}{N-1}\sum_{i=1}^{N}\left(\frac{(Q_{\text{exp}} - Q_{\text{cal}})}{Q_{\text{exp}}}\right)_i^2 \tag{4.13}$$

Table 4.5 Parameters of Langmuir and Freundlich isotherm models for caffeine adsorption onto starting solids and magnetic carbonaceous composites at 293 K

	Langmuir model			Freundlich model		
Sample	Qmax (mg g^{-1})	K_L (L mg^{-1})	R^2	Kf (L g^{-1})	1/n	R^2
GAC	232.9	0.138	0.981	81.83	0.177	0.920
GAC –M	18.71	6.891	0.609	15.51	0.066	0.617
GACoxN	150.9	0.168	0.955	64.79	0.135	0.916
GACoxN-M	36.49	0.080	0.809	11.76	0.179	0.872
GACoxP	159.6	0.101	0.968	47.87	0.196	0.942
GACoxP-M	37.59	0.753	0.847	18.65	0.116	0.805
GACoxCl	154.7	0.071	0.908	70.62	0.122	0.896
GACoxCl-M	38.02	0.053	0.675	10.54	0.196	0.642
GACoxS	219.8	0.099	0.982	64.25	0.207	0.951
GACoxS-M	42.21	0.036	0.660	11.32	0.201	0.770
BC850	7.459	0.009	0.931	0.559	0.371	0.947
BC850-M	14.59	0.002	0.939	0.153	0.598	0.948
Magnetite	23.46	0.0073	0.908	0.012	1.119	0.748

Table 4.6 Parameters of Redlich-Peterson and Toth isotherm models for caffeine adsorption onto starting solids and magnetic carbonaceous composites at 293 K

	Redlich-Peterson				Toth			
Sample	K_{RP} (L g^{-1})	a_{RP} (L mg^{-1})	B	R^2	K_T (L g^{-1})	a_T (mg L^{-1})	t	R^2
GAC	26.91	0.080	1.067	0.985	222.1	4043	3.489	0.986
GAC-M	9143	588.2	0.934	0.617	24.55	0.089	0.189	0.616
GACoxN	7.644	0.028	1.094	0.911	149.7	342.1	1.807	0.910
GACoxN-M	1.524	0.027	1.061	0.677	41.88	3.381	0.597	0.683
GACoxP	17.00	0.046	1.151	0.968	148.0	1293	2.970	0.963
GACoxP-M	566.2	47.53	0.8228	0.872	100.64	0.398	0.145	0.877
GACoxCl	15.51	0.092	1.009	0.968	169.3	2.935	0.666	0.970
GACoxCl-M	37.08	1.186	0.9684	0.851	40.11	0.544	0.503	0.863
GACoxS	21.31	0.093	1.007	0.982	212.0	45.91	1.500	0.982
GACoxS-M	4200	3714	0.7992	0.770	317.1	0.365	0.090	0.765
BC850	0.451	0.612	0.6678	0.948	27.67	2.080	0.242	0.976
BC850-M	0.588	3.470	0.4143	0.948	117.8	7.481	0.284	0.946
Magnetite	0.028	0.551	0.0005	0.909	9436	245.3	0.420	0.743

$$\chi^2 = \sum_{i=1}^{N} \left(\frac{(Q_{\text{exp}} - Q_{\text{cal}})^2}{Q_{\text{cal}}} \right) \tag{4.14}$$

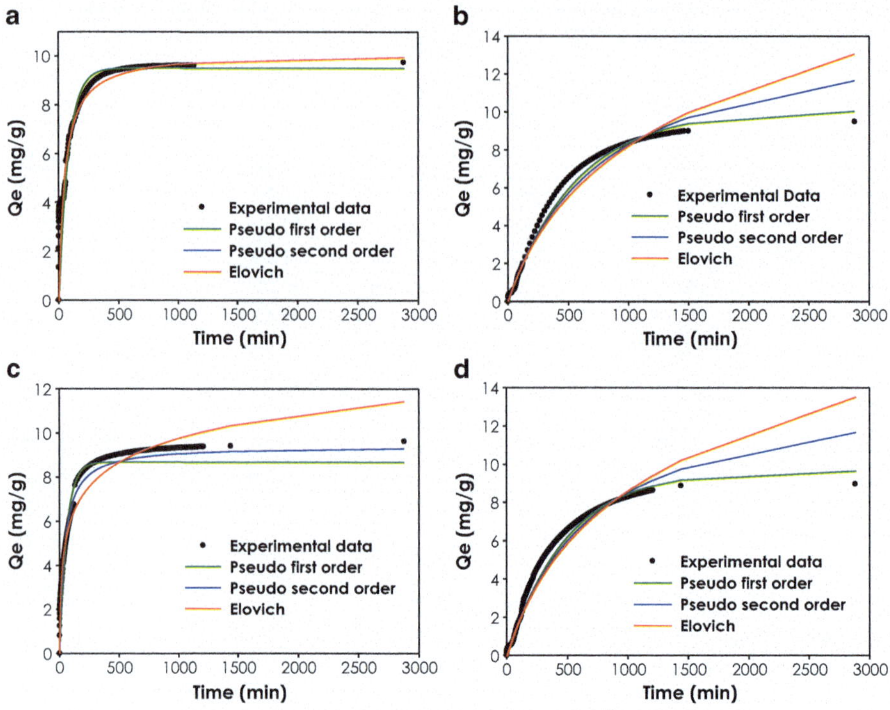

Fig. 4.10 Sorption kinetics of caffeine on (**a**) CAG, (**b**) CAG-M, (**c**) GACoxCl, and (**d**) GACoxCl-M

$$\text{HYBRID} = \frac{100}{N-p}\sum_{i=1}^{N}\left(\frac{(Q_{\text{cal}} - Q_{\text{exp}})}{Q_{\text{exp}}}\right)_i^2 \quad (4.15)$$

where Q_{exp} and Q_{cal}, respectively, are the experimental value and the calculated value of adsorption capacity of carbonaceous magnetic materials evaluated at time "t" or equilibrium concentration C_e, N is the number of measurements, and p is the number of parameters in each model. The results are displayed in Table 4.7.

Figure 4.10 shows that adsorption rate is faster in carbonaceous materials than in magnetic carbon materials. This may be due to the higher number of pores and active sites in the magnetite-free materials. However, for both types of materials, adsorption capacity increased rapidly and then slowly, and then tended to equilibrate.

Statistical error analysis showed that the pseudo-first-order and pseudo-second-order models described the kinetic data well for GAC-M, GACoxCl-M, GAC, and GACoxCl, respectively. This suggests that adsorption mechanisms were different between carbonaceous materials and their magnetic derivatives. The pseudo-first-order model assumed physical adsorption, then caffeine retention rate with time is directly proportional to caffeine concentration and the amount of magnetic

Table 4.7 Parameters of different kinetic models and statistical indices for adsorption of CFN onto GAC, GAC-M, GACoxCl, and GACoxCl-M at 293 K

	Caffeine				
Model	Parameters	GAC	GACoxCl	GAC-M	GACoxCl-M
Pseudo-first order	$q_{e\ exp}$ (mg g^{-1})	9.750	9.621	9.481	8.972
	$q_{e\ cal}$ (mg g^{-1})	9.280	8.973	9.483	9.173
	k (1 min^{-1})	0.016	0.015	0.002	0.002
	R^2	0.930	0.910	0.990	0.990
	Δq(%)	33.74	19.82	25.90	20.57
	ARE (%)	16.80	12.25	16.90	12.42
	χ^2	287.7	134.1	6.362	19.34
	Hybrid	31.11	12.00	4.161	3.134
Pseudo-second order	$q_{e\ cal}$ (mg g^{-1})	9.980	9.710	12.77	12.82
	k_2 (1 min^{-1})	0.002	0.002	0.001	0.001
	R^2	0.980	0.990	0.840	0.850
	Δq(%)	25.01	14.59	26.30	20.98
	ARE (%)	14.62	7.253	18.02	13.51
	χ^2	210.2	54.86	14.00	40.17
	Hybrid	22.95	5.501	5.631	4.162
Elovich	$q_{e\ cal}$ (mg g^{-1})	11.23	11.44	13.08	13.54
	α (mg g^{-1} min^{-1})	1.075	0.757	0.019	0.021
	β (g mg^{-1})	0.681	0.632	0.185	0.187
	R^2	0.90	0.960	0.860	0.860
	Δq(%)	16.63	7.571	26.82	21.23
	ARE (%)	12.38	5.342	18.86	14.45
	χ^2	130.23	53.37	29.18	73.77
	Hybrid	11.90	2.563	7.562	5.491

carbonaceous materials. While the pseudo-second-order model assumed that the rate-limiting step may be chemical sorption, involving valence forces through sharing or exchange of electrons between caffeine and carbonaceous materials (Sun et al. 2013). These differences could be caused by magnetite deposition on carbonaceous materials; as a result, the surface was modified and there were fewer functional surface chemical groups available. In addition, calculated qe values for GAC, GACoxCl, GAC-M, and GACoxCl-M were similar to experimental data in the two best models evaluated. Analogous results were obtained with biochar of *Eichhornia crassipes*, activated carbon fibers (ACFs) from pineapple plant leaves, chars from gasification of coal and pine activated with K_2CO_3, activated carbons from dende coco and babassu coco, and a commercially available activate carbon (NO: Norit1 GAC 1240 plus), among other researchers who determined that pseudo-second-order kinetic model described better caffeine kinetic adsorption (Galhetas et al. 2014a; Couto et al. 2015; Ngeno et al. 2016).

Nickel Adsorption

All the carbonaceous and magnetized samples obtained were tested as adsorbents of nickel from aqueous solution; the isotherms are presented in Fig. 4.11a for carbonaceous materials and Fig. 4.11b for the magnetized materials. The experimental data were also fitted to the Langmuir, Freundlich, Redlich-Peterson, and Toth models.

The Langmuir model fits well with experimental data of nickel adsorption; the mechanism by this model suggested the formation of the monolayer on a homogenous surface of the solid, and the lateral interactions between adsorbed particles are unlikely. On the other hand, Freundlich model described a multilayer adsorption on heterogeneous surface (Allen et al. 2004). Freundlich model fits better for the nickel adsorption on GACoxP, GACoxS-M, BC850, and magnetite samples (Table 4.7). These results suggested that the nickel ions are preferably adsorbed in monolayer on the surface of the solid. The maximum capacity evaluated by Langmuir model shows a better efficiency of the magnetic solids only for GAC-M, GACoxP-M, and GACoxS-M.

Three parameter models have been developed to describe the adsorption process from a better mathematical approach; Redlich-Peterson and Toth models are equations closely related to conventional Langmuir and Freundlich models. Toth model describes an adsorption mechanism which implies the formation of heterogeneous monolayer on a surface with finite number of energetically different sites; Toth model fits better to all experimental data except to GACoxCl (Table 4.8). From this model, it is also possible to compare the maximum adsorption capacity with the characteristic parameter of the Toth model K_T (Mukherjee et al. 2019); when comparing this parameter, higher adsorption capacity for solids is observed: GACoxS-M > GACoxCl-M > Magnetite > GAC-M > BC850 > GACoxN-M > GACoxP-M > GACoxP > GACoxS > GAC > BC850-M and finally GACoxN-M. In the next section, the proposed adsorption mechanism is discussed (Table 4.9).

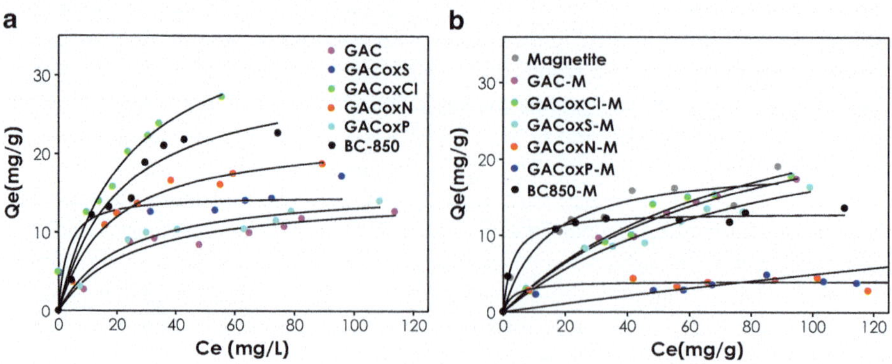

Fig. 4.11 Equilibrium data of Ni(II) adsorption on (**a**) CAG series and BC-850, (**b**) Magnetic modified solids. The solid line correspond to Langmuir model fitting

Table 4.8 Parameters of Langmuir and Freundlich isotherm models for nickel adsorption onto starting solids and magnetic carbonaceous composites at 293 K

	Langmuir model			Freundlich model		
Sample	Qmax (mg g^{-1})	K_L (L mg^{-1})	R^2	Kf (L g^{-1})	1/n	R^2
GAC	14.91	0.039	0.975	1.955	0.400	0.965
GAC-M	32.77	0.012	0.995	0.978	0.641	0.993
GACoxN	22.77	0.055	0.994	5.210	0.292	0.991
GACoxN-M	4.061	0.320	0.926	2.712	0.080	0.915
GACoxP	14.67	0.342	0.964	8.566	0.153	0.969
GACoxP-M	35.18	0.012	0.991	4.036	0.333	0.971
GACoxCl	39.25	0.042	0.976	9.370	0.282	0.732
GACoxCl-M	29.44	0.012	0.989	0.884	0.634	0.989
GACoxS	15.93	0.045	0.984	5.527	0.113	0.948
GACoxS-M	40.07	0.001	0.847	0.191	0.704	0.859
BC850	30.80	0.047	0.984	3.142	0.604	0.991
BC850-M	13.08	0.329	0.994	5.725	0.189	0.977
Magnetite	19.52	0.072	0.968	4.036	0.333	0.971

Table 4.9 Parameters of Redlich-Peterson and Toth isotherm models for nickel adsorption onto starting solids and magnetic carbonaceous composites at 293 K

	Redlich-Peterson				Toth			
Sample	K_{RP} (L g^{-1})	a_{RP} (L mg^{-1})	B	R^2	K_T (L g^{-1})	a_T (mg L^{-1})	T	R^2
GAC	0.599	0.044	0.979	0.975	13.94	47.82	1.172	0.975
GAC-M	0.405	0.015	0.967	0.995	32.56	84.78	1.007	0.995
GACoxN	3.756	0.403	0.838	0.994	30.49	1.819	0.448	0.994
GACoxN-M	0.623	0.070	1.167	0.934	3.930	67.52	2.095	0.930
GACoxP	23.45	1.996	0.923	0.975	19.43	0.472	0.417	0.976
GACoxP-M	0.342	0.001	1.461	0.984	25.96	612.5	1.490	0.992
GACoxCl	6.088	0.355	0.844	0.731	51.35	1.556	0.426	0.732
GACoxCl-M	0.973	0.550	0.477	0.999	98.86	9.431	0.442	0.990
GACoxS	0.711	0.044	1.002	0.984	14.58	57.15	1.264	0.985
GACoxS-M	2.391	12.32	0.292	0.856	320.2	9.515	0.296	0.853
BC850	1.057	0.008	1.340	0.986	31.06	13.80	0.959	0.996
BC850-M	5.079	0.457	0.962	0.995	13.68	1.936	0.794	0.995
Magnetite	4.374	0.680	0.766	0.974	38.09	1.523	0.350	0.974

Adsorption Mechanism

Figure 4.12a, b shows the correlation between the available surface area and micropore volume on the started and modified solid on the maximum caffeine adsorption capacity; it is possible to establish that the greater the available area and micropore volume for the adsorption process, the adsorbate-adsorbent

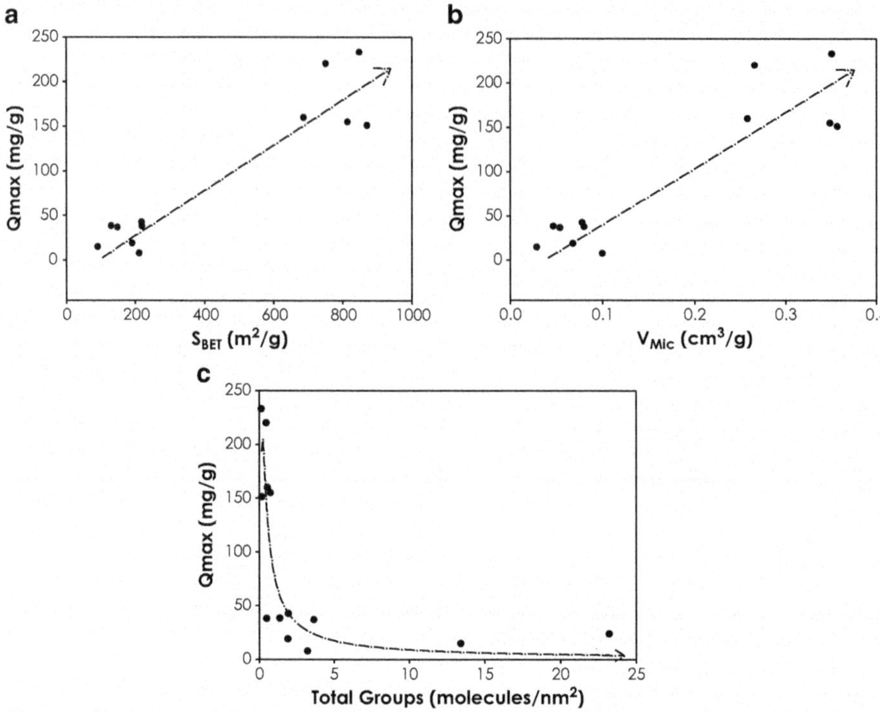

Fig. 4.12 Correlation between textural parameter: superficial area -SBET- (a) and micropore volume, -VMic- (b), and total superficial groups evaluated from potentiometric titration with the maximum caffeine adsorption capacity from Langmuir model (c)

interactions are favored. In this sense, the carbonaceous materials with high surface areas have a porous structure that allows to diffuse the caffeine molecules throughout the porous structure. However, restrictions on this diffusion must be considered due to the dimensions of the caffeine molecule (0.78 nm, 0.61 nm, and 0.21 nm for length, width, and height, respectively) (Pendolino 2014) and the pore size distribution of the materials. As previously discussed, the inclusion of surface groups oxygenated by oxidation processes affects the size distribution of porous structures, since the new groups are located mainly in the openings of the pores, blocking the diffusion of the molecules into the porous network.

On the other hand, the caffeine molecule is a weak electrolyte composed by pyrimidine and imidazole rings with 10 electrons, aromatic character, and zwitterionic resonance structure (Ptaszkowska-Koniarz et al. 2018; Oliveira et al. 2019). Figure 4.12c shows the behavior of the caffeine adsorption with respect to the concentration of total surface groups evaluated by potentiometric titration; in this figure, it is possible to observe that caffeine adsorption capacity decreased with increase of total chemical groups on the surface. The affinity between caffeine

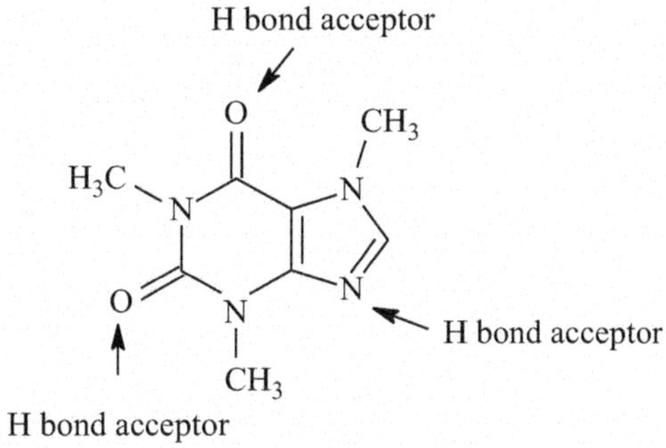

Fig. 4.13 Chemical structure of caffeine

molecules and the solids can be related to the specific adsorbate–adsorbent interactions (Galhetas et al. 2014b; Álvarez-Torrellas et al. 2016).

It has been widely reported that organic compound adsorption onto activated carbons is caused by a combination of physical properties such as specific surface area and pore volume, and chemical properties of the surface (Feiqiang et al. 2018). It is suggested that caffeine adsorption onto the carbonaceous materials evaluated was caused by hydrogen bond interactions between hydroxyl group of CAGs and the two oxygen atoms and the H bond acceptor nitrogen atom of CFN. In addition, aromatic ring structure of CFN could serve as a π–π electron-donor-acceptor complex interaction (Fig. 4.13). Also it is possible the establishment of interactions between hydroxyl moieties on iron oxide surface and the functional groups of CFN (Reguyal et al. 2017; Dong et al. 2018; Lompe et al. 2018; Reguyal and Sarmah 2018).

On the other hand, the nickel ion adsorption on carbonaceous surfaces and magnetite nanoparticles can be described by different mechanisms according to the amphoteric nature of the carbonaceous and Fe_xO_y surfaces; therefore, these solids can develop surface charges by the protonation and deprotonation reactions of the oxygen groups—carboxylic acids, lactone, phenolic, among others—of the carbonaceous solids, and the hydroxo and oxo sites on the magnetite surface; so the surface can develop negative, positive, and neutral functional groups that can coexist on the surface; in this way, the solid can be charged differently and can interact with anions or cations.

The driving forces of the metal ion adsorption onto solid surface mainly involved the next mechanism: (1) physisorption, (2) chemical adsorption, (3) precipitation on the surface, (4) complex formation, and (5) ion exchange (Tian et al. 2011). In the adsorption process on magnetite, nickel ions can replace hydroxyl ions on the surfaces; also the nickel ions can form inner-sphere and outer-sphere complexes, including monodentate and bidentate; electrostatic interaction can occur between negatively charged surface groups and positively charged nickel ions, so the -OH and -oxo groups of the magnetite surface increase the functional groups on the magnetite nanoparticles, and thus the adsorption ability increases (El-Dib et al. 2019).

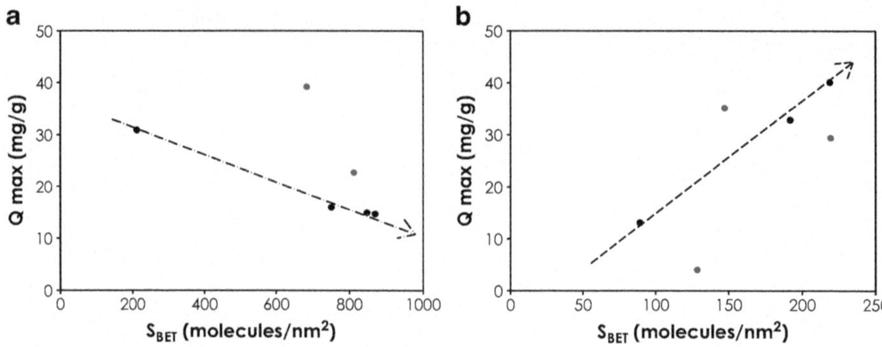

Fig. 4.14 Correlation between textural and chemical parameters with the maximum nickel adsorption capacity from Langmuir model. S_{BET} vs. Qmax in carbonaceous solids (**a**) and magnetic modified solids (**b**)

Figures 4.14 and 4.15 show the correlation between textural and chemical parameters with the maximum nickel adsorption capacity from Langmuir model. For carbonaceous, it is possible to see that the increase in the total groups on the surface of the solids studied favors the nickel ion adsorption due to the increased active sites for the interactions and development of the mechanism previously mentioned. In contrast to caffeine, the area available for adsorption of nickel ions seems to have less effect on their removal.

Finally, Fig. 4.16 describes the adsorption mechanisms for the caffeine molecules and nickel ions mentioned above.

4.4 Conclusion

Magnetic nanoparticles in combination with carbonaceous materials and their subsequent use in removal of organic and inorganic contaminants have been studied. Results of this research showed that carbonaceous material precursors had higher adsorption capacity of caffeine than their magnetic derivatives. CAG was the best precursor material, and the best derivate magnetic material was CAGoxS-M with values of 232.9 mg g^{-1} and 42.21 mg g^{-1} for adsorption capacity of caffeine, respectively. This performance is caused by greater surface area and appropriate microporosity of the activated carbon precursors. In addition, decreased surface area in magnetic materials is related to iron nanoparticle deposition onto carbonaceous material surface employed. On the other hand, FTIR spectra showed hydroxyl and carbonyl groups on the aromatic carbon skeleton of the carbonaceous material precursors; therefore, it is possible to assume hydrogen bonds and π–π interactions between caffeine and the adsorbent solids evaluated. In contrast, the adsorption of

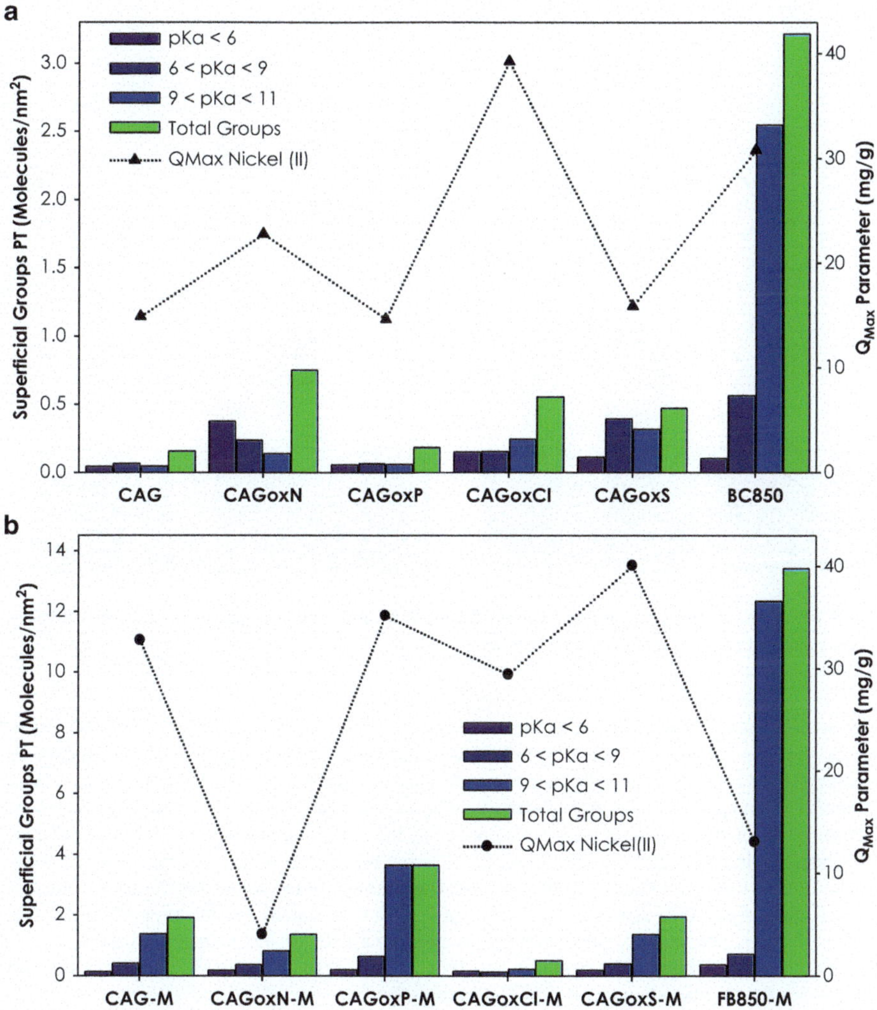

Fig. 4.15 Correlation between total superficial groups evaluated by potentiometric titration with the maximum nickel adsorption capacity from Langmuir model in carbonaceous solids (**a**) and magnetic modified solids (**b**)

nickel ions on the carbonaceous surface seems to have a close relationship between the adsorption capacity of the solid and the concentration of surface groups and a low effect of the surface area available for the adsorption process. For the magnetized solids, it is observed that the solids GAC, GACoxP, and GACoxS present a greater adsorption capacity after the modification process with the magnetite nanoparticles, between 2.1 and 2.5 times greater.

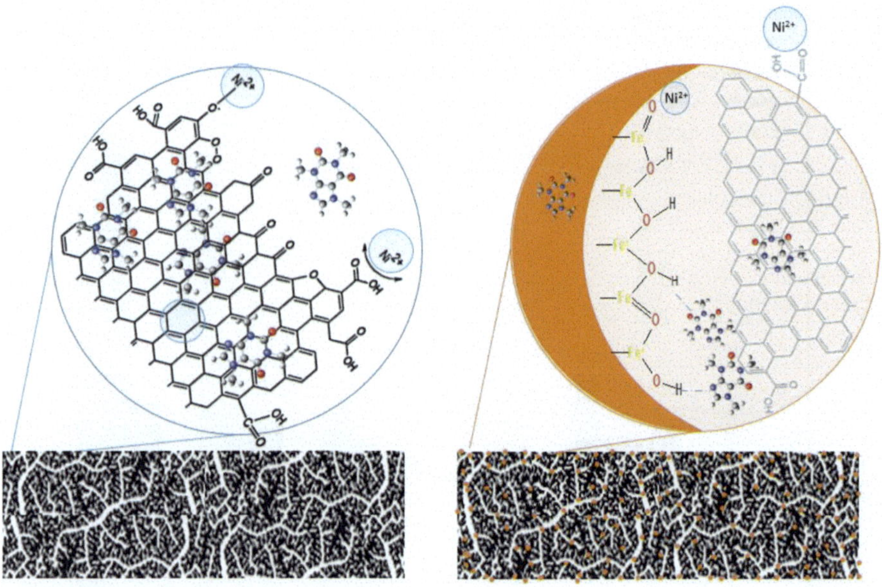

Fig. 4.16 Proposed adsorption mechanisms for caffeine and nickel ion molecules on the carbonaceous surface and on magnetite nanoparticles

Acknowledgments The authors thank the Framework Agreement between the Universidad de Los Andes and the Universidad Nacional de Colombia-Sede Bogotá and the act of agreement established between the chemistry departments of both universities. The Prof. Dr. Juan Carlos Moreno-Piraján appreciate the grant for the funding of research programs for associate professors, full professors, and emeritus professors announced by the Faculty of Sciences of the Universidad de los Andes (Colombia), 20-01-2020, 20-01-2022, according to the project number INV-2019-84-1786. Authors thank the Faculty of Sciences of Universidad de los Andes for the partial funding through the project INV-2019-86-1819 and Universidad de Caldas for the support to doctoral studies. Finally, the authors wish to express their gratitude to the Ministry of Science and Technology of Colombia (Minciencias) and its Call 811-2018 "Postdoctoral stay program for Colciencias training beneficiaries in SNCTeI entities."

References

Alahabadi A, Hosseini-Bandegharaei A, Moussavi G et al (2017) Comparing adsorption properties of NH4Cl-modified activated carbon towards chlortetracycline antibiotic with those of commercial activated carbon. J Mol Liq 232:367–381. https://doi.org/10.1016/j.molliq.2017.02.077

Allen SJ, Mckay G, Porter JF (2004) Adsorption isotherm models for basic dye adsorption by peat in single and binary component systems. J Colloid Interface Sci 280:322–333. https://doi.org/10.1016/j.jcis.2004.08.078

Álvarez-Torrellas S, Rodríguez A, Ovejero G et al (2016) Removal of caffeine from pharmaceutical wastewater by adsorption: influence of NOM, textural and chemical properties of the adsorbent. Environ Technol (UK) 37:1618–1630. https://doi.org/10.1080/09593330.2015.1122666

Asasian Kolur N, Sharifian S, Kaghazchi T (2019) Investigation of sulfuric acid-treated activated carbon properties. Turkish J Chem 43:663–675. https://doi.org/10.3906/kim-1810-63

Bandosz TJ, Ania CO (2006) Surface chemistry of activated carbons and its characterization. In: Bandosz TJ (ed) Activated carbon surfaces in environmental remediation. Elsevier, New York, pp 159–229

Bandosz TJ, Jagiello J, Contescu C, Schwarz JA (1993) Characterization of the surfaces of activated carbons in terms of their acidity constant distributions. Carbon N Y 31:1193–1202. https://doi.org/10.1016/0008-6223(93)90072-I

Baruah A, Chaudhary V, Malik R, Tomer VK (2019) Nanotechnology based solutions for wastewater treatment. In: Ahsan A, Ismail AF (eds) Micro and nano technologies, nanotechnology in water and wastewater treatment. Elsevier, Amsterdam, pp 337–368

Basheer AA (2018) New generation nano-adsorbents for the removal of emerging contaminants in water. J Mol Liq 261:583–593. https://doi.org/10.1016/j.molliq.2018.04.021

Beltrame KK, Cazetta AL, de Souza PSC et al (2018) Adsorption of caffeine on mesoporous activated carbon fibers prepared from pineapple plant leaves. Ecotoxicol Environ Saf 147:64–71. https://doi.org/10.1016/j.ecoenv.2017.08.034

Boruah PK, Borthakur P, Das MR (2019) Magnetic metal/metal oxide nanoparticles and nanocomposite materials for water purification. In: Thomas S, Pasquini D, Leu S-Y, Gopakumar DA (eds) Micro and nano technologies, nanoscale materials in water purification. Elsevier, Amsterdam, pp 473–503

Chung YT, Wang CK, Wang KS et al (2017) Facile modification of graphite sheet by novel electrochemical exfoliation/oxidant method and its adsorption of caffeine from water. J Taiwan Inst Chem Eng 80:747–753. https://doi.org/10.1016/j.jtice.2017.09.012

Contescu A, Contescu C, Putyera K, Schwarz JA (1997) Surface acidity of carbons characterized by their continuous pK distribution and Boehm titration. Carbon N Y 35:83–94. https://doi.org/10.1016/S0008-6223(96)00125-X

Correa-Navarro YM, Giraldo L, Moreno-Piraján JC (2019) Dataset for effect of pH on caffeine and diclofenac adsorption from aqueous solution onto fi que bagasse biochars. Data Br 25:1–6. https://doi.org/10.1016/j.dib.2019.104111

Couto OM, Matos I, da Fonseca IM et al (2015) Effect of solution pH and influence of water hardness on caffeine adsorption onto activated carbons. Can J Chem Eng 93:68–77. https://doi.org/10.1002/cjce.22104

de Andrade JR, Oliveira MF, da Silva MGC, Vieira MGA (2018) Adsorption of pharmaceuticals from water and wastewater using nonconventional low-cost materials: a review. Ind Eng Chem Res 57:3103–3127. https://doi.org/10.1021/acs.iecr.7b05137

de Morais LC (2019) Surface modifications of magnetic nanoparticles for water purification. In: Thomas S, Pasquini D, Leu S-Y, Gopakumar DA (eds) Micro and nano technologies, nanoscale materials in water purification. Elsevier, Amsterdam, pp 505–519

Dong X, He L, Hu H et al (2018) Removal of 17B-estradiol by using highly adsorptive magnetic biochar nanoparticles from aqueous solution. Chem Eng J 352:371–379. https://doi.org/10.1016/j.cej.2018.07.025

dos Santos Lins PV, Henrique DC, Ide AH et al (2019) Evaluation of caffeine adsorption by MgAl-LDH/biochar composite. Environ Sci Pollut Res 26:31804–31811. https://doi.org/10.1007/s11356-019-06288-3

El-Dib FI, Mohamed DE, El-Shamy OAA, Mishrif MR (2019) Study the adsorption properties of magnetite nanoparticles in the presence of different synthesized surfactants for heavy metal ions removal. Egypt J Pet. https://doi.org/10.1016/j.ejpe.2019.08.004

Feiqiang G, Xiaolei L, Xiaochen J et al (2018) Characteristics and toxic dye adsorption of magnetic activated carbon prepared from biomass waste by modified one-step synthesis. Colloids Surf A Physicochem Eng Asp 555:43–54. https://doi.org/10.1016/j.colsurfa.2018.06.061

Figueiredo J, Pereira MF, Freitas MM, Órfão JJ (1999) Modification of the surface chemistry of activated carbons. Carbon N Y 37:1379–1389. https://doi.org/10.1016/S0008-6223(98)00333-9

Fleet ME (1981) The structure of magnetite. Acta Crystallogr B37:917–920. https://doi.org/10.1107/S0567740881004597
Freundlich H (1906) Adsorption in solution. Phys Chem Soc 40:1361–1368
Galhetas M, Mestre AS, Pinto ML et al (2014a) Chars from gasification of coal and pine activated with K2CO3: acetaminophen and caffeine adsorption from aqueous solutions. J Colloid Interface Sci 433:94–103. https://doi.org/10.1016/j.jcis.2014.06.043
Galhetas M, Mestre AS, Pinto ML et al (2014b) Chars from gasification of coal and pine activated with K2CO3: acetaminophen and caffeine adsorption from aqueous solutions. J Colloid Interface Sci 433:94–103. https://doi.org/10.1016/j.jcis.2014.06.043
Jagiełło J, Bandosz TJ, Putyera K, Schwarz JA (1995) Determination of proton affinity distributions for chemical Systems in Aqueous Environments Using a stable numerical solution of the adsorption integral equation. J Colloid Interface Sci 172:341–346. https://doi.org/10.1006/jcis.1995.1262
Kajjumba GW, Emik S, Öngen A et al (2018) Modelling of adsorption kinetic processes—errors, theory and application. In: Edebali S (ed) Advanced sorption process applications, IntechOpen, London, pp 1–19
Kwon J, Wilson L, Sammynaiken R (2014) Synthesis and characterization of magnetite and activated carbon binary composites. Synth Met 197:8–17. https://doi.org/10.1016/j.synthmet.2014.08.010
Langmuir I (1916) The constitution and fundamental properties of solids and liquids, part I. solids. J Am Chem Soc:2221–2295
Li J, Zhang S, Chen C et al (2012a) Removal of Cu (II) and fulvic acid by graphene oxide Nanosheets decorated with Fe_3O_4 nanoparticles. Appl Mater Interfaces 4:4991–5000. https://doi.org/10.1021/am301358b
Li X, Liu L, Wang Y et al (2012b) Integrated assessment of heavy metal contamination in sediments from a coastal industrial basin, NE China. PLoS One 7. https://doi.org/10.1371/journal.pone.0039690
Lompe KM, Vo Duy S, Peldszus S et al (2018) Removal of micropollutants by fresh and colonized magnetic powdered activated carbon. J Hazard Mater 360:349–355. https://doi.org/10.1016/j.jhazmat.2018.07.072
Madrid L (2012) "Heavy metals": reminding a long-standing and sometimes forgotten controversy. Geoderma 155:128–129. https://doi.org/10.1016/j.geoderma.2009.11.031
Mahmoudiab M, Sant S, Wang B et al (2011) Superparamagnetic iron oxide nanoparticles (SPIONS): development, surface modification and applications in chemotherapy. Adv Drug Deliv Rev 63:24–46. https://doi.org/10.1016/j.addr.2010.05.006
Mansour F, Al-Hindi M, Yahfoufi R et al (2018) The use of activated carbon for the removal of pharmaceuticals from aqueous solutions: a review. Rev Environ Sci Biotechnol 17:109–145. https://doi.org/10.1007/s11157-017-9456-8
Martinez-Boubeta CSK (2019) Magnetic nanoparticles for water purification. In: Thomas S, Pasquini D, Leu S-Y, Gopakumar DA (eds) Micro and nano technologies, nanoscale materials in water purification. Elsevier, Amsterdam, pp 521–552
Mukherjee S, Mishra A, Samanta AN (2019) Amine-impregnated MCM-41 in post-combustion CO2 capture: synthesis, characterization, isotherm modelling. Adv Powder Technol 30:3231–3240. https://doi.org/10.1016/j.apt.2019.09.032
Ngeno EC, Orata F, Danstone Baraza L et al (2016) Adsorption of caffeine and ciprofloxacin onto Pyrolitically derived water hyacinth biochar: isothermal, kinetic and thermodynamic studies. J Chem Chem Eng 10:185–194. https://doi.org/10.17265/1934-7375/2016.04.006
Niu R, Li H, Ma Y et al (2015) An insight into the improved capacitive deionization performance of activated carbon treated by sulfuric acid. Electrochim Acta 176:755–762. https://doi.org/10.1016/j.electacta.2015.07.012
Oliveira MF, da Silva MGC, Vieira MGA (2019) Equilibrium and kinetic studies of caffeine adsorption from aqueous solutions on thermally modified Verde-lodo bentonite. Appl Clay Sci 168:366–373. https://doi.org/10.1016/j.clay.2018.12.011

Pal A, He Y, Jekel M et al (2014) Emerging contaminants of public health significance as water quality indicator compounds in the urban water cycle untreated water sewer system. Environ Int 71:46–62. https://doi.org/10.1016/j.envint.2014.05.025

Pendolino F (2014) Self-assembly of molecules on nanostructured graphene. Universidad Autónoma de Madrid, Madrid

Portinho R, Zanella O, Féris LA (2017) Grape stalk application for caffeine removal through adsorption. J Environ Manag 202:178–187. https://doi.org/10.1016/j.jenvman.2017.07.033

Ptaszkowska-Koniarz M, Goscianska J, Pietrzak R (2018) Synthesis of carbon xerogels modified with amine groups and copper for efficient adsorption of caffeine. Chem Eng J 345:13–21. https://doi.org/10.1016/j.cej.2018.03.132

Redlich O, Peterson DL (1958) A useful adsorption isotherm. J Am Chem Soc 63:1024–1025

Reguyal F, Sarmah AK (2018) Site energy distribution analysis and influence of Fe3O4 nanoparticles on sulfamethoxazole sorption in aqueous solution by magnetic pine sawdust biochar. Environ Pollut 233:510–519. https://doi.org/10.1016/j.envpol.2017.09.076

Reguyal F, Sarmah AK, Gao W (2017) Synthesis of magnetic biochar from pine sawdust via oxidative hydrolysis of FeCl2 for the removal sulfamethoxazole from aqueous solution. J Hazard Mater 321:868–878. https://doi.org/10.1016/j.envpol.2016.07.040

Rodríguez-Estupiñán P, Giraldo L, Moreno-Piraján JC (2018) Carbonaceous porous materials for the adsorption of heavy metals: chemical characterization of oxidized activated carbons. In: Crini G, E. L (eds) Green adsorbents for pollutant removal. Environmental chemistry for a sustainable world. Springer Cham pp 163–191

Santhosh C, Malathi A, Dhaneshvar E et al (2019) Iron oxide nanomaterials for water purification. In: Thomas S, Pasquini D, Leu S-Y, Deepu G (eds) Micro and nano technologies, nanoscale materials in water purification. Elsevier, Amsterdam, pp 431–446

Satheesh R, Vignesh K, Suganthi A, Rajarajan M (2014) Visible light responsive photocatalytic applications of transition metal (M=Cu, Ni and Co) doped α-Fe2O3 nanoparticles. J Environ Chem Eng 2:1956–1968. https://doi.org/10.1016/j.jece.2014.08.016

Soto-Jimenez MF (2011) Transferencia de elementos traza en tramas tróficas acuáticas. Hidrobiológica 21:239–248

Sotomayor F, Cychosz K, Thommes M (2018) Characterization of micro/Mesoporous materials by Physisorption: concepts and case studies. Accounts Mater Surf Res 3:34–50

Srivastava VC, Mall ID, Mishra IM (2009) Competitive adsorption of cadmium(II) and nickel (II) metal ions from aqueous solution onto rice husk ash. Chem Eng Process Process Intensif 48:370–379. https://doi.org/10.1016/j.cep.2008.05.001

Sun L, Wan S, Luo W (2013) Biochars prepared from anaerobic digestion residue, palm bark, and eucalyptus for adsorption of cationic methylene blue dye: characterization, equilibrium, and kinetic studies. Bioresour Technol 140:406–413. https://doi.org/10.1016/j.biortech.2013.04.116

Tian G, Geng J, Jin Y et al (2011) Sorption of uranium(VI) using oxime-grafted ordered mesoporous carbon CMK-5. J Hazard Mater 190:442–450. https://doi.org/10.1016/j.jhazmat.2011.03.066

Toth J (1971) State equations of the solid gas interface layer. Acta Chem Acad Hung:311–317

Tran NH, Reinhard M, Gin KYH (2018) Occurrence and fate of emerging contaminants in municipal wastewater treatment plants from different geographical regions-a review. Water Res 133:182–207. https://doi.org/10.1016/j.watres.2017.12.029

UNICEF and WHO (2015) Progress on sanitation and drinking water – 2015 update and MDG assessment. UNICEF, Geneva

Vidojkovic SM, Rakin MP (2017) Surface properties of magnetite in high temperature aqueous electrolyte solutions: a review. Adv Colloid Interf Sci 245:108–129. https://doi.org/10.1016/j.cis.2016.08.008

World Health Organization, UN-Water (2014) UN-water global analysis and assessment of sanitation and drinking-water (GLAAS) 2014-report. Investing in water and sanitation: increasing access, reducing inequalities. World Health Organization, Geneva

Yoon S-Y, Lee C-G, Park J-A et al (2014) Kinetic, equilibrium and thermodynamic studies for phosphate adsorption to magnetic iron oxide nanoparticles. Chem Eng J 236:341–347. https://doi.org/10.1016/j.cej.2013.09.053

Zebardast HR, Pawlik M, Rogak S, Asselin E (2014) Potentiometric titration of hematite and magnetite at elevated temperatures using a ZrO2-based pH probe. Colloids Surf A Physicochem Eng Asp 444:144–152. https://doi.org/10.1016/j.colsurfa.2013.12.039

Chapter 5
Magnetic Biochar Fibers for Copper Removal

Ioannis Anastopoulos and Ioannis Pashalidis

Contents

Abstract The present chapter gives an overview of the environmental chemistry of divalent copper ions (Cu(II)), including analysis, chemical behavior, and removal of Cu(II) using magnetic biochar fibers. After a short overview of preparation/modification procedures and characterization techniques associated with biochar adsorbents, a comprehensive compilation of data related to Cu(II) adsorption by magnetic biochar materials is presented and the data discussed. The discussion involves Cu (II) binding by the biochar surface, thermodynamic and kinetic modelling, and surface speciation. Generally, the removal of Cu(II) ions from (waste)waters by magnetic biochars is based on different sorption mechanisms such as ion exchange, surface complexation, cation π-interaction, and surface polynucleation and precipitation, depending on the solution pH, the surface modification, and the Cu(II) ion concentration. The adsorption capacity ranges from 2 to 141 mg/g and the adsorption data are best fitted by the *Langmuir* adsorption isotherm and the pseudo second-order kinetic model. Future work has to be focused on the characterization of surface species by sophisticated spectroscopic techniques (e.g., EXAFS, XPS, and RAMAN) and studies related to Cu(II) redox reactions occurring on the biochar surface.

I. Anastopoulos (✉) · I. Pashalidis
Department of Chemistry, University of Cyprus, Nicosia, Cyprus

L. Meili, G. L. Dotto (eds.), *Advanced Magnetic Adsorbents for Water Treatment*, Environmental Chemistry for a Sustainable World 61,
https://doi.org/10.1007/978-3-030-64092-7_5

Keywords Magnetic biochars · Equilibrium modelling · Characterization techniques · Sorption mechanism

5.1 Introduction

5.1.1 Hydrolysis and Complexation Reactions of Cu(II) Ions

Copper (Cu), which is a transition metal, is an abundant trace element found in many rocks and minerals on earth's crust. The oxidation states of copper found in natural environments are three: zero-valent copper (Cu^{o}, pure/native metal); monovalent copper (Cu(I), cuprous ion); and divalent copper (Cu(II), cupric ion). Cu is one of the essential trace elements, which is necessary for a wide range of metabolic processes in bacteria and higher organisms. However, at elevated levels, copper becomes toxic and therefore Cu concentrations in environmental systems and its biological availability is of particular interest. Under oxic conditions, Cu(II) is the predominant oxidation state of copper in aqueous solutions, which is primarily present in the form of complexes with naturally occurring ligands (e.g., OH^{-}, Cl^{-}, SO_4^{2-}, and CO_3^{2-}) or bound to organic (e.g., humic acid) or inorganic particulate matter (e.g., clay minerals). Bioavailability and toxicity of copper depends on its physical and chemical speciation in solution and the free bivalent copper (Cu^{2+}) is one of the most bioavailable species of the metal (Sauvé et al. 1997).

The Cu^{2+} ion is actually hydrated in solution and can be regarded as aquo-complex ion, which gives to its concentrated aqueous solutions the blue color. This is the case in acidic solutions and when the pH increases, the hydrolysis of the Cu^{2+} ion takes place, resulting in the gradual formation of Cu(II)-hydroxo complexes. Assuming formation of only mononuclear species the relative species distribution of Cu(II) complexes as a function of pH formed under ambient conditions in a dilute aqueous solution is given in Fig. 5.1. The species distribution diagram has been calculated using the complex formation constants from Baes and Mesmer (1977). From Fig. 5.1, it is obvious that below pH 6, the free copper ion (Cu^{2+}) is the predominant species, whereas in the alkaline pH range, the Cu(II)-carbonato complexes are becoming the dominating copper species in solution.

In seawater, where a large excess of Cl^{-} ions exists, one would expect the formation of Cu(II)-chloro complexes (e.g., $CuCl_4^{2-}$) to become predominant. However, in seawater Cu(II) is predominantly complexed by thiourea-type thiols and humic substances, because these ligands form very strong complexes with Cu (II) (Whitby et al. 2017). Moreover, in the presence of excess NH_3, Cu(II) forms very readily complexes with ammonia ($[Cu(NH_3)_4]^{2+}$), which have deep blue color and are often used to detect the presence of the metal ion in solution.

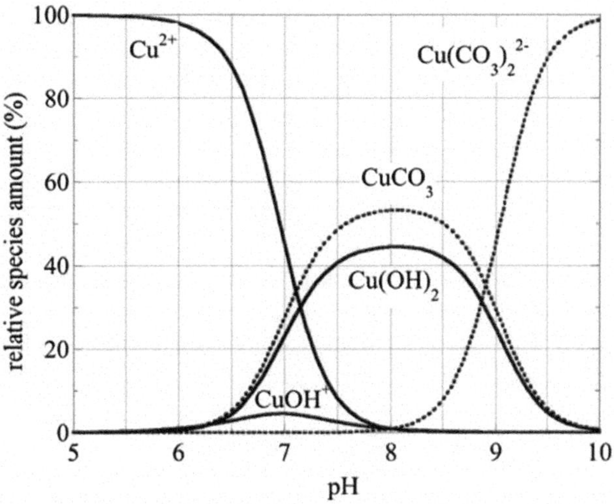

Fig. 5.1 Relative species distribution of Cu (II) complexes as a function of pH, formed under ambient conditions in a dilute aqueous solution

5.1.2 Environmental Chemistry of Cu(II)

Copper is abundant and is found in all water systems (e.g., surface, ground-, and seawaters) including drinking water. The copper levels in surface waters varies strongly between 0.0005 and 1 mg/l with a median value about 0.01 mg/l. In unpolluted areas, the copper concentration in the respective waters is significantly lower than in waters in contact with contaminated soils or even waters downstream of sewage treatment plants. Copper levels in drinking waters vary widely depending mainly on the water characteristics such as pH, hardness, and copper availability from the water distribution system, since the corrosion of interior copper plumbing has been found to be often the primary source of copper (Flemming and Trevors 1989).

Elevated Cu levels in surface waters and sediments caused by anthropogenic activities have been widely reported (Flemming and Trevors 1989). Anthropogenic activities such as mining, smelting, industrial (metal plating, steelworks, etc.) and agricultural applications (fertilizers, fungicides, etc.), domestic waste, and sewage sludge have led to increased levels of Cu in certain areas, which may adversely affect living organism and the quality of life. As an element, copper is persistent and bioaccumulative in the environment, and is not easily metabolized and broken down (Flemming and Trevors 1989). Copper is relatively nontoxic to mammals and is of vital importance for human life and health. However, above certain levels it can be toxic like all heavy metals. The suggested safe level of Cu in drinking water for humans is set by the European Drinking Water Directive 98/83/EC to 2 mg/L (European Union 1998).

Nevertheless, copper is remarkably toxic to aquatic organisms (e.g., fish, algae) and the toxicological tolerance limits for fish and crustaceans are generally one to two orders of magnitude lower than those of mammals. The toxicological tolerance limits for algae are even three orders of magnitude lower and strongly depend on dimensions and respiratory rates of the aquatic biota (Flemming and Trevors 1989).

Copper pollution of waters and sediments/soils, as well as activated sludge may have adverse effects on native microorganisms present in natural environments (Flemming and Trevors 1989). The ecological importance of microorganisms as primary decomposers can be dramatically affected due to alterations in microbial numbers and diversity, and microbial activities involved in C- and N-mineralization processes. Because of its potential toxicity to algae, copper in the form of copper sulfate ($CuSO_4$) has been used as an algicide since the early 1900s and is still widely used. Addition of copper sulfate to surface waters at concentration up to 0.01 mg/l results in substantial reduction in algal biomass and a shift in algal community structure as well as bacterial population (Flemming and Trevors 1989). Therefore, it is clear that (waste)waters with increased copper load must be treated before discharge to surface and groundwater reservoirs.

Copper can be removed by different processes such electrochemical separation, coagulation/flocculation (Carvalho Barros et al. 2018; Hargreaves et al. 2018; Xu et al. 2017; Yang et al. 2013), ion exchange, membrane filtration (Arbabi and Golshani 2016), and adsorption (Ince and Ince 2017; Wan Ngah et al. 2011), including biosorption (Abbas et al. 2014; Bilal et al. 2013; McKay et al. 1999). In recent years, adsorption methods have received considerable attention because of their low operational cost, ease in processing as compared to other processes, environmental compatibility and technically feasible operation, and possible reuse of adsorbents (Ince and Ince 2017). Specifically, biochar materials being produced from biomass by-products have gained increased attention, because they possess the properties of activated carbon produced from fossil fuel materials and at the same time are low-cost adsorbents of easy availability (Hadjittofi et al. 2014; Trakal et al. 2014). In addition, the affinity of biochars can be easily modified toward polar (e.g., ionic species) or nonpolar compounds (e.g., organic pollutants) upon surface modification. Surface modification can be performed upon surface oxidation (Hadjittofi et al. 2014; Liatsou et al. 2017) and following derivatization (Liatsou et al. 2019), as well surface deposition or coating with metal oxides (Hadjiyiannis and Pashalidis 2019), and may increase the selectivity, adsorption capacity, and physical properties of the material (e.g., magnetization) (Nicolaou et al. 2019; Reddy and Yun 2016).

5.1.3 Analysis of Cu(II) by Cu(II)-ISE Electrode

The most common analytical techniques for copper quantification in environmental samples are inductively coupled plasma mass spectrometry (ICP-MS) and electro thermal atomic absorption (ETAAS), because of the high sensitivity and low sample quantity required by these methods. Nevertheless, prober sampling and sample preparation procedures are needed for obtaining reliable data. On the other hand, the copper ion-selective electrode (Cu-ISE) is one of few analytical means for determining the activity (effective concentration, a) of the free metal species

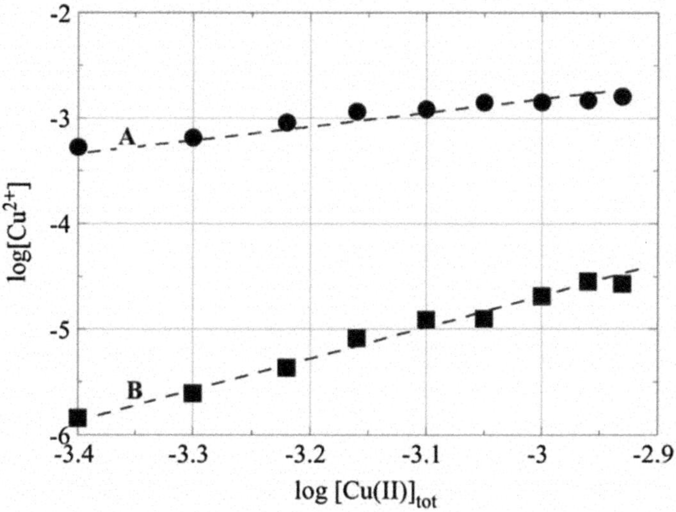

Fig. 5.2 Free copper ion concentration ([Cu^{2+}]) as a function of the total Cu(II) concentration in two aqueous solutions (in absence (A) and the presence (B) of an adsorbent) under ambient conditions at pH 5. (Konstantinou et al. 2007. Reprinted with permission by Springer)

(Cu^{2+}) (Rachou et al. 2007). The activity can be related to the molar concentration (C) using the activity coefficient (g) as follows:

$$a_{Cu^{2+}} = \gamma_{Cu^{2+}} \cdot C_{Cu^{2+}} \tag{5.1}$$

The value of γ depends on the ionic strength of the solution and its value is expected close to unity for dilute aqueous solutions. It is noteworthy to mention that when determining the Cu^{2+} activity by ISE to keep the ionic strength and subsequently the activity coefficient constant to allow easy evaluation of the molar concentration from the activity. The latter is calculated from the Nernst potential ($E_{Cu}{}^{2+}$) determined using a Cu-ISE and Cu^{2+} standard solutions. The correlation between the Nernst potential and copper ion activity is given by following equation:

$$E_{Cu^{2+}} = E^{o}_{Cu^{2+}} - \frac{0.059}{2} \cdot \log_{10} \frac{1}{a_{Cu^{2+}}} \tag{5.2}$$

Figure 5.2 shows the free copper ion concentration ([Cu^{2+}]) as a function of the total Cu(II) concentration in two aqueous solutions under ambient conditions at pH 5. The curve A corresponds to dilute aqueous solutions, whereas the curve B corresponds to a similar solution in which a certain amount of olive cake, which acts as an adsorbent for Cu(II), has been added. According to Fig. 5.2, in the case of curve A, the free copper ion concentration ([Cu^{2+}]) is equal to total Cu (II) concentration ([Cu(II)]$_{tot}$). This is because under the given conditions, the total amount of copper exists in the form of the free copper ion, which can be detected by the Cu-ISE. On the other hand, in the presence of the adsorbent, a significant amount of Cu(II) is adsorbed and hence the free copper ion concentration in solution decreases dramatically (Konstantinou et al. 2007). This is an example which shows the usefulness of Cu-ISE in the investigation of Cu(II) adsorption by different adsorbent materials.

5.2 Preparation of Magnetic Biochar Fibers

Generally, biochar fibers are produced by pyrolysis of plant fibers under restricted oxygen conditions at moderate temperatures (350–700 °C) and have significantly higher content of oxygen containing surface groups (e.g., C-O, C=O, COOH, OH, etc.) than activated carbons. In addition, the tubular structure of the biochar fibers provides to the material robustness and large external surface area, which allows fast fluid transport and material exchange (Hadjittofi et al. 2014; Liatsou et al. 2017). Moreover, the surface of the biochar fibers can be easily modified by oxidation to dramatically increase the oxygen-containing surface moieties (e.g., -COOH), by chemical derivatization, which allows specific chelating molecules to be attached on the biochar surface, and surface coating or deposition of metal oxides (e.g., MnO_2) to increase the materials capacity and selectivity toward certain metal ions (Hadjiyiannis and Pashalidis 2019; Nicolaou et al. 2019). Specifically, deposition of magnetite particles (Fe_3O_4) on the biochar's surface results in both magnetization of the biochar fibers, which allows easy separation of the adsorbent from the liquid phase, as well as dramatic increase of the surface area due to the deposition of iron oxide nanoparticles on the biochar surface (Nicolaou et al. 2019) and enhanced chemical affinity towards certain metal ions (e.g., U(VI)). Magnetic biochars as adsorbent materials for various contaminants including (radio)toxic metals (Hu et al. 2018; Philippou et al. 2019), POPs (Yang et al. 2019), and dyes (Akbarnezhad and Safa 2018) have been extensively studied. The main purpose for using magnetized biochar adsorbents are not only the magnetic properties of the materials but also in some cases the increased affinity of iron oxides for certain metal ions such as As (III/V) (Wang et al. 2015) and U(VI) (Li et al. 2019).

Methods to produce magnetic adsorbents include coprecipitation, thermal decomposition, hydrothermal processes, and matrix-assisted approaches (e.g., using polyols, micelles, and organometallic compounds), chemical reduction, and pyrolysis. The combination of coprecipitation and pyrolysis is a favored method to produce magnetic biochar. Usually, first the inorganic phase is coprecipitated in the presence of biomass and following the latter is carbonized at temperatures below 800 °C in the absence of oxygen to avoid burn-up and improve the surface characteristics of the biochar composite (Noor et al. 2017).

Coprecipitation is an easy and simple way to synthesize magnetite from aqueous Fe(II/III) salt solution by increasing the pH under inert atmosphere to avoid carbonate formation. However, the experimental parameters such as pH and mixing rate of the base solution, ionic strength, temperature, Fe(II)/Fe(III) ratio, which affect the physicochemical properties (e.g., composition, particle size, and surface characteristics) of the magnetic adsorbents, are difficult to control and usually the materials prepared consist of polydispersed particles.

The thermal decomposition of organometallic precursors, which are dissolved in organic solvents and in the presence of stabilizing micelles, is a simple way to prepare magnetic adsorbents consisting of nanoparticles with controlled size and morphology. However, the use of organic solvents and the use of relatively high temperatures (>300 °C) to remove organic residues (e.g., solvents and surfactants) from the final product are the main disadvantages of the thermal decomposition method. On the other hand, the hydrothermal process is the method of choice for the preparation magnetic particles with increased crystallinity. However, the slow reaction kinetics even at elevated temperatures and pressures (autoclave) is one of the main drawbacks of the method.

The sol-gel process is a relatively simple wet-chemical technique for the preparation of various materials of high purity and uniform nanostructures with novel properties, at relatively low temperatures. During the sol-gel a liquid (sol) is transformed into a gel, which is subsequently thermally treated to obtain the metal oxide. The method has been successfully applied for the preparation of magnetic mesoporous silica-coated carbon nanotubes (Tong et al. 2016).

The chemical reduction of metal salts is another simple method for producing magnetic nanoparticles among the various solution-phase chemistry routes. The method has been used for the preparation of manganese ferrite nanoparticles starting from manganese and iron nitrate salts and ethylene diamine in an alkaline aqueous solution (Mahmoodi 2014). Similarly, the polyol process, which results in the preparation of fine powdered solids, is based on the reduction of metal compounds in liquid polyols (e.g., 1,2-ethanediol, 1,2-propanediol, etc.). The metal compounds (e.g., nitrates, chlorides, and acetates) are suspended or even dissolved in the polyol, which acts as mild reducing agent, and the reactions take place at a given temperature, which depends on the reducibility of the metal, and may vary between 0 °C and the boiling temperature of the polyol.

Pyrolysis is very often used to produce magnetic biochars by heating the material via conduction, radiation, and convection. The pyrolysis can be carried out in two steps, carbonization step, which involves the formation of the biochar at pyrolysis temperatures between 600 and 900 °C under inert gas atmosphere to avoid burn-up, and the activation step, which involves biochar treatment with CO_2 or steam at 600–1200 °C to partly oxidize the biochar surface and form a well-developed porous structure (Suhas et al. 2007). Using microwave-assisted pyrolysis/heating usually results in reduced preparation times and improved surface characteristics of the adsorbent (Ania et al. 2005).

The successful modification/magnetization of the biochar fibers can be easily followed by using an external magnetic field or performing advanced magnetic measurements (Reddy and Yun 2016). In addition, spectroscopic measurements such as FTIR, Raman, and XPS (Philippou et al. 2019) can be carried out to identify the magnetite phases. Particle size determination and element identification on the biochar surface, as well as crystal phase recognition and primary crystal size evaluation can be carried out by means of SEM-EDX and XRD measurements (Philippou et al. 2019).

5.3 Adsorption of Cu(II) by Magnetic Biochars

The main sorption mechanisms on which the removal process of Cu(II) by magnetic biochars is based are: (1) ion exchange (particularly at low pH) and surface complexation of the Cu^{2+} cations by the carboxylic moieties of the oxidized biochar surface and the hydroxo groups present on the magnetite surface, (2) cation π-interaction, and (3) surface precipitation, which occurs at near neutral and alkaline pH and increased Cu(II) concentration.

1. Ion exchange

Ion exchange is the surface reaction which involves the reversible exchange of cations between the adsorbent surface and water. The cations are electrostatically attracted by the negative surface charges, which are mainly attributed to oxygen ions generating a negative electrostatic potential on the adsorbent's surface. Depending on the pH and the ionic strength of the solution, the negative surface charges can be neutralized by interaction with protons (H^+) or other cations such as Na^+. At low pH values, the negative surface charges are mainly balanced by protons (H^+), suggesting that cation exchange reactions are pH sensitive. Moreover, because of the weak electrostatic attraction, the process is reversible and H^+ or Na^+ cations can be displaced by other cations, which are present in solution and may have greater electrostatic charge or molar concentration. Figure 5.3 shows schematically the displacement of H^+ from the surface by Cu^{2+} cations, which are electrostatically attached to the surface and finally form inner-sphere complexes with the oxygen ions present on the adsorbent's surface in the form of carboxylic moieties (-COOH) on the biochar and as hydroxyl groups (-OH) on the magnetite particle surface.

The formation of inner-sphere complexes can be proved with FTIR spectroscopic measurements as well as by investigating the effect of ionic strength (Hadjiyiannis and Pashalidis 2019). The formation of inner-sphere complexes results in dramatic changes of the corresponding absorption peaks due to changes in binding affinity and symmetry. In addition, increasing ionic strength does insignificantly affect the adsorption efficiency since inner-sphere complexes are based on specific interactions

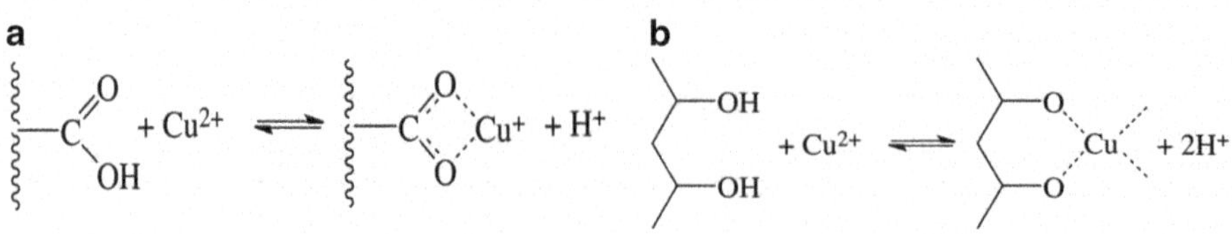

Fig. 5.3 Schematic illustration of the cation exchange reaction and formation of inner-sphere complexes of Cu^{2+} with (**a**) biochar carboxylic and (**b**) magnetite hydroxyl groups

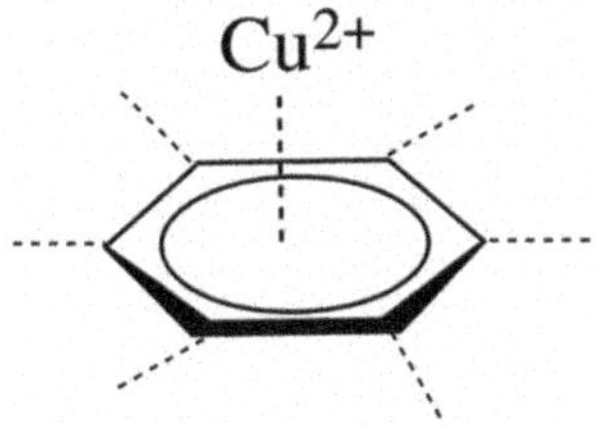

Fig. 5.4 Cation π-interaction between a Cu^{2+} cation and a benzene π-system

between the surface oxygen anions and the Cu^{2+} cations. On the other hand, outer-sphere complexation, which occurs basically in the diffuse layer close to the surface and is based solely on electrostatic attraction, is strongly affected by ionic strength changes (Konstantinou et al. 2006; Philippou et al. 2019).

2. Cation π-interaction

Due to the fact that biochar adsorbents are carbonaceous materials consisting of graphite phases, pi-system Cu^{2+} cation interactions are also possible and may govern the Cu(II) adsorption by biochar, particularly when other interaction modes are restricted due to the absence of oxygen containing moieties (e.g., carboxy- and hydoxy groups) on the biochar surface.

Figure 5.4 shows an example of a cation π-interaction between a Cu^{2+} cation and a benzene π-system, which is a noncovalent molecular interaction between the face of an electron-rich π system and an adjacent Cu^{2+} cation (Rimola et al. 2006).

3. Surface precipitation

The formation of a solid phase consisting of several sorption layers on the adsorbents surface or the formation of a solid solution of the adsorbing cation with the adsorbent initiated via dissolution/precipitation processes is defined as surface precipitation (Lützenkirchen and Behra 1995). Generally, surface precipitation occurs in near neutral and alkaline solutions and is indicated by a steep increase of the adsorption capacity (q_e) in the adsorption isotherm curve (Liatsou et al. 2017). Identification and characterization of the formed solid phase, including crystal size evaluations, can be performed by XRD measurements. Figure 5.5 shows an XRD diffractogram of a Cu(II) solid phase precipitated in an Cu(II)-biochar adsorption system at pH 6 and increased total copper concentration ($[Cu(II)]_{tot} = 5 \times 10^{-3}$ M) under ambient conditions.

5.3.1 *Adsorption Modeling and Thermodynamic Studies*

For understanding the mechanism of Cu(II) adsorption and for developing an effective and accurate design model, the use of adsorption isotherm models, kinetics models, and thermodynamic studies is obligatory (Anastopoulos and Kyzas 2014).

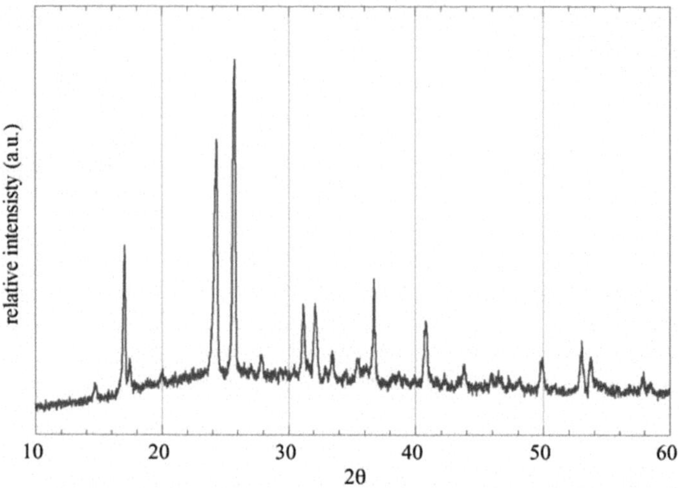

Fig. 5.5 Copper solid phase precipitated during adsorption batch-type investigations on the Cu (II) adsorption by biochar fibers performed at pH 5.5 and ($[Cu(II)]_{tot} = 5 \times 10^{-3}$ M), under ambient conditions. (Liatsou 2019. Retrieved from the library of the University of Cyprus)

Adsorption isotherm models provide information about the adsorption capacities (Chen 2015) and describe also interactions among the adsorbent and adsorbate (Foo and Hameed 2010). The most widely isotherm models are the *Langmuir* (Langmuir 1918) and *Freundlich* (Freundlich 1906) adsorption models.

The *Langmuir* isotherm model describes monolayer sorption at specific homogenous sites on the adsorbent. It can be expressed as follows:

$$q_e = q_m \frac{K_L C_e}{1 + K_L C_e} \tag{5.3}$$

where C_e is the equilibrium concentration of the adsorbate in solution (mg/L), q_e is the amount of the adsorbate adsorbed per gram of adsorbent at equilibrium (mg/g), q_m is the maximum monolayer adsorption capacity (mg/g), and K_L is an equilibrium constant related to energy of adsorption.

The *Freundlich* isotherm is applicable to nonideal adsorption on heterogeneous surfaces as well as to multilayer sorption, and it can be represented as:

$$q_e = K_F {C_e}^{1/n} \tag{5.4}$$

where C_e is the equilibrium concentration of the adsorbate in solution (mg/L), q_e is the amount of the adsorbate adsorbed per gram of adsorbent at equilibrium (mg/g), and K_F and n are the Freundlich constants.

Generally, the experimental data related to Cu(II) adsorption by magnetic biochars (Table 5.1) are well described by the *Langmuir* isotherm and the corresponding q_{max} values vary between 2 mg/g and 141 mg/g.

Table 5.1 Isotherm, kinetic, and thermodynamic data obtained from the studies focused on the magnetic biochar for Cu(II) adsorption

Adsorbent	Isotherm	Kinetic	Maximum monolayer adsorption capacity (mg/g)	Thermodynamic	References
Raw biochar (RBC) obtained from rice straw	L	Ps2	65.40	–	Yin et al. (2018)
Activated biochar (ABC) obtained from rice straw	L	Ps2	75.25	–	Yin et al. (2018)
Magnetized biochar (MBC)	L	Ps2	70.59	–	Yin et al. (2018)
Activated magnetic biochars obtained from rice straw (carbonization temperature 300 °C, AMBCs-300 °C)	L	Ps2	85.93	–	Yin et al. (2018)
Activated magnetic biochars obtained from rice straw (carbonization temperature 500 °C, AMBCs-500 °C)	L	Ps2	80.33	–	Yin et al. (2018)
Activated magnetic biochars obtained from rice straw (carbonization temperature 700 °C, AMBCs-700 °C)	L	Ps2	79.59	–	Yin et al. (2018)
Magnetic biochar obtained from gasification process of wastes. (MBC1: Prepared at the molar ratio $FeSO_4$: $NaBH_4$ 1:1 by stirring the suspension for 30 min)	L	Ps2	57.22	T: 295 to 335 K. ΔG^0 in kJ/moL: −14.19, −14.82 and − 14.83 at 295, 315and 335 K, respectively. ΔH^0 in kJ/moL: −9.38 ΔS^0 in J/moL K: −40.70	Kołodyńska and Bąk (2018)
Magnetic biochar obtained from gasification process of wastes. (MBC2: Prepared at the	L	Ps2	58.80	T: 295 to 335 K. ΔG^0 in kJ/moL: −14.14, −14.93 and − 15.07 at 295, 315and	Kołodyńska and Bąk (2018)

(continued)

Table 5.1 (continued)

Adsorbent	Isotherm	Kinetic	Maximum monolayer adsorption capacity (mg/g)	Thermodynamic	References
molar ratio $FeSO_4$: $NaBH_4$ 1:2 by stirring the suspension for 30 min)				335 K, respectively. ΔH^0 in kJ/moL: −7.22 ΔS^0 in J/moL K: −33.50	
Magnetic biochar obtained from gasification process of wastes. (MBC3: Prepared by dissolving the same amounts of $FeSO_4$ and $NaBH_4$ as MBC1 and stirring for 360 min)	L	Ps2	57.79	T: 295 to 335 K. ΔG^0 in kJ/moL: −13.50, −14.11 and − 14.37 at 295, 315and 335 K, respectively. ΔH^0 in kJ/moL: −7.14 ΔS^0 in J/moL K: −35.60	Kołodyńska and Bąk (2018)
Magnetic biochars derived from kelp marine macro-algae (KBCmag−0.05, where 0.05 is the molar concentration of $FeCl_3$ solution)	L	–	55.86[a]	–	Son et al. (2018)
Magnetic biochars derived from hijikia marine macro-algae (HBCmag−0.05, where 0.05 is the molar concentration of $FeCl_3$ solution)	L	–	47.75[a]	–	Son et al. (2018)
Biochar obtained from corn straw	L	–	117.63 ± 2.95	–	Song et al. (2016)
Chitosan-modified magnetic biochar obtained from corn straw (CC-Fe)	L	Ps2	131.19 ± 5.26	–	Song et al. (2016)
Chitosan-modified biochar obtained from corn straw (CC)	L	–	141.18 ± 1.510	–	Song et al. (2016)

(continued)

Table 5.1 (continued)

Adsorbent	Isotherm	Kinetic	Maximum monolayer adsorption capacity (mg/g)	Thermodynamic	References
Biochar obtained from water hyacinth (*Eichhornia Crassipes*) EC	L	Ps2	2.06	T: 298.15 to 323.15 K. ΔG^0 in kJ/moL: −9.571, −8.044, −6.517, −4.991, −3.464, and − 1.937 at 298.15, 303.15, 308.15, 313.15, 318.15, and 323.15 K, respectively. ΔH^0 in kJ/moL: −100.608 ΔS^0 in kJ/moL K: −0.305	Nyamunda et al. (2019)
Magnetic biochar derived from water hyacinth (*Eichhornia Crassipes*) (Fe_2O_3-EC)	L	Ps2	3.53	T: 298.15 to 323.15 K. ΔG^0 in kJ/moL: −6.370, −8.018, −9.667, −11.315, −12.964, and −14.612 at 298.15, 303.15, 308.15, 313.15, 318.15, and 323.15 K, respectively. ΔH^0 in kJ/moL: −91.928 ΔS^0 in kJ/moL K: −0.330	Nyamunda et al. (2019)

The adsorption capacity was found to be in a range from 2 to 141 mg/g. In all cases the adsorption was found to be spontaneous and exothermic. Langmuir and pseudo second-order kinetic model had the best fit.

Note: [a]Obtained from experimental results. *L* Langmuir, *Ps2* Pseudo second-order

For the effective design and potential scale-up of an adsorption processes from lab to industrial scale, the estimation of the rate of adsorption is a crucial issue (Hubbe et al. 2019). The most widely used kinetic models are the pseudo first-order (Eq. 5.5) (Lagergren 1898) and the pseudo second-order kinetic (Eq. 5.6) models (Blanchard et al. 1984; Ho and McKay 1999):

$$q_t = q_e\left(1 - e^{-k_1 t}\right) \tag{5.5}$$

$$q_t = \frac{k_2 {q_e}^2 t}{1 + k_2 q_e t} \tag{5.6}$$

where q_e (mg/g) and q_t are the amounts of adsorbed adsorbate on the adsorbents at the equilibrium and at any time t, respectively; k_1 (min^{-1}) is rate constant of the first-order adsorption; and k_2 (g/mg min) is the rate constant of pseudo second-order adsorption, respectively.

Based on Table 5.1, the adsorption kinetics follow generally the pseudo second-order kinetics.

The thermodynamic parameters, Gibbs free energy change ΔG^0 (kJ/mol), the enthalpy change ΔH^0 (kJ/mol), and the entropy change ΔS^0 (kJ/mol K) indicate if the adsorption is spontaneous, enthalpy-, or entropy-driven and reveal the trend (increase or decrease) of randomness at the solid–liquid interface (Anastopoulos and Kyzas 2016). They were estimated using following equations:

$$\Delta G^0 = -RT \ln K \tag{5.7}$$

$$\Delta G^0 = \Delta H^0 - T\Delta S^0 \tag{5.8}$$

where R is the gas constant (8.314 J/mol K), T is the absolute temperature (Kelvin), and K is thermodynamic equilibrium constant. It should be noted that the K equilibrium constant must be unitless as clearly described in a recent review article by Lima et al. (2019). Otherwise, the use of inappropriate values for the thermodynamic equilibrium constant in the van't Hoff equation results in reporting erroneous thermodynamic parameters. It should be noted that the values of ΔG°, ΔH°, and ΔS° reported in Table 5.1 might be wrongly calculated, using erroneous data as thermodynamic equilibrium constants.

Nevertheless, the thermodynamic data compiled in Table 5.1 reveal that both ΔH^0 and ΔS^0 have negative values suggesting that the adsorption of Cu(II) by magnetic biochars is an enthalpy-driven process. It has to be noted that the Cu (II) adsorption on magnetic biochars is associated with more than one type of reaction and different binding modes and hence the observed thermodynamic parameters are cumulative quantities and the corresponding data average values.

5.4 Conclusions and Outlook

Magnetic biochars are capable of removing copper ions from (waste)waters and the main removal mechanisms are ion exchange, surface complexation cation π-interaction, and surface polynucleation and precipitation. The predominance of a specific adsorption mechanism depends on the solution pH, the species distribution on the biochar surface, and the metal ion concentration. The adsorption capacity ranges from 2 to 141 mg/g and the adsorption data are best fitted by the Langmuir adsorption isotherm and pseudo second-order kinetic model. Generally, the Cu (II) adsorption by magnetic biochar materials was found to be spontaneous and basically an enthalpy-driven process.

However, to better understand and describe the adsorption mechanisms at a molecular level, the characterization of surface species by sophisticated spectroscopic techniques (e.g., EXAFS, XPS, and RAMAN) is compulsory. Moreover, Cu (II) redox reactions, which may occur on the biochar surface, have to be studied to have a deeper understanding of the chemical behavior of Cu(II) ions on magnetic biochar composites.

References

Abbas SH, Ismail IM, Mostafa TM, Sulaymon AH (2014) Biosorption of heavy metals: a review. J Chem Sci Technol 3:74–102

Akbarnezhad AA, Safa F (2018) Biochar-based magnetic nanocomposite for dye removal from aqueous solutions: response surface modeling and kinetic study. Russ J App Chem 91:1856–1866. https://doi.org/10.1134/S1070427218110174

Anastopoulos I, Kyzas GZ (2014) Agricultural peels for dye adsorption: a review of recent literature. J Mol Liq 200:381–389. https://doi.org/10.1016/j.molliq.2014.11.006

Anastopoulos I, Kyzas GZ (2016) Are the thermodynamic parameters correctly estimated in liquid-phase adsorption phenomena? J Mol Liq 218:174–185. https://doi.org/10.1016/j.molliq.2016.02.059

Ania CO, Parra JB, Menéndez JA, Pis JJ (2005) Effect of microwave and conventional regeneration on the microporous and mesoporous network and on the adsorptive capacity of activated carbons. Microporous Mesoporous Mater 85:7–15. https://doi.org/10.1016/j.micromeso.2005.06.013

Baes CF, Mesmer RS (1977) The hydrolysis of cations. Ber Bunsen-Ges 81:245–246. https://doi.org/10.1002/bbpc.19770810252

Arbabi M, Golshani N (2016) Removal of copper ions cu (II) from industrial wastewater: a review of removal methods. Int J Epidemiol Res 3:283–293

Bilal M, Shah JA, Ashfaq T, Gardazi SMH, Tahir AA, Pervez A, Haroon H, Mahmood Q (2013) Waste biomass adsorbents for copper removal from industrial wastewater—a review. J Hazard Mater 263:322–333. https://doi.org/10.1016/j.jhazmat.2013.07.071

Blanchard G, Maunaye M, Martin G (1984) Removal of heavy metals from waters by means of natural zeolites. Water Res 18:1501–1507. https://doi.org/10.1016/0043-1354(84)90124-6

Carvalho Barros GKG, Melo RPF, ELd BN (2018) Removal of copper ions using sodium hexadecanoate by ionic flocculation. Sep Purif Technol 200:294–299. https://doi.org/10.1016/j.seppur.2018.01.062

Chen X (2015) Modeling of experimental adsorption isotherm data. Information 6:14–22. https://doi.org/10.3390/info6010014

Flemming CA, Trevors JT (1989) Copper toxicity and chemistry in the environment: a review. Water Air Soil Poll 44:143–158. https://doi.org/10.1007/BF00228784

European Union (1998) Council Directive 98/83/EC of 3 November 1998 on the quality of water intended for human consumption. Official Journal of the European Communities L 330/32 05/12/1998. The Council of the European Union, Brussels

Freundlich H (1906) Over the adsorption in solution. J Phys Chem 57:385–470

Foo KY, Hameed BH (2010) Insights into the modeling of adsorption isotherm systems. Chem Eng J 56:2–10. https://doi.org/10.1016/j.cej.2009.09.013

Hadjittofi L, Prodromou M, Pashalidis I (2014) Activated biochar derived from cactus fibres–preparation, characterization and application on Cu (II) removal from aqueous solutions. Bioresour Technol 159:460–464. https://doi.org/10.1016/j.biortech.2014.03.073

Hadjiyiannis P, Pashalidis I (2019) Copper (II) adsorption by *Opuntia ficus-indica* biochar fiber-MnO_2 composites. Desalin Water Treat 159:60–65. https://doi.org/10.5004/dwt.2019.23719

Hargreaves AJ, Vale P, Whelan J, Alibardi L, Constantino C, Dotro G, Cartmell E, Campo P (2018) Coagulation–flocculation process with metal salts, synthetic polymers and biopolymers for the removal of trace metals (Cu, Pb, Ni, Zn) from municipal wastewater. Clean Techn Environ Policy 20:393–402. https://doi.org/10.1007/s10098-017-1481-3

Ho Y-S, McKay G (1999) Pseudo-second order model for sorption processes. Process Biochem 34:451–465. https://doi.org/10.1016/S0032-9592(98)00112-5

Hu Q, Zhu Y, Hu B, Lu S, Sheng G (2018) Mechanistic insights into sequestration of U(VI) toward magnetic biochar: batch, XPS and EXAFS techniques. J Environ Sci (China) 70:217–225. https://doi.org/10.1016/j.jes.2018.01.013

Hubbe MA, Azizian S, Douven S (2019) Implications of apparent pseudo-second-order adsorption kinetics onto cellulosic materials: a review. BioRes 14:7582–7626

Ince M, Ince OK (2017) An overview of adsorption technique for heavy metal removal from water/wastewater: a critical review. Int J Pure Appl Sci 3:10–19

Kołodyńska D, Bąk J (2018) Use of three types of magnetic biochar in the removal of copper (II) ions from wastewaters. Sep Sci Technol 53:1045–1057. https://doi.org/10.1080/01496395.2017.1345944

Konstantinou M, Kasseta G, Pashalidis I (2006) Boron adsorption on alumina (Al_2O_3) and magnesia (MgO) in aqueous solutions: a comparative study. Int J Environ Technol Manag 6:466–479. https://doi.org/10.1504/ijetm.2006.010478

Konstantinou M, Kolokassidou K, Pashalidis I (2007) Sorption of Cu (II) and Eu (III) ions from aqueous solution by olive cake. Adsorption 13:33–40. https://doi.org/10.1007/s10450-007-9007-7

Lützenkirchen J, Behra P (1995) On the surface precipitation model for cation sorption at the (hydr)oxide water interface. Aquat Geochem 1:375–397. https://doi.org/10.1007/BF00702740

Lagergren S (1898) About the theory of so-called adsorption of soluble substances. Sven Vetenskapsakad Handingarl 24:1–39

Langmuir I (1918) The adsorption of gases on plane surfaces of glass, mica and platinum. J Am Chem Soc 40:1361–1403. https://doi.org/10.1021/ja02242a004

Li M, Liu H, Chen T, Dong C, Sun Y (2019) Synthesis of magnetic biochar composites for enhanced uranium(VI) adsorption. Sci Total Environ 651:1020–1028. https://doi.org/10.1016/j.scitotenv.2018.09.259

Liatsou I, Constantinou P, Pashalidis I (2017) Copper binding by activated biochar Fibres derived from *Luffa cylindrica*. Water Air Soil Poll 228:255. https://doi.org/10.1007/s11270-017-3411-8

Liatsou I (2019) Adsorption of metal ions by chemically modified biochar fibers obtained from *Luffa Cylindrica*. Doctoral dissertation

Liatsou I, Pashalidis I, Dosche C (2019) Cu(II) adsorption on 2-thiouracil-modified Luffa Cylindrica biochar fibres from artificial and real samples, and competition reactions with U (VI). J Hazard Mater 5:120950. https://doi.org/10.1016/j.jhazmat.2019.120950

Lima EC, Hosseini-Bandegharaei A, Moreno-Piraján JC, Anastopoulos I (2019) A critical review of the estimation of the thermodynamic parameters on adsorption equilibria. Wrong use of equilibrium constant in the Van't Hoff equation for calculation of thermodynamic parameters of adsorption. J Mol Liq 273:425–434. https://doi.org/10.1016/j.molliq.2018.10.048

Mahmoodi NM (2014) Synthesis of core–shell magnetic adsorbent nanoparticle and selectivity analysis for binary system dye removal. J Ind Eng Chem 20:2050–2058. https://doi.org/10.1016/j.jiec.2013.09.030

McKay G, Ho Y, Ng J (1999) Biosorption of copper from waste waters: a review. Sep Purif Methods 28:87–125. https://doi.org/10.1080/03602549909351645

Nicolaou E, Philippou K, Anastopoulos I, Pashalidis I (2019) Copper adsorption by magnetized pine-needle biochar. PRO 7:903. https://doi.org/10.3390/pr7120903

Noor NM, Othman R, Mubarak NM, Abdullah EC (2017) Agricultural biomass-derived magnetic adsorbents: preparation and application for heavy metals removal. J Taiwan Inst Chem E 78:168–177. https://doi.org/10.1016/j.jtice.2017.05.023

Nyamunda BC, Chivhanga T, Guyo U, Chigondo F (2019) Removal of Zn (II) and Cu (II) ions from industrial wastewaters using magnetic biochar derived from water hyacinth. J Eng 2019, Article ID 5656983. https://doi.org/10.1155/2019/5656983

Philippou K, Anastopoulos I, Dosche C, Pashalidis I (2019) Synthesis and characterization of a novel Fe3O4-loaded oxidized biochar from pine needles and its application for uranium removal. Kinetic, thermodynamic, and mechanistic analysis. J Environ Manag 252:109677. https://doi.org/10.1016/j.jenvman.2019.109677

Rachou J, Gagnon C, Sauvé S (2007) Use of an ion-selective electrode for free copper measurements in low salinity and low ionic strength matrices. Environ Chem 4:90–97. https://doi.org/10.1071/EN06036

Reddy DHK, Yun Y-S (2016) Spinel ferrite magnetic adsorbents: alternative future materials for water purification? Coord Chem Rev 315:90–111. https://doi.org/10.1016/j.ccr.2016.01.012

Rimola A, Rodríguez-Santiago L, Sodupe M (2006) Cation−π interactions and oxidative effects on Cu^{+} and Cu^{2+} binding to Phe, Tyr, Trp, and his amino acids in the gas phase. Insights from first-principles calculations. J Phys Chem B 110:24189–24199. https://doi.org/10.1021/jp0649571

Sauvé S, McBride MB, Norvell WA, Hendershot WH (1997) Copper solubility and speciation of in situ contaminated soils: effects of copper level, pH and organic matter. Water Air Soil Poll 100:133–149. https://doi.org/10.1023/A:1018312109677

Son E-B, Poo K-M, Chang J-S, Chae K-J (2018) Heavy metal removal from aqueous solutions using engineered magnetic biochars derived from waste marine macro-algal biomass. Sci Total Environ 615:161–168. https://doi.org/10.1016/j.scitotenv.2017.09.171

Song Q, Yang B, Wang H, Xu S, Cao Y (2016) Effective removal of copper (II) and cadmium (II) by adsorbent prepared from chitosan-modified magnetic biochar. J Residuals Sci Technol 13:197–205. https://doi.org/10.12783/issn.1544-8053/13/3/3

Suhas, PJM C, Ribeiro Carrott MML (2007) Lignin – from natural adsorbent to activated carbon: a review. Bioresour Technol 98:2301–2312. https://doi.org/10.1016/j.biortech.2006.08.008

Tong Y, Zhang M, Xia P, Wang L, Zheng J, Li W, Xu J (2016) Progammed synthesis of magnetic mesoporous silica coated carbon nanotubes for organic pollutant adsorption. J Magn Magn Mater 406:35–41. https://doi.org/10.1016/j.jmmm.2015.12.085

Trakal L, Šigut R, Šillerová H, Faturíková D, Komárek M (2014) Copper removal from aqueous solution using biochar: effect of chemical activation. Arab J Chem 7:43–52. https://doi.org/10.1016/j.arabjc.2013.08.001

Wan Ngah WS, Teong LC, Hanafiah MAKM (2011) Adsorption of dyes and heavy metal ions by chitosan composites: a review. Carbohyd Polym 83:1446–1456. https://doi.org/10.1016/j.carbpol.2010.11.004

Wang S, Gao B, Zimmerman AR, Li Y, Ma L, Harris WG, Migliaccio KW (2015) Removal of arsenic by magnetic biochar prepared from pinewood and natural hematite. Bioresour Technol 175:391–395. https://doi.org/10.1016/j.biortech.2014.10.104

Whitby H, Hollibaugh JT, van den Berg CMG (2017) Chemical speciation of copper in a salt marsh estuary and bioavailability to Thaumarchaeota. Front Mar Sci 4:178. https://doi.org/10.3389/fmars.2017.00178

Xu T, Lei X, Sun B, Yu G, Zeng Y (2017) Highly efficient and energy-conserved flocculation of copper in wastewater by pulse-alternating current. Environ Sci Pollut Res 24:20577–20586. https://doi.org/10.1007/s11356-017-9280-2

Yang X, Zhang S, Ju M, Liu L (2019) Preparation and modification of biochar materials and their application in soil remediation. Appl Sci 9(7):1365. https://doi.org/10.3390/app9071365

Yang YG, Yang YL, Kang XH (2013) Removal of copper ion in industrial wastewater by magnetic flocculation experimental research. Appl Mech Mater 295–298:1231–1234. https://doi.org/10.4028/www.scientific.net/AMM.295-298.1231

Yin Z, Liu Y, Liu S, Jiang L, Tan X, Zeng G, Li M, Liu S, Tian S, Fang Y (2018) Activated magnetic biochar by one-step synthesis: enhanced adsorption and coadsorption for 17β-estradiol and copper. Sci Total Environ 639:1530–1542. https://doi.org/10.1016/j.scitotenv.2018.05.130

Chapter 6
Synthesizing Magnetic Adsorbents for Landfill Leachate Remediation

Mir Amir Mohammad Reshadi, Alireza Bazargan, and Gordon McKay

Contents

Abstract Leachate is a complex mixture of chemicals and pollutants formed in the landfill. This wastewater stream has come into contact with the waste and the biological activity within the landfill, and can cause potentially severe environmental impacts if not managed and treated properly. For the treatment of landfill leachate, various processes have been proposed and studied. In this chapter, after providing an overview of leachate characteristics and identifying humic acid, fulvic acid, and ammonia nitrogen as the most important components, the use of magnetic adsorbents for leachate treatment has been reviewed. In particular, the synthesis processes for the adsorbents have been investigated. It was concluded that the best performing adsorbents are not necessarily the ones with the most expensive and complicated synthesis process. On the contrary, two adsorbents were highlighted as those requiring further study due to their simplicity and high adsorption capacities, namely,

M. A. M. Reshadi
Department of Civil Engineering, K. N. Toosi University of Technology, Tehran, Iran

A. Bazargan (✉)
School of Environment, College of Engineering, University of Tehran, Tehran, Iran
e-mail: alireza.bazargan@ut.ac.ir

G. McKay
Division of Sustainable Development, College of Science and Engineering, Hamad Bin Khalifa University, Doha, Qatar

L. Meili, G. L. Dotto (eds.), *Advanced Magnetic Adsorbents for Water Treatment*, Environmental Chemistry for a Sustainable World 61,
https://doi.org/10.1007/978-3-030-64092-7_6

magnetic mesoporous activated carbon and magnetic Fe_3O_4 nanoparticles. The best performing adsorbent (in terms of capacity) was a combined maghemite and nanosized hydroxyapatite adsorbent.

Keywords Nanoparticle · Sorbent · Sorption · Regeneration · Separation processes

6.1 Introduction

Adsorption can be generally defined as a surface phenomenon in which a highly porous material removes species from a solution through the formation of chemical and physical bonds. In general, this technique has been very common for removing organic and inorganic substances in water and wastewater treatment (Mubita et al. 2019; Zhang et al. 2019b), and has been used in broad applications for the treatment of gases (Abbasi and Sardroodi 2019; Sepahvand et al. 2020).

Although adsorption has been used for hundreds (if not thousands) of years, the term itself became popular at a much more recent date (Dąbrowski 2001). Since then, adsorbents have been derived from a plethora of feedstocks and their adsorption capacities (defined as the milligram of pollutant removed per gram of adsorbent) as well as their specific surface areas (reported as m^2 of unit area per gram of adsorbent) have improved markedly (Zhang et al. 2019a).

Although these two properties are related, it should not be assumed that the increase in one necessitates an increase in the other and vice versa. For example, in one study porous crumpled graphene has shown much better adsorption performance of rhodamine B (exceeding 300 mg/g) compared to activated carbon, even though the specific surface area of the active carbon was nearly double that of the crumpled graphene (Zhao et al. 2019a).

Nowadays, research is focused on the synthesis of adsorbents with higher adsorption capacities, lower environmental impact, better regeneration and reusability, and more diverse chemical and molecular forms, geometries and surface structures. Some examples of this diverse area of research include hierarchical C/NiO–ZnO nanocomposites for the adsorption for Congo red (Chen et al. 2019), fluffy honeycomb-like activated carbon from popcorn for organic dye removal (Yu et al. 2019), nitrogen-doped graphitic porous carbon for capacitive deionization (Gao et al. 2019), and graphene aerogels for cadmium ion adsorption (Trinh et al. 2019), just to name a few.

On the other hand, the growing trend of waste generation has become a problem on both local and global scales, which is due to the world's large population as well as the increase in a materialistic lifestyle. Previous studies have predicted an annual waste production of 2.2 billion tons on a global scale by 2025 (Hoornweg and Bhada-Tata 2012). Landfilling, composting, incineration, and recycling are the major methods proposed for waste management.

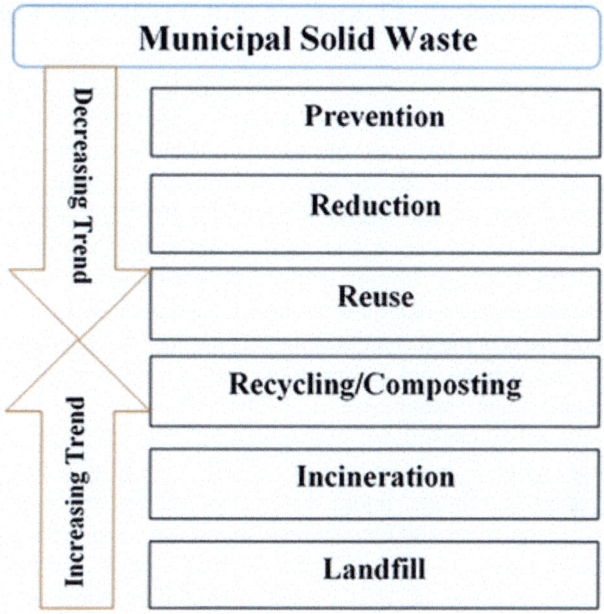

Fig. 6.1 Waste management hierarchy ordered from most suitable (top) to least suitable (bottom) (reproduced with permission from Yakubu and Zhou (2019))

Although it is the last resort based on the waste management hierarchy (Fig. 6.1), landfilling is the most widespread technique on a global scale. Unfortunately, this does not always occur in engineered "sanitary" landfills, but often, in open dumps with little or no control. This is the case especially in developing countries, owing to the preference for cheaper methods. It should also be noted that the lack of public awareness, the absence of proper regulations as well as the low acceptance of such regulations are some of the main reasons hampering the implementation of more novel and efficient methods (Oakley and Jimenez 2012; Torretta et al. 2016). Not to mention the other – and perhaps more pressing – issues in society demanding a share of the national budget.

One of the major environmental issues associated with landfilling and/or dumping waste is the generation of leachate which can be defined as the wastewater resulting from the aerobic and anaerobic decomposition of waste as well as the percolation of water through the waste deposit. Landfill leachate is strongly contaminated, which can cause serious adverse effects by finding its way into water resources and subsequently entering the food chain (Bashir et al. 2012; Negi et al. 2020). That, among other reasons, is why when planning suitable locations for landfills, complex decision-making processes must be considered as reviewed previously by our team, including the consideration of groundwater depth, surface water vicinity, elevation, land slope, soil permeability, soil stability, flooding susceptibility, lithology and stratification, faults, land use type, nearby settlements and urbanization, cultural and protected site vicinity, wind direction, roads, railways, proximity to building materials, pipelines and powerlines, and proximity to airports (Rezaeisabzevar et al. 2020).

6.2 Leachate Properties and Treatment Options

The generation of contaminated landfill leachate, for long periods of time spanning decades, is one of the major issues associated with landfills (Zhang et al. 2019c). A previous toxicity study on dozens of landfills reported that at least 32 chemicals found in leachate are directly carcinogenic, 21 induce genetic damage, and 10 can cause birth defects (Achankeng 2004). Besides, some emerging contaminants as by-products of newer and more complicated industrial processes are detected in landfill leachate (Stuart et al. 2012). Stricter pollution control obligations and discharge standards, as well as increased awareness of the associated hazards have motivated many researches toward new landfill leachate treatment (Top et al. 2019).

The volume and composition are two major factors when considering leachate management (Mukherjee et al. 2015; Renou et al. 2008). There are many factors influencing landfill leachate generation rate, namely, landfill cover, groundwater level, rainfall, and the type of landfilled waste as well as other factors, such as soil properties, climate, liquid waste disposal, waste compaction level, landfill design, waste moisture, and biochemical interactions inside the landfill (Aquino and Stuckey 2004; Li et al. 2010; Mukherjee et al. 2015; Tsarpali et al. 2012; Umar et al. 2010a, b; Xu et al. 2010). Also, it should be noted that leachate generation rate is inversely proportionate with landfill age, meaning that leachate generation decreases as a landfill gets older (Brennan et al. 2016; Renou et al. 2008). Landfill age is often cited as one of the most important factors influencing leachate composition and volume, and therefore, many studies have given ranges for various leachate characteristics based on the landfill age. As leachate becomes older, the BOD_5 content decreases which also causes a decrease in BOD_5/COD ratio. The relative percentage of refractory compounds (fulvic and humic acids) also increases compared to other organic compounds (Ahmed and Lan 2012; Bashir et al. 2013). In addition, as the leachate becomes older, it becomes more transparent and clear.

Since the composition of the leachate is heavily dependent on numerous factors, different studies have provided various values. Tables 6.1, 6.2, and 6.3 provided from the literature show the discrepancy and differences. Therefore, when choosing treatment options, it is very important to know the actual composition of the leachate for each site, while also acknowledging the composition of each site will also change based on the seasons, climatic conditions, waste composition, etc. Additionally, Table 6.4 shows the liters of leachate produced in various Chinese landfills per ton of waste (wet waste) each day, showing extreme variability. Some of this variability has been ascribed to the different covers employed over the landfills, allowing or not allowing rainwater to enter (Yang et al. 2015).

Table 6.1 Landfill leachate composition for young, middle-aged (intermediate), and stabilized (old) leachate (reproduced with permission from (Foo and Hameed 2009))

Composition	Young	Intermediate	Stabilized
Age (years)	<5	5–10	>10
pH	<6.5	6.5–7.5	>7.5
COD (mg/L)	>10,000	4000–10,000	<4000
BOD5/COD	0.5–1.0	0.1–0.5	<0.1
Organic compounds	80% volatile fatty acids (VFA)	5–30% VFA+ humic and fulvic acids	Humic and fulvic acids
Ammonia nitrogen (mg/L)	<400	NA	>400
TOC/COD	<0.3	0.3–0.5	>0.5
Kjeldahl nitrogen (g/L)	0.1–0.2	NA	NA
Heavy metals (mg/L)	Low to medium	Low	Low
Biodegradability	Important	Medium	Low

6.2.1 Ammonia

Since there is no mechanism to degrade ammonia under methanogenic conditions inside the landfill, ammonia remains as one of the most significant component in leachate (Kruempelbeck and Ehrig 1999; Robinson 1995). The literature indicates that amino acid deamination which is the result of organic compound decomposition, is the main reason leading to the formation of ammoniacal nitrogen (Tatsi and Zouboulis 2002). Hydrolysis and fermentation of fractions of biodegradable substrates having nitrogen in their structure have been identified as a source of ammonia concentration in landfill leachate (Carley and Mavinic 1991).

6.2.2 Humic and Fulvic Acids

Humic acid (HA), fulvic acid (FA), and humin are three main fractions of humic substances which can mainly be characterized by their different sizes and solubility in different pH values (Doulia et al. 2009). Humic acid, having a molecular weight of 2000–5000 Daltons, is soluble in basic solutions and insoluble in acidic water ($pH < 2$) (Esmaeili et al. 2012; Jones and Bryan 1998) while FA is soluble in water at any pH value. On the other hand, unlike fulvic acid, humin is insoluble in both basic and acidic water and has the highest molecular weight and largest molecular size (Liu et al. 2011). Humic and fulvic acids are produced as a result of continuous decomposition of the organic fraction of waste. The presence of humic substances makes the treatment of leachate extremely more complicated than municipal wastewater (Ghaedi et al. 2013).

Table 6.2 Composition of leachates from different locations and ages (reproduced with permission from (Renou et al. 2008))

Location	Age	COD	BOD	BOD/COD	pH	SS	TKN	NH_3-N
Canada	Y	13,800	9660	0.70	5.8	–	212	42
Canada	Y	1870	90	0.05	6.58	–	75	10
China, Hong Kong	Y	15,700	4200	0.27	7.7	–	–	2260
China, Hong Kong	Y	17,000	7300	0.43	7.0–8.3	>5000	3200	3000
	Y	13,000	5000	0.38	6.8–9.1	2000	11,000	11,000
	Y	50,000	22,000	0.44	7.8–9.0	2000	13,000	13,000
China, Mainland	Y	1900–3180	3700–8890	0.36–0.51	7.4–8.5	–	–	630–1800
Greece	Y	70,900	26,800	0.38	6.2	950	3400	3100
Italy	Y	19,900	4000	0.20	8	–	–	3917
Italy	Y	10,540	2300	0.22	8.2	1666	–	5210
South Korea	Y	24,400	10,800	0.44	7.3	2400	1766	1682
Turkey	Y	16,200–20,000	10,800–11,000	0.55–0.67	7.3–7.8	–	–	1120–2500
	Y	35,000–50,000	21,000–25,000	0.5–0.6	5.6–7.0	–	–	2020
Turkey	Y	35,000–50,000	21,000–25,000	0.5–0.6	5.6–7.0	2630–3930	2370	2020
Turkey	Y	10,750–18,420	6380–9660	0.52–0.59	7.7–8.2	1013–1540	–	1946–2002
Canada	I	3210–9190	–	–	6.9–9.0	–	–	–
China	I	5800	430	0.07	7.6	–	–	–
China, Hong Kong	I	7439	1436	0.19	8.22	784	–	–
Germany	I	3180	1060	0.33	–	–	1135	884
Germany	I	4000	800	0.20	–	–	–	800
Greece	I	5350	1050	0.20	7.9	480	1100	940
Italy	I	5050	1270	0.25	8.38	–	1670	1330
Italy	I	3840	1200	0.31	8	–	–	–
Poland	I	1180	331	0.28	8	–	–	743
Taiwan	I	6500	500	0.08	8.1	–	–	5500
Turkey	I	9500	–	–	8.15	–	1450	1270
Brazil	S	3460	150	0.04	8.2	–	–	800
Estonia	S	2170	800	0.37	11.5	–	–	–
Finland	S	556	62	0.11	–	–	192	159
Finland	S	340–920	84	0.09–0.25	7.1–7.6	–	–	330–560
France	S	500	7.1	0.01	7.5	130	540	430
France	S	100	3	0.03	7.7	13–1480	5–960	0.2
France	S	1930	–	–	7	–	–	295
Malaysia	S	1533–2580	48–105	0.03–0.04	7.5–9.4	159–233	–	–
South Korea	S	1409	62	0.04	8.57	404	141	1522
Turkey	S	10,000	–	–	8.6	1600	1680	1590

Y Young, *I* Intermediate (Middle-aged), *S* Stabilized (Old), *SS* Suspended Solids, *TKN* Total Kjeldahl Nitrogen

Table 6.3 Landfill leachate composition in different periods of landfill age (reproduced with permission from (Costa et al. 2019))

	Landfill age (years)			
Parameters (mg/L) other than pH	0–5	5–10	10–20	>20
pH	3–6	6–7	7–7.5	7.5
BOD	10,000–25,000	1000–4000	50–1000	<50
COD	15,000–40,000	10,000–20,000	1000–5000	<1000
TKN	1500–4500	400–800	75–300	<50
$NH_4^+ - N$	1500–4250	250–700	50–200	<30
Cl^-	1000–3000	500–2000	100–500	<100
P	100–300	10–100	–	<10
Alkalinity	8000–18,000	4500–6000	–	–
Conductivity*	15,000–41,500	6000–14,000	–	–
SO_4^{2-}	500–2000	200–1000	50–200	<50
Fe^{2+}	500–1500	500–1000	100–500	<100
Zn^{2+}	100–200	50–100	10–50	<10
Total dissolved solids (TDS)	10,000–25,000	5000–10,000	2000–5000	<1000

Table 6.4 Landfill leachate production rate among different Chinese landfills (reproduced with permission from (Yang et al. 2015))

City	Operating period (years)	Leachate volume produced ($L \cdot t^{-1}ww$)
Yi'ning	5	70
Beijing (Liulitun)	3	400
Beijing (Beishenshu)	11	167
Beijing (Anding)	3	490
Tangshan (Jianzigu)	2	280
Xi'an (Jiangcungou)	8	230
Suzhou (Qizishan)	1	830
Changzhou	8	380
Shanghai (Laogang)	5	400
Ningbo (Yinzhou)	4	400
Lishui (Wulinggen)	2	950
Nanchang (Maiyuan)	14	320
Changsha (Qiaoyi)	8	290
Guiyang (Gaoyan)	10	190
Chongqing (Changshengqiao)	3	800
Chengdu (Chang'an)	18	400
Kunming (Xijiao)	5	154
Guangzhou (Xingfeng)	10	330
Shenzhen (Xiaping)	15	454

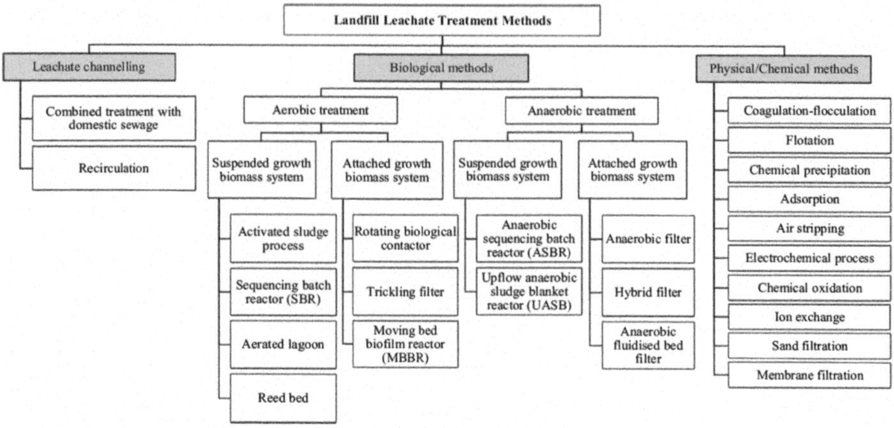

Fig. 6.2 Hierarchy of different landfill leachate treatment methods (reproduced with permission from Kamaruddin et al. (2017))

6.2.3 Leachate Treatment Options

In addition to adverse environmental impacts, landfill leachate can also cause other problems such as structural damages to concrete structures (Agamuthu and Fauziah 2008). Since landfill leachate composition is highly variable among different landfills, there is no magic bullet with which the leachate can be treated. In other words, there are various options from which the proper treatment option according to the site-specific considerations should be chosen. Many different landfill leachate treatment methods have been proposed. For example, recirculation of leachate is a good option when good control of liquid flow is possible and moisture control within the landfill is wanted. Sending the leachate to conventional wastewater treatment plants, to be diluted and treated with domestic wastewater is also an option. However, it requires severe dilution based on leachate quality and discharge quality standards. Biological methods are more suitable for younger leachate (with higher BOD/COD ratio) whereas physiochemical processes are suited for older leachate with more recalcitrant components. Figure 6.2 provides a comprehensive hierarchy of the methods proposed for landfill leachate treatment. Each technique has its benefits and drawbacks. For this reason, employing a combination of the options is the best strategy in many instances.

6.3 Magnetic Adsorbents for Leachate Treatment

From among various leachate treatment methods, adsorption is one of the more widely used. Adsorption has gained acceptance in part due to its ease of design and operation, effectiveness at the removal of various pollutants such as COD, and low capital expenditure costs (although the running costs can be high due to the

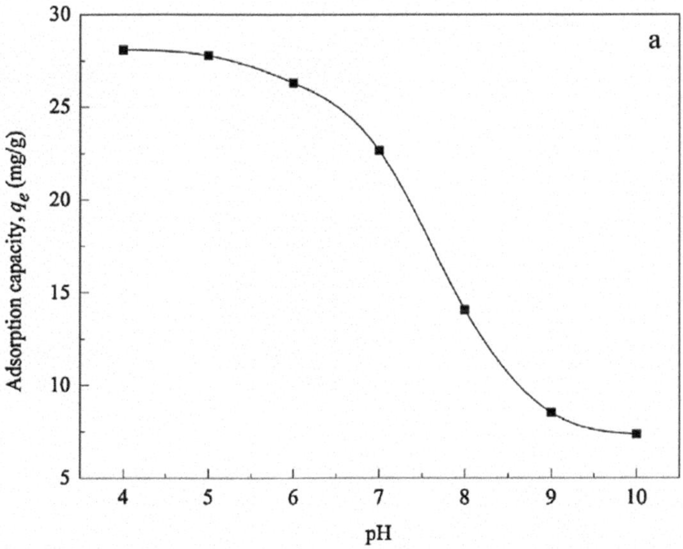

Fig. 6.3 Effect of different pH values on adsorption capacity of HA on magnetic chitosan particles (reproduced with permission from Dong et al. (2014a))

exhaustion and replacement of the adsorbent). Magnetic adsorption is a subcategory of adsorption processes with its own particular advantages and features. For example, magnetic adsorbents require shorter separation time, have higher chemical stability, and have the possibility of improving the process capacity (Dong et al. 2014b; Parsonage 1992). Another important feature of magnetic adsorbents is their relatively easier regeneration and reuse compared to activated carbon.

This section investigates studies which have focused on the adsorption of humic acid, fulvic acid, and ammonia nitrogen onto magnetic adsorbents as indicators of the adsorbent's applicability for landfill leachate treatment. This is because so far, very few studies have focused on employing magnetic adsorbents in the treatment of real landfill leachate. A previous review article by our team has investigated this topic to some extent (Reshadi et al. 2020).

For example, the removal of HA using magnetic chitosan nanoparticles was investigated in one study (Dong et al. 2014b). The maximum adsorption capacity was slightly more than 30 mg/g for HA, which was under the conditions of pH = 7 and 25 °C. It was observed that the adsorption capacity was adversely affected by higher pH values. An increase of solution pH from 4 to 10 appeared to reduce the adsorption capacity from just under 30 mg/g to 7.4 mg/g. This is particularly important because younger leachate has lower pH values and as the leachate becomes older, the pH increases. The relationship between pH variation and adsorption capacity for the said adsorbent is depicted in Fig. 6.3. Similarly, there was a similar relationship between adsorption efficiency and ionic strength, meaning that adsorption efficiency was enhanced at lower ionic strength.

Fig. 6.4 Gathering amino-functionalized magnetic cellulose composite loaded with hexavalent chromium. (**a**) Before the magnetic field, (**b**) less than 1 min after the magnetic field (reproduced with permission from (Sun et al. 2014))

As previously mentioned, one of the important advantages of magnetic adsorbents is the relative ease of their collection for regeneration and reuse. Figure 6.4 depicts magnetic adsorbent gathering using a magnet.

A variety of different magnetic adsorbents ranging from magnetic activated carbon to more advanced adsorbents have been tested. Here, a total of 18 relevant studies have been comprehensively summarized in Table 6.5. The table provides information regarding the adsorbent name, the target pollutant to be removed, the synthesis process, and the adsorption capacity for the target pollutant. By careful examination of the synthesis processes, it can be seen that the production processes need not be overtly expensive or complicated to achieve good results. Actually, it is most favorable if the process for synthesizing the adsorbent is as simple and inexpensive as possible. Particularly, two studies stand out:

The first article that demands attention is the one in which, magnetic mesoporous activated carbon (Ogata and Miura 2013) was simply synthesized by adding iron nitrate to commercial activated carbon (from coconut shells) under vacuum conditions followed by heating under nitrogen flow at 800 °C and at 850 °C under CO_2 flow. The product was grinded to powder form, and was able to remove impressive amounts in excess of 400 mg of humic acid per gram of adsorbent.

In the second noteworthy article, magnetic Fe_3O_4 nanoparticles were recently synthesized (Zhao et al. 2019b) by dissolving $FeCl_3{\cdot}6H_2O$ and $FeCl_2{\cdot}4H_2O$ in deionized water under nitrogen flow, followed by the addition of 1.5 M sodium hydroxide with continuous stirring. The black Fe_3O_4 nanoparticles were able to remove upwards of 100 mg of fulvic acid per gram of adsorbent.

These two examples show that in order to obtain favorable results, researchers need not look for overtly complicated methods. Considering that simplicity is an extremely important feature for industrial application of the product, such studies could be used as starting points for further investigation. The best performing adsorbent in those reviewed is combined magnetic maghemite with nanosized hydroxyapatite (Shi et al. 2016) exhibiting a tremendous adsorption capacity

Table 6.5 A list of magnetic adsorbents used for the removal of landfill leachate pollutants, reproduced with significant modification from Reshadi et al. (2020)

#	Adsorbent name and reference	Target pollutant	Preparation method	Adsorption capacity and/or removal %
1	Oxidized magnetic activated carbon (Ogata and Miura 2013)	AN[a]	3 and 1.5 g of $FeCl_3$ and $FeCl_2.4H_2O$ (molar ratio of 1:2.45) were added to 100 ml of DI water with a nitrogen flow and with vigorous stirring. Ten milliliters of ammonia water and a mixture of different coating agents (hexanoic and oleic acid) were added to the first solution. Magnetic nanoparticles were filtered, magnetically separated, and washed with ethanol several times. After being dried for 24 h at 80 °C under vacuum conditions, the particles were stored. Then, "magnetic seeding" was used, in which the magnetite nanoparticles were coprecipitated with activated carbon allowing them to physically get adsorbed into activated carbon sites. CO_2 heat treatment increases magnetization as it expands pore radius. Further treatment with CO_2 for 90 min and adsorption for 120 min yields the final product of Ox-MAC.	5.3 mg/g, 80% removal
2	Magnetic zeolite (Sugawara et al. 2016)	AN	2.5 and 1.25 g of $FeCl_3$ and $FeCl_2$ respectively were dissolved in deionized water. The mixture was kept in a water bath (90 °C) for 15 min. Fifteen grams of fly-ash and 100 ml of 2 M sodium hydroxide were mixed at 95 °C and were refluxed for 24 h. The two prepared solutions were mixed to coprecipitate magnetic zeolite. Magnetic zeolite was separated magnetically and dried at 50 °C. Using pressure, the particles were pelletized and	83.4% removal

(continued)

Table 6.5 (continued)

#	Adsorbent name and reference	Target pollutant	Preparation method	Adsorption capacity and/or removal %
			annealed for 15 h at 700 °C under vacuum for crystallization. Finally, magnetic zeolite was powdered for activation.	
3	Magnetically modified zeolite (Nah et al. 2007, 2008)	AN	Synthetic zeolite and magnetite were provided commercially. Five grams of magnetite, zeolite, and urethane were mixed with 20 g of thinner acting as adhesive followed by vacuum drying for 5 h at 60 °C. Magnetically modified zeolite particles were separated using a magnet. The separated particles were grinded for 3 h and washed with DI water several times for further use.	22 mg/g
4	Aminopropyl functionalized silica-coated Fe_3O_4 nanoparticles (Wang et al. 2015)	HA[b]	0.02 and 0.04 mol of $FeCl_2.4H_2O$ and $FeCl_3.6H_2O$, were added respectively into a 0.5 HCL solution which was then added to 500 ml of 1.5 M solution of sodium hydroxide under a nitrogen flow of 40 ml/min at 80 °C with vigorous stirring. This resulted in Fe_3O_4 nanoparticles which were subsequently washed with deionized water as well as drying under vacuum at 50 °C for 4 h. Silica-coated Fe_3O_4 nanoparticles were prepared by adding 2 g of the nanoparticles to 400 ml of deionized water, which was heated to 80 °C under the same nitrogen flow. Forty milliliters of 1 M sodium silicate solution was added to the suspension under stirring. 2 M HCL solution was added to adjust the suspension pH to 6. After stirring for 3 h at 80 °C, the resulting silica-coated nanoparticles were washed with deionized water	181.82 mg/g

(continued)

Table 6.5 (continued)

#	Adsorbent name and reference	Target pollutant	Preparation method	Adsorption capacity and/or removal %
			and then separated using a magnet followed by drying for 12 h under vacuum at 50 °C. To functionalize the silica-coated nanoparticle surface, 3-aminopropyl trimethoxysilane (APTMS) was used as the silylation agent. Two grams of silica-coated nanoparticles were ultrasonically dispersed with 50 ml of toluene for 15 min in a 250-ml three-necked flask. Then, 4 ml of APTMS was added to the flask and the mixture was refluxed with continuous stirring at 110 °C under the same nitrogen flow for 12 h. Functionalized silica-coated nanoparticles were filtered followed by washing using acetone and ethanol several times, and then dried for 12 h under vacuum at 50 °C.	
5	Combined magnetic maghemite with nanosized hydroxyapatite (Shi et al. 2016)	HA	A 0.05 M solution of calcium nitrate containing 200 mg of maghemite was added to a 0.03 M solution of ammonium hydrogen phosphate while stirring. The solution pH was adjusted to 11 by adding ammonia. The final step consisted of heating the solution to 90 °C for 2 h and then cooling down to room temperature without stirring. The resulting precipitates were separated magnetically and washed by deionized water. After drying at 90 °C, grinding led to the final maghemite incorporated nanocomposites.	601.91 mg/g
6	Iron oxide magnetic nanoparticles (Esmaeili et al. 2012)	HA	Under ambient conditions, Iron (II) and Iron (III) chlorides were added to 2 mol/L HCL solution to reach the final	3.7 mg/g, 98% removal

(continued)

Table 6.5 (continued)

#	Adsorbent name and reference	Target pollutant	Preparation method	Adsorption capacity and/or removal %
			concentration of 250 mM. Citrate ions were chosen as nucleation stabilizers. The reactor was filled with argon and 5.5 M ammonia solution was added dropwise in order to precipitate the nanoparticles. The resulting product was precipitated with acetone, followed by centrifuging at 4000 rpm for 5 min. Then, argon-bubbled water was used to disperse the pellets. Under vacuum conditions, acetone residues were evaporated. The final product was filtered through a 0.22μm filter, resulting in nanoparticles that could be stored in an aqueous solution or in lyophilized form.	
7	Magnetic chitosan nanoparticle (Dong et al. 2014a, b)	HA	0.5 g of chitosan was dissolved in 200 mL of acetic acid solution with vigorous stirring. 4.7 and 2.4 g of $FeCl_3 \cdot 6H_2O$ and $FeSO_4 \cdot 7H_2O$, respectively, were dissolved in 22 mL of distilled water for 5 min using ultrasonication, and were added to the first solution by stirring for 20 min at 1000 rpm at 40 °C. Then, 40 mL of concentrated ammonia (28%) was added to the system. After 20 min, the reaction temperature was set to 60 °C. Six milliliters of C_3H_5ClO (epichlorohydrin) was added to the reaction system with stirring for 3 h at 1000 rpm. The resulting magnetic chitosan nanoparticles (MCNP) were separated magnetically, washed several times with acetic acid, alcohol and distilled water and then dried at 60 °C.	32.6 mg/g

(continued)

Table 6.5 (continued)

#	Adsorbent name and reference	Target pollutant	Preparation method	Adsorption capacity and/or removal %
8	Magnetic cucurbituril (Yang et al. 2014)	HA	1.5 g of cucurbituril, 2.51 g of $(NH_4)Fe(SO_4)_3 \cdot 12H_2O$, and 1.7 g of $(NH_4)Fe(SO_4)_3 \cdot 6H_2O$ were added to 200 ml of deionized water. After sonication for 10 min, 10 mL of $NH_3.H_2O$ (8 M) was heated to 50 °C and added to the suspension while stirring at 1000 rpm under nitrogen. The reaction pH was adjusted to 11–12. Finally, the precipitate was magnetically separated and washed with deionized water and ethanol several times and dried at 60 °C under vacuum.	9.52 mg/g
9	Magnetic montmorillonite–Cu (II)/Fe(III) oxides (Peng et al. 2006)	HA	0.5 M sodium hydroxide and 0.2 M aluminum chloride solutions were mixed dropwise with vigorous stirring at 85 °C. The pillaring solution was added to 1.5% Na-montmorillonite suspension under ambient conditions with continuous stirring. Meanwhile, NaOH was added to a solution of 2.525 g of iron (III) nitrate $Fe(NO_3) \cdot 9H_2O$, 0.533 g of copper(II) chloride dihydrate $CuCl_2 \cdot 2H_2O$, and 0.375 g of polyvinyl alcohol (PVA) in 100 mL water to reach a pH value of 11. Then, the magnetic dispersion was mixed with pillared montmorillonite and stirred for 2 h at 95–100 °C to prepare the magnetic adsorbent. After centrifuging and washing with deionized water, the solids were dried, milled, and calcinated for 2 h at 500 °C to produce the final product.	96% removal
10	Magnetic multiwalled carbon nanotubes decorated with calcium (Li et al. 2017)	HA	The addition of 0.3 M NaOH and 0.5 M Na_2CO_3 comprised the base solution. Meanwhile,	90.27%

(continued)

Table 6.5 (continued)

#	Adsorbent name and reference	Target pollutant	Preparation method	Adsorption capacity and/or removal %
			calcium chloride was dissolved in 20 ml of distilled water. A specific amount of magnetic multiwalled carbon nanotubes (MWCNT) was suspended in 100 ml of water containing $NH_4Fe(SO_4)_2 \cdot 12H_2O$ and $(NH_4)_2Fe(SO_4)_2 \cdot 6H_2O$ and ultrasonicated for 10 min. After keeping the above solution into a bath and adjusting the temperature at 60 °C, the solution was stirred at 400 rpm and the as-prepared base solution was added dropwise. The pH value of the solution was maintained at 11–12. After churning for 30 min (while maintaining the temperature), the solution was left to age for 30 min. The final product was filtered, washed, dried and grinded to obtain magnetic multiwalled carbon nanotubes decorated with calcium.	
11	Magnetically separable polyaniline (Wang et al. 2014)	HA	To prepare Fe_3O_4 nanoparticles, 0.02 g and 0.04 g of iron(II) and iron(III) chlorides were dissolved in 0.5 M HCl. This mixture was then added to 500 mL of 1.5 M NaOH solution at 80 °C with continuous stirring under the presence of nitrogen flow. Using a magnet, the Fe_3O_4 nanoparticles were separated and stored for further stages. After being washed with HCl solution, the particles were suspended in a mixture of deionized water and ethanol (60 and 240 mL respectively). The mixture was sonicated at 25 °C for 30 min. Then, ammonia water was added to adjust the pH value to 11.	36.36 mg/g

(continued)

Table 6.5 (continued)

#	Adsorbent name and reference	Target pollutant	Preparation method	Adsorption capacity and/or removal %
			While stirring continuously, 1 mL of tetraethyl orthosilicate was added dropwise to the mixture. Silica-coated nanoparticles ($Fe_3O_4@SiO_2$) were separated using a magnet and subsequently washed with deionized water and dried for 12 h at 50 °C. To prepare magnetic polyaniline (PANI) adsorbents, 0.3 g of silica-coated nanoparticles were suspended in deionized water followed by sonication at ambient conditions for 30 min. After addition of 0.2 mL of aniline, the pH was adjusted to 3 using 1 M HCl. After stirring for 1 h, ammonium persulfate was added to the solution. This mixture was kept in an ice bath followed by vigorous stirring for 6 h. The resulting solids were gathered magnetically and washed with deionized water and dried under vacuum for 12 h.	
12	Magnetic graphene oxide (Zhang et al. 2016)	HA	A suspension was made by adding 1 g of graphite oxide to a mixed solution and sonicated for 30 min. Meanwhile 10.7 g and 5.8 g of $NH_4Fe(SO_4)_2.12H_2O$ and $(NH_4)_2Fe(SO_4)_2.6H_2O$, respectively, were mixed in deionized water followed by addition of 10 ml of 25% ammonia solution under nitrogen flow. After that, the graphite oxide suspension was mixed with this solution with continuous stirring for 45 min at 250 rpm. The resulting suspension was cooled down to room temperature and the produced magnetic graphene oxide (MGO)	98.82%

(continued)

Table 6.5 (continued)

#	Adsorbent name and reference	Target pollutant	Preparation method	Adsorption capacity and/or removal %
			was separated magnetically and washed with deionized water and alcohol several times, followed by vacuum drying at 60 °C.	
13	Amine functionalized magnetic mesoporous composite microspheres (Tang et al. 2012)	HA	First, for preparing magnetic microspheres 0.86 g and 2.16 g of $FeCl_3{\cdot}6H_2O$ and CH_3COONa (sodium acetate) were added to 30 mL of ethylene glycol with continuous stirring. The obtained solution was heated for 8 h at 200 °C. Then, the product was left to cool down to room temperature, and washed with deionized water and ethanol several times. Next, the produced magnetic microspheres were dispersed in deionized water to prepare a Fe_3O_4 solution (100 mg/mL). Four ml of this solution was mixed with 0.1 M HCl for 20 min under sonication. The treated microspheres were dispersed in a mixture of water and ethanol (v/v = 30/70). For adjusting the solution pH to 9.5, NH_4OH was added to the solution. Under ambient conditions, 0.12 mL of tetraethyl orthosilicate was added to the reactor with continuous stirring and the reaction was continued for 24 h, followed by several water and ethanol washes. 7.2 mL of NH_4OH was added to a mixture of deionized water and ethanol (v/v = 12/88). Five milliliters of $C_{16}TMS$ and ethanol TEOS (molar ratio of 1:4.7) was added slowly. After stirring for 6 h at room temperature, the reaction stopped and the product was washed with	128.64 mg/g

(continued)

Table 6.5 (continued)

#	Adsorbent name and reference	Target pollutant	Preparation method	Adsorption capacity and/or removal %
			deionized water and ethanol several times followed by calcination for 6 h at 550 °C. This step produced the Fe_3O_4@mesoporous silica composite particles. As the final step, polyethylenimine (PEI) was added to the aqueous solution of the produced particles from the second step under ultrasonication and stirring for 30 min. The final product was then obtained after several washes with deionized water.	
14	Fe_3O_4-chitosan hybrid nanoparticles (Zulfikar et al. 2016)	HA	First, Fe_3O_4 nanoparticles were prepared by mixing $FeSO_4 \cdot 7H_2O$ and $FeCl_3$ with 150 ml of deionized water in a three-necked flask under nitrogen for 3 min. Then, 20 mL of 25% NH_4OH was added to the solution. After stirring for 30 min, 3 mL of C_3H_5ClO (epichlorohydrin) was added, followed by heating at 75 °C for 1 h, under nitrogen atmosphere. After cooling to room temperature, the nanoproduct was extracted using a magnet and washed with ethanol and deionized water several times. After drying, in order to prepare the Fe_3O_4-chitosan nanoparticles, 1.5 of the nanoparticles were mixed with 0.5 g of chitosan in 100 ml of acetic acid in a three-necked flask, and refluxed at 65 °C for 4 h at pH = 6. Finally, the Fe_3O_4-chitosan hybrid nanoparticles were collected, washed, and dried.	44.84 mg/g

(continued)

Table 6.5 (continued)

#	Adsorbent name and reference	Target pollutant	Preparation method	Adsorption capacity and/or removal %
15	Magnetic mesoporous activated carbon (Ogata and Miura 2013)	HA	A specific amount of commercial activated carbon (from coconut shells) was added to iron nitrate solution under vacuum condition. Then, the mixture was heated under nitrogen flow at 800 °C followed by heating at 850 °C under CO_2 flow. The product was grinded to powder form.	420 mg/g, 90% removal
16	Magnetic Fe_3O_4 nanoparticles (Zhao et al. 2019b)	FA[c]	First, 5.2 g and 2 g of $FeCl_3{\cdot}6H_2O$ and $FeCl_2{\cdot}4H_2O$ were respectively dissolved in deionized water under nitrogen flow. Then, this solution was added into 250 mL of 1.5 M sodium hydroxide with continuous stirring under nitrogen. The black Fe_3O_4 nanoparticles were rinsed with deionized water several times, and dispersed in deionized water for further use.	128.6 mg/g
17	Magnetic graphene oxide (Zhang et al. 2016)	FA	0.7 g and 5.8 g of $NH_4Fe(SO_4)_2.12H_2O$ and $(NH_4)_2Fe(SO_4)_2.6H_2O$, respectively (molar ratio 1:1.5) were mixed in 100 ml of deionized water followed by addition of 10 ml of ammonia solution under nitrogen flow. After that, this suspension was added to a graphite oxide suspension in deionized water at stirred for 45 min at 250 rpm. The magnetic graphene oxide was then separated magnetically, washed with deionized water and alcohol several times, and dried in a vacuum oven at 60 °C.	72.38 mg/g
18	Graphene oxide nanosheets decorated with Fe_3O_4 nanoparticles (Li et al. 2012)	FA	40 mL of sulfuric acid, 1 g $NaNO_3$ and 1 g of graphite were mixed in a three-necked flask inside an acid bath followed by slow addition of 6 g $KMnO_4$ with vigorous	19.09 mg/g

(continued)

Table 6.5 (continued)

#	Adsorbent name and reference	Target pollutant	Preparation method	Adsorption capacity and/or removal %
			stirring. The solution was then moved to a water bath (35 °C) and stirred for 1 h. After addition of 80 mL of Milli-Q water, the solution was stirred for 30 min at 90 °C. After that, another 150 mL of Milli-Q water was added slowly along with 6 mL of H_2O_2. The solution was filtered, rinsed, and vacuum dried to obtain graphene oxide particles. 0.96 g and 1.86 g of $FeSO_4{\cdot}7H_2O$ and $FeCl_3{\cdot}6H_2O$ respectively were added to the graphene oxide solution under nitrogen flow at 80 °C. A mixture of $FeSO_4$ and $FeCl_3$ was added to the graphene oxide suspension, followed by the quick addition of $NH_4.OH$. The temperature was set to 85 °C and the solution pH was adjusted to 10 by adding ammonia water. After stirring for 45 min, the solution was left to cool down to room temperature. The solution was filtered followed by washing with Milli-Q water and ethanol. After being dried under vacuum at 70 °C, the final GO/Fe_3O_4 product was obtained.	

[a]Ammonia Nitrogen (AN)
[b]Humic Acid (HA)
[c]Fulvic Acid (FA)

exceeding 600 mg/g for humic acid. Incidentally, this is also one of the more simple processes, particularly attractive because it does not employ high temperatures during the synthesis, thereby having reduced energy usage and environmental impact.

6.4 Conclusion

Landfill leachate potentially has severe environmental impacts, and is a threat to the ecosystem, including humans. The variety of pollutants in the leachate emerging from landfills that have not been constructed based on appropriate guidelines are sources of possible contamination. For the treatment of landfill leachate, various processes have been used which can be generally divided into the three categories as (1) leachate channeling and recirculation, (2) biological treatment, and (3) physical/ chemical treatment. Adsorption has been one of the commonly used successful methods for leachate treatment.

Magnetic adsorbents are specific types of adsorbents with some important advantages. If properly modified, magnetic adsorbents are capable of easy separation after contaminant removal. The possibility of regeneration is another critical point that enables the successive application of magnetic adsorbents and avoids additional costs and environmental impacts.

In this chapter, humic acid, fulvic acid, and ammonia nitrogen were chosen as the three indicative pollutants in landfill leachate, and the articles pertaining to their removal with magnetic adsorbents were summarized. In particular, the synthesis processes for the adsorbents was reviewed. It was concluded that the best adsorbents in terms of performance are not necessarily the ones with the most expensive and complicated synthesis processes. On the contrary, two of the best performing adsorbents were singled out as those requiring further study due to their simplicity. They were magnetic mesoporous activated carbon (Ogata and Miura 2013) and magnetic Fe_3O_4 nanoparticles (Zhao et al. 2019b). The study using combined magnetic maghemite with nanosized hydroxyapatite (Shi et al. 2016) was also very promising, particularly because it exhibited a tremendous adsorption capacity exceeding 600 mg/g for humic acid.

References

Abbasi A, Sardroodi JJ (2019) Adsorption of O_3, SO_2 and SO_3 gas molecules on MoS_2 monolayers: a computational investigation. Appl Surf Sci 469:781–791

Achankeng E (2004) Sustainability in municipal solid waste management in Bamenda and Yaounde. University of Bamenda, Cameroon

Agamuthu P, Fauziah SH (2008) Solid waste landfilling: environmental factors and health. In: Proceedings of the EU-Asia solid waste management conference, Malaysia

Ahmed FN, Lan CQ (2012) Treatment of landfill leachate using membrane bioreactors: a review. Desalination 287:41–54

Aquino SF, Stuckey DC (2004) Soluble microbial products formation in anaerobic chemostats in the presence of toxic compounds. Water Res 38:255–266

Bashir MJK, Aziz HA, Yusoff MS, Aziz SQ (2012) Investigation of color and COD eliminations from mature semi-aerobic landfill leachate using anion-exchange resin: equilibrium and kinetic study. Environ Eng Sci 29:297–305

Bashir MJ, Aziz HA, Aziz SQ, Abu Amr SS (2013) An overview of electro-oxidation processes performance in stabilized landfill leachate treatment. Desalin Water Treat 51:2170–2184

Brennan RB, Healy MG, Morrison L, Hynes S, Norton D, Clifford E (2016) Management of landfill leachate: the legacy of European Union Directives. Waste Manag 55:355–363

Carley BN, Mavinic DS (1991) The effects of external carbon loading on nitrification and denitrification of a high-ammonia landfill leachate. Res J Water Pollut Control Fed 63:51–59

Chen H, Wageh S, Al-Ghamdi AA, Wang H, Yu J, Jiang C (2019) Hierarchical C/NiO-ZnO nanocomposite fibers with enhanced adsorption capacity for Congo red. J Colloid Interface Sci 537:736–745

Costa AM, Alfaia RGDSM, Campos JC (2019) Landfill leachate treatment in Brazil–an overview. J Environ Manag 232:110–116

Dąbrowski A (2001) Adsorption—from theory to practice. Adv Colloid Interf Sci 93:135–224

Dong C, Chen W, Liu C (2014a) Preparation of novel magnetic chitosan nanoparticle and its application for removal of humic acid from aqueous solution. Appl Surf Sci 292:1067–1076

Dong C, Chen W, Liu C, Liu Y, Liu H (2014b) Synthesis of magnetic chitosan nanoparticle and its adsorption property for humic acid from aqueous solution. Colloids Surf A Physicochem Eng Asp 446:179–189

Doulia D, Leodopoulos C, Gimouhopoulos K, Rigas F (2009) Adsorption of humic acid on acid-activated Greek bentonite. J Colloid Interface Sci 340:131–141

Esmaeili H, Ebrahimi A, Hajian M, Pourzamani HR (2012) Kinetic and isotherm studies of humic acid adsorption onto iron oxide magnetic nanoparticles in aqueous solutions. Int J Environ Health Eng 1:33

Foo KY, Hameed BH (2009) An overview of landfill leachate treatment via activated carbon adsorption process. J Hazard Mater 171:54–60

Gao T, Du Y, Li H (2019) Preparation of nitrogen-doped graphitic porous carbon towards capacitive deionization with high adsorption capacity and rate capability. Sep Purif Technol 211:233–241

Ghaedi M, Ansari A, Sahraei R (2013) ZnS: Cu nanoparticles loaded on activated carbon as novel adsorbent for kinetic, thermodynamic and isotherm studies of Reactive Orange 12 and Direct yellow 12 adsorption. Spectrochim Acta Part A Mol Biomol Spectrosc 114:687–694

Hoornweg D, Bhada-Tata P (2012) What a waste: a global review of solid waste management. World Bank 15:116

Jones MN, Bryan ND (1998) Colloidal properties of humic substances. Adv Colloid Interf Sci 78:1–48

Kamaruddin MA, Yusoff MS, Rui LM, Isa AM, Zawawi MH, Alrozi R (2017) An overview of municipal solid waste management and landfill leachate treatment: Malaysia and Asian perspectives. Environ Sci Pollut Res 24:26988–27020

Kruempelbeck I, Ehrig H-J (1999) Long-term behavior of municipal solid waste landfills in Germany. In: Sardinia 99, seventh international waste management and landfill symposium, p 51

Li W, Hua T, Zhou Q, Zhang S, Li F (2010) Treatment of stabilized landfill leachate by the combined process of coagulation/flocculation and powder activated carbon adsorption. Desalination 264:56–62

Li J, Zhang S, Chen C, Zhao G, Yang X, Li J, Wang X (2012) Removal of Cu (II) and fulvic acid by graphene oxide nanosheets decorated with Fe_3O_4 nanoparticles. ACS Appl Mater Interfaces 4:4991–5000

Li S, He M, Li Z, Li D, Pan Z (2017) Removal of humic acid from aqueous solution by magnetic multi-walled carbon nanotubes decorated with calcium. J Mol Liq 230:520–528

Liu Z, Zu Y, Meng R, Xing Z, Tan S, Zhao L, Sun T, Zhou Z (2011) Adsorption of humic acid onto carbonaceous surfaces: atomic force microscopy study. Microsc Microanal 17:1015–1021

Mubita TM, Dykstra JE, Biesheuvel PM, Van Der Wal A, Porada S (2019) Selective adsorption of nitrate over chloride in microporous carbons. Water Res 164:114885

Mukherjee S, Mukhopadhyay S, Hashim MA, Sen Gupta B (2015) Contemporary environmental issues of landfill leachate: assessment and remedies. Crit Rev Environ Sci Technol 45:472–590

Nah IW, Hwang K-Y, Shul Y-G (2007) A simple synthesis of magnetically modified zeolite. Powder Technol 177:99–101

Nah IW, Hwang KY, Shul YG, Jeon C (2008) Removal of ammonium ion from aqueous solution using magnetically modified zeolite. Environ Technol 29:633–639

Negi P, Mor S, Ravindra K (2020) Impact of landfill leachate on the groundwater quality in three cities of North India and health risk assessment. Environ Dev Sustain 22:1455–1474

Oakley SM, Jimenez R (2012) Sustainable sanitary landfills for neglected small cities in developing countries: the semi-mechanized trench method from Villanueva, Honduras. Waste Manag 32:2535–2551

Ogata T, Miura O (2013) Removal of humic substances and ammonia nitrogen in water by superconducting magnetic separation and magnetic activated carbon. IEEE Trans Appl Supercond 24:1–4

Parsonage P (1992) Coating and carrier methods for enhancing magnetic and flotation separations. Dev Miner Process 12:331–359

Peng X, Luan Z, Zhang H (2006) Montmorillonite–Cu (II)/Fe (III) oxides magnetic material as adsorbent for removal of humic acid and its thermal regeneration. Chemosphere 63:300–306

Renou S, Givaudan JG, Poulain S, Dirassouyan F, Moulin P (2008) Landfill leachate treatment: review and opportunity. J Hazard Mater 150:468–493

Reshadi MAM, Bazargan A, McKay G (2020) A review of the application of adsorbents for landfill leachate treatment: focus on magnetic adsorption. Sci Total Environ 731:138863. https://doi.org/10.1016/j.scitotenv.2020.138863

Rezaeisabzevar Y, Bazargan A, Zohourian B (2020) Landfill site selection using multi criteria decision making: focus on influential factors for comparing locations. J Environ Sci 93:170–184. https://doi.org/10.1016/j.jes.2020.02.030

Robinson HD (1995) The technical aspects of controlled waste management. A review of the composition of leachates from domestic wastes in landfill sites. Academic, London

Sepahvand S, Jonoobi M, Ashori A, Gauvin F, Brouwers HJH, Oksman K, Yu Q (2020) A promising process to modify cellulose nanofibers for carbon dioxide (CO2) adsorption. Carbohydr Polym 230:115571

Shi M, Yang L, Wei Z, Zhong W, Li S, Cui J, Wei W (2016) Humic acid removal by combining the magnetic property of maghemite with the adsorption property of nanosized hydroxyapatite. J Dispers Sci Technol 37:1724–1737

Stuart M, Lapworth D, Crane E, Hart A (2012) Review of risk from potential emerging contaminants in UK groundwater. Sci Total Environ 416:1–21

Sugawara T, Matsuura Y, Anzai T, Miura O (2016) Removal of ammonia nitrogen from water by magnetic zeolite and high-gradient magnetic separation. IEEE Trans Appl Supercond 26:1–4

Sun X, Yang L, Li Q, Zhao J, Li X, Wang X, Liu H (2014) Amino-functionalized magnetic cellulose nanocomposite as adsorbent for removal of Cr (VI): synthesis and adsorption studies. Chem Eng J 241:175–183

Tang Y, Liang S, Yu S, Gao N, Zhang J, Guo H, Wang Y (2012) Enhanced adsorption of humic acid on amine functionalized magnetic mesoporous composite microspheres. Colloids Surf A Physicochem Eng Asp 406:61–67

Tatsi AA, Zouboulis AI (2002) A field investigation of the quantity and quality of leachate from a municipal solid waste landfill in a Mediterranean climate (Thessaloniki, Greece). Adv Environ Res 6:207–219

Top S, Akkaya GK, Demir A, Yıldız Ş, Balahorli V, Bilgili MS (2019) Investigation of leachate characteristics in field-scale landfill test cells. Int J Environ Res 13:829–842

Torretta V, Ferronato N, Katsoyiannis I, Tolkou A, Airoldi M (2016) Novel and conventional technologies for landfill leachates treatment: a review. Sustainability 9:9

Trinh TTPNX, Quang DT, Tu TH, Dat NM, Linh VNP, Loan TT, Hang PT, Phuong NTL, Phong MT, Nam HM et al (2019) Fabrication, characterization, and adsorption capacity for cadmium ions of graphene aerogels. Synth Met 247:116–123

Tsarpali V, Kamilari M, Dailianis S (2012) Seasonal alterations of landfill leachate composition and toxic potency in semi-arid regions. J Hazard Mater 233:163–171
Umar M, Aziz HA, Yusoff MS (2010a) Trends in the use of Fenton, electro-Fenton and photo-Fenton for the treatment of landfill leachate. Waste Manag 30:2113–2121
Umar M, Aziz HA, Yusoff MS (2010b) Variability of parameters involved in leachate pollution index and determination of LPI from four landfills in Malaysia. Int J Chem Eng 2010:56–61
Wang J, Bi L, Ji Y, Ma H, Yin X (2014) Removal of humic acid from aqueous solution by magnetically separable polyaniline: adsorption behavior and mechanism. J Colloid Interface Sci 430:140–146
Wang J, Tian H, Ji Y (2015) Adsorption behavior and mechanism of humic acid on aminated magnetic nanoadsorbent. Sep Sci Technol 50:1285–1293
Xu ZY, Zeng GM, Yang ZH, Xiao Y, Cao M, Sun HS, Ji LL, Chen Y (2010) Biological treatment of landfill leachate with the integration of partial nitrification, anaerobic ammonium oxidation and heterotrophic denitrification. Bioresour Technol 101:79–86
Yakubu Y, Zhou J (2019) Novel approach to quantify municipal solid waste management hierarchy based on analytical hierarchy process. Int J Environ Sci Technol 16:1897–1908
Yang Q, Jiang Y, Li X, Yang Y, Hu L (2014) Magnetic-supported cucurbituril: a recyclable adsorbent for the removal of humic acid from simulated water. Bull Mater Sci 37:1167–1174
Yang N, Damgaard A, Kjeldsen P, Shao L-M, He P-J (2015) Quantification of regional leachate variance from municipal solid waste landfills in China. Waste Manag 46:362–372
Yu Y, Qiao N, Wang D, Zhu Q, Fu F, Cao R, Wang R, Liu W, Xu B (2019) Fluffy honeycomb-like activated carbon from popcorn with high surface area and well-developed porosity for ultra-high efficiency adsorption of organic dyes. Bioresour Technol 285:121340
Zhang J, Gong JL, Zenga GM, Ou XM, Jiang Y, Chang YN, Guo M, Zhang C, Liu HY (2016) Simultaneous removal of humic acid/fulvic acid and lead from landfill leachate using magnetic graphene oxide. Appl Surf Sci 370:335–350
Zhang G, Liu Y, Zheng S, Hashisho Z (2019a) Adsorption of volatile organic compounds onto natural porous minerals. J Hazard Mater 364:317–324
Zhang H, Lyu S, Zhou X, Gu H, Ma C, Wang C, Ding T, Shao Q, Liu H, Guo Z (2019b) Super light 3D hierarchical nanocellulose aerogel foam with superior oil adsorption. J Colloid Interface Sci 536:245–251
Zhang T, Shi J, Qian X, Ai Y (2019c) Temperature and gas pressure monitoring and leachate pumping tests in a newly filled MSW layer of a landfill. Int J Environ Res 13:1–19
Zhao J, Tang Z, Qiu Y, Gao X, Wan J, Bi W, Shen S, Yang J (2019a) Porous crumpled graphene with improved specific surface area based on hydrophilic pre-reduction and its adsorption performance. J Mater Sci 54:8108–8120
Zhao T, Fang M, Tang Z, Zhao X, Xie F, Wu F, Giesy JP (2019b) Effects of fulvic acid on aggregation, sedimentation, and adsorption of Fe 3 O 4 magnetic nanoparticles. Environ Sci Pollut Res 2019:1–12
Zulfikar MA, Afrita S, Wahyuningrum D, Ledyastuti M (2016) Preparation of Fe_3O_4-chitosan hybrid nano-particles used for humic acid adsorption. Environ Nanotechnology, Monit Manag 6:64–75

Chapter 7
Removal of Emerging Pollutants Using Magnetic Adsorbents

Julia Resende de Andrade, Giani de Vargas Brião, Meuris Gurgel Carlos da Silva, and Melissa Gurgel Adeodato Vieira

Contents

Abstract Adsorption is a preeminent advanced technology for effective environmental remediation of emerging pollutants and offers additional advantages with the use of state-of-the-art adsorbents with magneto-responsive features. This chapter highlights the adsorption of antibiotic, endocrine-disrupting chemical, and nonsteroidal anti-inflammatory drugs (NSAIDs) using magnetic adsorbents. The distinct magnetization and functionalization approaches highly influence the final structural and chemical properties of the adsorbents. The complexity of adsorbent matrices and the particularities of emerging pollutants led to miscellaneous adsorption performances. Of note, maximum Langmuir adsorption capacities ranged between 18.48 and 283.44 mg/g for tetracycline, between 9.13 and 253.8 mg/g for bisphenol A, and between 11.81 and 240.5 mg/g for NSAIDs. In all selected studies that inspected adsorption kinetics, the pseudo-second order was the most accurate fitting equation,

J. R. de Andrade · G. de V. Brião · M. G. C. da Silva · M. G. A. Vieira (✉)
School of Chemical Engineering, Department of Processes and Products Design, University of Campinas, Campinas, Sao Paulo, Brazil
e-mail: meuris@unicamp.br; melissag@unicamp.br

L. Meili, G. L. Dotto (eds.), *Advanced Magnetic Adsorbents for Water Treatment*, Environmental Chemistry for a Sustainable World 61,
https://doi.org/10.1007/978-3-030-64092-7_7

as indicative of the major role of chemical reactions. The isotherms of tetracycline and NSAIDs had good adjustments by Langmuir, Freundlich, Temkin, Sips, and Dubinin–Radushkevich models, while bisphenol A isotherms were mostly represented by Langmuir equation. The magnetic adsorbents showed high recoveries by external magnetic field over recycling cycles. Remarkably, some materials have been reused for up to 20 cycles and kept the original adsorptive performances. The durability of magnetic adsorbents substantiates their potential application for the uptake of emerging pollutants.

Keywords Adsorption · Magnetic adsorbents · Emerging pollutants · Antibiotics · Endocrine-disrupting chemicals · Nonsteroidal anti-inflammatory drugs

Abbreviations

ΔG	Gibbs free energy change
ΔH	Enthalpy change
AC	Activated carbon
AOPs	Advanced oxidation processes
BPA	Bisphenol A
CAS	Conventional activated sludge reactor
CNT	Carbon nanotube
DIC	Diclofenac
E2	17β-estradiol
EDCs	Endocrine-disrupting chemicals
EPs	Emerging pollutants
GAC	Granular activated carbon
GO	Graphene oxide
IBU	Ibuprofen
K_F ($mg^{1-n}L^n/g$)	Freundlich constant related to the adsorption capacity
MNPs	Magnetic nanoparticles
MOFs	Metal-organic frameworks
MS	Municipal sewage sludge
Ms emu/g	Saturation magnetization
n (–)	Freundlich constant related to the adsorption intensity
NAP	Naproxen
NOM	Natural organic matter
NP	Nanoparticle
NSAIDs	Nonsteroidal anti-inflammatory drugs
PAC	Powdered activated carbon
PC	Porous carbon
q_{max} (mg/g)	Maximum adsorption capacity from Langmuir isotherm
r-GO	Reduced graphene oxide
S_{BET} (m^2/g)	Brunauer–Emmett–Teller (BET) surface area obtained from N_2 isotherms at 77 K
TC	Tetracycline

UF	Ultrafiltration membrane separation
WHO	World Health Organization
WWTF	Wastewater treatment facility
β-CD	β-Cyclodextrin

7.1 Introduction

7.1.1 Depletion of Water Resources

There is an unsustainable pressure on global water resources led by population expansion, socioeconomic improvements, and progressing consumption patterns. According to the United Nations (UNESCO), the world demand for water has been augmenting by approximately 1% each year since the 1980s, impelled by domestic and industrial activities. As a result, water stress reaches more than 2 billion people and approximately 30% of global population has no access to safe drinking water. If environmental degradation and water pressures persist, more than half of population will be possibly threatened by 2050. Equitable access to sustainable water supply depends on the availability, accessibility, and appropriate treatment of water. The latter precondition refers to the purification, disinfection, and protection of water resources (UNESCO/WWAP 2019).

Around 3928 km^3 of freshwater is globally withdrawn per year, of which around 44% is consumed and 56% is dispersed as wastewater, including run-off from rural areas, industrial effluents, and municipal effluents. An estimated 80% of the wastewater generated worldwide is dropped into the environment lacking satisfactory treatment. Thus, the dumping of untreated or inadequately treated effluents is a paramount source of water contamination. A broad spectrum of natural, organic, and inorganic substances has been traced in water compartments and some of them present acute and chronic toxic effects, and may be poisonous, carcinogenic, and mutagenic. The widespread water pollution is harmful to human health, puts environment at risk, and negatively affects the global economy (UNESCO/WWAP 2017).

7.1.2 Contamination by Emerging Contaminants

In recent years, there is great interest on the so-called emerging pollutants (EPs), which consist of synthetic or naturally occurring chemicals or microorganisms that are not historically regulated or monitored and which fate, behavior, and ecotoxicological consequences are generally unknown. Up to now, over 1000 emerging substances have been recognized as EPs, including pharmaceuticals, surfactants, steroids, hormones, personal care products, pesticides, herbicides, and flame retardants (NORMAN Network 2016).

Advances in modern analytical methods have allowed the detection and quantification of EPs in water and effluents (Gros et al. 2009; Al Aukidy et al. 2012). EPs and their metabolites have traced levels varying between ng/L and mg/L in treated or untreated effluents from urban and industrial wastewater treatment facilities (WWTFs), and agricultural drainage waters (Rosal et al. 2010; Loos et al. 2013). EPs have also been encountered in surface water, ground water, and drinking water across the globe (Ebele et al. 2017). If in one hand the ubiquity of EPs in water is well recognized, the latent impacts of such chemicals on people's well-being and the environment still instigate scientific research and have become an important global concern (Ginebreda et al. 2010).

Environmental persistence, bioaccumulation, and toxicity are some of the problems arising from chemical contaminants in water matrix. For example, trace levels of pharmaceuticals and metabolites can be biologically active and influence nontarget organisms and all aquatic biota that are accidentally exposed to them, that is, they represent great risks of ecotoxicological effects (Ebele et al. 2017). A distinct kind of toxicity is presented by endocrine-disrupting EPs, that is, substances that modify the function(s) of the endocrine system and thus cause adverse health effects, such as hormone-related cancers, reproductive dysfunctions, neurodevelopmental, and metabolic disorders. Endocrine disruptors encompass various natural substances (e.g., phytoestrogens and natural hormones) as well as anthropogenic chemicals (e.g., pharmaceuticals, personal care product ingredients, pesticides, plasticizers, and product additives) (WHO/UNEP 2013). The list of endocrine disruptors is being constantly updated due to discoveries on new potential sources, synthesis of novel chemicals, and changes in consumption patterns.

Another important issue regarding EPs contamination is the expansion of bacterial pathogen resistance. Recently, the World Health Organization (WHO) prepared a global priority list of antibiotic-resistant pathogens, which includes 12 families of bacteria that threaten human health (WHO 2017a, b). The input and consequent accumulation of numerous antimicrobial agents can trigger the multiplication and dissemination of nonpathogenic bacteria, which serve as reservoirs of genes in the environment. These genes may convert into antibiotic-resistance genes in pathogenic bacteria (Baquero et al. 2008). Therefore, preventing antibiotic release is essential to curb the proliferation of resistant pathogenic bacteria and their associated resistance genes (Martinez 2009).

The mixtures of pharmaceutical residues that are dumped into aquatic matrices are quite complex because of the regular and simultaneous use of a wide range of medicines, their particular metabolism, their partial abatement in WWTFs, and the formation of metabolites and transformation products. Although low individual levels (ng/L) of pharmaceuticals can induce no or nonsignificant ecotoxicological effects, multichemical mixtures may have adverse combined outcomes due to synergistic, additive, or antagonistic interactions (Fent et al. 2006; Vasquez et al. 2014). The potential effects of exposure to pharmaceutical cocktails on living organisms and the environment are barely known; however, some initiatives have been placed to develop guidelines for assessing the risks of mixtures of pharmaceutical (Benfenati et al. 2003; Kortenkamp et al. 2009; European Commission 2012; Michael et al. 2014; WHO 2017a, b).

7.1.3 Wastewater Treatment Approaches

The ubiquitous water pollution demands appropriate management to palliate the degradation of water quality, which compromises the availability and exacerbates water scarcity scenario. WWTFs are chiefly conceived to treat compounds that are continually received in average levels of mg/L, including particulate matter, carbonaceous substances, nutrients, and pathogens. These macropollutants are substantially and systematically removed, but the same fate does not happen to microcontaminants found in lower concentrations (ng/L or μg/L). Most of existing traditional treatment options are not effective in completely removing compounds such as pharmaceuticals and hormones (Luo et al. 2014; Krzeminski et al. 2019). Thus, the outflows from WWTFs are important sources of EPs in aquatic environment. Up-to-date approaches are imperative to complement current water treatment methods to mitigate microcontaminant loads.

WWTFs traditionally comprise preliminary treatments (bar screening and grit removal), primary treatment (sedimentation tank), and secondary biological suspended mass reactor (conventional activated sludge, CAS) (Verlicchi et al. 2012). Investments on advanced tertiary treatments, such as membrane separation, activated carbon adsorption, and advanced oxidation processes, are subordinated to public or environmental demands for higher quality water, such as for reuse purposes (Luo et al. 2014).

The preliminary and primary treatment processes aim to retain suspended solids and have shown minimal eliminations of microcontaminants from effluents (Radjenović et al. 2009; Behera et al. 2011; Gao et al. 2012a; Gracia-Lor et al. 2012). Better outcomes are observed for secondary treatment by CAS, which can reduce microcontaminant load either through abiotic (sorption onto sludge) or biotic processes (biotransformation/biodegradation) (Radjenović et al. 2009; Luo et al. 2014). Previous studies revealed that biodegradation has a predominant mechanistic role over sorption in removing pharmaceuticals (Gao et al. 2012a; Verlicchi et al. 2012). However, the removal rates vary noticeably due to seasonal fluctuations in plant performance and inlet concentrations (Gracia-Lor et al. 2012). In certain cases, even negative pharmaceutical removal efficiencies occur due to increasing levels of persistent compounds, which can be formed by deconjugation or accumulate in the treatment system. Moreover, the overall effluent quality can be aggravated in the event of partial biotransformations that generate intermediates more toxic than the parent substances (Krzeminski et al. 2019).

The worrisome issue of WWTFs as relevant sources of EPs can be potentially addressed by the implementation of novel effective tertiary processes, redesign, and improvement of current treatment systems and/or optimization of operational parameters of existing biological treatments in WWTFs (Krzeminski et al. 2019). Under this canopy, developing new and effective advanced wastewater treatment methods is essential to complement the conventional ones, in order to sustain long-term improvements in mitigating EPs (Dubey et al. 2017). Activated carbon (AC) adsorption, ozonation, membrane technology, and advanced oxidation processes (AOPs) are among the most preeminent options (Loos et al. 2013; Plakas et al. 2015). AC treatment and ozonation can substantially diminish the loads of EPs in effluents (Lee et al. 2013; Kårelid et al. 2017; Real et al. 2017) and have already been gradually exploited at full scale in Switzerland. The Swiss National Policy conjectures upgrading 100 municipal WWTFs by 2040 to halve the micropollutant load in surface waters (Logar et al. 2014; FOEN 2015).

Nevertheless, an important drawback of ozonation technology is the generation of oxidation/transformation by-products, which removal requires additional posttreatment steps (e.g., biological slow sand filter). The generation of by-products also impairs AOPs, such as UV/H_2O_2, photo-Fenton, and UV/TiO_2 systems (Rizzo et al. 2019). Recent papers deal with AOPs for EPs elimination (Jallouli et al. 2018; Kang et al. 2018; Miklos et al. 2018), but full-scale performance still needs research. Membrane technology, such as nanofiltration and reverse osmosis, offers great prospects, but requires treatments for solid removal and disposal of concentrate waste stream, and generally has high-energy consumption and variable rejection percentages (Rizzo et al. 2019).

Adsorption does not form by-products, unlike ozonation and AOPs, and has lower energy demands. However, it provides null disinfection, which can be partially overcome by coupling powdered AC (PAC) with ultrafiltration membrane separation (UF) (Rizzo et al. 2019). The PAC/UF system displayed superior sustainability performance for environmental, economic, and social factors in comparison to other approaches, including reverse osmosis, ozone/ultraviolet-light oxidation, and photocatalytic membrane reactor (Plakas et al. 2015). It is worth mentioning that adsorption is a separation process, and thereby, the pollutants are not actually removed but rather transferred to the solid adsorbent, which entails adequate disposal and so extra costs (Adeleye et al. 2016).

7.1.4 Adsorption Technology

Adsorption is defined as the mass transference of molecules (adsorbate) from a fluid phase to the surface of a solid phase (adsorbent) (Sing et al. 1985). The adsorption method relies on several factors, primarily the properties of the adsorbent (e.g., features and chemistry of the surface) and of the target pollutants (e.g., hydrophobicity, chemical configuration, and ionization constant). The adsorption capacity also relies on pH and temperature, wastewater quality (competitive adsorption), and operational parameters.

On applying adsorption, it is imperative to relate the adsorbed amounts and the characteristic process parameters. Information about adsorption equilibrium, adsorption kinetics, and adsorption dynamics compose the basis of adsorption theory (Worch 2012). The adsorption equilibrium relates the adsorbed amounts and the liquid-phase concentration at certain temperature. Adsorption equilibrium data have been traditionally described by isotherm models, such as those from Langmuir (1918), Freundlich (1906), Dubinin (1960), Temkin and Pyzhev (1940), and Sips (1948). Adsorption equilibrium is not attained immediately, and the time dependence of the process is denoted as adsorption kinetics. The kinetic equations of pseudo-first order (Lagergren 1898) and pseudo-second order (Ho and McKay 1998) have been employed to represent kinetic data. Adsorption equilibrium and kinetics are determined in batch tests, followed by continuous tests on fixed-bed columns for the adsorption dynamics (Worch 2012).

Most traditional adsorption processes employ ACs, which are endowed with well-developed porosity and surface functionalities that confer great trapping capacities for a myriad of contaminants (Yang 2003). ACs are generally applied in the form of powdered AC (PAC) in contact reactors or as granular AC (GAC) in fixed-bed columns (Snyder et al. 2007). PAC cannot be regenerated and have to be withdrawn from solution after saturation. Prolonged lifetimes of PAC can be attained by recycling into CAS tanks of WWTF, which can also boost contaminant removal. The recirculated PAC can be burned up jointly with the waste sludge, as executed in Switzerland (Boehler et al. 2012). Dissimilar to PAC, GAC can be regenerated for removing the adsorbed compounds. If the adsorbent is not regenerated on site, it has to be dealt as hazardous waste (Adeleye et al. 2016). Another relevant matter of ACs is that, while their raw materials can be low cost, the fabrication involves a high primary energy demand (Rizzo et al. 2019).

There is a great scientific effort in searching and developing alternative adsorbents to AC, such as zeolites (Garcia et al. 2019), clay minerals (Oliveira et al. 2018; Maia et al. 2019), and inactive biomass and waste (Araujo et al. 2018; Bazarin et al. 2019). Even though these adsorbents display utmost potential for effective remediation of pharmaceuticals, they suffer from hard separation in continuous water/wastewater treatment systems. In recent times, the use of magnetism has been acknowledged as a promising approach to overcome such limitation in adsorption processes.

7.1.5 Adsorption Technology Using Magnetic Adsorbents

Magnetic adsorbents comprise materials with embedded magnetic particles, mainly metal oxides that are more stable to oxidation than pure metals (Ni, Co, and Fe) (Philippova et al. 2011). Ascribable to the metal component, those complex adsorbents can be directly gathered from the fluid by magnetism, what improves the efficiency of water/wastewater treatments and cut extra costs with separation steps.

Iron oxides [MFe_2O_4, where M is a transition metal, such as Mn, Fe, Co, Zn] have been chosen as magnetic carriers owing to high saturation magnetization, small electrical loss, physicochemical stability, and low inherent toxicity (Pereira et al. 2012; Luo et al. 2017). The magnetic characteristics of metal oxides are primarily governed by particle size, shape, crystallinity, and chemical composition (Philippova et al. 2011). The size of iron oxide particles is a key factor that influences the coercivity, which measures the material resistance to become demagnetized after reaching saturation magnetization. Coercivity is null only for tiny particles with single-domain structure, in which the groups of spins all point to a common direction and act cooperatively. These particles are denoted superparamagnetic. Conversely, larger single-domain particles, in addition to particles with multidomain structure, own some coercivity, so that magnetization reversal occurs by nucleation and motion of the domain walls. They are called ferromagnetic. The nano-sized particles of iron oxides (5–15 nm size) are superparamagnetic, whereas micron-sized particles are ferromagnetic (Philippova et al. 2011).

Aiming at adsorption processes with adsorbent separation by means of external magnetic field, the adsorbent particles should normally present superparamagnetic features, so that they can be easily magnetized without remanent magnetization (free from magnetic memory). In other words, superparamagnetic particles magnetize and tend to promptly agglomerate under the action of magnetic field. After removing the field, these particles easily redisperse, because they lack magnetic memory. This fact favors rapid sorption–desorption processes (Franzreb et al. 2006).

First records of applying magnetism for water purification purposes date back to 1870 (Mehta et al. 2015). In the 1940s, pure magnetic iron oxides started being employed for adsorption and uptake of dissolved and colloidal biological compounds from wastewater (Franzreb et al. 2006). Since then, different synthesis process of magnetic particles has been experimented, including coprecipitation, hydrothermal, and solvothermal approaches (Kharissova et al. 2015). The new generation of magnetic adsorbents target enhanced adsorption efficiencies for a myriad of contaminants. This chapter focuses on discussing the application of magnetic materials for the adsorption of emerging organic contaminants.

7.2 Application of Magnetic Adsorbents in the Removal of Emerging Contaminants

This section aims at examining the adsorption of EPs, namely antibiotics, endocrine-disrupting chemicals, and nonsteroidal anti-inflammatory drugs, using different advanced magnetic adsorbents.

7.2.1 Antibiotics

Antibiotics are complex molecules with specific antimicrobial action on bacteria or fungi in humans and animals. Antibiotic pollution is of high concern because of latent effects on the structure and function of microorganism populations. The direct impacts include bactericidal and bacteriostatic activities with consequent decline of some environmental microorganisms. Indirectly, antibiotics can perform as selective force and develop multiresistant bacteria, which pose serious risks to human and veterinary health (Grenni et al. 2018). Hence, cleaning-up antibiotic compounds from water is fundamental.

In this chapter, we chose tetracyclines as model antibiotic pollutants because they are among the most regularly used, especially in livestock and poultry industries (Ji et al. 2009; Zhu et al. 2014a). Tetracyclines own a broad spectrum of activity, have lower costs than other antibiotics, and are employed not only for therapeutic purposes but also as growth promoters. More than 20 compounds from tetracycline family are available, and tetracycline, chlortetracycline, oxytetracycline, and doxycycline are the most commonly consumed in agriculture (Granados-Chinchilla and Rodríguez 2017).

Tetracycline (TC, $C_{22}H_{24}N_2O_8$) has high water solubility and prolonged half-life in the environment (Li et al. 2010). Triggered by intense use, TC levels in water tend to increase with time. Thus, the adsorption of this antibiotic is a hot topic of research, although it is a complex process given the amphoteric nature of TC molecules, which can occur as cation ($pH < 3.3$), zwitterion ($3.3 < pH < 7.3$), and anion ($pH > 7.3$) (Kulshrestha et al. 2004). TC adsorption has been studied using carbon-based solids, like powdered activated carbons (Choi et al. 2008), biochar (Jing et al. 2014), graphene oxide (Gao et al. 2012b), carbon nanotubes (Zhang et al. 2011a), and porous carbons (Zhu et al. 2014b; Zhang et al. 2019). Strong interactions have been verified between the TC molecules and the surface of carbonaceous adsorbents, such as van der Waals forces, π–π electron donor–acceptor interactions, and cation–π bonding (Yang et al. 2011). To overcome the limitation of difficult collection of the carbon-based adsorbents, aggravated by their hydrophilic nature, the introduction of magnetic properties has been pursued (Yang et al. 2011). Table 7.1 presents some magnetic adsorbents described in literature for TC removal.

The benefit of magnetic character in TC removal can be demonstrated by comparing pure and modified porous carbon (PC) produced from hydrochar, which is a residue of hydrothermal carbonization of biomass. The raw waste hydrochar-derived PC showed a surface area from Brunauer–Emmett–Teller (BET) method as $S_{BET} = 316\ m^2/g$ and Freundlich adsorption capacity of TC as $K_F = 8.57\ mg^{1-n}L^n/g$ ($1/n = 0.404$) (Zhu et al. 2014b). However, the insertion of $\gamma\text{-}Fe_2O_3$ particles into the waste hydrochar-derived PC (MPC) raised S_{BET} to $349\ m^2/g$ and improved K_F to $11.78\ mg^{1-n}L^n/g$ ($1/n = 0.23$) (Zhu et al. 2014a). The enhanced surface area and porosity, as well as the graphite layers of MPC, might have favored TC uptake. Although MPC exhibited relatively low saturation magnetization of 0.7 emu/g, it could be isolated by an external magnetic field. In addition,

Table 7.1 Tetracycline antibiotic removal by distinct magnetic adsorbents (S_{BET} = BET surface area; Ms = saturation magnetization) in equilibrium tests (T = temperature; pH; contact time; agitation; C_i = initial concentration range; C_{ads} = adsorbent dosage) for Langmuir maximum adsorption capacity (q_{max}) and adsorbent reusability in cycles

Magnetic adsorbent	Magnetization method	S_{BET} (m^2/g)	Ms (emu/g)	Experimental conditions	q_{max} (mg/g)	Reusability	References
Fe_2O_3/hydrochar-derived porous carbon, MPC	Thermal pyrolysis at 700 °C of $FeCl_3$ pretreated hydrochar	349	0.7	T = 298 K; pH 3; 120 h; C_i = 5–80 mg/L; C_{ads} = 50 g/L	25.44	–	Zhu et al. (2014a)
Ni/auricularia-derived porous carbon, AMPC	Simultaneous magnetization and activation to encapsulate Ni and impregnate KOH, respectively	823.2	11	T = 298 K; pH natural; 12 h; 120 rpm; C_i = 40–280 mg/L	136.12	6 cycles q_e = 243 mg/g (1st) q_e = 114 mg/g (6th) 0.2 M NaOH	Xie et al. (2019)
$MnFe_2O_4$/powdered activated carbon, $MnFe_2O_4$/PAC	Coprecipitation reaction with $MnCl_2$ and $FeCl_3$ in the presence of coconut powdered activated carbon	512	10.5	T = 298 K; pH 5; 24 h; C_i = 48–960-mg/L; C_{ads} = 1 g/L	283.44	–	Shao et al. (2012)
Alkali-acid combined method modified sewage sludge biochar	Municipal sewage sludge containing polyferric sulfate flocculant (Fe source) pyrolyzed at 800 °C	202.52	~1	T = 298 k; pH 7; 160 rpm; C_i = 50–800-mg/L; C_{ads} = 1 g/L	187.27	5 cycles q_e ~ 97 mg/g (1st and 5th) NaOH	Tang et al. (2018)
Fe_3O_4/thiourea-dioxide–reduced graphene oxide, r-MGO	Coprecipitation reaction with $FeCl_3$ and $FeSO_4$	–	4.38	T = 303 K; pH = 4; 24 h; C_i = 5–100 mg/L; C_{ads} = 70 g/L	980	5 cycles q_e = 148 mg/g (1st) q_e = 55 mg/g (5th)	Yang et al. (2017)
α-Fe_2O_3/thermally reduced graphene oxide, α-Fe_2O_3/r-GO	Simultaneous thermal reduction of GO at 700 °C and synthesis of α-Fe_2O_3 nanoparticles with $FeCl_3.6H_2O$ as Fe precursor	35	7.15	T = 298 K; pH 4; 2 h; C_i = 2–25 mg/L; C_{ads} = 0.7 g/L	18.48	–	Huízar-Félix et al. (2019)

DMSA-Fe_3O_4/ polyethylenimine-grafted–reduced graphene oxide, Fe_3O_4-rGO	Covalent bonding of DMSA-modified Fe_3O_4 nanoparticles onto the surface of polyethylenimine-reduced GO	–	28	T = 298 K; 24 h; C_i = 12.5–75 mg/L; C_{ads} = 0.53 g/L	–	–	Zhang et al. (2011b)
NH_2-Fe_3O_4/graphene oxide	Covalent bonding of NH_2-modified Fe_3O_4 nanoparticles onto the surface of GO	–	–	T ~ 298 K; 10 min; C_i = 0–100 mg/L; C_{ads} = 2 g/L	39.1	–	Lin et al. (2013)
Fe_3O_4/macro-reticulated and cross-linked chitosan, MRC	Coprecipitation reaction with $FeCl_3$ and $FeSO_4$ and incorporation of Fe_3O_4 particles into chitosan matrix	47.95	–	2 h; C_i = 25–200 mg/L; C_{ads} = 2 g/L	20.7	–	Oladoja et al. (2014)
Fe3O4@SiO_2-chitosan/GO, MSCG	Solvothermal synthesis of Fe_3O_4 particles, coating with SiO_2, grafting Fe_3O_4@SiO_2 particles into chitosan matrix, and assembling on the surface of graphene oxide	–	10.5	T = 298 K; pH 6; 24 h; C_i = 0–96 mg/L; C_{ads} = 0.4 g/L	119	5 cycles q_e ~ 67 mmol/kg (1st) q_e ~ 59 mmol/kg (5th) 0.05 M NaOH	Huang et al. (2017)
NH_2-Fe_3O_4/coordination complex modified polyoxometalate, NH2-Fe_3O_4/ $CuSiW_{12}$NP	Colloidal blending of $CuSiW_{12}$ and NH_2-modified Fe_3O_4 nanoparticles	–	8.19	T = 298 K; pH = 6.8; 24 h; C_i = 10–300-mg/L; C_{ads} = 1 g/L	–	5 cycles 91% (1st) 80% (5th) Ethanol	Ou et al. (2016)

TC kinetics onto MPC was consistently fitted by pseudo-second-order model. The negative values of Gibbs free energy change (ΔG) corroborated a spontaneous adsorption, while the value of positive enthalpy change (ΔH) revealed that the process is favored at higher temperatures (endothermic).

The recent paper by Xie et al. (2019) proposes the synthesis of a magnetic PC from the fungus Auricularia, using $NiCl_2.6H_2O$ as magnetic precursor and KOH as activation agent. The impregnation of KOH significantly improved the surface area and porosity of the material. As a consequence, its adsorption ability for TC is higher than many other magnetic materials (Table 7.1). The satisfactory fitting of Langmuir model indicated a mono-layer adsorption process, in which adsorbate molecules on adjacent sites do not interact with each other. Moreover, the process was verified to be endothermic and spontaneous.

Shao et al. (2012) prepared a magnetic coconut PAC with joint advantages of high adsorbability and effective magnetic separability. Despite the fact that magnetic $MnFe_2O_4$/PAC composite owed S_{BET} around 36% smaller than pure PAC, its adsorption capacity for TC was not expressively affected. Indeed, the maximum adsorption capacity from Langmuir equation (q_{max}) for the composite was almost 5% higher than that of pure PAC. The enhanced performance, especially at high concentrations, was associated to the greater heterogeneity of the adsorbent surface after the introduction of $MnFe_2O_4$ particles. This fact not only enabled the complexation of TC molecules with the surface metals but may also have shifted the uptake mechanism from monolayer to multilayer. The good correlation of Freundlich isotherm model ($R^2 > 0.99$) endorses the presence of adsorption sites with different energies in the heterogeneous surface of composite. The adsorption of TC over $MnFe_2O_4$/PAC was thermodynamically spontaneous, endothermic, and accompanied an increased degree of freedom.

Municipal sewage sludge (MS) is rich in organic matter and so a potential precursor of biochar. Worth noting, MS from WWTFs is generally fully mixed with flocculants like polyferric sulfate, which acts as Fe source and endows intrinsic magnetism to MS-derived biochars. Tang et al. (2018) tested TC removal using a magnetic MS-derived biochar, modified with soda lye and nitric acid to preserve the magnetism. Analyzes showed that Fe was well wrapped within the adsorbent, what prevents fall off or corrosion. On comparing pseudo-first-order model and pseudo-second-order model, the latter offered better correlation to TC kinetics. As both Langmuir and Freundlich models displayed good fittings to the isotherms, physisorption and chemisorption mechanisms were speculated to have relevant roles on TC adsorption onto MS-derived biochar. The thermodynamic behavior suggested a spontaneous and endothermic adsorption process. The adsorbent showed constant adsorption ability in deionized water, tap water, or lake water and had good regeneration performance with insignificant losses after five consecutive cycles.

Graphene oxide (GO) is an engineered nanomaterial with characteristic single-layer sp^2-carbon network and great environmental remediation appeal due to hydrophobicity, relatively high specific area and numerous oxygen functionalities (Mukherjee et al. 2016). Concerning TC uptake, adsorption rates over 71% have been obtained using pristine GO adsorbent (Gao et al. 2012b). The antibiotic was shown to easily adsorb onto GO via intense π–π stacking interactions owing to the four aromatic rings of TC molecule (Lin et al. 2013). Modifications of GO have been proposed for improved adsorptive properties (e.g., chemical selectiveness, solubility, thermal, and electronic conductivity) and reusability in consecutive cycles (Huízar-Félix et al. 2019). One of the changes is reducing GO by thermal or chemical treatments to obtain so-called reduced GO (r-GO). Song et al. (2016) prepared r-GO by adding hydrazine monohydrate and obtained $q_{max} = 219.1$ mg/g (298 K and pH = 6) for TC, which was verified to have stronger adsorbability than the antibiotic sulfamethazine ($q_{max} = 174.4$ mg/g).

As alternative to hydrazine reductant, which is unstable and carcinogenic, Yang et al. (2017) proposed thiourea dioxide to produce a reduced magnetic GO (TDMGO) which was compared to magnetic GO (MGO). The TDMGO exhibited higher adsorption capacity than MGO ascribable to the abundant nitrogen functionalities that were introduced into sp^2-C network by thiourea dioxide. The optimum pH for TC uptake was selected as 4, at which TC molecules are in the zwitterionic form. The adsorption kinetics was satisfactorily correlated ($R^2 > 0.98$) by pseudo-second-order equation, and film diffusion was identified as the major rate-controlling step. Langmuir isotherm best represented equilibrium data and q_{max} was 980 mg/g (303 K). The process spontaneity and endothermicity were unveiled by the thermodynamic parameters. The saturation magnetization of TDMGO was Ms = 4.38 emu/g. After five consecutive cycles, the adsorption capacity of TDMGO significantly diminished by almost 63%.

Recently, Huízar-Félix et al. (2019) fabricated r-GO by thermal reduction through heating GO up to 700 °C, what causes dramatic withdraw of oxygen-functional groups from GO surface. The remaining moieties on the edges of the adsorbent facilitate electrostatic interactions, π–π interactions, and cation–π binding to organic molecules. In parallel, the authors prepared a reduced magnetic GO by the simultaneous pyrolysis of GO and synthesis of α-Fe_2O_3 nanoparticles (NPs) using $FeCl_3.6H_2O$ as iron salt precursor. The as-synthesized α-Fe_2O_3/r-GO composite showed TC adsorption capacity as $q_{max} = 9.7$ mg/g, while r-GO had $q_{max} = 44.2$ mg/g. The abridged adsorbability of α-Fe_2O_3/r-GO in comparison to r-GO was associated to the less electrostatic adsorbent/adsorbate interactions. Moreover, the values of Freundlich heterogeneity factor n were in 1–2 range for r-GO, but surpassed 2 for the hybrid nanomaterial. The greater heterogeneity of the surface of α-Fe_2O_3/r-GO, owing to the coexistence of two phases with distinct chemistries in the layer, might have further disturbed TC adsorption.

Alternatively to in situ reduction of iron salt precursors, the assembly of magneto-responsive NPs on GO surface has been adopted to obtain magnetic GO composites. For instance, Zhang et al. (2011b) proposed an easy method for depositing Fe_3O_4 NPs of controlled size and morphology on r-GO matrix. In brief, functionalized r-GO was obtained by chemical reduction using polyethylenimine, and Fe_3O_4 NPs were prepared by thermal decomposition of Fe precursor and further functionalized by 2,3-dimercaptosuccinnic acid. The Fe_3O_4 NPs were assembled in the surface of r-GO through covalent bonding, and as-prepared Fe_3O_4-rGO composites were used for TC adsorption. The equilibrium pattern followed Freundlich profile, and the maximum amount adsorbed was 95 mg/g at 298 K. Sonication was not effective for TC desorption from GO sheets, suggesting a strong adsorbability over Fe_3O_4-rGO. The high Ms = 28 emu/g of Fe_3O_4-rGO enabled the easy retrieval from solution using magnetic field.

Interestingly, Lin et al. (2013) developed GO-based magnetic composites by the covalent attachment of amine-functionalized magnetic magneto-responsive NPs on GO surface. The study showed that equilibrium of TC adsorption was achieved remarkably fast (10 min) and the kinetics pattern well-followed pseudo-second-order model. The equilibrium was satisfactorily represented by Langmuir isotherm with q_{max} = 39.1 mg/g. The ability of the adsorbent in TC remediation was verified not only in deionized water but also in environmental waters (mineral water and river water samples).

Instead of covalent attachment of magnetic NPs to GO, Hazell et al. (2016) used the Coulombic attraction between GO and a paramagnetic electrolyte based on $FeCl_4^-$ to generate magnetic responses. In that case, GO was firstly dispersed into the contaminated solution for TC adsorption, followed by the insertion of the magnetically active polyelectrolyte, which induced a strong flocculation. In the end, GO could be collected by magnetic field due to the polyelectrolyte adsorbed on its surface. Apparently, the affinity of TC for GO was not undermined by the electrolyte co-presence in the system.

Surface modifications via covalent binding or physical coating have also been researched in order to overcome common limitations of raw magnetic Fe_3O_4 NPs, such as poor dispersion and oxidation. For instance, Oladoja et al. (2014) incorporated Fe_3O_4 into macro-reticulated and crosslinked chitosan frame and the adsorbent was referred as MRC. Gastropod shells were employed as the pore-forming agent and glutaraldehyde as the chitosan crosslinker. The multicomponent facet conferred MRC a heterogeneous surface, which had distinct affinities for TC molecules, as demonstrated by the good correlation of Freundlich model to equilibrium data ($R^2 > 0.99$). Complementarily, TC kinetics over MRC well-obeyed pseudo-second-order kinetics ($R^2 > 0.99$), as an indicative of chemisorption. Acidic conditions (pH 3) were verified for improved adsorption performance of TC due to concomitant interactions established by permanent and induced dipoles, hydrogen bonding, electrostatic forces, and cation exchange.

More recently, Huang et al. (2017) prepared magnetic Fe_3O_4@SiO_2 nanoparticles with developed core–shell structure, which were grafted into chitosan matrix and assembled on the surface of GO layers. The Fe_3O_4@SiO_2-chitosan/GO nanocomposite (MSCG) presented Ms = 10.5 emu/g and could be effectively magnetically collected. The impact of pH on TC uptake onto MSCG was examined over 3–10 range and pH 6 conferred the uppermost adsorption capacity (~32.4 mg/g). At pH = 6, there is a diminished electrostatic repulsion between the adsorbent's surface (negatively charged) and TC molecules (predominantly neutral). In the presence of copper, TC uptake massively increases to 88 mg/g due to the complexation of TC molecules with Cu(II). The formed complexes had higher affinity for the adsorbent. Moreover, the time required to reach equilibrium significantly halved from 480 min to 240 min with the coexistence of Cu(II). Relative to pseudo-first-order model, pseudo-second-order rate law represented kinetics more accurately. The isotherms were better described by Freundlich equation rather than Langmuir equation, which points to a multilayer adsorption process. After five consecutive cycles, MSCG nanocomposite still removed over 85% of TC from solution.

TC adsorption was studied by Ou et al. (2016) using a magnetic NPs of amino functionalized Fe_3O_4 loaded on coordination complex modified polyoxometalates (NH_2-Fe_3O_4/$CuSiW_{12}$NP). The new adsorbent presented Ms = 8.19 emu/g, which endows effective external magnetic separation. The pseudo-second-order model ($R^2 > 0.99$) was verified to best represent TC adsorption kinetics. The isotherms were best described by Temkin model ($R^2 > 0.99$), what points to the key role of electrostatic interactions. From the thermodynamic point of view, the process was characteristically spontaneous and endothermic.

In summary, all papers herein examined about TC removal using magnetic adsorbents showed the superior representativeness of pseudo-second-order model over pseudo-first-order model for the adsorption kinetics. So, it is deducible that TC adsorption processes onto magnetic adsorbents are controlled mostly by chemical reaction. Moreover, the spontaneous and endothermic character of the processes was identified without exception in all studies that performed thermodynamic studies. On the other hand, adsorption isotherms were well correlated not only by Langmuir equation but also by Freundlich and Temkin models.

7.2.2 *Endocrine-Disrupting Chemicals (EDCs)*

Endocrine-disrupting chemicals (EDCs) are natural or synthetic substances that can mimic or antagonize the biological activity of natural hormones. A myriad of papers describe exposure-related issues for endocrine-disrupting chemicals (EDCs) in wildlife (Cowin et al. 2010) and humans (Colborn 1995).

As representative EDC, bisphenol A (BPA) is extensively spent as manufacturing intermediate of epoxy resins, polycarbonate plastic, polysulfones, and others (Sohoni et al. 2001). BPA metabolites have estrogenic activity and have been linked to reduction on the number and activity of sperm (Li et al. 2015). Surface waters from Brazil have presented BPA levels ranging from 204 to 13,016 ng/L (Montagner and Jardim 2011). The rampant occurrence of BPA in the environment is worrisome. In response, research on BPA adsorption using innovative materials grows exponentially.

Table 7.2 lists magnetic adsorbents that have been tried for BPA uptake from aqueous solutions. First of all, there is a significant attention from the scientific community in carbon nanotubes (CNTs) for removing EDCs from water and/or effluents (Zaib et al. 2012; Sun and Zhou 2014; Senin et al. 2018). This is because CNTs contain a sp^2-hybridized carbon framework in tubular structure and unique physicochemical properties. Nonetheless, the release of pristine nano-level-sized

Table 7.2 Bisphenol A removal by distinct magnetic adsorbents (S_{BET} = BET surface area; Ms = saturation magnetization) in equilibrium tests (T = temperature; pH; contact time; agitation; C_i = initial concentration range; C_{ads} = adsorbent dosage) for Langmuir maximum adsorption capacity (q_{max}) and adsorbent reusability in cycles

Magnetic adsorbent	Magnetization method	S_{BET} (m^2/g)	Ms (emu/g)	Experimental conditions	q_{max} (mg/g)	Reusability	References
Carbon nanotubes (CNTs)/Fe_3O_4	Solvothermal synthesis of Fe_3O_4 particles in the presence of multiwalled CNTs (20 wt.% CNT:Fe_3O_4)	34.3	125	T = 298 K; pH = 6.2; 130 rpm; 2 h; C_i = 25–400 mg/L; C_{ads} = 5 g/L	45.31	5 cycles q_e ~ 22 mg/g (5th) Fe^{2+}/H_2O_2 oxidation	Li et al. (2015)
Fe_3O_4@Co/Ni-LDH	Solvothermal synthesis of Fe_3O_4 particles and coating with ZIF-67-derived layered double hydroxide	128.13	14.2	T = room temperature; pH = 7; 250 rpm; C_i = 0–300 mg/L; C_{ads} = 0.5 g/L	238.96	5 cycles ~67% (1st and 5th) Methanol	Li et al. (2020a)
Fe_3O_4@polyaniline (PANI)	Solvothermal synthesis of Fe_3O_4 particles and two-step polymerization of polyaniline	–	40.4	T = 298 K; pH natural; 270 rpm; 5 h; C_i = 1–20-mg/L; C_{ads} = 0.8 g/L	9.13	–	Zhou et al. (2016)
Fe_3O_4@2-Vinylpyridine, Mag-PVP	Solvothermal synthesis of Fe_3O_4 particles, modification with γ-methacryloxypropyltrimethoxysilane, and coating with 2-vinylpyridine	–	44.5	T = 293 K; 150 rpm; 4 h; C_i = 2.5–50 mg/L; C_{ads} = 0.2 g/L	115.87	12 cycles q_e ~ 9.9 mg/g (1st) q_e ~ 9.5 mg/g (12th) Methanol	Li et al. (2018)
Fe_3O_4/vermiculite modified with poly (trimesoyl chloride-melamine)	Coprecipitation reaction with $FeCl_3$ and $FeCl_2$ in the presence of vermiculite and interfacial polymerization with poly (trimesoyl chloride- melamine)	–	–	120 rpm	237.7	7 cycles 97% (1st) 60% (7th) Methanol	Saleh et al. (2019)

β-cyclodextrin capped graphene-magnetite nanocomposite, G-Fe_3O_4-βCD	Coprecipitation reaction with Fe^{3+} and Fe^{2+} ions into Fe_3O_4 on graphene oxide and capping with ethylene diamide-activated β-cyclodextrin	–	97	T = 298 K; pH = 6; 200 rpm; 240 min; C_i = 10–200 mg/L; C_{ads} = 0.5 g/L	66.01	6 cycles 90% (1st) 81% (6th) Methanol	Ragavan and Rastogi (2017)
Magnetic clusters (MMCs) functionalized with β-cyclodextrin, CD-MG	Mesoporous magnetite clusters prepared with $FeCl_3$, ethylene glycol, Na_3Cit, and NH_4Ac; modified with HMDI; and functionalized with β-cyclodextrin	–	26.4	T = 303 K; pH = 6.5–7; 200 rpm; C_i = 4–130 mg/L; C_{ads} = 1 g/L	52.7	4 cycles 98.5% (1st) 84.5% (4th) Ethanol	Lee and Kwak (2019)
Fe_3O_4/r-graphene oxide, rGO-MNPs-1	Coprecipitation reaction with $FeCl_3$ and $FeCl_2$ in the presence of reduced-graphene oxide (1:2, w(rGO)/w(ferric salt))	–	36.2	T = 298 K; pH = 6; 200 rpm; 6 h; C_i = 10–180 mg/L; C_{ads} = 0.2 g/L	125.0	5 cycles q_e ~ 78 mg/g (1st) q_e ~ 72 mg/g (5th) Methanol	Zhang et al. (2014)
Fe_3O_4/graphene aerogel (GA), FGA2	Coprecipitating method to obtain Fe_3O_4; ex situ method to form nanocomposites through self-assembly of reduced-graphene oxide in the presence of Fe_3O_4 (1:1, w(Fe_3O_4)/w(GO))	414.27	21.2	T = room temperature; equilibration pH; 4 h; C_i = 25–400 mg/L; C_{ads} = 1 g/L	253.8	–	Quan et al. (2019)
Ferrihydrite-impregnated powdered activated carbon, ferrihydrite/PAC	Mixture of powdered activated carbon with iron–salt solution; pH adjusted to 7–8 using 2 M NaOH; collection of ferrihydrite/PAC; washing with water and drying at room temperature	323	–	T = 298 K; pH = 7; 300 rpm; 200 min; C_i = 0.001–0.06 mg/L; C_{ads} = 0.17 g/L	–	–	Park et al. (2015)

CNTs into the environment can potentially harm living organisms and influence the fate of organic pollutants. To overcome this issue, Li et al. (2015) proposed the incorporation of Fe_3O_4 NPs into the CNTs to assure their magnetic recovery from solution. The CNTs/Fe_3O_4 nanocomposites with 20% CNT content presented Ms = 125 emu/g, which imply strong magnetic response. Increasing the additive content of CNTs led to diminished magnetization, but concomitant higher adsorption of BPA due to larger S_{BET}. BPA isotherms of CNTs/Fe_3O_4 obeyed Freundlich model better than Langmuir model, so a heterogeneous arrangement of the adsorption sites on the CNTs is deductible. The adsorption mechanism was attributed to π–π bond interactions among aromatic moieties of BPA and graphene sheets of CNTs/Fe_3O_4. The Fe^{2+}/H_2O_2 oxidation method was shown to be interesting not only for the regeneration of CNTs/Fe_3O_4 (99%) but also for complete in situ degradation of BPA postadsorption. CNTs/Fe_3O_4 maintained stable adsorption performance throughout five successive adsorption–deactivation–regeneration cycles.

Metal-organic frameworks (MOFs) are a category of functional porous compounds that are organized by metal ions and organic ligands. On the basis of well-developed surface area and porosity, tunable structure, and other advantages, MOFs have attracted attention for BPA adsorption (Park et al. 2013; Qin et al. 2015). Recently, Li et al. (2020a) developed a core–shell material with Fe_3O_4 as core and MOF (ZIF-67) derivative layered double hydroxide (LDH) as shell (Fe_3O_4@Co/Ni-LDH). The peculiar hierarchical rattle-like arrangement endowed the material with enhanced area (S_{BET} = 128.13 m^2/g), greater porosity (pore volume = 0.34 cm^3/g), and magnetic features (Ms = 14.2 emu/g). The adsorption rate of BPA over Fe_3O_4@Co/Ni-LDH was almost 67% after 60-min contact time. Pseudo-second-order equation well represented the kinetics pattern, and the removal mechanism was mainly related to hydrogen-bonding interactions between BPA and oxygen functionalities of Fe_3O_4@Co/Ni-LDH. Langmuir model described BPA isotherm with $R^2 > 0.99$, and the adsorption capacity, $q_{max} > 238$ mg/g (Table 7.2), demonstrates the utmost potential of Fe_3O_4@Co/Ni-LDH for BPA removal. Adsorbent performance showed a 10% decrease in real water compared to deionized water. Using methanol as regenerant agent, Fe_3O_4@Co/Ni-LDH was recycled five times without significant impairment of BPA uptake. In addition, the adsorbent was highly stable as the evolved layer prevented the dissolution of the Fe_3O_4.

Another core–shell nanocomposite, consisting of Fe_3O_4 cores and polyaniline shells (Fe_3O_4@PANI), was fabricated and tested for BPA removal from water matrices (Zhou et al. 2016). Characterizations verified the successful polymerization of polyaniline on the surface of Fe_3O_4, which provided higher stability and enhanced adsorption performance for BPA. Kinetic study revealed 300 min as enough time to attain equilibrium and better fittings of pseudo-second-order model in relation to pseudo-first-order or diffusion models. The BPA isotherms were more accurately represented by Freundlich model, suggesting multilayer adsorption over Fe_3O_4@PANI composite. Thermodynamic parameters determined the feasible, spontaneous, and endothermic character of the overall process.

Alternatively to polyaniline polymer to prepare functionalized magnetic NPs, Li et al. (2018) tested 2-vinylpyridine (PVP), which is a good hydrogen-bonding acceptor. The Mag-PVP adsorbent contained approximately 150 nm diameter Fe_3O_4 nanospheres coated by a 10-nm thick PVP shell. The polymer layers caused reduced magnetic forces to Mag-PVP (Ms = 44.5 emu/g), in comparison to pure Fe_3O_4 NPs (Ms = 88.3 emu/g); however, the composite could still be effectively collect by external magnetic field. Nevertheless, the adsorption capacity of BPA more than doubled using Mag-PVP (q_{max} = 116 mg/g) instead of Fe_3O_4 NPs (q_{max} = 42 mg/g) because of the numerous binding sites from the PVP layer. Studies with environmental matrices showed no obvious effects on BPA removal. In fact, Mag-PVP adsorption performance was higher in tap water and surface water than in deionized water. Mag-PVP was regenerated by methanol and reused over 12 times without compromising BPA adsorbability.

Saleh et al. (2019) adopted the strategy of combining magnetic properties with efficient polymers and cost-effective supports. In that case, vermiculite clay was used as support, magnetized using Fe^{3+} and Fe^{2+} precursors, and further modified by interfacial polymerization with poly(trimesoyl chloride-melamine) to form the adsorbent labeled as MV-MP. The polymeric chain introduction into the magnetized vermiculite expressively enhanced its performance for BPA removal by almost 57%. A factorial design analysis was used to scrutinize the effects of pH, contact time, adsorbent dosage, and BPA concentration. The highest BPA removal rate was obtained at the highest pH (8) and the longest contact time (90 min). The Langmuir adsorption capacity was predicted as q_{max} = 273.7 mg/g. Thermodynamic parameters unveiled that BPA adsorption is driven by spontaneous and exothermic processes. Methanol was tested for desorption, and the adsorption efficiency of MV-MP decreased from 97% to 60% after seven cycles.

β-Cyclodextrin (β-CD) is a kind of cyclic oligosaccharide originated from enzymatic degradation of starch, and that presents a toroid structure with a hydrophilic outer side and relatively nonpolar interior. This configuration grants β-CD a great capacity of entrapping other compounds (Martin et al. 2018), which is attractive for the functionalization of magnetic adsorbents. For example, Kang et al. (2011) prepared β-CD-modified $Fe_3O_4@SiO_2$ NPs, but the adsorbent provided a poor 24% uptake of BPA molecules. Alternatively, Ragavan and Rastogi (2017) synthesized magnetic nanocomposites capped by β-CD following three steps: (i) formation of graphene-Fe_3O_4 composite by coprecipitation and reduction; (ii) mixture with ethylene diamine-activated β-CD; (iii) purification by magnetic decantation to obtain the β-CD-graphene-Fe_3O_4 nanocomposites (G-Fe_3O_4-βCD). The adsorption of BPA onto G-Fe_3O_4-βCD reached equilibrium after 240 min for 25 mg/L BPA initial concentration. The kinetics obeyed pseudo-second-order equation ($R^2 > 0.99$). Isotherms followed Langmuir model (q_{max} = 66 mg/g) and the proceeding was verified to be spontaneous and endothermic. The G-Fe_3O_4-βCD was tested in adsorption/desorption cycles with methanol washing, and its adsorption efficiency for BPA reduced from initial 90% to 81% after six consecutive runs.

As a substitute to magnetic NPs, mesoporous magnetic clusters (MMCs) have been researched due to their larger cluster size, which presumably promotes more rapid adsorption kinetics. Indeed, the β-CD crosslinked-MMCs (CD-MG) fabricated by Lee and Kwak (2019) demonstrated faster BPA removal than the NPs from

Ragavan and Rastogi (2017) (G-Fe_3O_4-βCD). Using a greater BPA initial concentration of 130 mg/L, the equilibrium over CD-MG was obtained after 40 min, while G-Fe_3O_4-βCD took 240 min. However, the BPA adsorption ability of the MMCs CD-MG, q_{max} = 52.7 mg/g, was comparatively lower (Table 7.2). The kinetics and isotherms of BPA adsorption onto CD-MG obeyed pseudo-second rate law and Langmuir equation, respectively. Ethanol was adopted for BPA desorption and CD-MG regeneration, and after four cycles, the recovery efficiency of BPA diminished to from 98.5% to 84.5% (Lee and Kwak 2019).

Several graphene-based nanocomposites have been tested for BPA remediation. To exemplify, Zhang et al. (2014) fabricated r-GO nanosheets decorated with Fe_3O_4 NPs. The material obtained with rGO: ferric salt mass ratio of 1:2, referred as rGO-MNPs-1, presented the greatest saturated adsorption capacity of BPA. The pH was attested to have minor effects on BPA adsorption over the range of 3–6, suggesting the key role of π–π interactions between BPA and the r-GO skeleton. The good fit of Langmuir model indicated a monolayer BPA adsorption on rGO-MNPs-1. Equilibrium was reached after 4 h, and the kinetics agreed with pseudo-second-order model more than pseudo-first-order model. Higher temperatures hindered BPA adsorption, revealing the exothermic character of the process. Methanol, ethanol, toluene, and cyclohexane were tested as eluent for rGO-MNPs-1 regeneration, and the first one showed the highest desorption efficiency (98%). After five cycles, nearly 95% of BPA could still be adsorbed over rGO-MNPs-1.

The assembly of 2D graphene sheets into 3D structures has been researched aiming at better mechanical properties and rapid mass and electron diffusions due to the combined properties of graphene and the interconnected framework (Liu et al. 2014). Quan et al. (2019) developed a 3D graphene aerogel (GA) functionalized with Fe_3O_4 NPs by ex situ method (self-assembly of rGO in the presence of Fe_3O_4). The anchoring of Fe_3O_4 on GA surface inhibited the restacking of the r-GO sheets and structural shrinkage, expanding the interlayer space and pore dimensions. Characterizations of the magnetic nanocomposite with Fe_3O_4: GO ratio of 1:1 (called FGA2) showed an interconnected porous framework with S_{BET} = 414.27 m^2/g, which is much larger than other graphene-based adsorbents. The great surface area of the Fe_3O_4/GA granted a superior adsorption capacity for BPA molecules, estimated as q_{max} = 253.8 mg/g. The adsorption mechanisms were associated to electrostatic interactions, π–π interactions, and hydrogen bonds. The adsorption equilibration time was established as 8 h. On comparing pseudo-first-order and pseudo-second-order models, the second one best represented the kinetics. In relation to equilibrium data, Langmuir model was more consistent than Temkin or Freundlich model.

Researches have shown that BPA adsorption onto ACs can be expressively affected in the copresence of natural organic matter (NOM), due to the development of organic complexes (Zhu et al. 2012). Pure iron oxides are preeminent materials used for selective adsorption of NOM in water treatment; but are relatively poor at removing BPA (Quan et al. 2019). So, Park et al. (2015) proposed adsorbents of PAC and iron oxides for concurrent removal of BPA and NOM. Three types of iron-oxide-impregnated PAC materials were prepared: ferrihydrite/PAC, magnetite/PAC, and hematite/PAC. Despite the fact that the introduction of iron oxide NPs into the pores of PAC substantially reduced surface areas and pore volumes, the adsorption rates for BPA and NOM increased considerably. For exemplification, ferrihydrite/PAC showed single BA adsorption almost three times higher than bare PAC. The greater performances of the hybrid adsorbents were assigned to surface coordination between the functional moieties of iron oxides (hydroxyl groups) and BPA (phenolic groups) or NOM (carboxyl groups). Moreover, unlike bare PAC, ferrihydrite/PAC sustained the same BPA removal capacity in the presence of NOM in solution, even with the potential formation of BPA–NOM complexes.

Overall, this section indicates that BPA removal is chiefly a chemical bonding process between the adsorbate molecules and the various magnetic adsorbents, given the prevalence of pseudo-second-order model. Of the ten works detailed in Table 7.2, seven works report the accuracy of Langmuir model to BPA isotherms, suggesting monolayer adsorptions over energetically homogeneous sites of the magnetic materials. Only four of the papers conducted thermodynamic studies, of which both endo- and exothermic processes were verified.

Regarding natural EDCs, the natural estrogen 17β-estradiol (E2) owns the uppermost endocrine-disrupting effect. Comparatively to BPA, E2 was verified to have higher binding affinities for model estrogen receptors (Shyu et al. 2011). Therefore, the treatment of E2 is indispensable for safe water management. More recently, E2 uptake has been examined using magnetic adsorbents, including functionalized GO (Jiang et al. 2017), biomass-derived materials (Dong et al. 2018; Liu et al. 2019), and other nanocomposites (Fakhri and Behrouz 2015). These papers will not be detailed in this chapter for the sake of brevity.

7.2.3 *Nonsteroidal Anti-inflammatory Drugs*

Nonsteroidal anti-inflammatory drugs (NSAIDs) comprise those pharmaceuticals with analgesic, antipyretic, and anti-inflammatory effects. Naproxen (NAP), ibuprofen (IBU), and diclofenac (DIC) are major representatives of NSAIDs (Husein et al. 2019). NSAIDs have been increasingly used, and their detection in environment is widespread. NAP has been encountered in water compartments at alarming concentrations. Some of the detrimental effects of NAP include cardiovascular disease and endocrine disruption (Li et al. 2020b). With regard to IBU, its estimated annual production in 2015 was over 13,500 tons (Gilevska et al. 2015). The removal rates of IBU in WWTFs vary between 75% and 90%. IBU is biologically active and its metabolites are toxic (Kaur et al. 2016). Finally, around 31% of an oral dose of DIC is excreted in urine of subjects with normal renal function (Sawchuk et al. 1995). The residual DIC and its metabolites can reach environmental compartments and endanger bacteria, invertebrates, algae, and animals (Sotelo et al. 2014). In common, NAP, IBU, and DIC are recalcitrant to conventional treatment approaches, thereby the adsorption of these NSAIDs is a research hotspot.

Examples of magnetic adsorbents that have been explored for NSAID uptake from water and/or effluents are listed in Table 7.3.

Plethora chitosan-based magnetic adsorbents have been explored for NSAID adsorption due to the renewable and environmental-friendly nature of chitosan. To upgrade the adsorption ability for microcontaminants, chemical modification by grafting co-polymerization has been suggested to obtain adsorbents with "core-brush" topology. In this case, the raw chitosan-magnetic composite acts as the core and polymeric branches on the outer surface compose one-end-fixed brushes. The flexibility of these brushes promotes greater chances of contact between the functionalities of the biopolymer and the pollutant molecules. Zhang et al. (2016) tested three different polymeric grafting branches (polycations, polyanions, and neutral polymer) for brush-like structure in chitosan-Fe_3O_4 composites. The S_{BET} of the material increased by up to 22.6 times with the introduction of the extended brushes. The adsorbent modified with polycations (2-methyl acryloyloxyethyl trimethyl ammonium chloride), referred as CD-MCP, presented the greatest adsorption performance for DIC. This was ascribable to greater attractive forces between the cationic functionalities of CD-MCP and the predominantly anionic species of DIC at pH $= 4$. The pivotal role of electrostatic interactions presupposes a chemisorptive process, endorsed by the satisfactory adjustment of pseudo-second-order kinetic pattern. In addition, the superior fitting of Langmuir isotherm reveals that a DIC monolayer is supposedly adsorbed on the brushes. Competitive assays in the presence of TC antibiotic did not impair CD-MCP performance for DIC removal. Based on thermodynamic parameters, DIC removal is spontaneous and exothermic. Desorption and reusability tests showed that CD-MCP has acceptable reusability after five cycles.

Recently, Soares et al. (2019) inspected DIC adsorption using a magnetic hybrid material functionalized with highly charged cationic chitosan. The authors employed a one-step strategy to encapsulate Fe_3O_4 NPs within siliceous network containing quaternary chitosan, which was obtained by introducing N-(2-hydroxypropyl)-3-trimethylammonium into the polymeric chain. The hybrid adsorbent, Fe_3O_4@SiO_2/SiHTCC, had 60% removal efficiency for DIC, which surpassed that obtained using bare Fe_3O_4@SiO_2 (1%) or Fe_3O_4@SiO_2/Si-chitosan (4%). So, the surface modification by quaternary chitosan was identified a key factor for improved uptake. DIC adsorption onto Fe_3O_4@SiO_2/SiHTCC had optimal pH as 6 and equilibrium time of 5 h. The monolayer capacity from Langmuir isotherm was estimated as $q_{max} = 240.4$ mg/g, which is superior to that of other magnetic adsorbents (Table 7.3).

The study from Chahm and Rodrigues (2017) inspected the adsorption of IBU using Fe_3O_4 NPs containing the amphiphilic chitosan derivative O-carboxymethylchitosan-N-lauryl. This adsorbent, labeled O-carboxymethyl-N-laurylchitosan/γ-Fe_2O_3 (OCh-LM), was firstly tested with success by Demarchi et al. (2015) for treating an anionic reactive dye. The OCh-LM was found to have combined magnetic properties, for easy collection, and functional groups from chitosan derivative, for improved adsorption ability. The adsorption of IBU over OCh-LM was optimized by response surface methodology for the effects of initial

Table 7.3 Nonsteroidal anti-inflammatory drugs (NSAIDs) removal by distinct magnetic adsorbents (S_{BET} = BET surface area; Ms = saturation magnetization) in equilibrium tests (T = temperature; pH; contact time; agitation; C_i = initial concentration range; C_{ads} = adsorbent dosage) for Langmuir maximum adsorption capacity (q_{max}) and adsorbent reusability in cycles

Magnetic adsorbent	NSAID	Magnetization method	S_{BET} (m^2/g)	Ms (emu/g)	Experimental conditions	q_{max} (mg/g)	Reusability	References
Chitosan-Fe_3O_4 composite with core-brush topology, CD-MCP	Diclofenac	Coprecipitation reaction with $FeCl_3$ and $FeCl_2$ in the presence of chitosan and grafting co-polymerization	8.48	–	T = 298 K; pH = 6; 140 rpm; 12 h; C_i = 100 mg/L; C_{ads} = 1 g	196	5 cycles ~94% (1st) ~85% (5th) Ethanol: water	Zhang et al. (2016)
Fe_3O_4@SiO_2/Si-quaternary chitosan, Fe_3O_4@SiO_2/SiHTCC	Diclofenac	Alkaline oxidative hydrolysis of $FeSO_4$ under N_2 stream to form Fe_3O_4 cores and coating with siliceous network containing quaternary chitosan	5.25	–	T = 298 K; pH = 6; 30 rpm; 5 h; C_i = 40–670 mg/L; C_{ads} = 0.5 g/L	240.5	–	Soares et al. (2019)
O-carboxymethyl-N-laurylchitosan/γ-Fe_2O_3, OCh-LM	Ibuprofen	Coprecipitation with Fe^{3+} and Fe^{2+} to form Fe_2O_3 and encapsulation in O-carboxymethyl-N-laurylchitosan matrix	–	18.4	T = 298 K; pH = 7; 120 min; C_i = 5–75-mg/L; C_{ads} = 1.25 g/L	99 [S]	6 cycles q_e ~ 35 mg/g (1st) q_e ~ 33 mg/g (6th) Ethanol	Demarchi et al. (2015), Chahm and Rodrigues (2017)
Fe_3O_4/CNT	Naproxen	Coprecipitation reaction of Fe^{3+} and Fe^{2+} in the presence of multiwalled CNT	–	–	T = 303 K; 120 rpm; 240 min; C_i = 1–30 mg/L; C_{ads} = 0.3 g	11.81	–	İlbay et al. (2015)
Fe_3O_4/AC	Naproxen	Coprecipitation reaction of Fe^{3+} and Fe^{2+} in the presence of AC	–	–	T = 303 K; 120 rpm; 240 min; C_i = 1–30 mg/L; C_{ads} = 0.3 g	75.02	–	İlbay et al. (2015)

(continued)

Table 7.3 (continued)

Magnetic adsorbent	NSAID	Magnetization method	S_{BET} (m^2/g)	Ms (emu/g)	Experimental conditions	q_{max} (mg/g)	Reusability	References
$NiFe_2O_4$/activated carbon, NiAC	Ibuprofen	Hydrothermal method by mixing aqueous solutions of AC, $NiSO_4$, $FeCl_3$, and NaOH	564.4	6.25	T = 298 K; pH = 2; 125	97.53	–	Fröhlich et al. (2019)
	Ketoprofen				rpm; 4 h; C_i = 0–100 mg/L; C_{ads} = 0.5 g/L	87.91		
Fe_3O_4@SiO_2/GO, GO-MNPs-SiO_2	Naproxen	Solvothermal synthesis of Fe_3O_4 particles, coating with SiO_2 by sol-gel synthesis and assembling on the surface of graphene oxide sheets	–	–	T = room temperature; 60 min; C_i = 10–200 mg/L; C_{ads} = 0.030 g	31.25	25 cycles	Mohammadi Nodeh et al. (2018)
							~90% (1st)	
							~78% (20th)	
							Methanol: water	
Recycled rusted iron particles from disposable heat pads, RRIP	Ibuprofen	The Fe^0 core confers magnetic feature to Fe@FeOOH core-shelled particles	136.45	0.6	T = 298 K; pH = 7; 10 rpm; $C_i \sim 2.1$ mg/L; C_{ads} = 0.25 g/L	3.47	–	Yin et al. (2018)
γ-Fe_2O_3-coated zeolite, MNCZ	Diclofenac	Coprecipitation with $FeCl_3$ and $FeCl_2$ in the presence of natural zeolite, oxidation at 300 °C	126	–	T = 303 K; pH = 7; 200 rpm; 300 min; C_i = 0.01–0.5 mg/L; C_{ads} = 0.2 g/L	–		Salem Attia et al. (2013)
	Naproxen							
	Gemfibrozil							
	Ibuprofen							
Poly(4-vinylpyridine)-functionalized magnetic aluminum-based MOF, Al-MOF-Fe_3O_4@P4VP	Naproxen	Coprecipitation with $FeCl_3$ and $FeCl_2$ in the presence aluminum-based MOFs to form Al-MOF-Fe_3O_4 and coating with poly (4-vinylpyridine)	123.68	34.35	T = 293 K; pH natural; 180 rpm; 240 min; C_i = 10–40 mg/L; C_{ads} = 0.001 g/L	226.24	10 cycles	Li et al. (2020b)
							$q_e \sim 11.8$ mg/g (1st)	
							$q_e \sim 9.8$ mg/g (10th)	
							Methanol	

pH, IBU concentration, and OCh-LM dosage. The most relevant parameters were IBU concentration and square IBU concentration. On comparing the isotherm models of Langmuir, Freundlich, and Sips, the latter best represented the equilibrium data of IBU adsorption onto OCh-LM. Regarding kinetics, pseudo-second-order equation had better fittings than pseudo-first-order model. The IBU removal was mainly motivated by exothermic process ($\Delta H < 0$), but is nonspontaneous ($\Delta G > 0$). Ethanol was used as desorption agent, and OCh-LM was shown to keep adsorption capacity around 30 mg/g even after six consecutive runs.

The adsorption of NSAIDs was also examined by carbon-based magnetic adsorbents. İlbay et al. (2015) prepared both magnetically magnetite-modified multiwalled CNTs (M-MWCNT) and ACs (M-AC) by coprecipitation method for NAP adsorption. The magnetite ratios were about 51.5% in M-MWCNT and 50.6% in M-AC. The uptake of NAP over both adsorbents took place by concomitant hydrogen bonding and van der Waals forces. The processes using either M-MWCNT or M-AC were spontaneous and exothermic and conformed pseudo-second-order kinetics pattern. The greater adsorption capacity of M-AC (q_{max} = 75 mg/g at 303 K) in comparison to M-MWCNT (q_{max} = 12 mg/g at 303 K) was associated to the greater surface acidity.

More recently, Fröhlich et al. (2019) developed a composite material from AC and $NiFe_2O_4$ magnetic particles by a straightforward hydrothermal strategy. The magnetic adsorbent (NiAC) owed $S_{BET} = 564.4$ m^2/g and Ms = 6.25 emu/g, which endows great surface area for adsorption and easy withdraw from solution after the process. NiAC was used for the adsorption not only of IBU but also of ketoprofen (KET), which is another NSAID. Both pharmaceuticals were better removed in acidic conditions, due to attractive electrostatic interactions. The pseudo-second-order model described the kinetic profiles. Among Langmuir, Freundlich, and Sips isotherm models, the latter better correlated equilibrium data, suggesting hybrid adsorption pathways. At 298 K, NiAC maximum adsorption capacity from Sips model was around 261 and 98 mg/g for IBU and KET, correspondingly. The hybrid NiAC adsorbent was verified to be more effective for pharmaceutical remediation than raw AC material, ascribable to the modified surface and chemical features. Either IBU or KET adsorptive processes were spontaneous, favorable, and endothermic. Finally, NiAC presented 85% removal rate of the pharmaceuticals from a simulated effluent containing a mixture of IBU, KET, and inorganic substances.

The removal of NAP by a GO adsorbent material was examined by Mohammadi Nodeh et al. (2018). The authors decorated GO sheets with silica-coated Fe_3O_4 NPs to form the nanocomposite GO-MNPs-SiO_2. The adsorption of NAP over GO-MNPs-SiO_2 attained equilibrium after about 60 min. Freundlich model ($R^2 > 0.99$) was more suitable than Langmuir or Dubinin–Radushkevich model to describe the isotherm. So, it is deductible that NAP follows multilayer adsorption over the heterogeneous surface of the composite. In real wastewater conditions, the matrix interferences did not hamper the removal performance of GO-MNPs-SiO2, which remained between 83% and 95%. The adsorbent showed formidable reusability as 78% of NAP could still be removed by GO-MNPs-SiO_2 after

20 consecutive cycles. Besides batch tests, fixed-bed column tests were performed and the removal efficiency was around 72% at 1 mL/min flow rate.

Based on what has been discussed in this chapter so far, iron-based materials have wide applicability in the removal of EPs. The recycling-and-reuse of iron particles from waste materials is a sustainable alternative approach. The study by Yin et al. (2018) proposes the recycle of rusted iron particles (RRIPs) from exhausted disposable skin heat pads. Characterization analyzes depicted the core–shell structure of RRIP, with Fe^0 core and surface FeOOH groups. The Fe^0 component confers magnetic features to RRIP, which was classified as soft magnetic due to low coercivity (<100 Oe) and remanence (<0.05 emu/g). The saturation magnetization of RRIP was determined as 0.6 emu/g. IBU was chosen as model pharmaceutical contaminant, and nearly 20% removal rate was obtained using 0.25 g/L RRIP. Comparatively, the same dosage of synthetic α-FeOOH offered 10% IBU removal. The IBU adsorption over both adsorbents was satisfactorily correlated by Langmuir isotherm. The monolayer adsorption coverage of RRIP and α-FeOOH was appraised as 3.47 and 0.72 mg/g, respectively. The superior adsorption potential of RRIP was ascribable to the greater S_{BET} and porosity in relation to α-FeOOH. The RRIP was also used for the catalytic activation of persulfate oxidant aiming the total degradation of IBU molecules by both $SO_4^{-\bullet}$ and $OH^{\bullet}$ radicals.

Salem Attia et al. (2014) firstly synthesized an adsorbent with combined maghemite NPs and natural zeolite. Briefly, iron chloride and iron sulfate were mixed together with zeolite to form α-Fe_3O_4-coated zeolite, which was oxidized at 300 °C to form γ-Fe_3O_4-coated zeolite (MNCZ). The successful application of MNCZ in arsenic removal has encouraged studies to treat pharmaceutical contaminants. Salem Attia et al. (2013) examined the removal of IBU, NAP, and DIC (NSAIDs), as well as of gemfibrozil (GEM) (lipid regulator) by adsorption using MNCZ. The material presented extraordinary adsorption ability over a wide pH range (2–9), which is positive for extensive usage in WWTPs. The removal efficiency of all examined pharmaceuticals was greater than 95%; however, while DIC took only 10 min to reach equilibrium, the others required an average of 300 min. The faster uptake of DIC was related to the fact that its molecule has both N–H and O–H groups, whereas IBU, NAP, and GEM have O–H as the only main group. Isotherms were obtained for initial concentrations between 10 and 500 μg/L. On the basis that saturation adsorption capability was not reached for any of the compounds, MNCZ could be potentially used to treat waters with higher pollutant loads. Freundlich model described the isotherms better than Langmuir model. Moreover, the adsorption kinetics features chemisorption mechanism due to the good agreement of better pseudo-second-order pattern. The adsorption study in fixed-bed column packed with MNCZ was performed and offered removal efficiencies for IBU, NAP, DIC, and GEM higher than other conventional treatment methods (biofiltration treatment, activated carbon, ozonation, sand filtration, coagulation, flocculation, and sedimentation).

Lastly, it is important to cite that the adsorption of NSAIDs has also been inspected using MOF-derived magnetic materials. Li et al. (2020b) examined NAP uptake over an adsorbent with synergistic properties of aluminum-based MOFs (Al-MOF), magnetite (Fe_3O_4), and poly(4-vinylpyridine) (P4VP). The as-fabricated adsorbent was designated Al-MOF-Fe_3O_4@P4VP. The combination of Al-MOFs and Fe_3O_4 NPs confers magnetic property (Ms = 34.35 emu/g) for easy separation/recycling/reuse, and the polymer encapsulation with P4VP provides greater physicochemical properties and structural stability. P4VP was proved as effective stabilizer of the magnetic composite by comparing the leaching of Fe ions from pure Al-MOF-Fe_3O_4 (4.75 ug/L) and encapsulated Al-MOF-Fe_3O_4@P4VP (1.29 ug/L). Moreover, the Al-MOF-Fe_3O_4@P4VP featured high $S_{BET} = 124\ m^2/g$, mesoporous structure, and plethora functional moieties, which boosted the adsorption ability for NAP molecules. Langmuir model had low correlation coefficient ($R^2 < 0.67$) for NAP isotherm. The better agreement of Freundlich model ($R^2 > 0.99$) supported a multilayer NAP adsorption over the energetically heterogeneous surface of Al-MOF-Fe_3O_4@P4VP. The NAP adsorption kinetics was rapid and obeyed pseudo-second-order model. Thermodynamically, the process is spontaneous and endothermic. Methanol acted as desorption agent of NAP molecules from Al-MOF-Fe_3O_4@P4VP, and almost 94% removal rates were attained. The Al-MOF-Fe_3O_4@P4VP was reused in ten adsorption–desorption cycles and sustained over 61% of the initial adsorption capacity in the last run.

Summarizing, similar to TC (Sect. 7.2.1) and BPA (Sect. 7.2.2), the NSAIDs examined also presented kinetic profiles over magnetic adsorbents better correlated by pseudo-second-order equation, assuming that the rate-controlling step involves chemisorption mechanisms. Conversely, the adsorption isotherms were described by various models, namely Langmuir, Sips, Dubinin–Radushkevich, and Freundlich. Finally, NSAID removal processes were not unanimously spontaneous and motivated by both exo- and endothermic pathways.

7.3 Conclusions

This chapter presented that magneto-responsive adsorbents can be produced by the magnetization of innumerous potential materials and biomaterials, such as activated carbon, carbon nanotubes, graphene oxide, chitosan, and zeolites. The magnetization mainly occurs by coprecipitation, solvothermal, and hydrothermal procedures, which provides the production of magnetic particles mostly from iron sources.

Tetracycline antibiotic, bisphenol A, and nonsteroidal anti-inflammatory drugs (naproxen, ibuprofen, and diclofenac) were selected as representative emerging pollutants (EPs) of concern. The adsorption processes using magnetic adsorbents exhibited joint advantages of high adsorption capacities and effective magnetic separation. The kinetic model of pseudo-second order offered good correlation of all the examined cases, corroborating that chemical mechanism has a key role in the uptake of EPs from solution. On the other hand, equilibrium data were well represented not only by Langmuir isotherm model but also by Freundlich, Temkin, Sips, and Dubinin–Radushkevich models. This fact could be related to the mostly multicomponent and heterogeneous surface of the as-prepared magnetic adsorbents, which endows more complex interactions with the organic molecules.

Overall, the process using magnetic adsorbents for treating water/wastewater contaminated with EPs present the following advantages: (i) easy collection from solution by external magnetic field, requiring no additional separation steps, such as centrifugation or filtration is not required, decreasing the operational costs; (ii) minimal losses of adsorbent mass in recyclability tests due to high recovery efficiency by magnetic separation; and (iii) the covering of magnetic particles with protecting layers (e.g., polymeric) endows enhanced resistance to corrosion and so higher durability. Some restrictive points about the adsorbent magnetization may be listed: (i) the doping of adsorbent surfaces with magnetic particles may reduce the available binding sites for adsorption, and thereby decrease the adsorption capacity of the adsorbent material; (ii) the synthesis methods of magnetic adsorbents tend to be complex, which represents an obstacle reaches large-scale applications; (iii) the adsorbent separation is just required in batch adsorption systems, which are recommended to treat low volumes of contaminated water/wastewaters. For the treatment of large volumes, fixed-bed adsorption systems are preferred, in which there is no need for adsorbent separation (magnetic features are dispensable).

The implementation of adsorption as effective technology for EP removal aiming at environmental remediation still requires more detailed studies using complex matrices (multicomponent mixtures and real effluents) and further research in pilot scale to verify magnetic separation efficiency for large-scale applications.

Acknowledgments The authors are grateful for the financial support from São Paulo Research Foundation (FAPESP) (Proc. 2017/18236-1), National Council for Scientific and Technological Development (CNPq) (406193/2018-5), and Coordination for the Improvement of Higher Education Personnel (CAPES).

References

Adeleye AS, Conway JR, Garner K, Huang Y, Su Y, Keller AA (2016) Engineered nanomaterials for water treatment and remediation: costs, benefits, and applicability. Chem Eng J 286:640–662. https://doi.org/10.1016/j.cej.2015.10.105

Al Aukidy M, Verlicchi P, Jelic A, Petrovic M, Barcelò D (2012) Monitoring release of pharmaceutical compounds: occurrence and environmental risk assessment of two WWTP effluents and their receiving bodies in the Po Valley, Italy. Sci Total Environ 438:15–25. https://doi.org/10.1016/j.scitotenv.2012.08.061

Araujo LA, Bezerra CO, Cusioli LF, Silva MF, Nishi L, Gomes RG, Bergamasco R (2018) Moringa oleifera biomass residue for the removal of pharmaceuticals from water. J Environ Chem Eng 6:7192–7199. https://doi.org/10.1016/j.jece.2018.11.016

Baquero F, Martínez J-L, Cantón R (2008) Antibiotics and antibiotic resistance in water environments. Curr Opin Biotechnol 19:260–265. https://doi.org/10.1016/j.copbio.2008.05.006

Bazarin G, Módenes AN, Vieira MGA, Borba CE, Espinoza-Quiñones FR, Scariotto MC (2019) Tilapia scales: characterization and study of Cu(II) removal by ion exchange with Ca(II). Sep Sci Technol 55:1–13. https://doi.org/10.1080/01496395.2019.1577260

Behera SK, Kim HW, Oh J-E, Park H-S (2011) Occurrence and removal of antibiotics, hormones and several other pharmaceuticals in wastewater treatment plants of the largest industrial city of Korea. Sci Total Environ 409:4351–4360. https://doi.org/10.1016/j.scitotenv.2011.07.015

Benfenati E, Barcelò D, Johnson I, Galassi S, Levsen K (2003) Emerging organic contaminants in leachates from industrial waste landfills and industrial effluent. TrAC Trends Anal Chem 22:757–765. https://doi.org/10.1016/S0165-9936(03)01004-5

Boehler M, Zwickenpflug B, Hollender J, Ternes T, Joss A, Siegrist H (2012) Removal of micropollutants in municipal wastewater treatment plants by powder-activated carbon. Water Sci Technol 66:2115–2121. https://doi.org/10.2166/wst.2012.353

Chahm T, Rodrigues CA (2017) Removal of ibuprofen from aqueous solutions using O-carboxymethyl-N-laurylchitosan/γ-Fe_2O_3. Environ Nanotechnol Monit Manage 7:139–148. https://doi.org/10.1016/j.enmm.2017.03.001

Choi K-J, Kim S-G, Kim S-H (2008) Removal of tetracycline and sulfonamide classes of antibiotic compound by powdered activated carbon. Environ Technol 29:333–342. https://doi.org/10.1080/09593330802102223

Colborn T (1995) Commentary: environmental estrogens: health implications for humans and wildlife. Environ Health Perspect 103:135–136. https://doi.org/10.2307/3432522

Cowin PA, Gold E, Aleksova J, O'Bryan MK, Foster PMD, Scott HS, Risbridger GP (2010) Vinclozolin exposure in utero induces postpubertal prostatitis and reduces sperm production via a reversible hormone-regulated mechanism. Endocrinology 151:783–792. https://doi.org/10.1210/en.2009-0982

Demarchi CA, Debrassi A, de Campos BF, Nedelko N, Ślawska-Waniewska A, Dłużewski P, Dal Magro J, Scapinello J, Rodrigues CA (2015) Adsorption of the dye Remazol Red 198 (RR198) by O-carboxymethylchitosan-N-lauryl/γ-Fe_2O_3 magnetic nanoparticles. Arabian J Chem 12:3444. https://doi.org/10.1016/j.arabjc.2015.08.028

Dong X, He L, Hu H, Liu N, Gao S, Piao Y (2018) Removal of 17β-estradiol by using highly adsorptive magnetic biochar nanoparticles from aqueous solution. Chem Eng J 352:371–379. https://doi.org/10.1016/j.cej.2018.07.025

Dubey S, Banerjee S, Upadhyay SN, Sharma YC (2017) Application of common nano-materials for removal of selected metallic species from water and wastewaters: a critical review. J Mol Liq 240:656–677. https://doi.org/10.1016/j.molliq.2017.05.107

Dubinin MM (1960) The potential theory of adsorption of gases and vapors for adsorbents with energetically nonuniform surfaces. Chem Rev 60:235–241

Ebele AJ, Abou-Elwafa Abdallah M, Harrad S (2017) Pharmaceuticals and personal care products (PPCPs) in the freshwater aquatic environment. Emerging Contam 3:1–16. https://doi.org/10.1016/j.emcon.2016.12.004

European Commission (2012) Communication from the Commission to the Council. The combination effects of chemicals – chemical mixtures. COM/2012/252 final. EC, Brussels, pp 1–10

Fakhri A, Behrouz S (2015) Improved uptake of steroid hormone from aqueous solution using γ-Fe_2O_3/NiO nanocomposites. J Ind Eng Chem 26:61–66. https://doi.org/10.1016/j.jiec.2014.10.042

Fent K, Weston AA, Caminada D (2006) Ecotoxicology of human pharmaceuticals. Aquat Toxicol 76:122–159. https://doi.org/10.1016/j.aquatox.2005.09.009

Franzreb M, Siemann-Herzberg M, Hobley TJ, Thomas ORT (2006) Protein purification using magnetic adsorbent particles. Appl Microbiol Biotechnol 70:505–516. https://doi.org/10.1007/s00253-006-0344-3

Freundlich HMF (1906) Over the adsorption in solution. J Phys Chem 57:385–470

Fröhlich AC, Foletto EL, Dotto GL (2019) Preparation and characterization of $NiFe_2O_4$/activated carbon composite as potential magnetic adsorbent for removal of ibuprofen and ketoprofen pharmaceuticals from aqueous solutions. J Cleaner Prod 229:828–837. https://doi.org/10.1016/j.jclepro.2019.05.037

Gao P, Ding Y, Li H, Xagoraraki I (2012a) Occurrence of pharmaceuticals in a municipal wastewater treatment plant: mass balance and removal processes. Chemosphere 88:17–24. https://doi.org/10.1016/j.chemosphere.2012.02.017

Gao Y, Li Y, Zhang L, Huang H, Hu J, Shah SM, Su X (2012b) Adsorption and removal of tetracycline antibiotics from aqueous solution by graphene oxide. J Colloid Interface Sci 368:540–546. https://doi.org/10.1016/j.jcis.2011.11.015

Garcia JJM, Nuñez JAP, Salapare HS, Vasquez MR (2019) Adsorption of diclofenac sodium in aqueous solution using plasma-activated natural zeolites. Results Phys 15:102629. https://doi.org/10.1016/j.rinp.2019.102629

Gilevska T, Gehre M, Richnow HH (2015) Multidimensional isotope analysis of carbon, hydrogen and oxygen as tool for identification of the origin of ibuprofen. J Pharm Biomed Anal 115:410–417. https://doi.org/10.1016/j.jpba.2015.07.030

Ginebreda A, Muñoz I, de Alda ML, Brix R, López-Doval J, Barceló D (2010) Environmental risk assessment of pharmaceuticals in rivers: relationships between hazard indexes and aquatic macroinvertebrate diversity indexes in the Llobregat River (NE Spain). Environ Int 36:153–162. https://doi.org/10.1016/j.envint.2009.10.003

Gracia-Lor E, Sancho JV, Serrano R, Hernández F (2012) Occurrence and removal of pharmaceuticals in wastewater treatment plants at the Spanish Mediterranean area of Valencia. Chemosphere 87:453–462. https://doi.org/10.1016/j.chemosphere.2011.12.025

Granados-Chinchilla F, Rodríguez C (2017) Tetracyclines in food and feedingstuffs: from regulation to analytical methods, bacterial resistance, and environmental and health implications. J Anal Methods Chem 2017:1315497–1315497. https://doi.org/10.1155/2017/1315497

Grenni P, Ancona V, Barra Caracciolo A (2018) Ecological effects of antibiotics on natural ecosystems: a review. Microchem J 136:25–39. https://doi.org/10.1016/j.microc.2017.02.006

Gros M, Petrović M, Barceló D (2009) Tracing pharmaceutical residues of different therapeutic classes in environmental waters by using liquid chromatography/quadrupole-linear ion trap mass spectrometry and automated library searching. Anal Chem 81:898–912. https://doi.org/10.1021/ac801358e

Hazell G, Hinojosa-Navarro M, McCoy TM, Tabor RF, Eastoe J (2016) Responsive materials based on magnetic polyelectrolytes and graphene oxide for water clean-up. J Colloid Interface Sci 464:285–290. https://doi.org/10.1016/j.jcis.2015.11.029

Ho YS, McKay G (1998) A comparison of chemisorption kinetic models applied to pollutant removal on various sorbents. Process Saf Environ Prot 76:332–340. https://doi.org/10.1205/095758298529696

Huang B, Liu Y, Li B, Liu S, Zeng G, Zeng Z, Wang X, Ning Q, Zheng B, Yang C (2017) Effect of Cu(II) ions on the enhancement of tetracycline adsorption by Fe_3O_4@SiO_2-Chitosan/graphene oxide nanocomposite. Carbohydr Polym 157:576–585. https://doi.org/10.1016/j.carbpol.2016.10.025

Huízar-Félix AM, Aguilar-Flores C, Martínez-de-la Cruz A, Barandiarán JM, Sepúlveda-Guzmán S, Cruz-Silva R (2019) Removal of tetracycline pollutants by adsorption and magnetic separation using reduced graphene oxide decorated with α-Fe_2O_3 nanoparticles. Nanomater 9:313. https://doi.org/10.3390/nano9030313

Husein DZ, Hassanien R, Al-Hakkani MF (2019) Green-synthesized copper nano-adsorbent for the removal of pharmaceutical pollutants from real wastewater samples. Heliyon 5:e02339. https://doi.org/10.1016/j.heliyon.2019.e02339

İlbay Z, Şahin S, Kerkez Ö, Bayazit ŞS (2015) Isolation of naproxen from wastewater using carbon-based magnetic adsorbents. Int J Environ Sci Technol 12:3541–3550. https://doi.org/10.1007/s13762-015-0775-4

Jallouli N, Pastrana-Martínez LM, Ribeiro AR, Moreira NFF, Faria JL, Hentati O, Silva AMT, Ksibi M (2018) Heterogeneous photocatalytic degradation of ibuprofen in ultrapure water, municipal and pharmaceutical industry wastewaters using a TiO2/UV-LED system. Chem Eng J 334:976–984. https://doi.org/10.1016/j.cej.2017.10.045

Ji L, Chen W, Duan L, Zhu D (2009) Mechanisms for strong adsorption of tetracycline to carbon nanotubes: a comparative study using activated carbon and graphite as adsorbents. Environ Sci Technol 43:2322–2327. https://doi.org/10.1021/es803268b

Jiang L, Liu Y, Liu S, Hu X, Zeng G, Hu X, Liu S, Liu S, Huang B, Li M (2017) Fabrication of β-cyclodextrin/poly (l-glutamic acid) supported magnetic graphene oxide and its adsorption behavior for 17β-estradiol. Chem Eng J 308:597–605. https://doi.org/10.1016/j.cej.2016.09.067

Jing X-R, Wang Y-Y, Liu W-J, Wang Y-K, Jiang H (2014) Enhanced adsorption performance of tetracycline in aqueous solutions by methanol-modified biochar. Chem Eng J 248:168–174. https://doi.org/10.1016/j.cej.2014.03.006

Kang Y, Zhou L, Li X, Yuan J (2011) β-Cyclodextrin-modified hybrid magnetic nanoparticles for catalysis and adsorption. J Mater Chem 21:3704–3710. https://doi.org/10.1039/c0jm03513k

Kang J, Zhou L, Duan X, Sun H, Wang S (2018) Catalytic degradation of antibiotics by metal-free catalysis over nitrogen-doped graphene. Catal Today 357:341. https://doi.org/10.1016/j.cattod.2018.12.002

Kårelid V, Larsson G, Björlenius B (2017) Pilot-scale removal of pharmaceuticals in municipal wastewater: comparison of granular and powdered activated carbon treatment at three wastewater treatment plants. J Environ Manage 193:491–502. https://doi.org/10.1016/j.jenvman.2017.02.042

Kaur A, Umar A, Kansal SK (2016) Heterogeneous photocatalytic studies of analgesic and non-steroidal anti-inflammatory drugs. Appl Catal A 510:134–155. https://doi.org/10.1016/j.apcata.2015.11.008

Kharissova OV, Dias HVR, Kharisov BI (2015) Magnetic adsorbents based on micro- and nano-structured materials. RSC Adv 5:6695–6719. https://doi.org/10.1039/c4ra11423j

Kortenkamp A, Backhaus T, Faust M (2009) State of the art report on mixture toxicity – final report, executive summary, study contract no. 070307/2007/485103/ETU/D.1

Krzeminski P, Tomei MC, Karaolia P, Langenhoff A, Almeida CMR, Felis E, Gritten F, Andersen HR, Fernandes T, Manaia CM, Rizzo L, Fatta-Kassinos D (2019) Performance of secondary wastewater treatment methods for the removal of contaminants of emerging concern implicated in crop uptake and antibiotic resistance spread: a review. Sci Total Environ 648:1052–1081. https://doi.org/10.1016/j.scitotenv.2018.08.130

Kulshrestha P, Giese RF, Aga DS (2004) Investigating the molecular interactions of oxytetracycline in clay and organic matter: insights on factors affecting its mobility in soil. Environ Sci Technol 38:4097–4105. https://doi.org/10.1021/es034856q

Lagergren S (1898) Zur theorie der sogenannten adsorption gelöster stoffe, Kungliga Svenska Vetenskapsakademiens. Handlingar 24:1–39

Langmuir I (1918) The adsorption of gases on plane surfaces of glass, mica and platinum. J Am Chem Soc 40:1361–1403

Lee JH, Kwak S-Y (2019) Rapid adsorption of bisphenol A from wastewater by β-cyclodextrin-functionalized mesoporous magnetic clusters. Appl Surf Sci 467–468:178–184. https://doi.org/10.1016/j.apsusc.2018.10.054

Lee Y, Gerrity D, Lee M, Bogeat AE, Salhi E, Gamage S, Trenholm RA, Wert EC, Snyder SA, von Gunten U (2013) Prediction of micropollutant elimination during ozonation of municipal wastewater effluents: use of kinetic and water specific information. Environ Sci Technol 47:5872–5881. https://doi.org/10.1021/es400781r

Li Z, Schulz L, Ackley C, Fenske N (2010) Adsorption of tetracycline on kaolinite with pH-dependent surface charges. J Colloid Interface Sci 351:254–260. https://doi.org/10.1016/j.jcis.2010.07.034

Li S, Gong Y, Yang Y, He C, Hu L, Zhu L, Sun L, Shu D (2015) Recyclable CNTs/Fe_3O_4 magnetic nanocomposites as adsorbents to remove bisphenol A from water and their regeneration. Chem Eng J 260:231–239. https://doi.org/10.1016/j.cej.2014.09.032

Li Q, Pan F, Li W, Li D, Xu H, Xia D, Li A (2018) Enhanced adsorption of bisphenol A from aqueous solution with 2-vinylpyridine functionalized magnetic nanoparticles. Polymers 10:1136. https://doi.org/10.3390/polym10101136

Li G, Zhang X, Sun J, Zhang A, Liao C (2020a) Effective removal of bisphenols from aqueous solution with magnetic hierarchical rattle-like Co/Ni-based LDH. J Hazard Mater 381:120985. https://doi.org/10.1016/j.jhazmat.2019.120985

Li Y, Wang Y, He L, Meng L, Lu H, Li X (2020b) Preparation of poly(4-vinylpyridine)-functionalized magnetic Al-MOF for the removal of naproxen from aqueous solution. J Hazard Mater 383:121144. https://doi.org/10.1016/j.jhazmat.2019.121144

Lin Y, Xu S, Li J (2013) Fast and highly efficient tetracyclines removal from environmental waters by graphene oxide functionalized magnetic particles. Chem Eng J 225:679–685. https://doi.org/10.1016/j.cej.2013.03.104

Liu X, Cui J, Sun J, Zhang X (2014) 3D graphene aerogel-supported SnO2 nanoparticles for efficient detection of NO2. RSC Adv 4:22601–22605. https://doi.org/10.1039/c4ra02453b

Liu S, Li M, Liu Y, Liu N, Tan X, Jiang L, Wen J, Hu X, Yin Z (2019) Removal of 17β-estradiol from aqueous solution by graphene oxide supported activated magnetic biochar: adsorption behavior and mechanism. J Taiwan Inst Chem Eng 102:330–339. https://doi.org/10.1016/j.jtice.2019.05.002

Logar I, Brouwer R, Maurer M, Ort C (2014) Cost-benefit analysis of the Swiss National Policy on reducing micropollutants in treated wastewater. Environ Sci Technol 48:12500–12508. https://doi.org/10.1021/es502338j

Loos R, Carvalho R, António DC, Comero S, Locoro G, Tavazzi S, Paracchini B, Ghiani M, Lettieri T, Blaha L, Jarosova B, Voorspoels S, Servaes K, Haglund P, Fick J, Lindberg RH, Schwesig D, Gawlik BM (2013) EU-wide monitoring survey on emerging polar organic contaminants in wastewater treatment plant effluents. Water Res 47:6475–6487. https://doi.org/10.1016/j.watres.2013.08.024

Luo Y, Guo W, Ngo HH, Nghiem LD, Hai FI, Zhang J, Liang S, Wang XC (2014) A review on the occurrence of micropollutants in the aquatic environment and their fate and removal during wastewater treatment. Sci Total Environ 473–474:619–641. https://doi.org/10.1016/j.scitotenv.2013.12.065

Luo T, Qu L, Hou X, Liu X, Wang S, Wu Y (2017) Preparation of 3-dimensional flower-like $NiFe_2O_4$ with enhanced adsorptive performance for water contaminants. J Alloys Compd 727:484–490. https://doi.org/10.1016/j.jallcom.2017.07.234

Maia GS, de Andrade JR, da Silva MGC, Vieira MGA (2019) Adsorption of diclofenac sodium onto commercial organoclay: kinetic, equilibrium and thermodynamic study. Powder Technol 345:140–150. https://doi.org/10.1016/j.powtec.2018.12.097

Martin J, Díaz-Montaña EJ, Asuero AG (2018) Cyclodextrins: past and present. In: Arora P, Dhingra N (eds) Cyclodextrin: a versatile ingredient. IntechOpen, London. Available online at: www.intechopen.com/books/cyclodextrin-a-versatile-ingredient/cyclodextrins-past-and-present

Martinez JL (2009) Environmental pollution by antibiotics and by antibiotic resistance determinants. Environ Pollut 157:2893–2902. https://doi.org/10.1016/j.envpol.2009.05.051

Mehta D, Mazumdar S, Singh SK (2015) Magnetic adsorbents for the treatment of water/wastewater—a review. J Water Process Eng 7:244–265. https://doi.org/10.1016/j.jwpe.2015.07.001

Michael I, Vasquez MI, Hapeshi E, Haddad T, Baginska E, Kümmerer K, Fatta-Kassinos D (2014) Metabolites and transformation products of pharmaceuticals in the aquatic environment as contaminants of emerging concern. In: Lambropoulou DA, Nollet LM (eds) Transformation products of emerging contaminants in the environment. Wiley, New York, pp 413–458. Available online at: https://onlinelibrary.wiley.com/doi/abs/10.1002/9781118339558.ch14

Miklos DB, Hartl R, Michel P, Linden KG, Drewes JE, Hübner U (2018) UV/H_2O_2 process stability and pilot-scale validation for trace organic chemical removal from wastewater treatment plant effluents. Water Res 136:169–179. https://doi.org/10.1016/j.watres.2018.02.044

Mohammadi Nodeh MK, Radfard M, Zardari LA, Rashidi Nodeh H (2018) Enhanced removal of naproxen from wastewater using silica magnetic nanoparticles decorated onto graphene oxide; parametric and equilibrium study. Sep Sci Technol 53:2476–2485. https://doi.org/10.1080/01496395.2018.1457054

Montagner CC, Jardim WF (2011) Spatial and seasonal variations of pharmaceuticals and endocrine disruptors in the Atibaia River, São Paulo State (Brazil). J Braz Chem Soc 22:1452–1462. https://doi.org/10.1590/S0103-50532011000800008

Mukherjee R, Bhunia P, De S (2016) Impact of graphene oxide on removal of heavy metals using mixed matrix membrane. Chem Eng J 292:284–297. https://doi.org/10.1016/j.cej.2016.02.015

NORMAN Network (2016) List of Emerging Substances, Network of reference laboratories, research centres and related organisations for monitoring of emerging environmental substances. Available online at www.norman-network.com. Accessed June 2019

Oladoja NA, Adelagun ROA, Ahmad AL, Unuabonah EI, Bello HA (2014) Preparation of magnetic, macro-reticulated cross-linked chitosan for tetracycline removal from aquatic systems. Colloids Surf B 117:51–59. https://doi.org/10.1016/j.colsurfb.2014.02.006

Oliveira MF, de Souza VM, da Silva MGC, Vieira MGA (2018) Fixed-bed adsorption of caffeine onto thermally modified verde-lodo bentonite. Ind Eng Chem Res 57:17480–17487. https://doi.org/10.1021/acs.iecr.8b03734

Ou J, Mei M, Xu X (2016) Magnetic adsorbent constructed from the loading of amino functionalized Fe_3O_4 on coordination complex modified polyoxometalates nanoparticle and its tetracycline adsorption removal property study. J Solid State Chem 238:182–188. https://doi.org/10.1016/j.jssc.2016.03.021

Park EY, Hasan Z, Khan NA, Jhung SH (2013) Adsorptive removal of bisphenol-A from water with a metal-organic framework, a porous chromium-benzenedicarboxylate. J Nanosci Nanotechnol 13:2789–2794. https://doi.org/10.1166/jnn.2013.7411

Park H-S, Koduru JR, Choo K-H, Lee B (2015) Activated carbons impregnated with iron oxide nanoparticles for enhanced removal of bisphenol A and natural organic matter. J Hazard Mater 286:315–324. https://doi.org/10.1016/j.jhazmat.2014.11.012

Pereira C, Pereira AM, Fernandes C, Rocha M, Mendes R, Fernández-García MP, Guedes A, Tavares PB, Grenèche J-M, Araújo JP, Freire C (2012) Superparamagnetic MFe_2O_4 ($M = Fe$, Co, Mn) nanoparticles: tuning the particle size and magnetic properties through a novel one-step coprecipitation route. Chem Mater 24:1496–1504. https://doi.org/10.1021/cm300301c

Philippova O, Barabanova A, Molchanov V, Khokhlov A (2011) Magnetic polymer beads: recent trends and developments in synthetic design and applications. Eur Polym J 47:542–559. https://doi.org/10.1016/j.eurpolymj.2010.11.006

Plakas KV, Georgiadis AA, Karabelas AJ (2015) Sustainability assessment of tertiary wastewater treatment technologies: a multi-criteria analysis. Water Sci Technol 73:1532–1540. https://doi.org/10.2166/wst.2015.630

Qin F-X, Jia S-Y, Liu Y, Li H-Y, Wu S-H (2015) Adsorptive removal of bisphenol A from aqueous solution using metal-organic frameworks. Desalin Water Treat 54:93–102. https://doi.org/10.1080/19443994.2014.883331

Quan LD, Dang NH, Tu TH, Phuong Linh VN, Mong Thy LT, Nam HM, Phong MT, Hieu NH (2019) Preparation of magnetic iron oxide/graphene aerogel nanocomposites for removal of bisphenol A from water. Synth Met 255:116106. https://doi.org/10.1016/j.synthmet.2019.116106

Radjenović J, Petrović M, Barceló D (2009) Fate and distribution of pharmaceuticals in wastewater and sewage sludge of the conventional activated sludge (CAS) and advanced membrane bioreactor (MBR) treatment. Water Res 43:831–841. https://doi.org/10.1016/j.watres.2008.11.043

Ragavan KV, Rastogi NK (2017) β-Cyclodextrin capped graphene-magnetite nanocomposite for selective adsorption of Bisphenol-A. Carbohydr Polym 168:129–137. https://doi.org/10.1016/j.carbpol.2017.03.045

Real FJ, Benitez FJ, Acero JL, Casas F (2017) Adsorption of selected emerging contaminants onto PAC and GAC: equilibrium isotherms, kinetics, and effect of the water matrix. J Environ Sci Health, Part A 52:727–734. https://doi.org/10.1080/10934529.2017.1301751

Rizzo L, Malato S, Antakyali D, Beretsou VG, Đolić MB, Gernjak W, Heath E, Ivancev-Tumbas I, Karaolia P, Lado Ribeiro AR, Mascolo G, McArdell CS, Schaar H, Silva AMT, Fatta-Kassinos

D (2019) Consolidated vs new advanced treatment methods for the removal of contaminants of emerging concern from urban wastewater. Sci Total Environ 655:986–1008. https://doi.org/10.1016/j.scitotenv.2018.11.265

Rosal R, Rodríguez A, Perdigón-Melón JA, Petre A, García-Calvo E, Gómez MJ, Agüera A, Fernández-Alba AR (2010) Occurrence of emerging pollutants in urban wastewater and their removal through biological treatment followed by ozonation. Water Res 44:578–588. https://doi.org/10.1016/j.watres.2009.07.004

Saleh TA, Tuzen M, Sarı A (2019) Magnetic vermiculite-modified by poly(trimesoyl chloride-melamine) as a sorbent for enhanced removal of bisphenol A. J Environ Chem Eng 7:103436. https://doi.org/10.1016/j.jece.2019.103436

Salem Attia TM, Hu XL, Yin DQ (2013) Synthesized magnetic nanoparticles coated zeolite for the adsorption of pharmaceutical compounds from aqueous solution using batch and column studies. Chemosphere 93:2076–2085. https://doi.org/10.1016/j.chemosphere.2013.07.046

Salem Attia TM, Hu XL, Yin DQ (2014) Synthesised magnetic nanoparticles coated zeolite (MNCZ) for the removal of arsenic (As) from aqueous solution. J Exp Nanosci 9:551–560. https://doi.org/10.1080/17458080.2012.677549

Sawchuk RJ, Maloney JA, Cartier LL, Rackley RJ, Chan KKH, Lau HSL (1995) Analysis of diclofenac and four of its metabolites in human urine by HPLC. Pharm Res 12:756–762. https://doi.org/10.1023/a:1016276012891

Senin RM, Ion I, Ion AC (2018) A sorption study of bisphenol A in aqueous solutions on pristine and oxidized multi-walled carbon nanotubes. Pol J Environ Stud 27:2245–2257. https://doi.org/10.15244/pjoes/78677

Shao L, Ren Z, Zhang G, Chen L (2012) Facile synthesis, characterization of a $MnFe_2O_4$/activated carbon magnetic composite and its effectiveness in tetracycline removal. Mater Chem Phys 135:16–24. https://doi.org/10.1016/j.matchemphys.2012.03.035

Shyu C, Cavileer TD, Nagler JJ, Ytreberg FM (2011) Computational estimation of rainbow trout estrogen receptor binding affinities for environmental estrogens. Toxicol Appl Pharmacol 250:322–326. https://doi.org/10.1016/j.taap.2010.11.005

Sing KSW, Everett DH, Haul RAW, Moscou L, Pierotti RA, Rouquérol J, Siemieniewska T (1985) Reporting physisorption data for gas/solid systems with special reference to the determination of surface area and porosity (Recommendations 1984). Pure Appl Chem 57:603–619

Sips R (1948) On the structure of a catalyst surface. J Chem Phys 16:490–495

Snyder SA, Adham S, Redding AM, Cannon FS, DeCarolis J, Oppenheimer J, Wert EC, Yoon Y (2007) Role of membranes and activated carbon in the removal of endocrine disruptors and pharmaceuticals. Desalination 202:156–181. https://doi.org/10.1016/j.desal.2005.12.052

Soares SF, Fernandes T, Sacramento M, Trindade T, Daniel-da-Silva AL (2019) Magnetic quaternary chitosan hybrid nanoparticles for the efficient uptake of diclofenac from water. Carbohydr Polym 203:35–44. https://doi.org/10.1016/j.carbpol.2018.09.030

Sohoni P, Tyler CR, Hurd K, Caunter J, Hetheridge M, Williams T, Woods C, Evans M, Toy R, Gargas M, Sumpter JP (2001) Reproductive effects of long-term exposure to bisphenol A in the fathead minnow (*Pimephales promelas*). Environ Sci Technol 35:2917–2925. https://doi.org/10.1021/es000198n

Song W, Yang T, Wang X, Sun Y, Ai Y, Sheng G, Hayat T, Wang X (2016) Experimental and theoretical evidence for competitive interactions of tetracycline and sulfamethazine with reduced graphene oxides. Environ Sci: Nano 3:1318–1326. https://doi.org/10.1039/c6en00306k

Sotelo JL, Ovejero G, Rodríguez A, Álvarez S, Galán J, García J (2014) Competitive adsorption studies of caffeine and diclofenac aqueous solutions by activated carbon. Chem Eng J 240:443–453. https://doi.org/10.1016/j.cej.2013.11.094

Sun W, Zhou K (2014) Adsorption of 17β-estradiol by multi-walled carbon nanotubes in natural waters with or without aquatic colloids. Chem Eng J 258:185–193. https://doi.org/10.1016/j.cej.2014.07.087

Swiss Federal Office for the Environment – FOEN (2015) Water quality: revision of the water protection act. Available online at www.bafu.admin.ch/bafu/fr/home/themes/formation/communiques.msg-id-59323.html. Accessed Nov 2019

Tang L, Yu J, Pang Y, Zeng G, Deng Y, Wang J, Ren X, Ye S, Peng B, Feng H (2018) Sustainable efficient adsorbent: Alkali-acid modified magnetic biochar derived from sewage sludge for aqueous organic contaminant removal. Chem Eng J 336:160–169. https://doi.org/10.1016/j.cej.2017.11.048

Temkin MJ, Pyzhev V (1940) Recent modifications to Langmuir isotherms. Acta Physiochim. URSS 12:217–222

United Nations World Water Assessment Programme – UNESCO/WWAP (2017) The United Nations World Water Development Report 2017. Wastewater: the untapped resource. UNESCO, Paris. ISBN 978-92-3-100201-4

United Nations World Water Assessment Programme – UNESCO/WWAP (2019) The United Nations World Water Development Report 2019: leaving no one behind. UNESCO, Paris. ISBN 978-92-3-100309-7

Vasquez MI, Lambrianides A, Schneider M, Kümmerer K, Fatta-Kassinos D (2014) Environmental side effects of pharmaceutical cocktails: what we know and what we should know. J Hazard Mater 279:169–189. https://doi.org/10.1016/j.jhazmat.2014.06.069

Verlicchi P, Al Aukidy M, Zambello E (2012) Occurrence of pharmaceutical compounds in urban wastewater: removal, mass load and environmental risk after a secondary treatment—a review. Sci Total Environ 429:123–155. https://doi.org/10.1016/j.scitotenv.2012.04.028

Worch E (2012) Adsorption technology in water treatment: fundamentals, processes, and modeling. De Gruyter, Berlin. ISBN 9783110240221

World Health Organization – WHO (2017a) Chemical mixtures in source water and drinking-water. WHO, Geneva. ISBN 978-92-4-151237-4

World Health Organization – WHO (2017b) Global priority list of antibiotic-resistant bacteria to guide research, discovery, and development of new antibiotics. WHO, Geneva, p 7

World Health Organization – WHO, United Nations Environment Programme – UNEP (2013) State of the science of endocrine disrupting chemicals – 2012. WHO/UNEP, Geneva/Nairobi. 296 p. ISBN 978 92 4 150503 1

Xie A, Cui J, Chen Y, Lang J, Li C, Yan Y, Dai J (2019) Simultaneous activation and magnetization toward facile preparation of auricularia-based magnetic porous carbon for efficient removal of tetracycline. J Alloys Compd 784:76–87. https://doi.org/10.1016/j.jallcom.2018.12.375

Yang RT (2003) Adsorbents: fundamentals and applications. Wiley, Hoboken. ISBN 978-0-471-29741-3

Yang W, Zheng F, Lu Y, Xue X, Li N (2011) Adsorption interaction of tetracyclines with porous synthetic resins. Ind Eng Chem Res 50:13892–13898. https://doi.org/10.1021/ie202166g

Yang Y, Hu X, Zhao Y, Cui L, Huang Z, Long J, Xu J, Deng J, Wu C, Liao W (2017) Decontamination of tetracycline by thiourea-dioxide–reduced magnetic graphene oxide: effects of pH, ionic strength, and humic acid concentration. J Colloid Interface Sci 495:68–77. https://doi.org/10.1016/j.jcis.2017.01.075

Yin R, Sun J, Xiang Y, Shang C (2018) Recycling and reuse of rusted iron particles containing core-shell Fe-FeOOH for ibuprofen removal: adsorption and persulfate-based advanced oxidation. J Cleaner Prod 178:441–448. https://doi.org/10.1016/j.jclepro.2018.01.005

Zaib Q, Khan IA, Saleh NB, Flora JRV, Park Y-G, Yoon Y (2012) Removal of bisphenol A and 17β-estradiol by single-walled carbon nanotubes in aqueous solution: adsorption and molecular modeling. Water, Air, Soil Pollut 223:3281–3293. https://doi.org/10.1007/s11270-012-1109-5

Zhang L, Song X, Liu X, Yang L, Pan F, Lv J (2011a) Studies on the removal of tetracycline by multi-walled carbon nanotubes. Chem Eng J 178:26–33. https://doi.org/10.1016/j.cej.2011.09.127

Zhang Y, Chen B, Zhang L, Huang J, Chen F, Yang Z, Yao J, Zhang Z (2011b) Controlled assembly of Fe_3O_4 magnetic nanoparticles on graphene oxide. Nanoscale 3:1446–1450. https://doi.org/10.1039/c0nr00776e

Zhang Y, Cheng Y, Chen N, Zhou Y, Li B, Gu W, Shi X, Xian Y (2014) Recyclable removal of bisphenol A from aqueous solution by reduced graphene oxide–magnetic nanoparticles: adsorption and desorption. J Colloid Interface Sci 421:85–92. https://doi.org/10.1016/j.jcis.2014.01.022

Zhang S, Dong Y, Yang Z, Yang W, Wu J, Dong C (2016) Adsorption of pharmaceuticals on chitosan-based magnetic composite particles with core-brush topology. Chem Eng J 304:325–334. https://doi.org/10.1016/j.cej.2016.06.087

Zhang M, Xu L, Qi C, Zhang M (2019) Highly effective removal of tetracycline from water by hierarchical porous carbon: batch and column adsorption. Ind Eng Chem Res 58:20036–20046. https://doi.org/10.1021/acs.iecr.9b03547

Zhou Q, Wang Y, Xiao J, Fan H (2016) Adsorption and removal of bisphenol A, α-naphthol and β-naphthol from aqueous solution by Fe_3O_4@polyaniline core–shell nanomaterials. Synth Met 212:113–122. https://doi.org/10.1016/j.synthmet.2015.12.008

Zhu F-D, Choo K-H, Chang H-S, Lee B (2012) Interaction of bisphenol A with dissolved organic matter in extractive and adsorptive removal processes. Chemosphere 87:857–864. https://doi.org/10.1016/j.chemosphere.2012.01.026

Zhu X, Liu Y, Qian F, Zhou C, Zhang S, Chen J (2014a) Preparation of magnetic porous carbon from waste hydrochar by simultaneous activation and magnetization for tetracycline removal. Bioresour Technol 154:209–214. https://doi.org/10.1016/j.biortech.2013.12.019

Zhu X, Liu Y, Zhou C, Luo G, Zhang S, Chen J (2014b) A novel porous carbon derived from hydrothermal carbon for efficient adsorption of tetracycline. Carbon 77:627–636. https://doi.org/10.1016/j.carbon.2014.05.067

Chapter 8
Magnetically Modified Biological Materials for Dye Removal

Ivo Safarik, Eva Baldikova, Jitka Prochazkova, and Kristyna Pospiskova

Contents

Abstract Various types of biological materials including agriculture and food waste and byproducts can be used as biosorbents for dye removal from contaminated waters. Magnetic modification of biosorbents enables to prepare smart materials

I. Safarik (✉)
Department of Nanobiotechnology, Biology Centre, ISB, CAS, Ceske Budejovice, Czech Republic

Regional Centre of Advanced Technologies and Materials, Palacky University, Olomouc, Czech Republic

Department of Magnetism, Institute of Experimental Physics, SAS, Kosice, Slovakia

E. Baldikova · J. Prochazkova
Department of Nanobiotechnology, Biology Centre, ISB, CAS, Ceske Budejovice, Czech Republic

K. Pospiskova (✉)
Regional Centre of Advanced Technologies and Materials, Palacky University, Olomouc, Czech Republic
e-mail: kristyna.pospiskova@upol.cz

L. Meili, G. L. Dotto (eds.), *Advanced Magnetic Adsorbents for Water Treatment*, Environmental Chemistry for a Sustainable World 61,
https://doi.org/10.1007/978-3-030-64092-7_8

exhibiting response to external magnetic field. Due to the presence of magnetic particles on or within the modified biosorbents, they can be rapidly, easily, and selectively separated from desired environments by means of magnetic separators. Magnetically responsive biomaterials represent very interesting, progressive, and easily obtainable biosorbents for potential environmental technology applications.

Keywords Biosorbents · Magnetic modification · Iron oxides · Waste biomaterials · Plant-based materials · Microbial cells · Microalgae · Marine algae · Dyes · Adsorption

Abbreviations

P-2-O	Pseudo-second-order
q_m	Maximum adsorption capacity

8.1 Introduction

Contamination of water resources by various types of pollutants represents currently one of the most important problems which human society has to solve. Huge amount of contaminated wastewater is produced and poured into the water environment at any moment. Enormous amount of both organic and inorganic pollutants is present in water resources. Synthetic dyes represent very significant contaminants and a large amount of dye containing wastewater is produced every year. Many different dye types are produced, including direct, acid, basic, reactive, mordant, metal complex, vat, sulfur, and disperse dyes. Dyes are often employed chemical compounds and are widely used in many fields of industry to color their products, such as textile, leather, paper, rubber, printing, plastics, and cosmetics. The total worldwide production of synthetic dyes is nearly 800,000 tons per year (Hassaan and El Nemr 2017; Katheresan et al. 2018; Saravanan et al. 2020).

Among different types of wastewaters, wastewater containing dissolved dyes and their metabolites deserves significant attention. Especially textile industries produce large amounts of liquid wastes containing both organic and inorganic compounds; dyes unfixed during the dyeing process are found in high concentrations. Many dyes are nonbiodegradable and carcinogenic and pose a major threat to health and the environment (Hassaan and El Nemr 2017; Katheresan et al. 2018; Saravanan et al. 2020).

Various procedures including flocculation, coagulation, chemical oxidation, membrane separation, electrodialysis, and aerobic and anaerobic microbial degradation have been developed for the removal of synthetic dyes. However, each of these methods has inherent limitations (Yagub et al. 2014).

Adsorption is characterized by the accumulation of a substance (adsorbate) to the adsorbent at the liquid–solid interface or gas–solid interface. Adsorption can be classified into two types: chemical sorption and physical sorption. Chemical adsorption (chemisorption) is characterized by the formation of strong chemical associations between the adsorbate and the adsorbent surface; thus chemisorption is usually irreversible. On the contrary, physical adsorption (physisorption) is characterized by weak interactions between adsorbate and adsorbent; that is why this interaction is usually reversible. Physisorption on most (bio)sorbents employs several physical forces including van der Waals forces, hydrogen bonds, polarity and dipole–dipole interactions, as well as π–π interactions (Yagub et al. 2014).

Adsorption is an efficient procedure for the treatment of polluted waters, especially in case of inexpensive adsorbents. Contaminants practically unaffected by conventional biological wastewater treatments can be efficiently removed. Adsorption is usually a flexible, simple, and low-cost procedure which does not produce harmful substances. Several factors including adsorbent surface area, adsorbent to adsorbate ratio, adsorbent particle size, temperature, pH, and contact time influence the adsorption process (Katheresan et al. 2018; Forgacs et al. 2004; Zhou et al. 2019; Yagub et al. 2014; Ahmad et al. 2015).

Magnetically modified biological materials can be efficiently used as adsorbents for dye removal. In this chapter, we summarize basic information about these materials and processes. Magnetic biosorbents used for dye adsorption are presented in table form.

8.2 Biological Materials for Dye Adsorption

Recently, a large number of papers have been published describing the applications of various biological materials that can serve as biosorbents for the removal of huge amount of organic and inorganic xenobiotics from contaminated water resources. Biological materials, such as peat, chitosan, alginate, plant gums, yeast, fungi, bacterial and algae biomass, sawdust, spent coffee grounds, and peanut husks, have already been used as biosorbents to remove different types of pollutants. In many cases, biosorbents are low-cost biological materials including wastes from food industry and agriculture (Safarik et al. 2011b). For the first time, the term "biosorbent" appeared in the 1970s (Jilek et al. 1976). Since that time enormous amount of scientific publications has appeared; currently (end of March 2020), there are 18,811 items covering the terms biosorbent*, bioadsorbent*, biosorption, or bioadsorption in Web of Science, and 13,910 items in Scopus.

Biosorption is usually defined as the removal of target substances from solutions by biological materials. Biosorption is a physicochemical process independent of metabolism that involves different types of mechanisms including absorption, adsorption, ion exchange, surface complexation, and precipitation. Biosorption is a property of both living and dead organisms, other biological materials, and their components. Biosorption of pollutants on biological materials usually includes several mechanisms based on the presence of many functional groups (e.g., hydroxyl, amino, carboxyl, phosphate, sulfate, amido, thiol, imidazole, and acetamido), which can interact with target pollutants. In many cases, the binding capacities of certain biosorbents are fully comparable with those of the commercial synthetic cation exchange resins. However, despite continuing dramatic increases in published research on biosorption and biosorbents, there has been little exploitation in an industrial scale (Safarik et al. 2011b).

Enormous amount of possible biosorbents is available for the potential xenobiotic removal. The renewable character of biomass makes it an inexhaustible pool of biosorbents of all kinds. Biosorption has several advantages compared with conventional separation techniques, namely low cost of the biosorbents (majority of them is made from abundant or even waste material). In addition, large variability of biological materials and their pretreatment enables to select an appropriate biosorbent exhibiting at least partial selectivity in xenobiotic adsorption. In specific cases, it is possible to regenerate the biosorbents (especially when this process is economically feasible), enabling their reuse (Srinivasan and Viraraghavan 2010; Abdolali et al. 2014; Adeniyi and Ighalo 2019; Safarik et al. 2011b).

8.3 Magnetically Modified Biological Materials for Dye Adsorption

Magnetically responsive materials are usually described as smart (stimuli-responsive) materials, exhibiting several specific responses to the external magnetic field. That is why magnetically responsive materials are used in many applications. These materials can be selectively separated from complex and difficult-to-handle environments including wastewater by means of an appropriate magnetic separator, and that is why they can be efficiently applied in environmental technologies. Possibility of magnetic separation is the basis of the most often used applications of magnetic particles in biosciences and environmental technologies. Absolute majority of biological materials have diamagnetic properties, decoration of their surface with magnetic particles results in preparation of magnetically modified materials. Selective efficient separation of these composites enables important broadening of application potential (Safarik et al. 2011b).

8.3.1 Magnetic Materials for Modification of Biological Materials

Many procedures to produce various types of magnetic iron oxide nano- and micromaterials such as classical coprecipitation, hydrothermal reactions, reactions in microemulsions, sol-gel syntheses, sonochemical reactions, electrospray and flow injection syntheses, hydrolysis and thermolysis of precursors, microwave-assisted synthesis, and mechanochemical processes have been developed (Majidi et al. 2016; Frey et al. 2009; Hasany et al. 2013; Laurent et al. 2008).

The most widely used process to obtain magnetite (Fe_3O_4) and maghemite (γ-Fe_2O_3) (nano)particles is based on the use of the coprecipitation reaction. Both magnetic iron oxides are prepared by aging a stoichiometric mixture of Fe^{2+} and Fe^{3+} salts in aqueous alkaline medium. Magnetite formation follows the simple chemical reaction (reaction 8.1):

$$Fe^{2+} + 2Fe^{3+} + 8OH^- \rightarrow Fe_3O_4 + 4H_2O \quad (8.1)$$

Formed magnetite (Fe_3O_4) is usually converted to maghemite (γ-Fe_2O_3) because it is not very stable and is sensitive to oxidation. The main advantage of precipitation reactions is that large quantities of particles can be obtained. The standard coprecipitation process generates particles with a broad size distribution. The size and shape of the iron oxide particles depend on the type of salts used, the ratio of ferric and ferrous ions, the pH value, the reaction temperature, the ionic strength of the media, and the other reaction parameters including stirring rate and the dropping speed of alkaline solution (Majidi et al. 2016; Laurent et al. 2008).

Hydrothermal synthesis refers to the heterogeneous reactions for synthesizing inorganic materials in aqueous media above ambient temperature and pressure. High pressure and temperature hydrothermal syntheses in autoclaves or reactors have also been used for synthesis of magnetite nanoparticles; the pressure is often higher than 13.8 MPa and the temperature exceeds 200 °C. Reaction conditions can influence the magnetite particle size (Laurent et al. 2008; Torres-Gomez et al. 2019).

Magnetic nanoparticles can also be synthesized by thermal decomposition of organometallic compounds at elevated temperatures by cleavage of chemical bonds. This method is mostly employed for organometallic compounds (e.g., acetyl acetonates) in organic solvents such as benzyl ether, containing also surface active agent such as oleic acid or oleyl amine. Morphology and size of particles can be controlled by varying the precursor composition (Khan et al. 2019).

The sol-gel process enables production of solid materials from small molecules. Hydroxylation and condensation of molecular precursors in solution lead to the formation of a "sol" of nanometer-sized particles. In the next step, the condensation and inorganic polymerization lead to the formation of a three-dimensional metal oxide network. These reactions are carried out at room temperature; that is why further heat treatments are needed to acquire the final crystalline state (Laurent et al. 2008). Using this procedure, 8 nm magnetite nanoparticles have been prepared (Lemine et al. 2012).

The polyol process employs polyethylene glycol or other selected polyols as solvents capable to dissolve inorganic compounds and also exhibiting high dielectric constants and high boiling points. Polyols can also serve as both reducing agents and stabilizers to control particle growth and prevent interparticle aggregation (Laurent et al. 2008). Facile synthesis of superparamagnetic magnetite nanoparticles in liquid polyols was described (Cai and Wan 2007).

Fig. 8.1 Scanning electron microscopy of magnetic iron oxide nano- and microparticles prepared by microwave-assisted synthesis from ferrous sulfate at high pH

Laser and spray pyrolysis are typical examples of aerosol technologies which enable the production of magnetic nanoparticles at a high rate. During spray pyrolysis, a solution of Fe^{3+} salt and a reducing agent in organic solvent is sprayed into a series of reactors, enabling aerosol solute condensation and solvent evaporation. Maghemite nanoparticles with different shapes and sizes (5–60 nm) were obtained using various iron precursor salts in alcoholic solution (Laurent et al. 2008).

Mechanochemical process has been used for magnetic iron oxide particle synthesis. Grinding in a mortar or ball milling of ferrous and ferric chlorides with sodium hydroxide led to the formation of magnetite particles which can be subsequently converted to maghemite during the milling (Lin et al. 2006; Safarik et al. 2014).

Microwave irradiation of an appropriate ferrous salt solution at high pH in a regular domestic microwave oven (700–750 W, 2450 MHz) for a suitable reaction time leads to the formation of magnetic iron oxide nano- and microparticles (Safarik and Safarikova 2014; Zheng et al. 2010); typical appearance of the formed magnetic label is shown in Fig. 8.1.

In addition to iron oxide particles, ferrite and chromium dioxide nano- and microparticles, nickel, or metallic cobalt have been used for specific magnetization applications; alternatively, paramagnetic cations (e.g., erbium ions) can be employed as magnetic labels (Safarik et al. 2016c).

8.3.2 Magnetic Modification of Biological Materials

Many already described procedures have been successfully used for magnetic modification of originally diamagnetic (nonmagnetic) biological materials (Safarik et al. 2012a, 2016c, 2018; Kanjilal and Bhattacharjee 2018). Magnetic modification is usually caused by the presence of magnetic labels (especially magnetic (nano)-particles) within the pores of the treated biomaterials, on the surface of the biomaterials, or within the biopolymer gels.

A simple and often used magnetic modification procedure is based on the standard chemical coprecipitation method, where magnetic iron oxide particles are prepared by aging a stoichiometric mixture of Fe^{2+} and Fe^{3+} salts in aqueous alkaline medium in the presence of modified diamagnetic biological materials, followed by heating (Safarik et al. 2012a; Taha and El-Maghraby 2016). Magnetic biocomposites containing different amount of iron oxide particles on their surface can be prepared (Khandanlou et al. 2014). Very often slightly modified procedures (e.g., use of inert gas during the magnetization process) have been described (Khandanlou et al. 2013). Alternatively, ferrites have been utilized as a magnetic label when ferrous ions are substituted by other divalent cations (e.g., Cu^{2+}, Mn^{2+}) (Hashemian and Salimi 2012; Wang et al. 2018b).

Laboratory-scale magnetic modification procedure employed perchloric acid-stabilized water-based magnetic fluid (ferrofluid). The modification of sawdust and peanut husks was performed in methanol (Safarik et al. 2007a; Safarik and Safarikova 2010). Alternatively, the modification was carried out by direct mixing of ferrofluid and the modified biological material, followed by drying; aggregates of magnetic iron oxides were deposited on the treated material (Safarik et al. 2012b).

Mechanochemical synthesis of various types of magnetically responsive biosorbents from nonmagnetic powdered precursors was performed by milling or grinding of ferrous and ferric salts with the modified biomaterial; after addition of powdered sodium or potassium hydroxide magnetic biosorbents were formed due to the deposition of magnetic iron oxide particles on the treated surfaces (Safarik et al. 2014).

Simple and rapid magnetic modification of diamagnetic materials can be performed using magnetic iron oxide nano- and microparticles prepared by microwave-assisted synthesis from ferrous sulfate at high pH. Direct one-pot synthesis is based on the microwave irradiation of the suspension of the treated material with iron(II) hydroxide prepared by alkalization of ferrous sulfate (Safarik et al. 2013). Indirect postmagnetization procedure employs magnetic iron oxide microparticles (prepared using a microwave-assisted procedure from ferrous sulfate at high pH) which are thoroughly mixed with the powder material to be modified; after drying, stable magnetically responsive materials are formed (Safarik and Safarikova 2014). In specific cases (magnetic modification of agglomerate forming materials), iron oxide particles were suspended in methanol or other water-miscible organic solvent (Safarik et al. 2016b). Typical appearance of native biosorbent and its

Fig. 8.2 Native and magnetically modified orange peel (left) and magnetic separation of magnetically modified biosorbent (right). Magnetic modification was performed by indirect postmagnetization procedure employing magnetic iron oxide nano- and microparticles prepared by microwave-assisted synthesis from ferrous sulfate at high pH

magnetic derivative prepared by indirect microwave postmagnetization procedure, and magnetic separation of modified biosorbent is presented in Fig. 8.2.

8.4 Magnetically Modified Plant-Based Materials for Dye Adsorption

Many studies are available on biosorption using naturally occurring and low-cost biosorbents, such as agricultural and related byproducts and wastes. Among them, sawdust, cereal byproducts (e.g., straw, husk, spent grain, maize cob), tea and coffee waste, and other waste biomaterials (e.g., litchi pericarps, oak acorns, peanut hulls and husks, pomelo peel, sugarcane bagasse, etc.) have been magnetically modified and used for dye adsorption (Safarik et al. 2018). The described magnetic plant-based biosorbents and further relevant information are summarized in Tables 8.1, 8.2, 8.3, and 8.4.

Sawdust belongs among the often-used low-cost biosorbents employed for the removal of acid and basic dyes, heavy metal ions, and some other unwanted materials from wastewater. Sawdust is both abundant and an efficient adsorbent for pollutant removal which is available in large quantities in lumber mills, and often presenting a disposal problem. The use of sawdust for pollutant removal can be useful for both the environmental technology and wood agriculture (Shukla et al. 2002; Saba et al. 2016). Literature reviewing the use of magnetically modified sawdust for dye adsorption is summarized in Table 8.1.

Cereal and maize are crops that are abundantly cultivated worldwide and are the most important staple foods for humans. Straw and husk produced by various cereals

Table 8.1 Literature review of magnetically modified sawdust for dye adsorption

Magnetic modification	Biological material treatment	Target dye	Analyzed material	Additional information	References
In situ coprecipitation of $CuFe_2O_4$ on material surface	Washing with water	Cyanine acid blue	Model solution	Maximum adsorption at pH 2 and 15 min, $q_m = 178.6$ mg/g; Langmuir and P-2-O models followed, exothermic process	Hashemian and Salimi (2012)
Treatment with water-based magnetic fluid stabilized with perchloric acid	None	Acridine orange, Bismarck brown Y, crystal violet, safranin O	Model solution	Spruce sawdust; magnetic and microscopy characterizations were carried out	Safarik et al. (2005)
Treatment with water-based magnetic fluid stabilized with perchloric acid	None	Acridine orange, Bismarck brown, crystal violet, malachite green, methyl green, Nile blue, safranin O, Saturn blue LBRR 200	Model solution	Spruce sawdust; q_m ranged between 34 and 59 mg/g	Safarik et al. (2007a)
Microwave irradiation in the presence of iron(II) sulfate at high pH	None	Bismarck brown Y, safranin O	Model solution	Spruce sawdust; $q_m = 50.1$ mg/g for Bismarck brown Y and 72.4 mg/g for safranin O	Safarik et al. (2013)
Treatment with Fe^{2+} and Fe^{3+} salts at high pH	Washing with water	Methylene blue	Model solution	*Tectona grandis* sawdust; q_m was 172.41 mg/g	Mashkoor and Nasar (2020)

(e.g., wheat, barley, rye, sorghum, or corn (maize)) are accessible in huge amounts and are mostly used as animal feed, animal bedding, fuel, and construction material. In some cases, straw is burnt in the field and wasted. There are many studies on the use of straw as a biosorbent for pollutant removal. Straw is typical of a large amount of carbohydrates (lignin and cellulose) containing functional groups, such as carboxyl, hydroxyl, and amino, which enable efficient pollutant adsorption. Straw modified with microwave-synthesized magnetic iron oxide nano- and microparticles is shown in Fig. 8.3. Chemical modification procedures employing acids, bases, or organic compounds can substantially increase straw adsorption properties (Baldikova et al. 2016a). Also spent grain and wheat bran can be employed for the same purpose (Saba et al. 2016). Literature reviewing the use of magnetically modified cereal materials for dye removal is summarized in Table 8.2.

Table 8.2 Literature review of magnetically modified cereal materials for dye removal

Biological material	Magnetic modification	Biological material treatment	Target dye	Analyzed material	Additional information	References
Barley straw	Postmagnetization with microwave-synthesized magnetic iron oxide particles	Citric acid-NaOH modification	Methylene blue, crystal violet, Bismarck brown Y, safranin O	Model solution	Chemical modification led to the four times increase in q_m	Baldikova et al. (2016a)
Corncobs	In situ coprecipitation of magnetic iron oxides	Ethanediamine modification	Congo red	Model solution	Langmuir and P-2-O kinetic models followed; $q_m = 198.2$ mg/g	Liu et al. (2016)
Corn straw	Impregnation by Fe_3O_4 nanoparticles in glutaraldehyde solution	Glutaraldehyde crosslinking and glutamic acid treatment	Methylene blue	Model solution	$q_m = 194.5$ mg/g at contact time 60 min, pH 6 and 60 °C	Zhao et al. (2014)
Corn straw	Impregnation by magnetic graphene oxide	Citric acid, graphene oxide	Methylene blue	Model solution	Freundlich and P-2-O kinetic models followed; $q_m = 315.5$ mg/g	Ge et al. (2016)
Maize cob	Impregnation by Fe_3O_4 nanoparticles (postmagnetization)	Activation with ammonium hydroxide	Methylene blue	Model solution	Optimal pH 6; P-2-O model followed	Tan et al. (2012)
Rice husk	Magnetic alginate/rice husk biocomposite beads	Ionotropic gelation method	Methylene blue	Model solution	$q_m = 274.9$ mg/g	Alver et al. (2020)
Rice straw	Branched polyethyleneimine-modified magnetic straw	Cutting into small pieces	Methyl orange	Model solution	$q_m = 510.2$ mg/g	Wang et al. (2018a)
Rye straw	Postmagnetization with microwave-synthesized magnetic iron oxide particles	Citric acid – NaOH modification	Acridine orange, methyl green	Model solution	More than four times increase in q_m after chemical modification	Baldikova et al. (2015)
Sorghum husk	In situ coprecipitation of magnetic iron oxides	Washing with water	Crystal violet, methylene blue	Model solution	$q_m = 18.87$ and 30.00 mg/g for CV and MB, resp.	Adeogun et al. (2019)

Spent grain	Treatment with water-based magnetic fluid stabilized with perchloric acid	Washing with water	Aniline blue, Bismarck brown Y, crystal violet, Nile blue	Model solution	Highest q_m for Bismarck brown (72.4 mg/g)	Safarik et al. (2011a)
Spent grain	Microwave irradiation in the presence of iron(II) sulfate at high pH	Washing with water	Bismarck brown Y, safranin O	Model solution	Also used as carrier for lipase immobilization	Safarik et al. (2013)
Wheat husk	Impregnation by magnetic particles	None	Methylene blue	Model solution	Optimal pH 5, Langmuir model followed, exothermic process	Shah et al. (2014)
Wheat straw	Impregnation by Fe_3O_4 nanoparticles (postmagnetization)	NaOH treatment	Methylene blue	Model solution	$q_m = 1374.6$ mg/g; P-1-O model followed	Ebrahimian Pirbazari et al. (2014)
Wheat straw	Impregnation by Fe_3O_4 nanoparticles (postmagnetization)	None	Basic Blue 9	Model solution	$q_m = 627.1$ mg/g; P-2-O model followed	Pirbazari et al. (2016)
Wheat straw	Postmagnetization with microwave-synthesized magnetic iron oxide particles	None	Crystal violet	Model solution	Experiment for secondary school students	Baldikova et al. (2016b)

Table 8.3 Literature review of magnetically modified coffee and tea derivatives for dye adsorption

Biological material	Magnetic modification	Biological material treatment	Target dye	Analyzed material	Additional information	References
Coffee particles	Treatment with water-based magnetic fluid stabilized with perchloric acid	Milling in a coffee mill	Methylene blue	Model solution	Langmuir model followed; $q_m = 88.49$ mg/g	Besharati et al. (2018)
Coffee silverskin, defective green coffee, spent coffee grounds	Treatment with ferrofluid stabilized with tetramethylammonium hydroxide	Washing with hot water	Methylene blue	Model solution	Regeneration six times, q_m highest for coffee silverskin, lowest for defective green coffee; Langmuir and P-2-O models followed	Zuorro et al. (2014)
Coffee silverskin	Treatment with ferrofluid stabilized with tetramethylammonium hydroxide	Washing with hot water	Methylene blue	Model solution	Langmuir and P-2-O models followed; $q_m = 556$ mg/g	Zuorro et al. (2013)
Green tea waste leaves	In situ coprecipitation of Fe_3O_4 on adsorbent surface	Boiling in hot water	Malachite green	Model solution	$q_m = 112.3$ mg/g	Besharati and Alizadeh (2018)
Spent coffee grounds	Treatment with water-based magnetic fluid stabilized with perchloric acid	Washing with hot water	Acridine orange, amido black 10B, Bismarck brown Y, Congo red, crystal violet, malachite green, safranin O	Model solution	Equilibrium in 90 min, also used for preconcentration of the target analytes from diluted solutions (MSPE)	Safarik et al. (2012c)
Spent coffee grounds	Microwave irradiation in the presence of iron (II) sulfate at high pH	Washing with hot water	Bismarck brown Y, safranin O	Model solution	$q_m = 49.3$ mg/g for Bismarck brown Y and 146.6 mg/g for safranin O	Safarik et al. (2013)

Waste rooibos tea	Postmagnetization with microwave-synthesized magnetite	Washing with hot water	Crystal violet, brilliant green, methyl green, Bismarck brown, acridine orange, methylene blue, Nile blue, safranin O	Model solution	q_m higher than 70 mg/g	Safarik et al. (2015a)
Waste tea leaves	In situ coprecipitation of Fe_3O_4 on the adsorbent surface	Washing with hot water	Janus green, methylene blue, thionine, crystal violet, Congo red, neutral red, reactive blue 19	Model solution	Adsorption higher for cationic dyes than for anionic ones	Madrakian et al. (2012)
Waste tea leaves	Microwave irradiation in the presence of iron (II) sulfate at high pH	Washing with hot water	Bismarck brown Y, safranin O	Model solution	Also used as carrier for lipase immobilization	Safarik et al. (2013)
Waste tea leaves	Treatment with ferrofluid stabilized with perchloric acid	Washing with boiling water	Acridine orange, Bismarck brown, crystal violet, malachite green, methyl green, Nile blue A, safranin O	Model solution	q_m up to 100 mg/g achieved	Safarik et al. (2012b)

Table 8.4 Literature review of other magnetically modified plant-derived materials for dye adsorption

Biological material	Magnetic modification	Biological material treatment	Target dye	Analyzed material	Additional information	References
Argan press cake nanocellulose	Modification with cobalt ferrite particles	Nanocellulose prepared from cellulose extracted from argan press cake	Sudan I, Sudan II, Sudan III, Sudan IV	Model solution, barbecue, and ketchup sauces	Efficient dye extraction from food samples	Benmassaoud et al. (2017)
Azolla filiculoides	In situ coprecipitation of magnetic iron oxides	Washing with water and drying	Crystal violet, methylene blue	Model solution	$q_m = 30.21$ and 25.0 mg/g for crystal violet and methylene blue resp.	Alizadeh et al. (2017)
Azolla filiculoides	In situ coprecipitation of magnetic iron oxides	Washing with water and drying	Malachite green	Model solution	$q_m = 23.1$ mg/g	Besharati and Alizadeh (2018)
Fig leaves	In situ coprecipitation of magnetic iron oxides	Washing with water and drying at 80 °C	Crystal violet, methylene blue	Model solution	$q_m = 53.47$ and 61.72 mg/g for crystal violet and methylene blue resp.	Alizadeh et al. (2017)
Fig leaves	In situ coprecipitation of magnetic iron oxides	Washing with water and drying at 80 °C	Malachite green	Model solution	$q_m = 73.2$ mg/g	Besharati and Alizadeh (2018)
Grapefruit peel	Solvothermal synthesis using grapefruit peel and ferric chloride	Washing with water	Congo red	Model solution	Adsorption followed P-2-O kinetic model and Freundlich isotherm	Inkoua et al. (2020)
Litchi pericarps	Impregnation by Fe_3O_4 nanoparticles	Washing with water	Malachite green	Model solution	Langmuir and P-2-O kinetic models followed; $q_m = 70.4$ mg/g	Zheng et al. (2015)
Oak acorns	Mechanochemical synthesis	None	Bismarck brown Y, safranin O	Model solution	$q_m = 20.9$ mg/g for Bismarck brown Y and 12.9 mg/g for safranin O	Safarik et al. (2014)

Peanut hulls	In situ coprecipitation of magnetic iron oxides	None	Methylene blue	Model solution	Temkin and P-2-O models followed	Taha and El-Maghraby (2016)
Peanut husk	Treatment with water-based magnetic fluid stabilized with perchloric acid	Washing with water	Malachite green	Model solution	q_m = 98.03 mg/g	Besharati and Alizadeh (2018)
Peanut husks	Treatment with water-based magnetic fluid stabilized with perchloric acid	None	Acridine orange, Bismarck brown Y, crystal violet, safranin O	Model solution	Equilibrium in 60–90 min; high q_m (71.4–95.7 mg/g)	Safarik and Safarikova (2010)
Peanut husks	Microwave irradiation in the presence of iron (II) sulfate at high pH	None	Bismarck brown Y, safranin O	Model solution	Also used as carriers for lipase immobilization	Safarik et al. (2013)
Peanut husks	Treatment with water-based magnetic fluid stabilized with perchloric acid	Washing with water	Methylene blue	Model solution	Langmuir model followed; q_m = 74.62 m/g	Besharati et al. (2018)
Pomelo peel	Postmagnetization with a suspension of magnetic particles in methanol	None	Acridine orange, methylene blue	Model solution	Langmuir model followed; q_m = 106.6 mg/g for acridine orange and 179.0 mg/g for methylene blue	Safarik et al. (2016b)
Sugarcane bagasse	In situ coprecipitation of Fe_3O_4 on adsorbent surface	Modification with pyromellitic dianhydride	Methylene blue, basic magenta	Model solution	q_m = 315.5 mg/g for methylene blue and 304.9 mg/g for basic magenta	Yu et al. (2012)
Sycamore (*Platanus occidentalis*) tree seed pod fibers	Reduction of adsorbed Fe^{3+} to Fe^0 using $NaBH_4$	Washing with water	Methylene blue, malachite green oxalate, methyl violet 2B	Synthetic wastewater	q_m = 92.59 mg/g for malachite green, 92.59 mg/g for methyl violet and 140.8 mg/g for methylene blue	Parlayici and Pehlivan (2019)

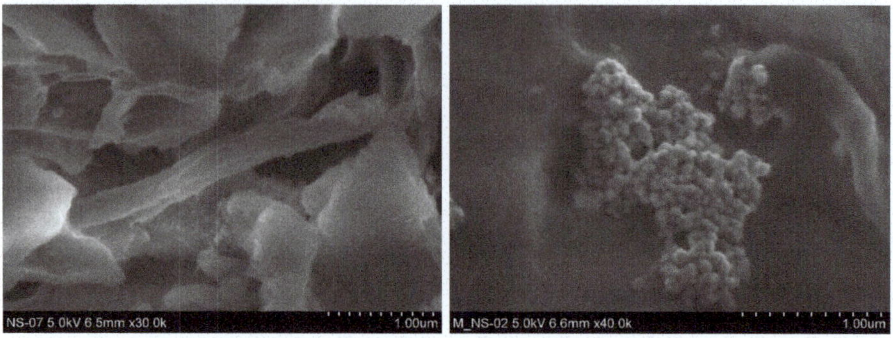

Fig. 8.3 Scanning electron microscopy images of native (left) and magnetically modified (right) barley straw. Magnetic modification was performed using microwave-synthesized iron oxide nano- and microparticles from ferrous sulfate at high pH. (Baldikova et al. 2016a)

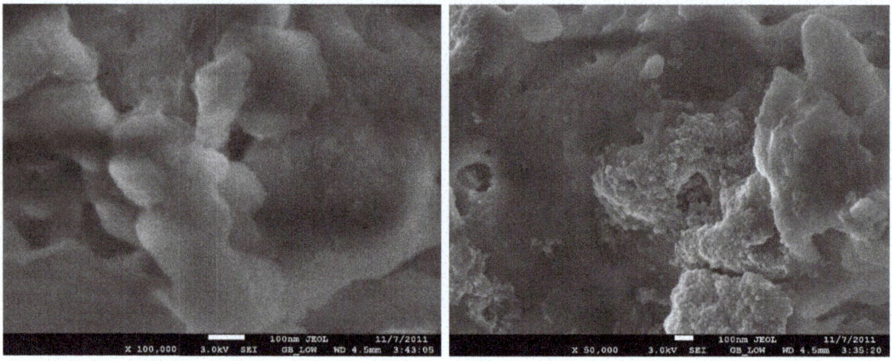

Fig. 8.4 Scanning electron microscopy images of native (left) and magnetically modified (right) tea. Magnetic modification was performed using perchloric acid-stabilized water-based magnetic fluid in methanol

Also coffee and tea derivatives are potential biosorbents for dye removal. Both coffee and tea are ones of the most popular and consumed beverages in the world. Several byproducts and waste biomaterials are generated during the processing of coffee fruits to the final products, namely defective green coffee, coffee silverskin, and spent coffee grounds. There are several studies showing that these biomaterials could be used as low-cost adsorbents for the removal of pollutants from wastewater (Zuorro et al. 2014). Also tea residue, a common waste obtained from households and beverage industries, has been employed as an efficient biosorbent for dye removal (Bommavaram et al. 2020). Magnetically responsive tea biosorbent can be easily prepared by modification with perchloric acid-stabilized water-based magnetic fluid (Safarik et al. 2016c); typical microscopy appearance of native and modified tea is shown in Fig. 8.4. In addition, spent rooibos (*Aspalathus linearis*) tea biomass has been employed for the same purpose (Safarik et al. 2015a). Literature reviewing the use of magnetically modified coffee and tea derivatives for dye adsorption is summarized in Table 8.3.

Also other types of waste plant-derived biomaterials including, for example, litchi pericarps, oak acorns, peanut husks, pomelo peel, or sugarcane bagasse have been magnetically modified and used for dye adsorption, as shown in Table 8.4.

8.5 Magnetically Modified Microbial Cells for Dye Adsorption

Diverse types of bacteria, fungi, and yeasts have been used as efficient adsorbents for the concentration and/or removal of different organic and inorganic pollutants. In addition to cultivated microbial cells, also low-cost waste microbial biomass can be efficiently used. Both living and dead microbial biomass can be employed as a biosorbent; the optimum form depends on the specific applications where both free and immobilized microorganisms can be used. The adsorption of xenobiotics on microbial cells depends on the composition of their cell walls. For instance, the cell walls of yeasts contain large number of complex organic compounds and their polymers such as glucan (28%), mannan (31%), proteins (13%), lipids (8%), chitin and chitosan (2%) (Safarik and Safarikova 2007; Safarik et al. 2011b).

Yeast biomass represents an important and promising material for dye biosorption. Yeast cells of the genus *Saccharomyces* and *Kluyveromyces* are non-pathogenic, are easily available, and enable simple manipulation. *Saccharomyces cerevisiae* cells (both baker's and brewer's yeasts), as well as *Kluyveromyces fragilis* (*marxianus*) cells (fodder yeast) can be magnetically modified by contact with perchloric acid-stabilized magnetic fluid (Safarikova et al. 2009; Uzun et al. 2011), as shown in Fig. 8.5. Very simple and inexpensive modification procedure is based on the use of microwave-synthesized magnetic iron oxide nano- and microparticles

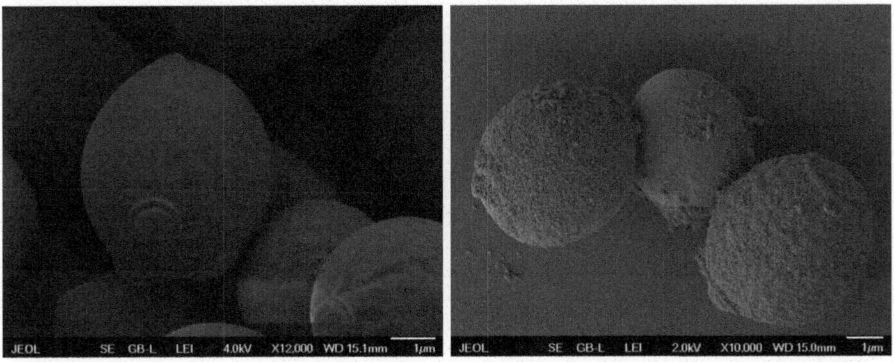

Fig. 8.5 Scanning electron microscopy of native cells of *Saccharomyces cerevisiae* (left) and cells modified with perchloric acid-stabilized water-based magnetic fluid in 0.1 M acetate buffer, pH 4.6 (right). (Safarikova et al. 2009)

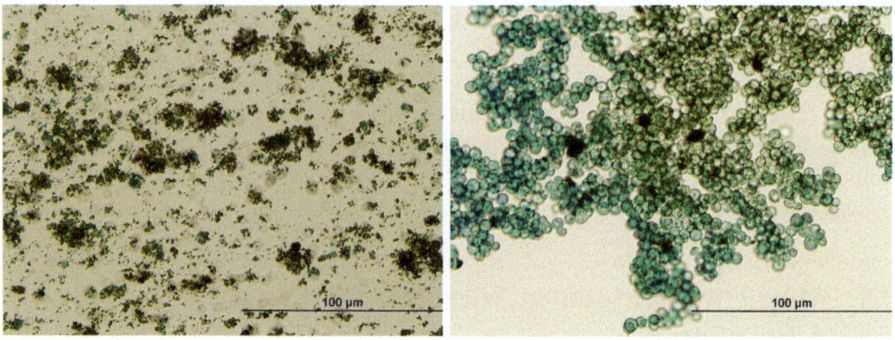

Fig. 8.6 Optical microscopy of microwave-synthesized magnetic iron oxide nano- and microparticles from ferrous sulfate at high pH (left) and stable magnetically responsive *Saccharomyces cerevisiae* cell aggregates prepared by the interaction of cells with magnetic particles (right). Perls' Prussian blue staining was applied to detect iron(III) compounds

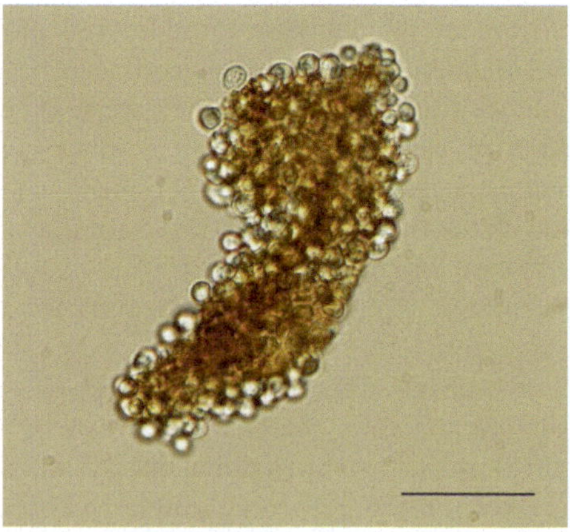

Fig. 8.7 Optical microscopy of *Saccharomyces cerevisiae* cells bound to microwave-synthesized magnetic chitosan particles. The bar corresponds to 25 µm

which enable flocculation of yeast cells and formation stable magnetically responsive yeast cell aggregates (Pospiskova et al. 2013) as shown in Fig. 8.6. Alternatively, yeast cells can be bound to microwave-synthesized magnetic chitosan particles (Safarik et al. 2015b), as shown in Fig. 8.7. Also selected bacteria have been used as biosorbents after their magnetic modification; Fig. 8.8 shows native and magnetically modified *Leptothrix* sheaths used for Amido black 10B removal (Safarik et al. 2017a; Angelova et al. 2017). Table 8.5 summarizes information about magnetically modified microbial cells used for dye adsorption.

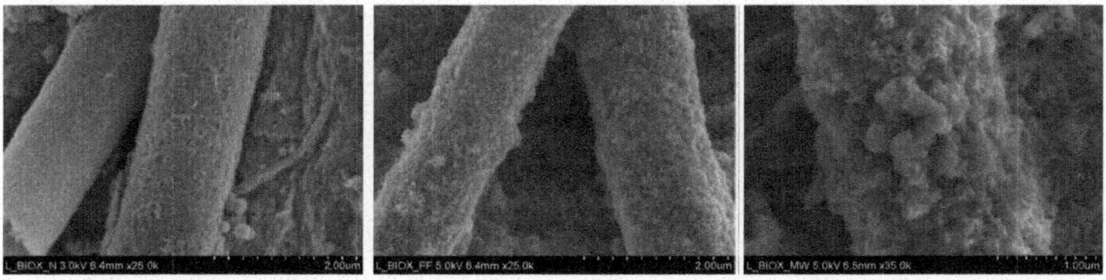

Fig. 8.8 Scanning electron microscopy of native *Lepthothrix* sp. sheaths (left), sheaths modified with perchloric acid-stabilized magnetic fluid (middle) and sheaths modified with microwave synthesized magnetic iron oxide nano- and microparticles (right)

8.6 Magnetically Modified Microalgae, Marine Algae, Seagrass and Related Organisms for Dye Adsorption

Biomass from unicellular algae, marine macroalgae and seagrass represents a typical example of low-cost, renewable natural biomaterial which can be obtained in large quantities. In many cases huge amounts of marine algae and seagrass can be found in beaches, thus causing problems to the tourist industry; the obtained biomass (often considered as a waste) can be efficiently used as adsorbents for the removal of specific pollutants including dyes or for the production of biochar. In addition to native algae biomass, also waste biomass obtained after selected industrial processes (e.g., solvent extraction of oil or colorants) can also be applied for the preparation of efficient biosorbents (Delrue et al. 2016; Safarik et al. 2020a, d).

Currently, only a few examples of magnetically modified microalgae, marine macroalgae, and seagrass have been described for dye separation and removal (Safarik et al. 2020a, d). *Chlorella vulgaris* in native and magnetically modified form is currently the only example of microalga used for the dye removal. Also marine macroalgae including *Cymopolia barbata* (see Fig. 8.9), *Cystoseira barbata, Sargassum horneri* (see Fig. 8.10), *Sargassum swartzii*, and seagrass *Posidonia oceanica* have been magnetically modified and subsequently employed as dye adsorbents. In addition, waste *Japonochytrium* sp. biomass after lipid extraction was employed as an efficient biosorbent for triphenylmethane dyes. More detailed information about these magnetically responsive biosorbents is presented in Table 8.6.

8.7 Magnetically Modified Marine Polysaccharides for Dye Adsorption

There are many papers describing dye adsorption on marine polysaccharides, especially chitosan, and alginate. Chitosan (polysaccharide composed of glucosamine and N-acetyl glucosamine) is prepared by deacetylation of chitin, which is produced

Table 8.5 Literature review of magnetically modified microbial cells for dye adsorption

Biological material	Magnetic modification	Biological material treatment	Target dye	Analyzed material	Additional information	References
Bacillus subtilis	Treatment with nano-sized magnetic silica	Washing with 0.9% NaCl	Methylene blue	Model solution	$q_m = 59$ mg/g; easy sorbent regeneration	Tural et al. (2017)
Beer yeast	In situ coprecipitation of magnetic iron oxides	Modification with pyromellitic dianhydride	Methylene blue, basic magenta	Model solution	q_m for methylene blue and basic magenta were 609.0 and 520.9 mg/g, resp.	Yu et al. (2013)
Kluyveromyces fragilis	Treatment with ferrofluid stabilized with perchloric acid	Washing with 0.1 M acetic acid	Crystal violet, amido black 10B, Congo red, Saturn blue LBRR, Bismarck brown, acridine orange, safranin O	Model solution	Langmuir model followed; q_m ranged between 29.9 (amido black 10B) and 138.2 (safranin O) mg/g	Safarik et al. (2007b)
Leptothrix sp.	Treatment with microwave synthesized iron oxide particles and perchloric acid stabilized ferrofluid	Washing with water	Amido black 10B	Model solution	$q_m = 339.2$ and 286.1 mg/g for ferrofluid modified and particles modified biosorbent, resp.	Angelova et al. (2017)
Neurospora sitophila immobilized on waste tea leaves	Modification with magnetite particles	Washing with water	Reactive violet 1	Model solution	Langmuir and P-2-O kinetic models followed; $q_m = 152.88$ mg/g	Divriklioglu et al. (2019)
Saccharomyces cerevisiae	Treatment with ferrofluid stabilized with perchloric acid	Suspension in 0.1 M acetate buffer, pH 4.6	Acridine orange, aniline blue, crystal violet, malachite green, safranin O	Model solution	Langmuir model followed; q_m ranged between 19.6 (malachite green) and 430.2 (aniline blue) mg/g	Safarik et al. (2002)
Saccharomyces cerevisiae subsp. *uvarum*	Treatment with ferrofluid stabilized with perchloric acid	Washing with saline	Aniline blue, Congo red, crystal violet, naphthol blue black, safranin O	Model solution	Langmuir model followed; $q_m = 228.0$ mg/g for aniline blue	Safarikova et al. (2005)

Fig. 8.9 Scanning electron microscopy of native (left) and magnetically modified (right) *Cymopolia barbata* biomass. Magnetic modification was performed using microwave-synthesized magnetic iron oxide nano- and microparticles. Presence of large amount of high aspect ratio calcium carbonate (aragonite) in both native and magnetically modified biomass is observed. (Mullerova et al. 2019)

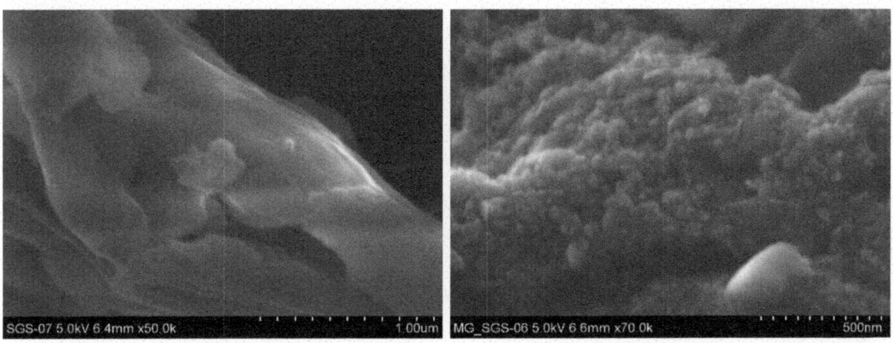

Fig. 8.10 Scanning electron microscopy images of native (left) and magnetically modified (right) *Sargassum horneri* biomass. Magnetic modification of native *Sargassum* biomass was performed using microwave-synthesized iron oxide nano- and microparticles in methanol. (Angelova et al. 2016)

by a great number of living creatures including crabs, other arthropods, and some fungi. Chitosan is soluble in acids and is chemically stable. Numerous dyes, mainly anionic ones, have been adsorbed by chitosan and its derivatives (Srinivasan and Viraraghavan 2010; Tran et al. 2015).

Several chitosan modification methods have been described to improve its sorption capacity and mechanical stability. Both physical and chemical modification procedures can be used. The physical modifications include the preparation of chitosan in various forms, such as powder, nanoparticles, gel beads, membranes, sponges, "honeycomb" type structure, and various types of fiber for subsequent biosorption applications. Chemical modifications involve crosslinking using glutaraldehyde or epichlorohydrin (Barbusinski et al. 2016). Also magnetic alginate and κ-carrageenan composite biosorbents have been prepared and tested. More detailed information about selected magnetically responsive marine polysaccharide-based biosorbents is presented in Table 8.7.

Table 8.6 Literature review of magnetically modified microalgae, marine algae, seagrass, and related organisms for dye adsorption

Biological material	Magnetic modification	Biological material treatment	Target dye	Analyzed material	Additional information	References
Cymopolia barbata	Modification with microwave-synthesized magnetic iron oxide particles	Washing with water	Safranin O	Model solution	Langmuir, Freundlich and P-2-O kinetic models followed; $q_m = 192.2$ mg/g	Mullerova et al. (2019)
Cystoseira barbata	Reaction of iron Fe(III) with NaOH at higher temperature	Washing with water	Methylene blue	Model solution	$q_m = 5.74$ and 1.08 mg/g at 25 °C and 45 °C, resp.	Ozudogru et al. (2016)
Chlorella vulgaris	Treatment with water-based magnetic fluid stabilized with perchloric acid	Washing with 0.1 M acetic acid	Aniline blue, Bismarck brown, Congo red, crystal violet, safranin O, Saturn blue LBRR	Model solution	q_m ranged between 24.2 (Saturn blue LBRR) and 257.9 (aniline blue) mg/g	Safarikova et al. (2008)
Pectin-*Chlorella vulgaris*	Precipitation of $FeSO_4$ by NaOH in the presence of pectin-*Chlorella vulgaris* followed by drying and heating	Composite formation	Methylene blue, methyl red, methyl orange, toluidine blue O, crystal violet, safranin O, bromothymol blue, bromophenol, malachite green	Model solution	Microwave-assisted regeneration of the used adsorbent. q_m ranged between 109.11 (methyl orange) and 287.62 (safranin O) mg/g	Khorasani and Shojaosadati (2019)

Japonochytrium sp.	Postmagnetization with a suspension of microwave-synthesized magnetic particles in methanol	Extraction of lipids	Crystal violet	Model solution	$q_m = 329.22$ mg/g at 294.15 K	Baldikova et al. (2019)
Posidonia oceanica	Modification with microwave-synthesized magnetic iron oxide particles	Washing with water and cutting into small fragments and milling	Acridine orange, Bismarck brown Y, brilliant green, crystal violet, methylene blue, Nile blue A sulfate, safranin O	Model solution	q_m ranged between 233.5 mg/g (Bismarck brown Y) and 88.1 mg/g (safranin O)	Safarik et al. (2016a)
Sargassum horneri	Postmagnetization with a suspension of microwave-synthesized magnetic particles in methanol	Washing with water	Acridine orange, crystal violet, malachite green, methylene blue, safranin O	Model solution	q_m ranged between 193.8 mg/g (acridine orange) and 110.4 mg/g (malachite green)	Angelova et al. (2016)

Table 8.7 Examples of magnetically modified marine polysaccharides for dye adsorption

Marine polysaccharide	Magnetic and other modification	Target dye	Analyzed material	Additional information	References
Alginate	In situ coprecipitation from $FeCl_3$ and $FeCl_2$ at high pH in presence of sodium alginate	Malachite green	Model solution	$q_m = 47.84$ mg/g	Mohammadi et al. (2014)
Alginate	Surface coating of nano cobalt ferrite with sodium alginate	Congo red, brilliant green, methylene blue	Model solution	$q_m = 93.0$, 92.1 and 95.8 mg/g for Congo red, brilliant green and methylene blue, resp.	Kumar et al. (2019)
Alginate	Sodium alginate @$CoFe_2O_4$-polydopamine beads	Methylene blue, crystal violet, malachite green	Model solution	$q_m = 466.60$ mg/g, 456.52 mg/g and 248.78 mg/g for methylene blue, crystal violet and malachite green, resp.	Li et al. (2016b)
κ-Carrageenan	Magnetic iron oxide nanoparticles coated with κ-carrageenan/silica organic/inorganic hybrid shells	Methylene blue	Model solution	$q_m = 530$ mg/g	Soares et al. (2017)
κ-Carrageenan	Magnetic beads composed of κ-carrageenan and Fe_3O_4 nanoparticles	Crystal violet	Model solution	$q_m = 84.7$ mg/g	Mahdavinia et al. (2014)
Chitosan	Glutaraldehyde crosslinked magnetic chitosan nanoparticles	Metanil Yellow, Reactive Black 5	Model solution	$q_m = 620$ mg/g for Metanil Yellow and 2549 mg/g for Reactive Black 5 at pH 3, 30 °C.	Tarhan et al. (2018)
Chitosan	Magnetic γ-Fe_2O_3/crosslinked chitosan composite	Methyl orange	Model solution	Adsorption kinetics followed the pseudo-second-order kinetic model	Zhu et al. (2010)
Chitosan	Magnetic chitosan-Fe(III) hydrogel	Acid Red 73	Model solution	$q_m = 294.5$ mg/g under pH = 12	Shen et al. (2011)
Chitosan	Ethylenediamine-modified magnetic chitosan nanoparticles	Acid Orange 7, Acid Orange 10	Model solution	$q_m = 3.47$ mmol/g for Acid Orange 7 and 2.25 mmol/g for Acid Orange 10	Zhou et al. (2011)

Chitosan	Chemically modified chitosan crosslinked using glutaraldehyde in the presence of magnetite	Reactive Black 5	Model solution	The best $q_m = 0.78$ mmol/g at 25 °C	Elwakeel (2009)
Chitosan	A quaternary ammonium salt modified chitosan magnetic composite adsorbent	Methyl orange	Model solution	Simultaneous adsorption of methyl orange and Cr(VI) ions studied	Li et al. (2016a)
Chitosan	$CoFe_2O_4$/chitosan magnetic composite	Indigotine blue	Model solution	$q_m = 224.81$ mg/g at 298 K and 289.87 mg/g at 328 K	Pang et al. (2020)
Chitosan	Templated crosslinked magnetic chitosan nanoparticles	Remazol Black 5, Remazol Brilliant Orange 3R	Model solution	Imprinting process used for biosorbent preparation	Chen et al. (2011)
Chitosan	Magnetic Fe_3O_4@chitosan@taurine adsorbent	Methylene blue	Model solution	$q_m = 204.1$ mg/g at pH 5 and 384.6 mg/g at pH 9	Huang et al. (2020)
Chitosan	Magnetic chitosan gel particles with covalently immobilized copper phthalocyanine dye	Acridine orange, Congo red, crystal violet, neutral red, safranin O	Model solution	q_m ranged between 17 and 54 mg/g	Safarik (1995)
Chitosan	A magnetic composite material composed of nano-magnetite, heulandite, and crosslinked chitosan	Methylene blue, methyl orange	Model solution	Adsorption followed the pseudo-second-order kinetics; $q_m = 45.1$ and 149.2 mg/g for methylene blue and methyl orange at pH 5.5, resp.	Cho et al. (2015)

8.8 Recyclability of Spent Magnetic Biosorbents

Biosorption is a very efficient procedure for the removal of organic pollutants (Safarik et al. 2018). However, the exhausted biosorbent after the adsorption process has to be handled in the proper way. Three main procedures can be considered, namely biosorbents regeneration and their further use for next biosorption cycles, exploitation of the exhausted biosorbents for potential (bio)technology applications, or the pyrolysis of exhausted biosorbents (usually for the formation of biochar) (Badescu et al. 2018).

From economic and environmental perspectives, reusability of saturated (spent) biosorbent is of great concern for potential application in large industrial scales through consecutive adsorption/desorption cycles. A successful operational desorption process can reduce the overall cost of as-used process. In general, an effective regeneration process depends on biosorbent type, sorption mechanism, and appropriate desorption scenario (Maksoud et al. 2020).

Pyrolysis conversion of magnetic biosorbents loaded with organic dyes into magnetically responsive biochar is a very interesting possibility, leading to the formation of new and efficient magnetic adsorbent for organic and inorganic pollutants removal (Safarik et al. 2016d; Trakal et al. 2016). In addition to adsorption properties, both native and magnetically modified biochars exhibit peroxidase-like catalytic activity which enabled the decolorization of crystal violet both in the model solution and the fish pond water containing suspended solids and dissolved organic matter. The observed biochar enzyme mimetic activity can thus find interesting applications in environmental technology for the degradation of selected xenobiotics (Safarik et al. 2020e).

Thorough experimental and economical evaluation is necessary to find the optimal way to use the spent low-cost magnetic biosorbents in accordance with the principles of the circular economy and sustainable biosorption technology (Baldikova et al. 2019).

8.9 Magnetically Modified Biomaterials for Dye Analysis

Magnetic solid phase extraction (MSPE) employing magnetically responsive adsorbents was developed by Safarikova and Safarik in 1999 to extract target analytes from large sample volumes or from difficult-to-handle samples (Safarikova and Safarik 1999). This approach has several advantages over traditional solid phase extraction performed in cartridges, because it provides a rapid and simple analyte separation that avoids the need for centrifugation or filtration steps; that is why it can be used for separations in samples containing suspended solids. Huge amount of MSPE applications has been described using various magnetically responsive adsorbents (Jiang et al. 2019; Capriotti et al. 2019). Currently, only a small amount of magnetically modified biomaterials has been applied for MSPE of dyes. Typical examples comprehend the determination of Sudan dyes (Sudan I, Sudan II, Sudan III, and Sudan IV) in food samples using MSPE – capillary liquid chromatography (Benmassaoud et al. 2017). Alternatively, magnetically responsive spent coffee grounds can also be used for the preconcentration of the target analytes from diluted solutions using the MSPE approach as shown by the extraction of crystal violet from diluted water solutions (Safarik et al. 2012c).

A new type of low-cost preconcentration method, based on the use of magnetically modified textile (magnetic textile solid phase extraction; MTSPE) has been developed recently (Safarik et al. 2017, 2019c, 2021). In this procedure, a piece of an appropriate textile (fabric) is used as a carrier for the immobilization of a specific affinity, ion exchange, or hydrophobic ligand. In order to prepare magnetically responsive adsorbent, an iron-based standard staple was inserted in the textile material using an office stapler. Magnetically modified pieces of textile can be moved using appropriate laboratory magnetic stirrers in a similar way as magnetic stirring bars. At the end of the extraction process, the piece of textile can be easily and rapidly separated magnetically. In the recent experiments, nonwoven acrylic felt modified with a positively charged polysaccharide chitosan was employed for the preconcentration of food acid dyes tartrazine, azorubine and indigo carmine (Safarik et al. 2019a, b) or blue fountain ink dye Acid blue 93 (Safarik and Pospiskova 2018). Adsorbed dyes were either eluted from the textile adsorbent and subsequently analyzed spectrophotometrically, or elution-free assay based on image analysis of stained textile photos was performed (Safarik et al. 2019d). Alternatively, magnetically responsive acrylic nonwoven textile impregnated with negatively charged κ-carrageenan was used as an adsorbent for the extraction of cationic dyes (Safarik et al. 2020b).

8.10 Conclusions

Different types of biological materials, including agro-industrial byproducts and wastes, microbial cells, and marine-based biomaterials, can be successfully employed as biosorbents for dye removal. Conversion of these biosorbents into "smart biosorbents" or "stimuli-responsive biosorbents" exhibiting response to external magnetic field is a progressive way how to improve biosorbents applicability, enabling their selective magnetic separation from wastewater after dye removal. Despite the fact that magnetically responsive biocomposites have currently been tested mainly in laboratory experiments, there is a great potential for their large-scale environmental technology applications in the near future.

Acknowledgments This research was supported by the ERDF/ESF projects "New Composite Materials for Environmental Applications" (No. CZ.02.1.01/0.0/0.0/17_048/0007399), "Development of pre-applied research in nanotechnology and biotechnology" (No. CZ.02.1.01/0.0/0.0/17_048/0007323) and by the Operational Program Research, Development and Education – European Regional Development Fund (project No. CZ.02.1.01/0.0/0.0/16_019/0000754) of the

Ministry of Education, Youth and Sports of the Czech Republic. Research was also supported by the project "Modified (nano)textile materials for health technologies" (No. ITMS 313011T548, Structural funds of EU, Ministry of Education, Slovakia) and bilateral project NASU-20-04 (Academy of Sciences, Czech Republic). This work was carried out in the frame of the COST Actions CA17107 (European Network to connect research and innovation efforts on advanced Smart Textiles), CA16215 (European network for the promotion of portable, affordable and simple analytical platforms), and CA18238 "European transdisciplinary networking platform for marine biotechnology".

References

Abdolali A, Guo WS, Ngo HH, Chen SS, Nguyen NC, Tung KL (2014) Typical lignocellulosic wastes and by-products for biosorption process in water and wastewater treatment: a critical review. Bioresour Technol 160:57–66. https://doi.org/10.1016/j.biortech.2013.12.037

Adeniyi AG, Ighalo JO (2019) Biosorption of pollutants by plant leaves: an empirical review. J Environ Chem Eng 7(3):103100. https://doi.org/10.1016/j.jece.2019.103100

Adeogun AI, Akande JA, Idowu MA, Kareem SO (2019) Magnetic tuned sorghum husk biosorbent for effective removal of cationic dyes from aqueous solution: isotherm, kinetics, thermodynamics and optimization studies. Appl Water Sci 9(7):160. https://doi.org/10.1007/s13201-019-1037-2

Ahmad A, Mohd-Setapar SH, Chuong CS, Khatoon A, Wani WA, Kumar R, Rafatullah M (2015) Recent advances in new generation dye removal technologies: novel search for approaches to reprocess wastewater. RSC Adv 5(39):30801–30818. https://doi.org/10.1039/C4RA16959J

Alizadeh N, Shariati S, Besharati N (2017) Adsorption of crystal violet and methylene blue on *Azolla* and fig leaves modified with magnetite iron oxide nanoparticles. Int J Environ Res 11:197–206. https://doi.org/10.1007/s41742-017-0019-1

Alver E, Metin AU, Brouers F (2020) Methylene blue adsorption on magnetic alginate/rice husk bio-composite. Int J Biol Macromol 154:104–113. https://doi.org/10.1016/j.ijbiomac.2020.02.330

Angelova R, Baldikova E, Pospiskova K, Maderova Z, Safarikova M, Safarik I (2016) Magnetically modified *Sargassum horneri* biomass as an adsorbent for organic dye removal. J Clean Prod 137:189–194. https://doi.org/10.1016/j.jclepro.2016.07.068

Angelova R, Baldikova E, Pospiskova K, Safarikova M, Safarik I (2017) Magnetically modified sheaths of *Leptothrix* sp. as an adsorbent for Amido black 10B removal. J Magn Magn Mater 427:314–319. https://doi.org/10.1016/j.jmmm.2016.10.094

Badescu IU, Bulgariu D, Ahmad I, Bulgariu L (2018) Valorisation possibilities of exhausted biosorbents loaded with metal ions – a review. J Environ Manag 224:288–297. https://doi.org/10.1016/j.jenvman.2018.07.066

Baldikova E, Safarikova M, Safarik I (2015) Organic dyes removal using magnetically modified rye straw. J Magn Magn Mater 380:181–185. https://doi.org/10.1016/j.jmmm.2014.09.003

Baldikova E, Politi D, Maderova Z, Pospiskova K, Sidiras D, Safarikova M, Safarik I (2016a) Utilization of magnetically responsive cereal by-product for organic dye removal. J Sci Food Agric 96(6):2204–2214. https://doi.org/10.1002/jsfa.7337

Baldikova E, Pospiskova K, Maderova Z, Safarikova M, Safarik I (2016b) Preparation of magnetic composite materials: experiments for secondary school students. Chem Listy 110(1):64–68

Baldikova E, Mullerova S, Prochazkova J, Rouskova M, Solcova O, Safarik I, Pospiskova K (2019) Use of waste *Japonochytrium* sp. biomass after lipid extraction as an efficient adsorbent for triphenylmethane dye applied in aquaculture. Biomass Convers Biorefinery 9(3):479–488. https://doi.org/10.1007/s13399-018-0362-2

Barbusinski K, Salwiczek S, Paszewska A (2016) The use of chitosan for removing selected pollutants from water and wastewater – Short review. Archit Civil Eng Environ 9 (2):107–115. https://doi.org/10.21307/acee-2016-026

Benmassaoud Y, Villaseñor MJ, Salghi R, Jodeh S, Algarra M, Zougagh M, Ríos Á (2017) Magnetic/non-magnetic argan press cake nanocellulose for the selective extraction of Sudan dyes in food samples prior to the determination by capillary liquid chromatography. Talanta 166 (Supplement C):63–69. https://doi.org/10.1016/j.talanta.2017.01.041

Besharati N, Alizadeh N (2018) Adsorption of malachite green dye on different natural absorbents modified with magnetite nanoparticles. J Nanoanal 5(3):143–155. https://doi.org/10.22034/jna.2018.551649.1048

Besharati N, Alizadeh N, Shariati S (2018) Removal of cationic dye methylene blue (MB) from aqueous solution by coffee and peanut husk modified with magnetite iron oxide nanoparticles. J Mex Chem Soc 62(3):110–124. https://doi.org/10.29356/jmcs.v62i3.433

Bommavaram K, Bhattacharjee A, Yadav D, Andra N, Pandey P, Ibrahim H (2020) Tea residue as a bio-sorbent for the treatment of textile industry effluents. Int J Environ Sci Technol 17:3351-3364. https://doi.org/10.1007/s13762-020-02628-w

Cai W, Wan J (2007) Facile synthesis of superparamagnetic magnetite nanoparticles in liquid polyols. J Colloid Interface Sci 305(2):366–370. https://doi.org/10.1016/j.jcis.2006.10.023

Capriotti AL, Cavaliere C, La Barbera G, Montone CM, Piovesana S, Lagana A (2019) Recent applications of magnetic solid-phase extraction for sample preparation. Chromatographia 82 (8):1251–1274. https://doi.org/10.1007/s10337-019-03721-0

Chen CY, Chang JC, Chen AH (2011) Competitive biosorption of azo dyes from aqueous solution on the templated crosslinked-chitosan nanoparticles. J Hazard Mater 185(1):430–441. https://doi.org/10.1016/j.jhazmat.2010.09.051

Cho DW, Jeon BH, Chon CM, Schwartz FW, Jeong Y, Song H (2015) Magnetic chitosan composite for adsorption of cationic and anionic dyes in aqueous solution. J Ind Eng Chem 28:60–66. https://doi.org/10.1016/j.jiec.2015.01.023

Delrue F, Alvarez-Diaz PD, Fon-Sing S, Fleury G, Sassi JF (2016) The environmental biorefinery: using microalgae to remediate wastewater, a win-win paradigm. Energies 9(3):132. https://doi.org/10.3390/en9030132

Divriklioglu M, Akar ST, Akar T (2019) A passively immobilized novel biomagsorbent for the effective biosorptive treatment of dye contamination. Environ Sci Pollut Res 26:25834–25843. https://doi.org/10.1007/s11356-019-05716-8

Ebrahimian Pirbazari A, Saberikhah E, Habibzadeh Kozani SS (2014) Fe_3O_4–wheat straw: preparation, characterization and its application for methylene blue adsorption. Water Resour Ind 7–8:23–37. https://doi.org/10.1016/j.wri.2014.09.001

Elwakeel KZ (2009) Removal of Reactive Black 5 from aqueous solutions using magnetic chitosan resins. J Hazard Mater 167(1–3):383–392. https://doi.org/10.1016/j.jhazmat.2009.01.051

Forgacs E, Cserháti T, Oros G (2004) Removal of synthetic dyes from wastewaters: a review. Environ Int 30(7):953–971. https://doi.org/10.1016/j.envint.2004.02.001

Frey NA, Peng S, Cheng K, Sun S (2009) Magnetic nanoparticles: synthesis, functionalization, and applications in bioimaging and magnetic energy storage. Chem Soc Rev 38(9):2532–2542. https://doi.org/10.1039/B815548H

Ge H, Wang C, Liu S, Huang Z (2016) Synthesis of citric acid functionalized magnetic graphene oxide coated corn straw for methylene blue adsorption. Bioresour Technol 221:419–429. https://doi.org/10.1016/j.biortech.2016.09.060

Hasany SF, Abdurahman NH, Sunarti AR, Jose R (2013) Magnetic iron oxide nanoparticles: chemical synthesis and applications review. Curr Nanosci 9(5):561–575. https://doi.org/10.2174/1573413711309990085

Hashemian S, Salimi M (2012) Nano composite a potential low cost adsorbent for removal of cyanine acid. Chem Eng J 188:57–63. https://doi.org/10.1016/j.cej.2012.02.008

Hassaan MA, El Nemr A (2017) Health and environmental impacts of dyes: mini review. Am J Environ Sci Eng 1(3):64–67. https://doi.org/10.11648/j.ajese.20170103.11

Huang L, Li D, Zhang D, Peng H, Ren Y (2020) Facile preparation of taurine modified magnetic chitosan nanocomposites as biodegradable adsorbents toward methylene blue. Environ Technol, in press. https://doi.org/10.1080/09593330.2020.1725140

Inkoua S, Maloko HL, Koko MM, Yan L (2020) Facile solvothermal synthesis of Fe_3O_4/magnetic grapefruit peel for adsorptive removal of congo red, humic acid and phosphate from aqueous solutions. Mater Express 10(1):37–44. https://doi.org/10.1166/mex.2020.1610

Jiang H-L, Li N, Cui L, Wang X, Zhao R-S (2019) Recent application of magnetic solid phase extraction for food safety analysis. TrAC Trends Anal Chem 120:115632. https://doi.org/10.1016/j.trac.2019.115632

Jilek R, Prochazka H, Stamberg K, Fuska J (1976) Application of fungal biomass in biosorbent preparation. Folia Microbiol 21(3):210–210

Kanjilal T, Bhattacharjee C (2018) Green applications of magnetic sorbents for environmental remediation. In: Inamuddin, Asiri AM, Mohammad A (eds) Organic pollutants in wastewater I, vol 29. Materials Research Foundations, pp 1–41

Katheresan V, Kansedo J, Lau SY (2018) Efficiency of various recent wastewater dye removal methods: a review. J Environ Chem Eng 6(4):4676–4697. https://doi.org/10.1016/j.jece.2018.06.060

Khan R, Rehman A, Hayat A, Andreescu S (2019) Magnetic particles-based analytical platforms for food safety monitoring. Magnetochemistry 5:Article No. 63. https://doi.org/10.3390/magnetochemistry5040063

Khandanlou R, Bin Ahmad M, Shameli K, Kalantari K (2013) Synthesis and characterization of rice straw/Fe_3O_4 nanocomposites by a quick precipitation method. Molecules 18(6):6597–6607. https://doi.org/10.3390/molecules18066597

Khandanlou R, Bin Ahmad M, Shameli K, Kalantari K (2014) Investigation of the role of reductant on the size control of Fe_3O_4 nanoparticles on rice straw. Bioresources 9(1):642–655

Khorasani AC, Shojaosadati SA (2019) Magnetic pectin-*Chlorella vulgaris* biosorbent for the adsorption of dyes. J Environ Chem Eng 7(3):103062. https://doi.org/10.1016/j.jece.2019.103062

Kumar M, Dosanjh HS, Singh H (2019) Surface modification of spinel ferrite with biopolymer for adsorption of cationic and anionic dyes in single and ternary dye system. Fibers Polym 20 (4):739–751. https://doi.org/10.1007/s12221-019-8462-6

Laurent S, Forge D, Port M, Roch A, Robic C, Vander Elst L, Muller RN (2008) Magnetic iron oxide nanoparticles: Synthesis, stabilization, vectorization, physicochemical characterizations, and biological applications. Chem Rev 108(6):2064–2110. https://doi.org/10.1021/cr068445e

Lemine OM, Omri K, Zhang B, El Mir L, Sajieddine M, Alyamani A, Bououdina M (2012) Sol–gel synthesis of 8 nm magnetite (Fe_3O_4) nanoparticles and their magnetic properties. Superlattice Microst 52(4):793–799. https://doi.org/10.1016/j.spmi.2012.07.009

Li R, Li P, Cai J, Xiao SJ, Yang H, Li AM (2016a) Efficient adsorption of both methyl orange and chromium from their aqueous mixtures using a quaternary ammonium salt modified chitosan magnetic composite adsorbent. Chemosphere 154:310–318. https://doi.org/10.1016/j.chemosphere.2016.03.100

Li X, Lu H, Zhang Y, He F, Jing L, He X (2016b) Fabrication of magnetic alginate beads with uniform dispersion of $CoFe_2O_4$ by the polydopamine surface functionalization for organic pollutants removal. Appl Surf Sci 389:567–577. https://doi.org/10.1016/j.apsusc.2016.07.162

Lin C-R, Chu Y-M, Wang S-C (2006) Magnetic properties of magnetite nanoparticles prepared by mechanochemical reaction. Mater Lett 60(4):447–450. https://doi.org/10.1016/j.matlet.2005.09.009

Liu D, Zhang ZP, Ding YG (2016) A simple method to prepare magnetic modified corncobs and its application for Congo red adsorption. J Dispers Sci Technol 37(1):73–79. https://doi.org/10.1080/01932691.2015.1027912

Madrakian T, Afkhami A, Ahmadi M (2012) Adsorption and kinetic studies of seven different organic dyes onto magnetite nanoparticles loaded tea waste and removal of them from wastewater samples. Spectrochim Acta A Mol Biomol Spectrosc 99:102–109. https://doi.org/10.1016/j.saa.2012.09.025

Mahdavinia GR, Iravani S, Zoroufi S, Hosseinzadeh H (2014) Magnetic and K^+-cross-linked kappa-carrageenan nanocomposite beads and adsorption of crystal violet. Iran Polym J 23 (5):335–344. https://doi.org/10.1007/s13726-014-0229-8

Majidi S, Zeinali Sehrig F, Farkhani SM, Soleymani Goloujeh M, Akbarzadeh A (2016) Current methods for synthesis of magnetic nanoparticles. Artif Cells Nanomed Biotechnol 44 (2):722–734. https://doi.org/10.3109/21691401.2014.982802

Maksoud MIAA, Elgarahy AM, Farrell C, Al-Muhtaseb AH, Rooney D, Osman AI (2020) Insight on water remediation application using magnetic nanomaterials and biosorbents. Coord Chem Rev 403:Article Number UNSP 213096. https://doi.org/10.1016/j.ccr.2019.213096

Mashkoor F, Nasar A (2020) Magnetized *Tectona grandis* sawdust as a novel adsorbent: preparation, characterization, and utilization for the removal of methylene blue from aqueous solution. Cellulose 27:2613-2635. https://doi.org/10.1007/s10570-019-02918-8

Mohammadi A, Daemi H, Barikani M (2014) Fast removal of malachite green dye using novel superparamagnetic sodium alginate-coated Fe_3O_4 nanoparticles. Int J Biol Macromol 69:447–455. https://doi.org/10.1016/j.ijbiomac.2014.05.042

Mullerova S, Baldikova E, Prochazkova J, Pospiskova K, Safarik I (2019) Magnetically modified macroalgae *Cymopolia barbata* biomass as an adsorbent for safranin O removal. Mater Chem Phys 225:174–180. https://doi.org/10.1016/j.matchemphys.2018.12.074

Ozudogru Y, Merdivan M, Goksan T (2016) Biosorption of methylene blue from aqueous solutions by iron oxide-coated *Cystoseira barbata*. J Turkish Chem Soc A 3:551–564. https://doi.org/10.18596/jotcsa.40601

Pang X, Bouzid M, dos Santos JMN, Gazzah M, Dotto GL, Belmabrouk H, Bajahzar A, Erto A, Li Z (2020) Theoretical study of indigotine blue dye adsorption on $CoFe_2O_4$/chitosan magnetic composite via analytical model. Colloids Surf A Physicochem Eng Asp 589:124467. https://doi.org/10.1016/j.colsurfa.2020.124467

Parlayici S, Pehlivan E (2019) Fast decolorization of cationic dyes by nano-scale zero valent iron immobilized in sycamore tree seed pod fibers: kinetics and modelling study. Int J Phytoremediation 21(11):1130–1144. https://doi.org/10.1080/15226514.2019.1606786

Pirbazari AE, Saberikhah E, Gorabi NGA (2016) Fe_3O_4 nanoparticles loaded onto wheat straw: an efficient adsorbent for Basic Blue 9 adsorption from aqueous solution. Desalin Water Treat 57 (9):4110–4121. https://doi.org/10.1080/19443994.2014.989918

Pospiskova K, Prochazkova G, Safarik I (2013) One-step magnetic modification of yeast cells by microwave-synthesized iron oxide microparticles. Lett Appl Microbiol 56(6):456–461. https://doi.org/10.1111/lam.12069

Saba B, Christy AD, Jabeen M (2016) Kinetic and enzymatic decolorization of industrial dyes utilizing plant-based biosorbents: a review. Environ Eng Sci 33(9):601–614. https://doi.org/10.1089/ees.2016.0038

Safarik I (1995) Removal of organic polycyclic compounds from water solutions with a magnetic chitosan based sorbent bearing copper phthalocyanine dye. Water Res 29(1):101–105. https://doi.org/10.1016/0043-1354(94)E0110-R

Safarik I, Pospiskova K (2018) A simple extraction of blue fountain ink dye (Acid blue 93) from water solutions using Magnetic Textile Solid-Phase Extraction. Sep Sci Plus 1(1):48–51. https://doi.org/10.1002/sscp.201700014

Safarik I, Safarikova M (2007) Magnetically modified microbial cells: a new type of magnetic adsorbents. China Particuology 5:19–25. https://doi.org/10.1016/j.cpart.2006.12.003

Safarik I, Safarikova M (2010) Magnetic fluid modified peanut husks as an adsorbent for organic dyes removal. Phys Procedia 9:274–278. https://doi.org/10.1016/j.phpro.2010.11.061

Safarik I, Safarikova M (2014) One-step magnetic modification of non-magnetic solid materials. Int J Mater Res 105(1):104–107. https://doi.org/10.3139/146.111009

Safarik I, Ptackova L, Safarikova M (2002) Adsorption of dyes on magnetically labeled baker's yeast cells. Eur Cells Mater 3(Suppl. 2):52–55

Safarik I, Safarikova M, Weyda F, Mosiniewicz-Szablewska E, Slawska-Waniewska A (2005) Ferrofluid-modified plant-based materials as adsorbents for batch separation of selected biologically active compounds and xenobiotics. J Magn Magn Mater 293(1):371–376. https://doi.org/10.1016/j.jmmm.2005.02.033

Safarik I, Lunackova P, Mosiniewicz-Szablewska E, Weyda F, Safarikova M (2007a) Adsorption of water-soluble organic dyes on ferrofluid-modified sawdust. Holzforschung 61(3):247–253. https://doi.org/10.1515/HF.2007.060

Safarik I, Rego LFT, Borovska M, Mosiniewicz-Szablewska E, Weyda F, Safarikova M (2007b) New magnetically responsive yeast-based biosorbent for the efficient removal of water-soluble dyes. Enzym Microb Technol 40(6):1551–1556. https://doi.org/10.1016/j.enzmictec.2006.10.034

Safarik I, Horska K, Safarikova M (2011a) Magnetically modified spent grain for dye removal. J Cereal Sci 53(1):78–80. https://doi.org/10.1016/j.jcs.2010.09.010

Safarik I, Horska K, Safarikova M (2011b) Magnetically responsive biocomposites for inorganic and organic xenobiotics removal. In: Kotrba P, Mackova M, Macek T (eds) Microbial biosorption of metals. Springer, Dordrecht, pp 301–320. https://doi.org/10.1007/978-94-007-0443-5_13

Safarik I, Horska K, Pospiskova K, Safarikova M (2012a) Magnetically responsive activated carbons for bio – and environmental applications. Int Rev Chem Eng 4(3):346–352

Safarik I, Horska K, Pospiskova K, Safarikova M (2012b) One-step preparation of magnetically responsive materials from non-magnetic powders. Powder Technol 229:285–289. https://doi.org/10.1016/j.powtec.2012.06.006

Safarik I, Horska K, Svobodova B, Safarikova M (2012c) Magnetically modified spent coffee grounds for dyes removal. Eur Food Res Technol 234(2):345–350. https://doi.org/10.1007/s00217-011-1641-3

Safarik I, Horska K, Pospiskova K, Maderova Z, Safarikova M (2013) Microwave assisted synthesis of magnetically responsive composite materials. IEEE Trans Magn 49(1):213–218. https://doi.org/10.1109/Tmag.2012.2221686

Safarik I, Horska K, Pospiskova K, Filip J, Safarikova M (2014) Mechanochemical synthesis of magnetically responsive materials from non-magnetic precursors. Mater Lett 126:202–206. https://doi.org/10.1016/j.matlet.2014.04.045

Safarik I, Maderova Z, Horska K, Baldikova E, Pospiskova K, Safarikova M (2015a) Spent rooibos (*Aspalathus linearis*) tea biomass as an adsorbent for organic dye removal. Biorem J 19 (3):183–187. https://doi.org/10.1080/10889868.2014.979279

Safarik I, Pospiskova K, Maderova Z, Baldikova E, Horska K, Safarikova M (2015b) Microwave-synthesized magnetic chitosan microparticles for the immobilization of yeast cells. Yeast 32 (1):239–243. https://doi.org/10.1002/yea.3017

Safarik I, Ashoura N, Maderova Z, Pospiskova K, Baldikova E, Safarikova M (2016a) Magnetically modified *Posidonia oceanica* biomass as an adsorbent for organic dyes removal. Mediterr Mar Sci 17(2):351–358. https://doi.org/10.12681/mms.1549

Safarik I, Baldikova E, Pospiskova K, Safarikova M (2016b) Magnetic modification of diamagnetic agglomerate forming powder materials. Particuology 29:169–171. https://doi.org/10.1016/j.partic.2016.05.002

Safarik I, Pospiskova K, Baldikova E, Safarikova M (2016c) Magnetically responsive biological materials and their applications. Adv Mater Lett 7(4):254–261. https://doi.org/10.5185/amlett.2016.6176

Safarik I, Maderova Z, Pospiskova K, Schmidt H-P, Baldikova E, Filip J, Krizek M, Malina O, Safarikova M (2016d) Magnetically modified biochar for organic xenobiotics removal. Water Sci Technol 74:1706–1715. https://doi.org/10.2166/wst.2016.335

Safarik I, Baldikova E, Safarikova M, Pospiskova K (2017) Magnetically responsive textile for a new preconcentration procedure: magnetic textile solid phase extraction. J Ind Text 48 (4):761–771. https://doi.org/10.1177/1528083717740767

Safarik I, Angelova R, Baldikova E, Pospiskova K, Safarikova M (2017a) *Leptothrix* sp. sheaths modified with iron oxide particles: Magnetically responsive, high aspect ratio functional material. Mater Sci Eng C 71:1342–1346. https://doi.org/10.1016/j.msec.2016.10.056

Safarik I, Baldikova E, Prochazkova J, Safarikova M, Pospiskova K (2018) Magnetically modified agricultural and food waste: preparation and application. J Agric Food Chem 66 (11):2538–2552. https://doi.org/10.1021/acs.jafc.7b06105

Safarik I, Mullerova S, Pospiskova K (2019a) Magnetically responsive textile for preconcentration of acid food dyes. Mater Chem Phys 232:205–208. https://doi.org/10.1016/j.matchemphys.2019.04.058

Safarik I, Mullerova S, Pospiskova K (2019b) Semiquantitative determination of food acid dyes by magnetic textile solid phase extraction followed by image analysis. Food Chem 274:215–219. https://doi.org/10.1016/j.foodchem.2018.08.125

Safarik I, Prochazkova J, Baldikova E, Pospiskova K (2019c) Magnetically responsive materials for solid phase extraction. Inženjerstvo okoliša – Environ Eng 6(1):15–20

Safarik I, Baldikova E, Prochazkova J, Pospiskova K (2019d) Smartphone-based image analysis for evaluation of magnetic textile solid phase extraction of colored compounds. Heliyon 5:e02995

Safarik I, Baldikova E, Prochazkova J, Pospiskova K (2020a) Magnetic particles in algae biotechnology: recent updates. J Appl Phycol 32:1743–1753. https://doi.org/10.1007/s10811-020-02109-0

Safarik I, Mullerova S, Pospiskova K (2020b) Magnetic textile solid phase extraction of cationic dyes from water solutions. Fibers Polym 21:2836–2841. https://doi.org/10.1007/s12221-020-9539-y

Safarik I, Prochazkova J, Baldikova E, Pospiskova K (2021) Magnetic textile solid phase extraction. In: Ehrmann A, Nguyen TA, Tri PN (eds) Nanosensors and nanodevices for smart multifunctional textiles. Elsevier, pp 149–161. https://doi.org/10.1016/B978-0-12-820777-2.00010-8

Safarik I, Prochazkova J, Baldikova E, Pospiskova K (2020d) Magnetically responsive algae and seagrass derivatives for pollutant removal. In: Vilarinho C, Castro F, Goncalves M, Fernando AL (eds) Wastes: solutions, treatments and opportunities III. CRC Press, Boca Raton, pp 131–136. https://doi.org/10.1201/9780429289798-21

Safarik I, Prochazkova J, Baldikova E, Schmidt HP, Kwapinski W, Medrik I, Jakubec P, Safarikova M, Pospiskova K (2020e) Biochars and their magnetic derivatives as enzyme-like catalysts mimicking peroxidases. Biochar 2:121–134. https://doi.org/10.1007/s42773-020-00035-5

Safarikova M, Safarik I (1999) Magnetic solid-phase extraction. J Magn Magn Mater 194(1):108–112. https://doi.org/10.1016/S0304-8853(98)00566-6

Safarikova M, Ptackova L, Kibrikova I, Safarik I (2005) Biosorption of water-soluble dyes on magnetically modified *Saccharomyces cerevisiae* subsp. *uvarum* cells. Chemosphere 59(6):831–835. https://doi.org/10.1016/j.chemosphere.2004.10.062

Safarikova M, Pona BMR, Mosiniewicz-Szablewska E, Weyda F, Safarik I (2008) Dye adsorption on magnetically modified *Chlorella vulgaris* cells. Fresenius Environ Bull 17(4):486–492

Safarikova M, Maderova Z, Safarik I (2009) Ferrofluid modified *Saccharomyces cerevisiae* cells for biocatalysis. Food Res Int 42(4):521–524. https://doi.org/10.1016/j.foodres.2009.01.001

Saravanan A, Senthil Kumar P, Yaashikaa PR (2020) Treatment of dye containing wastewater using agricultural biomass derived magnetic adsorbents. In: Naushad M, Lichtfouse E (eds) Green Materials for Wastewater Treatment. Environmental Chemistry for a Sustainable World, vol 38. Springer, Cham, pp 149–169. https://doi.org/10.1007/978-3-030-17724-9_7

Shah J, Jan MR, Jamil S, ul Haq A (2014) Magnetic particles precipitated onto wheat husk for removal of methyl blue from aqueous solution. Toxicol Environ Chem 96(2):218–226. https://doi.org/10.1080/02772248.2014.929690

Shen CS, Shen Y, Wen YZ, Wang HY, Liu WP (2011) Fast and highly efficient removal of dyes under alkaline conditions using magnetic chitosan-Fe(III) hydrogel. Water Res 45(16):5200–5210. https://doi.org/10.1016/j.watres.2011.07.018

Shukla A, Zhang Y-H, Dubey P, Margrave JL, Shukla SS (2002) The role of sawdust in the removal of unwanted materials from water. J Hazard Mater 95(1):137–152. https://doi.org/10.1016/S0304-3894(02)00089-4

Soares SF, Simões TR, Trindade T, Daniel-da-Silva AL (2017) Highly efficient removal of dye from water using magnetic carrageenan/silica hybrid nano-adsorbents. Water Air Soil Pollut 228(3):87. https://doi.org/10.1007/s11270-017-3281-0

Srinivasan A, Viraraghavan T (2010) Decolorization of dye wastewaters by biosorbents: a review. J Environ Manag 91(10):1915–1929. https://doi.org/10.1016/j.jenvman.2010.05.003

Taha NA, El-Maghraby A (2016) Magnetic peanut hulls for methylene blue dye removal: isotherm and kinetic study. Global NEST J 18(1):25–37. https://doi.org/10.30955/gnj.001730

Tan KA, Morad N, Teng TT, Norli I, Panneerselvam P (2012) Removal of cationic dye by magnetic nanoparticle (Fe_3O_4) impregnated onto activated maize cob powder and kinetic study of dye waste adsorption. Int Conf Environ Sci Dev (ICESD 2012) 1:83–89. https://doi.org/10.1016/j.apcbee.2012.03.015

Tarhan T, Tural B, Boga K, Tural S (2018) Adsorptive performance of magnetic nano-biosorbent for binary dyes and investigation of comparative biosorption. SN Appl Sci 1(1):8. https://doi.org/10.1007/s42452-018-0011-1

Torres-Gomez N, Nava O, Argueta-Figueroa L, Garcia-Contreras R, Baeza-Barrera A, Vilchis-Nestor AR (2019) Shape tuning of magnetite nanoparticles obtained by hydrothermal synthesis: effect of temperature. J Nanomater 2019:15. https://doi.org/10.1155/2019/7921273

Trakal L, Veselska V, Safarik I, Vitkova M, Cihalova S, Komarek M (2016) Lead and cadmium sorption mechanisms on magnetically modified biochars. Bioresour Technol 203:318–324. https://doi.org/10.1016/j.biortech.2015.12.056

Tran VS, Ngo HH, Guo WS, Zhang J, Liang S, Ton-That C, Zhang XB (2015) Typical low cost biosorbents for adsorptive removal of specific organic pollutants from water. Bioresour Technol 182:353–363. https://doi.org/10.1016/j.biortech.2015.02.003

Tural B, Ertaş E, Enez B, Fincan SA, Tural S (2017) Preparation and characterization of a novel magnetic biosorbent functionalized with biomass of *Bacillus subtilis*: Kinetic and isotherm studies of biosorption processes in the removal of Methylene Blue. J Environ Chem Eng 5 (5):4795–4802. https://doi.org/10.1016/j.jece.2017.09.019

Uzun L, Saglam N, Safarikova M, Safarik I, Denizli A (2011) Copper biosorption on magnetically modified yeast cells under magnetic field. Sep Sci Technol 46(6):1045–1051. https://doi.org/10.1080/01496395.2010.541400

Wang J, Cao M, Jiang C, Zheng Y, Zhang C, Wei J (2018a) Adsorption and coadsorption mechanisms of Hg^{2+} and methyl orange by branched polyethyleneimine modified magnetic straw. Mater Lett 229:160–163. https://doi.org/10.1016/j.matlet.2018.07.015

Wang HW, Chen YP, Dang BK, Shen XP, Jin CD, Sun QF, Pei JC (2018b) Ultrafine Mn ferrite by anchoring in a cellulose framework for efficient toxic ions capture and fast water/oil separation. Carbohydr Polym 196:117–125. https://doi.org/10.1016/j.carbpol.2018.05.031

Yagub MT, Sen TK, Afroze S, Ang HM (2014) Dye and its removal from aqueous solution by adsorption: a review. Adv Colloid Interf Sci 209:172–184. https://doi.org/10.1016/j.cis.2014.04.002

Yu J-X, Chi R-A, Zhang Y-F, Xu Z-G, Xiao C-Q, Guo J (2012) A situ co-precipitation method to prepare magnetic PMDA modified sugarcane bagasse and its application for competitive adsorption of methylene blue and basic magenta. Bioresour Technol 110:160–166. https://doi.org/10.1016/j.biortech.2012.01.134

Yu JX, Wang LY, Chi RA, Zhang YF, Xu ZG, Guo J (2013) A simple method to prepare magnetic modified beer yeast and its application for cationic dye adsorption. Environ Sci Pollut Res Int 20 (1):543–551. https://doi.org/10.1007/s11356-012-0903-3

Zhao Y, Xia YX, Yang HY, Wang YY, Zhao MJ (2014) Synthesis of glutamic acid-modified magnetic corn straw: equilibrium and kinetic studies on methylene blue adsorption. Desalin Water Treat 52(1–3):199–207. https://doi.org/10.1080/19443994.2013.782256

Zheng B, Zhang M, Xiao D, Jin Y, Choi MMF (2010) Fast microwave synthesis of Fe_3O_4 and Fe_3O_4/Ag magnetic nanoparticles using Fe^{2+} as precursor. Inorg Mater 46(10):1106–1111. https://doi.org/10.1134/S0020168510100146

Zheng H, Qi JQ, Jiang RX, Gao Y, Li XC (2015) Adsorption of malachite green by magnetic litchi pericarps: a response surface methodology investigation. J Environ Manag 162:232–239. https://doi.org/10.1016/j.jenvman.2015.07.057

Zhou LM, Jin JY, Liu ZR, Liang XZ, Shang C (2011) Adsorption of acid dyes from aqueous solutions by the ethylenediamine-modified magnetic chitosan nanoparticles. J Hazard Mater 185 (2–3):1045–1052. https://doi.org/10.1016/j.jhazmat.2010.10.012

Zhou Y, Lu J, Zhou Y, Liu Y (2019) Recent advances for dyes removal using novel adsorbents: a review. Environ Pollut 252:352–365. https://doi.org/10.1016/j.envpol.2019.05.072

Zhu HY, Jiang R, Xiao L, Li W (2010) A novel magnetically separable γ-Fe_2O_3/crosslinked chitosan adsorbent: preparation, characterization and adsorption application for removal of hazardous azo dye. J Hazard Mater 179(1–3):251–257. https://doi.org/10.1016/j.jhazmat.2010.02.087

Zuorro A, Di Battista A, Lavecchia R (2013) Magnetically modified coffee silverskin for the removal of xenobiotics from wastewater. 16th International Conference on Process Integration, Modelling and Optimisation for Energy Saving and Pollution Reduction 35:1375–1380. https://doi.org/10.3303/cet1335229

Zuorro A, Lavecchia R, Natali S (2014) Magnetically modified agro-industrial wastes as efficient and easily recoverable adsorbents for water treatment. Chem Eng Trans 38:349–354. https://doi.org/10.3303/CET1438059

Chapter 9
Regeneration of Magnetic Adsorbents Saturated by Organic Pollutants

Ye Xiao and Josephine M. Hill

Contents

Abstract Many processes generate wastewater that must be treated to remove contaminants, which range from the ppb to percentage level depending on the sources and pollutants, before being released. Approximately 2.5 billion people

Y. Xiao · J. M. Hill (✉)
Department of Chemical and Petroleum Engineering, University of Calgary, Calgary, AB, Canada
e-mail: jhill@ucalgary.ca

L. Meili, G. L. Dotto (eds.), *Advanced Magnetic Adsorbents for Water Treatment*, Environmental Chemistry for a Sustainable World 61,
https://doi.org/10.1007/978-3-030-64092-7_9

will be affected by water contaminated with organic pollutants, including dyes, pharmaceuticals, personal care products, flame retardants, and pesticides by 2050. The most common method for removing the organic contaminants is adsorption, which generally involves flowing the wastewater over a high surface area solid (the adsorbent) on which the pollutants will concentrate. Depending on the situation, regeneration of the used (spent) adsorbents may be a better environmental and economic choice than disposal. Here, we review the different methods of regeneration, and the corresponding challenges, for magnetic adsorbents used for the removal of organic species from wastewater. The major points covered include (1) the adsorption mechanisms and their impact on subsequent regeneration; (2) desorption-based regeneration methods; and (3) decomposition-based regeneration methods focusing on advanced oxidation technologies. Currently, the most effective advanced oxidation regeneration processes use electrochemistry to simultaneously regenerate the adsorbents and decompose the pollutants with efficiencies over 90%. Future research should focus on direct measurement of adsorption mechanisms, development of tailored adsorbents, and the scale-up of laboratory methods.

Keywords Adsorption · Regeneration · Magnetic adsorbent · Desorption · Decomposition · Advanced oxidation · Electrochemistry · Electro-Fenton · Microporosity · Mesoporosity

9.1 Introduction

With the current human activities in our global society, more and more organic chemicals are being synthesized, used, and inevitably released into the environment. The ubiquitous presence of various organic pollutants in surface water, groundwater, and wastewater has drawn tremendous attention from both the general public and researchers. In order to abate the influence of these organic pollutants, numerous technologies have been developed to treat water contaminated by toxic organic pollutants, such as Fenton oxidation (Pliego et al. 2015), electrochemical methods (Martínez-Huitle and Brillas 2009; Brillas and Martínez-Huitle 2015), photocatalysis (Malato et al. 2009; Chong et al. 2010), ultrasonic processes (Adewuyi 2005), and adsorption (Pan and Xing 2008; Gupta and Suhas 2009; Lin and Juang 2009). Among these technologies, adsorption processes have the advantages of low capital investment; simplicity in terms of design, installation, and operation; insensitivity to toxic substrates; and high efficiency (up to ~100%) for contaminant removal, even at low concentrations (such as <1 mg/L) (Foo and Hameed 2010; Fallou et al. 2016). Adsorption is normally carried out by packing the adsorbents in a column, but this process suffers from mass transfer limitations and significant pressure drops during water purification. An alternative is the suspension of the adsorbents in a stirred reactor to increase diffusion and eliminate

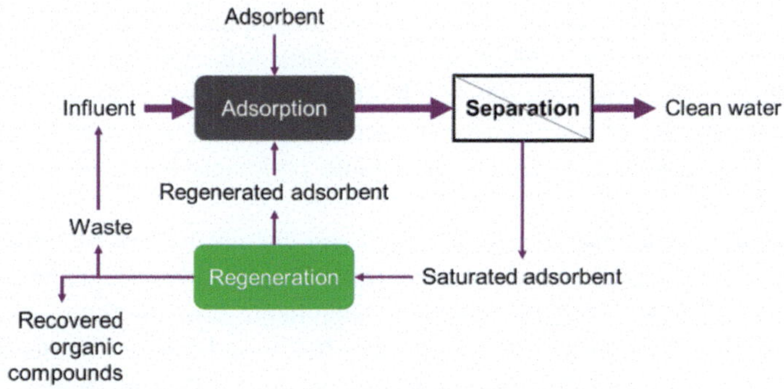

Fig. 9.1 Adsorption-regeneration cycle for water treatment

pressure drop. This process, however, requires separation of adsorbents after saturation (Fig. 9.1). Granular adsorbents (> ~ 0.3 mm diameter) will settle relatively quickly making the separation easier, but these adsorbents can cause abrasion of the reactor and have lower diffusion rates. The diffusion rates can be increased by using powder adsorbents (< ~ 0.2 mm diameter), but then the separation process is slower.

After use, it may be desirable to regenerate and recycle the saturated adsorbent, rather than to burn or dispose of it in a landfill because of the potential for secondary pollution. In addition, the process is expensive; for example, the estimated annual cost for a wastewater treatment plant using an activated carbon-based adsorption process with an average daily flow of 100,000 m^3/d is more than 10 million USD (USEPA 2000; Inyang and Dickenson 2015; Salvador et al. 2015b). Numerous regeneration studies of adsorbents saturated with organic pollutants have been done (see Salvador et al. 2015a, b and references therein). The regeneration methods studied include thermal, solvent extraction, pH variation, and regeneration by microwave, microbiology, chemical oxidation, and electrochemical oxidation. The following sections describe some of the more commonly used methods and their application in regeneration of magnetic adsorbents. Unless otherwise specified, all solutions are aqueous.

9.2 Organic Pollutants

With the wide application of organic chemicals in our daily life, numerous pollutants are released into the water systems through domestic wastewater, industrial wastewater, and agriculture activities. These organic pollutants include but are not limited to pesticides, surfactants, synthetic dyes, pharmaceuticals, personal care products, and flammable retardants. For instance, naphthenic acids are considered the main toxic contaminants in oil sands process-affected water in Alberta, Canada (Ahad et al. 2013); perfluorooctane sulfonate was detected at the level of 10–1000 pg/L in the ocean water (Yamashita et al. 2005); and pharmaceuticals and hormones were detected in the effluent of most wastewater treatment plants with concentrations up to tens of μg/L (Kolpin et al. 2002; Chen et al. 2006). Although most of the emerging pollutants are at low concentrations, the large volumes of water treated each day—the treatment capacity of a wastewater treatment plant can be larger than 1 million m^3/d—result in significant amounts of organic pollutants being released into the environment daily.

These organic pollutants, especially synthetic chemicals, can be toxic or cause reproductive, neurologic, endocrine, and immunologic adverse health effects on wild species (USEPA 2017). For instance, naphthenic acid solutions at concentrations in tens of mg/L had toxic and inhibitory effects on different organisms such as plants, fish, and bacteria (Clemente and Fedorak 2005). Pharmaceuticals and hormones at low concentration levels (<1 μg/L) can have chronic effects on the aquatic biota and increase the possibility that bacteria develop resistance (Kolpin et al. 2002). In addition, some of these chemicals can remain in the environment for a long time with a half-life ranging from years to >20 years (Terzaghi et al. 2018). Thus, these chemicals accumulate in the food chain. For example, perfluorooctane sulfonate concentrations as high as 4000 ng/g were detected in some marine mammal species and polar bears (Kannan et al. 2001; Yamashita et al. 2005).

Although the properties of the pollutants differ significantly in solubility, molecular size, polarity, and functional groups, adsorption processes are feasible for the removal of most types of organic pollutants. The adsorbents, however, must be designed according to these properties and the adsorption mechanism.

9.3 Magnetic Adsorbents

Magnetic adsorbents are a set of materials that can be separated from a solution using an external magnetic field and are generally ferromagnetic, ferromagnetic, or superparamagnetic materials (Opdyke and Channell 1996). The adsorbents may consist of these magnetic species by themselves or of these materials loaded on nonmagnetic materials (such activated carbon).

Table 9.1 includes the saturation magnetization, which is the highest obtainable magnetization, of several bulk materials. The values range from 55 emu/g for nickel to 218 emu/g for iron. When these materials are dispersed as nanoparticles on a support, the magnetization decreases. For example, the saturation magnetization value is 92 emu/g for bulk Fe_3O_4 but 40 emu/g for 80 nm particles, and 15 emu/g for 5 nm particles (Feng et al. 2012). The presence of other nonmagnetic materials also decreases the saturation magnetization values as seen in a Fe_3O_4/biochar composite for which the values decreased from 41 emu/g to 12 emu/g when the Fe content was reduced from 33% to 14% (Zhu et al. 2011). Thus, there is a balance between strength of magnetization and dispersion to maintain adsorption capacity. A saturation magnetization value of ~7 emu/g for Fe_3O_4 supported on a mesoporous carbon adsorbent was reported to be sufficient for magnetic separation (Kondo et al. 2010), but a value of > ~ 20 emu/g was generally required for fast (seconds using a

Table 9.1 Values of saturation magnetization of various magnetic materials used in adsorption processes

Materials	Saturation magnetization (emu/g)	References
Fe_3O_4	92	Cao et al. (2008)
γ-Fe_2O_3	76	Cao et al. (2016)
Fe	218	Crangle and Goodman (1971)
Ni	55	Crangle and Goodman (1971)
Co	163	I Schwerdt et al. (2012)
$CoFe_2O_4$	85	Singhal et al. (2010)
$MnFe_2O_4$	82	Aslibeiki et al. (2016)

conventional magnet with strength of ~0.01 T) separation (Dhoble et al. 2011; Hritcu et al. 2012).

9.4 Adsorption/Desorption Mechanism

The adsorption, and reverse desorption, mechanisms can vary significantly depending on the properties of the adsorbate and adsorbent, and the water chemistry (Moreno-Castilla 2004; Pan and Xing 2008). As interpreted in the literature (Moreno-Castilla 2004; Pan and Xing 2008), the mechanisms for the adsorption of organic pollutants included, but were not limited to, electrostatic interactions, hydrogen bonding, π–π interactions, van der Waals interactions, dipole–dipole interactions, and dipole–ion interactions. Almost all of these interactions are intermolecular without formation and/or dissociation of chemical bonds so that the adsorption of organic pollutants is generally physical adsorption. Among these mechanisms, however, electrostatic interactions and van der Waals interactions were most commonly reported as the main mechanisms of adsorption.

9.4.1 Electrostatic Interaction

Electrostatic interactions are affected by the pH of the water. In fact, adsorption experiments at different pH values were often conducted to determine if the main adsorption mechanism was electrostatic (Ma et al. 2012; Liu et al. 2015b; Yu et al. 2015; Mokhtari et al. 2016; Tajizadegan et al. 2016; Wang et al. 2016) and if so, how the adsorption capacity changed with pH (Cheah et al. 2013; Hassanzadeh-Tabrizi et al. 2016). An adsorbent surface with functional groups can be positively charged at low pH (reaction (9.1)) and negatively charged at high pH (reaction (9.2)), as shown in Fig. 9.2a (Worch 2012).

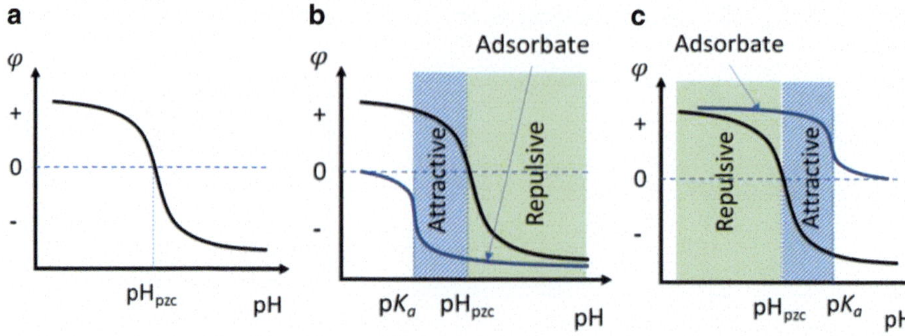

Fig. 9.2 Relationship between charge and pH for (**a**) an adsorbent, an adsorbent with (**b**) anionic or acidic adsorbates, and (**c**) cationic or basic adsorbates, where φ represents the potential of both the adsorbent surface (zeta potential) and the adsorbate (potential induced by molecular charge), pH_{pzc} is the point of zero charge, and p*K*a represents the acidic strength of the adsorbate

$$\equiv S - OH + H^+ \rightleftharpoons \equiv S - OH_2^+ \tag{9.1}$$

$$\equiv S - OH + OH^- \rightleftharpoons \equiv S - O^- + H_2O \tag{9.2}$$

where $\equiv S$ represents the adsorbent surface (not a sulphur group). When the surface charge of an adsorbent reaches zero, the corresponding pH value is referred to as the point of zero charge (pH_{pzc}).

Depending on the relationship between pH and p*K*a (*K*a is the acidity constant, and p*K*a $= -\log_{10}(Ka)$) of the adsorbate and the pH_{pzc} of the adsorbent, the interactions can be either attractive or repulsive (Worch 2012) as shown in Figs. 9.2b, c. This relationship allows the adsorption capacity and the extent of regeneration to be controlled by varying the pH.

The strength of the electrostatic interaction can be determined by the potential energy, which can be estimated by the zeta potential (φ) of the adsorbent according to Eq. (9.3).

$$E_e = \varphi \times e \times N_A = 96.5\varphi \text{ kJ/mol} \tag{9.3}$$

where E_e is the electrostatic potential energy in kJ/mol, φ is the zeta potential (V), e is the charge of one electron (1.602×10^{-19} C), N_A is Avogadro's number (6.022×10^{23} /mol). Published values for the zeta potentials of various adsorbents range from −100 mV to 100 mV as a function of pH, resulting in potential energies for the electrostatic interactions of less than 10 kJ/mol.

9.4.2 Van der Waals Interaction

If electrostatic interactions are a negligible part of the adsorption process, the adsorption capacity will not be significantly affected by the pH of the solution (Ma et al. 2012; Yu et al. 2015). van der Waals interactions include all the intermolecular attractions and repulsions resulting from correlations in the fluctuating polarizations of the molecules, ions, atoms, and surfaces. This broad definition includes dipole–dipole, ion–dipole, induced dipole, hydrogen bond, and dispersion forces.

The van de Waals interactions have different strengths with potential energies varying significantly from ~2.0 to >40 kJ/mol. Dispersion forces impact adsorption processes and can be described between an adsorbent surface and a molecule as follows (Israelachvili 2011; Bazhin 2012).

$$E_d = \alpha\varphi^2/2r^2 \tag{9.4}$$

where E_d is the dispersion interaction energy in kJ/mol, α is the electric polarizability (C^2 m^2/J), φ is the zeta potential (V), and r is the distance (m) between the center of the molecule and the adsorbent surface. As the electric polarizability of a molecule increases with the increase of volume occupied by the electrons (Anslyn and Dougherty 2006), dispersion forces will increase as the size of the adsorbate molecules increase.

9.4.3 Adsorption in Pores

Regardless of the type of interaction, the interaction strength (represented by potential energy) is affected by the pore size of the adsorbent as shown in Fig. 9.3. In pores that are several times the adsorbate size (so called big pores), the potential energy curve is similar to that on the external surface (Fig. 9.3a). In pores that are less than twice as large as the adsorbate (so called narrow pores), however, there are additional

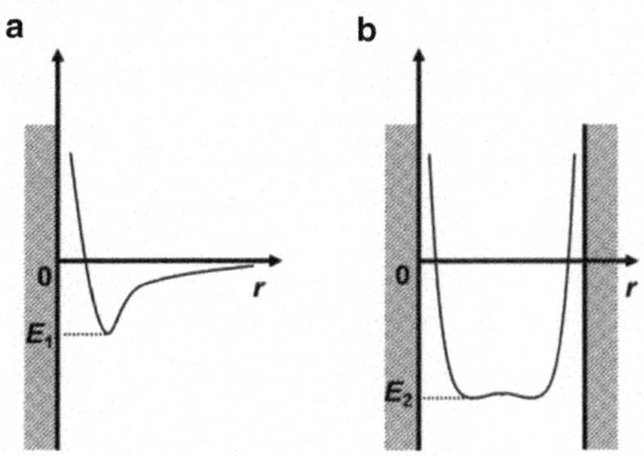

Fig. 9.3 Effect of pore size (r, radius) on the potential energy of adsorption (**a**) on the external surface, E_1, and (**b**) in narrow pores, E_2. Big pores refer to pores that are several times the adsorbate size while narrow pores are of similar size to the adsorbate

interactions between the adsorbate and adsorbent so that the potential energies may be up to twice as large as those in big pores or on the external surface. Thus, adsorption in smaller pores is thermodynamically favored and regeneration of these sites will be more difficult.

Understanding the adsorption mechanism for a specific adsorbate–adsorbent system is the first step for the design of an effective regeneration method. Unfortunately, measuring adsorption isotherms, adsorption kinetics, and thermodynamic parameters do not directly probe the mechanism, although many researchers have speculated based on these measurements. Techniques such as atomic force microscopy and surface plasmon resonance spectroscopy have been applied for the measurement of thermodynamic properties of peptide adsorption on certain surfaces (Wei and Latour 2010). These techniques, however, only provided information on adsorption energies and not the adsorption mechanism. Other researchers have used X-ray absorption near-edge structure spectroscopy but still direct evidence was not obtained (Yan et al. 2017). Techniques for the direct investigation of adsorption mechanisms have yet to be developed.

9.5 Regeneration

9.5.1 Regeneration Efficiency Calculation

Before discussing the details of the different regeneration methods, a discussion on the calculation of the regeneration efficiency is required as there are potentially several issues with the values reported in the literature. In particular, the efficiencies may be overestimated. In most cases, the regeneration efficiency, η, is calculated based on the adsorption capacity of the regenerated adsorbent, q_r, relative to that of the fresh adsorbent, q_0, under the same adsorption conditions, Eq. (9.5):

$$\eta = \frac{q_r}{q_0} \times 100\% \tag{9.5}$$

This method, however, may result in the overestimation of the regeneration efficiency if the adsorption experiments were not properly designed (Narbaitz and Cen 1997). As indicated in Fig. 9.4, if an adsorption process is operated according to the right operational line with a large initial concentration (indicated by the x-axis intercept) and a lower ratio of solid to liquid (indicated by the slope), the adsorption capacities of the fresh (q_0) and regenerated (q_r) adsorbents are in the plateau region of the isotherms. This situation will better represent the regeneration efficiency. When the equilibrium concentrations (C_e) are low as on the left operational line, the adsorbent will not be saturated with the adsorbate before regeneration resulting in lower adsorption capacities ($q_{0,\,1}$ and $q_{r,\,1}$) than at higher equilibrium concentrations. The difference in the value of q_r, however, will be smaller, and as a result, the calculated regeneration efficiency based on $q_{0,\,1}$ and $q_{r,\,1}$ will be larger than that

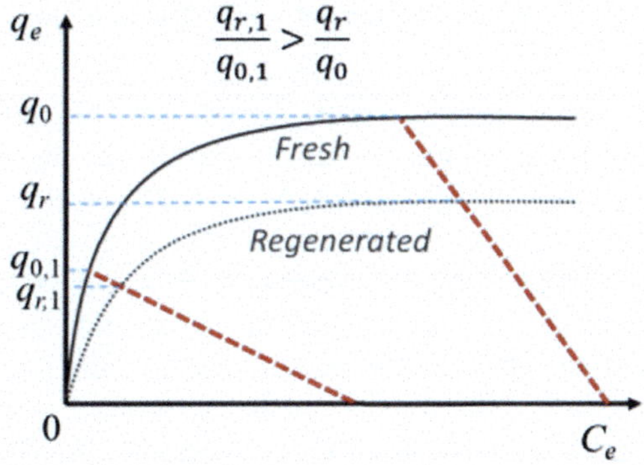

Fig. 9.4 Adsorption isotherms for fresh and regenerated adsorbents with different operational conditions (red dashed lines). The adsorption amount, q_e, is plotted versus equilibrium concentration, C_e. Values q_0 and $q_{0,1}$ are amounts adsorbed on the fresh adsorbent, while values q_r and $q_{r,1}$ are amounts adsorbed on the regenerated adsorbents

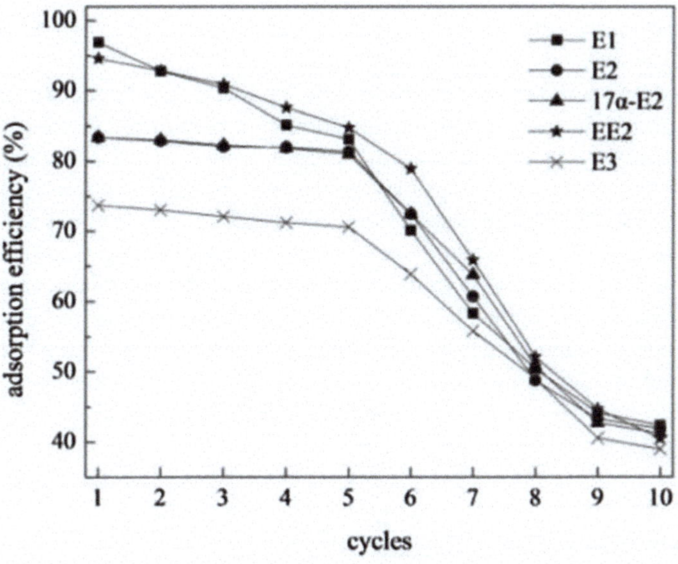

Fig. 9.5 The adsorption efficiency (removal efficiency of estrogens from solution by adsorption) of regenerated magnetic graphene oxide to the estrogens (Wang et al. 2018b). E1, estrone; E2, 17β-estradiol; 17α-E2, 17α-ethinylestradiol; EE2, synthetic estrogen; E3, estriol. (No special permission is required according to MDPI)

calculated based on q_0 and q_r, that is, there is an overestimation of the regeneration efficiency.

Regeneration before saturation may explain some of the trends reported in the literature. For example, Wang et al. (2018b) adsorbed various estrogen species on magnetic graphene oxide and observed a sharp decrease in the adsorption efficiency after cycle 5 as shown in Fig. 9.5. Because the adsorbate was not saturated, the regeneration efficiencies were likely overestimated for the first five cycles.

9.5.2 *Regeneration Classifications*

To regenerate adsorbents, the adsorbed organic pollutants must be removed through desorption and/or degradation, which are the main classifications for the regeneration methods (Zanella et al. 2014). Desorption-based regeneration processes include solvent extraction, pH variation, thermal, and pressure swing methods to remove the adsorbate from the adsorbent (Salvador et al. 2015b). The main goals of these methods are reduction of the interaction energies between the adsorbent surface and adsorbate and/or enhancement of the interaction energies between the solvent and adsorbate. Based on the adsorption mechanism, a suitable desorption-based regeneration method may be proposed. If electrostatic interactions (attractive) dominate, then changing the pH will regenerate most of the adsorption sites on the adsorbent. If, however, dispersion forces dominate, then application of a solvent with a high affinity for the adsorbed organic pollutants will be effective to regenerate the adsorption sites.

After a desorption-based technology, the pollutant has been transferred to a secondary phase that may then require additional treatment. If recovery of the pollutant is not of interest, simultaneous decomposition of the pollutant/adsorbate and regeneration of the adsorbent may be the best treatment option (Zanella et al. 2014). Decomposition-based regeneration processes include thermal regeneration, wet oxidation process, advanced oxidation processes, electrochemical regeneration, and biological regeneration. One issue in these methods can be the generation of byproducts that are more toxic than the original pollutant; for example, electrochemical regeneration in the presence of chlorine ions generated the more toxic chlorinated organic species (Hussain et al. 2013). Ideally, the pollutants are converted completely to minerals while fully regenerating the adsorption sites.

9.6 Desorption-Based Regeneration

Most of the studies involving the regeneration of magnetic materials have used desorption-based methods. Varying the solution pH is one of the easiest regeneration methods and can be applied if electrostatic interactions are involved in the adsorption mechanism (Qurrat Ul et al. 2019). The hydrophilicity/hydrophobicity of the adsorbates may also be a basis for the desorption method (Huang and Keller 2013). That is, the adsorbents may be mixed with organic solvents with different polarities to promote desorption of the adsorbates. Khan et al. (2019) tested various solvents (referred to as eluents in Fig. 9.6) for the regeneration of nanomagnetic copper ferrite/drumstick pod biomass composites with adsorbed malachite green. The adsorbent had a pH_{pzc} of ~5.8 with electrostatic interaction as the main adsorption mechanism for the cationic malachite green, so it was expected that an acidic solution would be able to regenerate the adsorbent. The regeneration (desorption), however, was not successful using hydrochloric acid and sulfuric acid. Therefore,

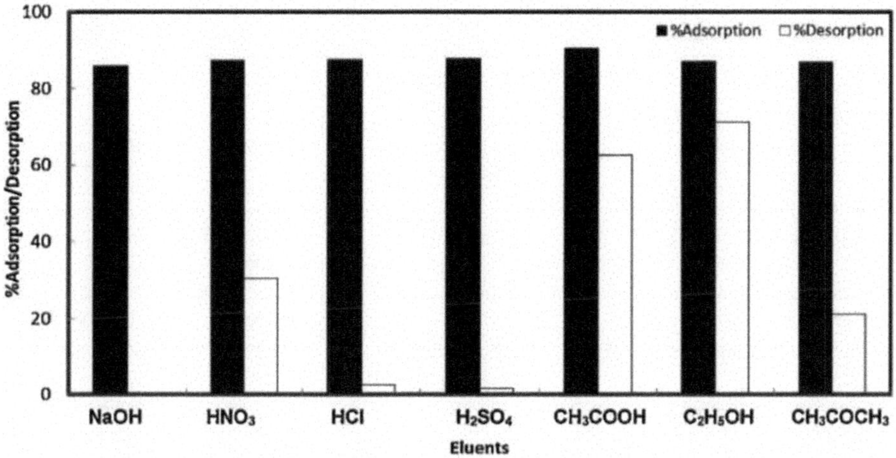

Fig. 9.6 Desorption of malachite green from a magnetic adsorbent.(Khan et al. (2019). Reprinted from Journal of Hazardous Materials, 365, Khan, M. A., M. Otero, M. Kazi, A. A. Alqadami, S. M. Wabaidur, M. R. Siddiqui, Z. A. Alothman and S. Sumbul, Unary and binary adsorption studies of lead and malachite green onto a nanomagnetic copper ferrite/drumstick pod biomass composite, 759–770, Copyright (2019), with permission from Elsevier)

knowledge of the adsorption mechanism is helpful but not sufficient to determine the appropriate regeneration method. As mentioned above, the complete adsorption mechanism generally cannot be determined with the currently available techniques. The following sections describe the use of specific solvents in more detail.

9.6.1 Regeneration in Alkali Solution

In the presence of hydroxyl ions, the adsorbent surface may become negatively charged. The surface will then repel negatively charged adsorbates, such as anionic dyes, including Congo red and methyl orange (Wang et al. 2018a; Chen et al. 2019a), and other pollutants, such as phenols and acids, that become anions at high pH (Eq. 9.6). Thus, regeneration in alkali solution has been successful for adsorption processes dominated by electrostatic interactions (Xu et al. 2018; Chen et al. 2019a; Qurrat Ul et al. 2019), and increasing the pH increased the amount of pollutant desorbed (Kumarasamy and Nachimuthu 2018).

$$\text{PhOH } or \text{ RCOOH} + \text{OH}^- \rightarrow \text{PhO}^- \; or \; \text{RCOO}^- + \text{H}_2\text{O} \tag{9.6}$$

Several studies in which magnetic adsorbents were regenerated in alkali solution are listed in Table 9.2. The regeneration efficiencies were > 90% in most cases even after several regeneration cycles for the anionic dyes, such as Congo red and methyl orange (Wang et al. 2018a, Chen et al. 2019a). There was no obvious trend for the

Table 9.2 Regeneration of magnetic adsorbents using alkali solutions of different strengths

Material (magnetic species)	Pollutant(s)	Regeneration efficiency	References
Regeneration in 0.01 M NaOH			
Activated carbon (Fe_3O_4)	p-chlorophenol and p-nitrophenol	>95%	Rong and Han (2019)
Regeneration in 0.1 M NaOH			
Camphor leaf biochar (γ-Fe_2O_3)	Ciprofloxacin	76% in third cycle	Hu et al. (2019)
Chitosan hydroxyapatite microspheres (Fe_3O_4)	Gallic acid	82% in fifth cycle	Fan et al. (2019)
Chitosan-based microspheres (Fe_3O_4)	Acid green 25 and reactive blue 19	>95%	Xu et al. (2018)
Amino-modified sepiolite composite (γ-Fe_2O_3)	Congo red	90% in tenth cycle	Wang et al. (2018a)
Willow tree-like functional groups modified nanoparticles (Fe_3O_4)	Congo red	88% in fifth cycle	Guo et al. (2019)
Regeneration in 0.5 M NaOH			
Reed biochar (Fe_3O_4)	Florfenicol	92% in fifth cycle	Zhao and Lang (2018)
Regeneration in 1 M NaOH			
Ionic liquid modified MoS_2-RGO (Fe_3O_4)	Methylene blue	~80%	Ni et al. (2019)
β-Cyclodextrin and polyethylenimine nanoadsorbent (Fe_3O_4)	Methyl orange	>95%	Chen et al. (2019a)
Multiwalled carbon nanotubes (Fe_3O_4)	Isonicotinic acid	>95%	Bhatia et al. (2019)
Amine-modified multiwalled carbon nanotubes (Fe_3O_4)	Bisphenol-A	>95%	Bhatia and Datta (2019)
Chitosan microspheres (Fe_3O_4)	Acid blue	88% in fourth cycle	Xu et al. (2019)

impact of alkali concentration or pollutant on regeneration. Other parameters such as the pore size of the adsorbents, stirring speed, time, and so on may have influenced the regeneration. The results reported in Table 9.2 are similar to those reported for nonmagnetic adsorbents. Based on the Pourbaix diagram of the iron–water system, magnetic materials including Fe_3O_4 and Fe_2O_3, which were mostly present in magnetic adsorbents (Table 9.2), are stable in alkali solution with pH lower than 13 (Beverskog and Puigdomenech 1996), which indicated that alkali regeneration had minimal impact on the magnetic adsorbents.

9.6.2 Regeneration in Acidic Solution

Just as regeneration in alkali solution is beneficial for anionic adsorbates, regeneration in acidic solution is beneficial for cationic adsorbates; that is, the adsorbents become positively charged in the presence of protons. As shown in Table 9.3, acidic solutions have been used to regenerate adsorbents saturated with cationic dyes, such as methylene blue and crystal violet (Çavuşoğlu et al. 2019; Zeng et al. 2019). During regeneration in an acidic solution, ion exchange and organic solvent extraction processes may occur in addition to the electrostatic repulsion process. For instance, cations in solution were reported to undergo ion exchange with the methylene blue cations from a nanofibrous metal-organic framework filter during regeneration (Li et al. 2020), while acetic acid may have extracted crystal violet from saturated adsorbents (Azari et al. 2019).

The main problem with acidic regeneration is the possibility of dissolving the magnetic species in the process. Regeneration with HCl resulted in decreased adsorption capacity for zinc ferrite–chitosan (Kumar et al. 2018) and $CuFe_2O_4$ adsorbents (Khan et al. 2019). Similarly, regeneration of γ-Fe_2O_3 loaded multiwalled carbon nanotubes saturated with methylene blue with 1 M HNO_3

Table 9.3 Regeneration of magnetic adsorbents using acidic solutions of different strengths

Material (magnetic species)	Pollutants	Regeneration efficiency	References
Regeneration in 0.1 M HCl			
Activated charcoal/β-Cyclodextrin/alginate polymer Nanocomposite (Fe_3O_4)	Methylene blue	>90% in fifth cycle	Yadav et al. (2020)
Amorphous carbon core-shell structure (Fe_3O_4)	Methylene blue	72.1% in fifth cycle	Zeng et al. (2019)
Spinel ferrite coated with sodium alginate ($CoFe_2O_4$)	Brilliant green and methylene blue	~75% in fifth cycle	Kumar et al. (2019)
Graphene oxide nanocomposites (Fe_3O_4 and γ-Fe_2O_3)	Methylene blue	>95%	Zhang et al. (2018)
Congo red functionalized SiO_2 nanoparticle (Fe_3O_4)	Methylene blue	60% in forth cycle	Yimin et al. (2018)
Regeneration in 0.1 M HNO_3			
Functionalized multi walled carbon nanotubes beads (γ-Fe_2O_3)	Methylene blue	>95%	Boukhalfa et al. (2019)
Alginate chitosan zeolite nanocomposite (Fe_3O_4)	Methylene blue	~75%	Kazemi and Javanbakht (2019)
Regeneration in acetic acid			
Activated carbon chitosan nanocomposite[a]	Crystal violet	75%	Çavuşoğlu et al. (2019)
Fe_3O_4/chitosan/Glutaraldehyde nanocomposites	Crystal violet	98%; 76% in tenth cycle	Azari et al. (2019)

Note: [a]Regenerated with 1 M acetic acid

reduced the regeneration efficiency to ~70% while regeneration with a less concentrated HNO_3 (0.1 M) solution resulted in regeneration efficiencies of >95% (Boukhalfa et al. 2019). Leaching of the magnetic species was suspected in each of these studies with evidence of adsorbent weight loss. Dissolution of magnetic species such as Fe_3O_4, $CuFe_2O_4$, $ZnFe_2O_4$ was observed in 1 M of HCl at room temperature with Fe_3O_4 being the most vulnerable magnetic species under acidic conditions (Lu and Muir 1988; Pang et al. 2011). In order to prevent loss during acidic regeneration, coated magnetic nanoparticles (polymer-coated Fe_3O_4 and γ-Fe_2O_3) were prepared and demonstrated to be stable (<1 wt% loss of Fe) under 1 M HCl conditions (Pang et al. 2011; Zheng et al. 2018).

9.6.3 Regeneration with Salt Solution

Addition of a salt changes the ionic strength of a solution and leads to the decrease in zeta potential of an adsorbent, which is called the screening effect. If an adsorption process is governed by electrostatic interactions, the presence of screening effects will reduce the adsorption capacity and result in regeneration because of the decreased potential energy of the interaction as indicated in Eq. (9.3) (Mittal et al. 2018; Bhatia and Datta 2019; Chen et al. 2019b). Besides the screening effect, the competitive adsorption and ion exchange of ions from the added salt were also proposed to contribute to the regeneration. For instance, Boukhalfa et al. (2019) found that the competitive adsorption of Na^+ ions reduced methylene blue (cation) adsorption capacity on a γ-Fe_2O_3-based adsorbent by 82%. Similarly, Liu et al. (2018) observed the regeneration of a γ-Fe_2O_3-based resin saturated with perfluorooctane sulfonate and perfluorooctanoate anions was aided by the ion exchange of Cl^- from the NaCl regeneration solution.

In terms of stability, magnetic species such as Fe_3O_4 and γ-Fe_2O_3 were stable under neutral conditions with various salts based on the Pourbaix diagram and laboratory experiments (Beverskog and Puigdomenech 1996; Kalska-Szostko et al. 2014). Salt solutions have been effective for the recovery of valuable adsorbates such as heparin (a pharmaceutical) from an Fe_3O_4-based adsorbent (Eskandarloo et al. 2019). Limited studies, however, have used salt solutions because of the high (10 wt% or higher) salt concentrations required (Shuang et al. 2012; Eskandarloo et al. 2019). The high concentrations introduced inorganic ions and increased the salinity of the treated water. Additional treatment of the water may be required to restore the original salinity.

9.6.4 Regeneration by Solvent

For adsorbates that are not charged or if dispersion interactions dominate, organic solvents can be used for regeneration. For aqueous adsorption systems, polar solvents such as methanol and ethanol often had higher regeneration efficiencies compared with nonpolar solvents including hexane or toluene (Reguyal et al. 2017; Minh et al. 2018). For instance, the desorption of bisphenol A from a carbon nanotube/Fe_3O_4 nanocomposite after adsorption in wastewater decreased from 98% to 59% when the regeneration solvent changed from methanol to cyclohexane. In contrast, nonpolar solvents were more favorable in organic adsorption systems. For example, the regeneration of a Ni-loaded mesoporous carbon saturated with dibenzothiophene from n-hexane was 97% regenerated in toluene compared to 23% regeneration in ethanol (Farzin Nejad et al. 2013).

During solvent regeneration with porous adsorbents, the diffusion rate of the solvent to the adsorption sites will be hindered if its polarity is different from that of the adsorption solution. In addition, the size of the solvent molecules relative to the pore size of the adsorbent is also a factor – higher regeneration efficiencies are achieved with smaller solvent molecules for a specific adsorbent (Alizadeh Fard and Barkdoll 2018). Thirdly, the nature of the surface functional groups will influence the regeneration. For example, the rate of desorption of methylene blue from a Fe_3S_4-loaded carbon adsorbent was several times lower in water than in ethanol despite the higher solubility of methylene blue in water than in ethanol (Wang et al. 2013). This result was attributed to the hydrophobicity of the adsorbent preventing the interaction of water with the adsorbent surface (Sun et al. 2006; Wang et al. 2013). Once the solvent reaches the adsorbate surface, the strength of interaction between the solvent and adsorbate may control the regeneration process. For instance, in the desorption of methyl violet from a Fe_3O_4-loaded polymer, ethanol disrupted and/or weakened the hydrophobic interaction (dispersion) for adsorption and regeneration was improved as the ethanol to water ratio increased (Liang et al. 2014).

In terms of the magnetic adsorbent stability during solvent regeneration, solvents had little effect as with salt solutions (Wang et al. 2012). Solvent and salt solutions have been used in combination for regeneration when multiple adsorption mechanisms are present. The organic solvent disrupted the dispersion interaction between magnetic adsorbents and organic pollutants (Ghaemi et al. 2014; Liang et al. 2014), while the salt reduced the electrostatic interaction through screening effects, as explained above (Mahdavinia and Mosallanezhad 2016).

9.7 Decomposition-Based Regeneration

Decomposition-based regeneration methods have been developed to regenerate magnetic adsorbents and to decompose adsorbed organic pollutants simultaneously. Importantly, if the organic pollutants are decomposed to water, CO_2, and minerals, these products will not compete with the target pollutants in subsequent adsorption cycles. A list of some recent studies focusing on thermal and advanced oxidation regeneration is given in Table 9.4.

Thermal regeneration is the only commercial method that has been used in a water treatment plant and the adsorbent was activated carbon. At an industrial scale, thermal treatment includes steaming, pyrolysis under inert atmosphere, and gasification and reactivation by oxidizing gases such as steam or carbon dioxide (Guo and Du 2012). High regeneration efficiencies (>90%) are generally achieved with thermal regeneration because complete decomposition of the adsorbed organic pollutants occurs with high temperature (>300 °C) calcination (Yang et al. 2013). The presence of magnetic species on the adsorbent was reported to catalyze the thermal decomposition process so that the regeneration temperature could be reduced. For example, the complete decomposition of azo-dye acid red B was achieved at 300 °C in the presence of $MnFe_2O_4$ compared to a self-degradation of over 600 °C (Wu et al. 2005). On the lab-scale, microwave heating has been used to reduce the regeneration time compared to that using conventional heating such as in a muffle furnace (Cui et al. 2015).

The high energy requirements—treatment temperatures up to 800 °C and drying of the wet adsorbents—increase the cost of thermal regeneration processes to ~50% of the material cost for the adsorbent (USEPA 2000). Additionally, the adsorbents can be damaged (collapse of the porous structure and subsequent reduction in surface area) during thermal regeneration resulting in reduced adsorption capacity after each regeneration cycle (Wu et al. 2005; Deng et al. 2016; Saroyan et al. 2017). Magnetic adsorbents consist of minerals (hydrotalcite or ferrites) that can be sintered resulting in reduced regeneration efficiencies (Wu et al. 2005; Deng et al. 2016). If the support is carbon-based, a reduction in the regeneration efficiency may be mainly a result of the decomposition of the carbon materials during thermal regeneration (Gao et al. 2014). Extra treatment steps, such as water washing, may be required to remove the generated minerals, such as sulfates and nitrates (Wu et al. 2003), before using the regenerated material for water treatment. The magnetism of the magnetic ferrites, such as $MnFe_2O_4$ and $CuFe_2O_4$, was not affected during thermal regeneration at temperatures up to 500 °C (Wu et al. 2005; Peng et al. 2006). Other magnetic species such as Fe_3O_4 and γ-Fe_2O_3 were likely converted to nonmagnetic α-Fe_2O_3 during thermal regeneration (Dinesen et al. 2001, Sandeep Kumar et al. 2019). Therefore, thermal regeneration of magnetic adsorbents requires a balance between adsorbate decomposition, adsorbent structure integrity, and magnetism maintenance.

Advanced oxidation processes are a set of chemical methods that can oxidize and mineralize organic pollutants in the presence of •OH (Andreozzi et al. 1999). The high standard reduction potential of •OH in acidic aqueous solutions

Table 9.4 Decomposition-based methods for the regeneration of magnetic adsorbents saturated with organic pollutants

Material (magnetic species)	Adsorbate	Regeneration conditions	Regeneration efficiency (%)[a]	References
Thermal regeneration				
Alumina/ferrite microfiber ($Ni_{0.5}Zn_{0.5}Fe_2O_4$)	Congo red and methylene blue	550 °C, 3 h	99% in fifth cycle	Yang et al. (2013)
Coal fly ash/ferrite composite ($CoFe_2O_4$)	Malachite green	500 °C, 30 min	92%	Zhang et al. (2016)
Ferrite/hydrotalcite composite ($CoFe_2O_4$)	Methyl orange	450 °C, 2 h	~50% in third cycle	Deng et al. (2016)
MnO–Fe_2O_3 ($MnFe_2O_4$)	ARB[b]	400 °C, 20 min	~90% in fifth cycle	Wu et al. (2005)
Hydroxyapatite/magnetite/zeolite composite (Fe_3O_4)	Congo red	400 °C, 30 min	97%	Fang et al. (2014)
Graphene oxide/magnetite composites (Fe_3O_4)	Methylene blue	250 °C 300 °C 350 °C	~88% 91% ~80%	Gao et al. (2014)
Magnetite/carbon nanotube (Fe_3O_4)	*p*-nitrophenol	Microwave, 850 W, 180 s	106%	Cui et al. (2015)
Advanced oxidation regeneration				
Ferrite material ($MnFe_2O_4$)	ARB	Excess H_2O_2, H_2O_2/Fe^{2+} = 50/1, pH 2–3, 3 h, 25 °C	166%	Wu and Qu (2005)
Magnetite adsorbent (Fe_3O_4)	ARB	2 mmol H_2O_2/mg ARB, H_2O_2/Fe^{2+} = 50/1, pH 2–3, 3 h, 25 °C	486%	Rongcheng and Jiuhui (2004)
Carbon nanotube/magnetite nanocomposite (Fe_3O_4)	Bisphenol A	H_2O_2/Fe^{2+} = 10/1, pH 4, 10 min	98.9%	Li et al. (2015)
Magnetite/graphene oxide nanoparticles (Fe_3O_4)	Methylene blue	0.1 mM H_2O_2	89%	Chang et al. (2014)
Magnetite/activated carbon (Fe_3O_4)	Methyl orange	0.28% H_2O_2	~65%	Do et al. (2011)
Maghemite-loaded mesoporous carbon (γ-Fe_2O_3)	Methyl orange	0.3% H_2O_2	93% in fifth cycle	Xiao and Hill (2017)
Magnetite@chitosan carbon microbeads (Fe_3O_4)	Doxycycline	5% H_2O_2	>80% in third cycle	Bai et al. (2018)

(continued)

Table 9.4 (continued)

Material (magnetic species)	Adsorbate	Regeneration conditions	Regeneration efficiency (%)[a]	References
Magnetite@yeast composite micro-spheres (Fe_3O_4)	Methylene blue	10% H_2O_2	84%	Song et al. (2015)
Reduced graphene oxide/iron oxide (Fe_3O_4)	Methylene blue	Anodic, 10 mA/cm^2, 30 min	~100%	Sharif et al. (2017)
Maghemite-loaded mesoporous carbon (γ-Fe_2O_3)	Methyl orange	Cathodic, −0.8 V, 8 h	96%	Xiao and Hill (2019)
Carbon nanotube/ magnetite nanocomposite (Fe_3O_4)	Bisphenol A	4.5 mM H_2O_2, 365 nm UV, 10 min	97.6%	Li et al. (2015)
Functional group-bridged bentonite adsorbent (Fe_3O_4)	Methylene blue	γ radiation, 1 kGy/h, 6 h	~96%	Lou et al. (2017)

Notes: [a]Efficiencies over 100% indicate that the adsorption capacity of the adsorbent increased after the regeneration treatment; [b]ARB is azo-dye acid red B

(– 2.80 V vs. standard hydrogen electrode) results in the decomposition of almost all organic pollutants to carbon dioxide and water, except simpler compounds such as short-chain carboxylic acids that are stable and difficult to decompose (Pera-Titus et al. 2004). There are numerous advanced oxidation processes that have been developed for water purification (Pera-Titus et al. 2004) and adsorbent regeneration (Toledo et al. 2003; Huling et al. 2007; Salvador et al. 2015b; Chen et al. 2017). Except conventional Fenton oxidation using H_2O_2 and Fe^{2+} (Table 9.4), some of these methods were operated using electrochemistry, such as electro-Fenton, anodic oxidation, and electro-peroxone (Narbaitz and Cen 1994; Brown et al. 2004; Salvador et al. 2015b; Zhan et al. 2016; Sharif et al. 2017), while others may use UV light or even γ radiation to generate •OH for the regeneration process (Li et al. 2015; Lou et al. 2017). Advanced oxidation processes had regeneration efficiencies varying from <50% to more than 90% (Lim et al. 2011; Salvador et al. 2015b; Chen et al. 2017). For some of the advanced oxidation processes, commercial chemicals are required either to degrade the organic pollutants or to adjust the solution chemistry. Fortunately, some of the chemicals can be generated in situ with electrochemistry and/or UV light (Li et al. 2015; Xiao and Hill 2019).

In comparison to thermal regeneration, the main advantage of advanced oxidation regeneration was the capability of regenerating adsorbents under ambient conditions. Also, no pretreatment (such as drying the adsorbent) was required before advanced oxidation regeneration because most of these processes were done in solution. As a result, advanced oxidation regeneration processes are promising technologies for the regeneration of magnetic adsorbents saturated with organic pollutants. The application of Fenton oxidation and electro-Fenton oxidation for adsorbent regeneration will be reviewed in the following Sections.

9.8 Advanced Oxidation Regeneration

9.8.1 Fenton Oxidation Mechanism

The Fenton oxidation process is initiated with the formation of •OH by the reaction between Fenton reagents, that is, H_2O_2 and Fe^{2+}, Eq. (9.7). The Fe^{2+} is then regenerated from the Fenton-like reaction between Fe^{3+} and H_2O_2, Eq. (9.8), which is much slower than the preceding reaction (9.7). The Fe^{2+} can be regenerated more quickly through reactions (9.9) and (9.10) by $HO_2^{\bullet}$ and $O_2^{\bullet -}$, which are generated from reactions (9.11) and (9.12) (Brillas et al. 2009).

$$Fe^{2+} + H_2O_2 + H^+ \rightarrow Fe^{3+} + H_2O + \bullet OH \tag{9.7}$$

$$Fe^{3+} + H_2O_2 \rightarrow Fe^{2+} + HO_2^{\bullet} + H^+ \tag{9.8}$$

$$Fe^{3+} + HO_2^{\bullet} \rightarrow Fe^{2+} + O_2 + H^+ \tag{9.9}$$

$$Fe^{3+} + O_2^{\bullet -} \rightarrow Fe^{2+} + O_2 \tag{9.10}$$

$$H_2O_2 + \bullet OH \rightarrow H_2O + HO_2^{\bullet} \tag{9.11}$$

$$HO_2^{\bullet} \rightleftharpoons H^+ + O_2^{\bullet -} \tag{9.12}$$

$$Fe^{2+} + \bullet OH \rightarrow Fe^{3+} + OH^- \tag{9.13}$$

Among these reactions, H_2O_2 can be decomposed by reactions (9.8) and (9.11), which are side reactions competing with Fenton's reaction. In addition, generated •OH is consumed through reactions (9.11) and (9.13) by Fenton's reagent, which will lead to a decrease in the oxidizing ability of the system (Brillas et al. 2009). Therefore, the concentrations or dosages of Fenton reagents should be adjusted for the specific application. The disadvantages of the conventional homogeneous Fenton oxidation process include the relatively high cost and potential risks for the transportation and handling of the commercial concentrated H_2O_2, a limited working pH range of 2.8–3.0, accumulation of iron sludge, and the impossible overall mineralization of organic pollutants because the formation of complexes of Fe(III) with generated carboxylic acid cannot be destroyed by bulk •OH (Brillas et al. 2009; Liu et al. 2015c).

Instead of using the soluble Fe ions as the catalysts, heterogeneous catalysts have been used in the Fenton oxidation process (referred to as heterogeneous Fenton oxidation). The catalysts were prepared by loading iron species onto different supports such as zeolite, alumina, silica, and carbon materials (Munoz et al. 2015). In order to facilitate separation of the catalysts from a reaction medium after the treatment, magnetic iron species including zero-valent iron, magnetite (Fe_3O_4), maghemite (γ-Fe_2O_3), and ferrites (such as $CuFeO_2$) were loaded on the supports through coprecipitation, hydrothermal, or thermal treatment methods (Zhang et al. 2014; Munoz et al. 2015). Application of heterogeneous catalysts can solve some problems related to the conventional homogeneous Fenton oxidation process, for instance broader pH range (neutral and basic pH), and no requirement for Fe recovery after treatment (Centi et al. 2000).

The generation of •OH during heterogeneous Fenton oxidation could happen in two ways: through the homogeneous reaction of dissolved Fe ions and through the surface catalytic reactions (Pignatello et al. 2006). Fe leaching can deactivate the catalyst by attrition (Ribeiro et al. 2016), as the dissolved Fe ions only contribute in a minor part to the whole oxidation process (Huling et al. 2007; Do et al. 2011). The mechanism of •OH generation on the heterogeneous catalyst surface with Fe(III) species was proposed and summarized by different researchers as shown by Eqs. (9.14) to (9.17) (Lin and Gurol 1998; Kwan and Voelker 2003; Luo et al. 2010), which could be well described by a Langmuir-Hinshelwood mechanism (Lin and Gurol 1998).

$$\equiv Fe^{III} + H_2O_2 \leftrightarrow \equiv Fe^{III}H_2O_2 \tag{9.14}$$

$$\equiv Fe^{III}H_2O_2 \rightarrow \equiv Fe^{II} + H^+ + HO_2^{\bullet} \tag{9.15}$$

$$\equiv Fe^{III} + HO_2^{\bullet} \rightarrow \equiv Fe^{II} + H^+ + O_2 \tag{9.16}$$

$$\equiv Fe^{II} + H_2O_2 \rightarrow \equiv Fe^{III} + OH^- + OH^{\bullet} \tag{9.17}$$

The oxidation of organic pollutants by •OH could occur on the surface of the catalyst or in the bulk solution. In most cases, the degradation kinetics were derived based on the concentration of •OH, indicating the importance of oxidation in solution especially for the catalysts having low adsorption capacity (Kwan and Voelker 2003, Luo et al. 2010).

9.8.2 Fenton Oxidation Regeneration

As illustrated in Fig. 9.7, several reactions can occur during regeneration with Fenton oxidation. First, adsorbed organic pollutants can desorb into solution to regenerate the adsorption sites directly (1). The desorbed organic pollutants and intermediates can experience homogeneous oxidation by •OH to induce more desorption (2). Then, the adsorbed organic pollutants may be oxidized to CO_2 and H_2O on the surface to realize regeneration (3). Finally, the adsorbed organic pollutants could be partially decomposed to intermediates (4), which can be desorbed from the surface to recover the adsorption sites (5). Based on the Fourier-transform infrared spectra of the regenerated adsorbent and the total organic carbon measurement of the solution during regeneration, Li et al. (2018) confirmed that part of the

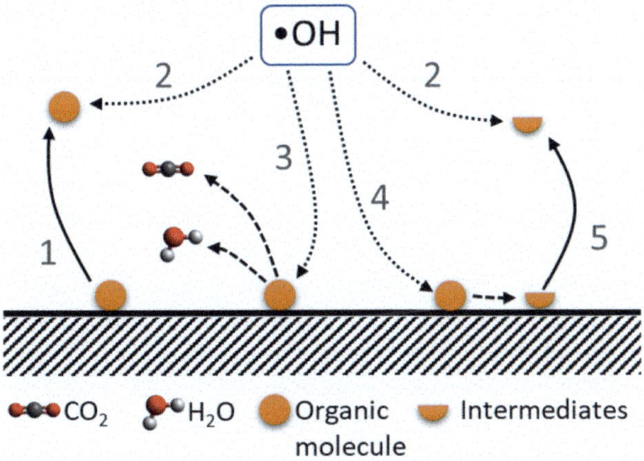

Fig. 9.7 Scheme of Fenton oxidation regeneration process: 1—desorption of adsorbed species; 2—oxidation of desorbed species; 3, 4—oxidation of adsorbed species; 5—desorption of intermediates

adsorbed pollutants were desorbed from the adsorbent surface before decomposition by Fenton-like oxidation. Both homogeneous and heterogeneous Fenton oxidation reactions were proposed to contribute to the oxidation of adsorbed acid red B during regeneration (Rongcheng and Jiuhui 2004).

Because of the effectiveness of this process, Fenton oxidation has been used to regenerate magnetic adsorbents saturated with organic pollutants. Both homogeneous and heterogeneous Fenton oxidation have been applied (Rongcheng and Jiuhui 2004; Xiao and Hill 2017) with the latter method only introducing H_2O_2 into the system. The regeneration efficiency was generally higher with homogeneous Fenton oxidation (Table 9.4). The higher regeneration efficiency of homogeneous Fenton oxidation may originate from the much higher •OH radical generation rate (~10^2 M^{-1} s^{-1}) compared with that of heterogeneous Fenton oxidation (~10^{-5} M^{-1} s^{-1} with a catalyst surface area of ~100 m^2/g) (Kwan and Voelker 2003; Brillas et al. 2009). In addition, the deposition of iron hydroxide from the Fenton reagents onto the adsorbent may have increased the regeneration efficiency and/or the adsorption capacity of the homogeneous Fenton oxidation regeneration (Rongcheng and Jiuhui 2004; Wu and Qu 2005).

In addition to the homogeneous Fenton oxidation regeneration with external Fe^{2+} salt as catalyst, heterogeneous Fenton oxidation regeneration of magnetic adsorbents has also been investigated with the advantage of using the magnetic species as catalysts for the production of •OH radicals. Some studies (Huling et al. 2009; Kan and Huling 2009) have suggested that only when the pollutants, H_2O_2, and the Fe species meet can Fe contribute to the regeneration and oxidation process. In general, loading Fe species on adsorbents increased regeneration efficiency. For example, the presence of Fe_3O_4 on an activated carbon based adsorbent was demonstrated to improve the regeneration efficiency from <10% to ~65% by promoting the heterogeneous Fenton oxidation reaction (Do et al. 2011). The improvement in regeneration efficiency was also observed for adsorbents loaded with other magnetic species such as γ-Fe_2O_3 and $MnFe_2O_4$ (Wan et al. 2014; Xiao 2018).

In some regeneration studies, the Fe_3O_4 species on the adsorbents were proposed to dissolve in solution as Fe^{3+} during Fenton oxidation regeneration leading to the decrease in regeneration efficiency over several regeneration cycles (Alizadeh Fard and Barkdoll 2018, 2019). This explanation seems unlikely because the loss of Fe ions was minimal (<1%) in various other studies for the regeneration of adsorbents containing Fe_3O_4 and γ-Fe_2O_3 at a pH of ~3.0 (Do et al. 2011; Xiao and Hill 2017; Li et al. 2018). The conversion of Fe_3O_4 into γ-Fe_2O_3, however, was possible during the regeneration and would result in lower regeneration efficiency due to the lower oxidation activity of γ-Fe_2O_3 (Rusevova et al. 2012). On the other hand, at typical Fenton oxidation conditions of pH of ~3.0, magnetic species such as ferrites may be dissolved during the regeneration process; for instance, the surface Fe and Mn contents on $MnFe_2O_4$ were reduced by 2% and 5%, respectively, after Fenton oxidation regeneration (Wu and Qu 2005).

The porosity of the support is also an important factor, with higher regeneration efficiencies obtained using mesoporous materials (Muranaka et al. 2010). As indicated in Table 9.4, most of the regeneration efficiencies were > ~ 90% when mesoporous adsorbents such as nanotubes and graphene oxides were regenerated, while the regeneration efficiency for an activated carbon (~80% microporosity)-based magnetic adsorbent was only ~65% under similar conditions (Do et al. 2011; Qin et al. 2014; Li et al. 2017). The impact of pore size was demonstrated for the regeneration of a series of carbon supports loaded with γ-Fe_2O_3 and then saturated with methyl orange (Xiao and Hill 2017). As shown in Fig. 9.8a, the regeneration efficiency decreased from almost 100% to less than 10% as the microporosity of the adsorbent increased from ~0% to 80%. The corresponding modeling indicated that the film diffusion and mesoporous diffusion rate constants were highly correlated with the regeneration efficiency (Figs. 9.8b, c). Although the regeneration efficiency was improved with larger pores, the adsorption capacity decreased from 362 mg/g to 26.5 mg/g.

9.8.3 Electro-Fenton Oxidation

Electro-Fenton oxidation combines the Fenton oxidation process with electrochemistry, in which the H_2O_2 is generated in situ electrochemically through the two-electron oxygen reduction reaction (ORR) on the cathode (Brillas et al. 2009) according to Eq. (9.18).

$$O_2 + 2H^+ + 2e^- \rightarrow H_2O_2 \tag{9.18}$$

The Fe^{3+} ion generated from Fenton's reaction (9.7) is regenerated to Fe^{2+} on the cathode (Sirés et al. 2014) through reaction (9.19).

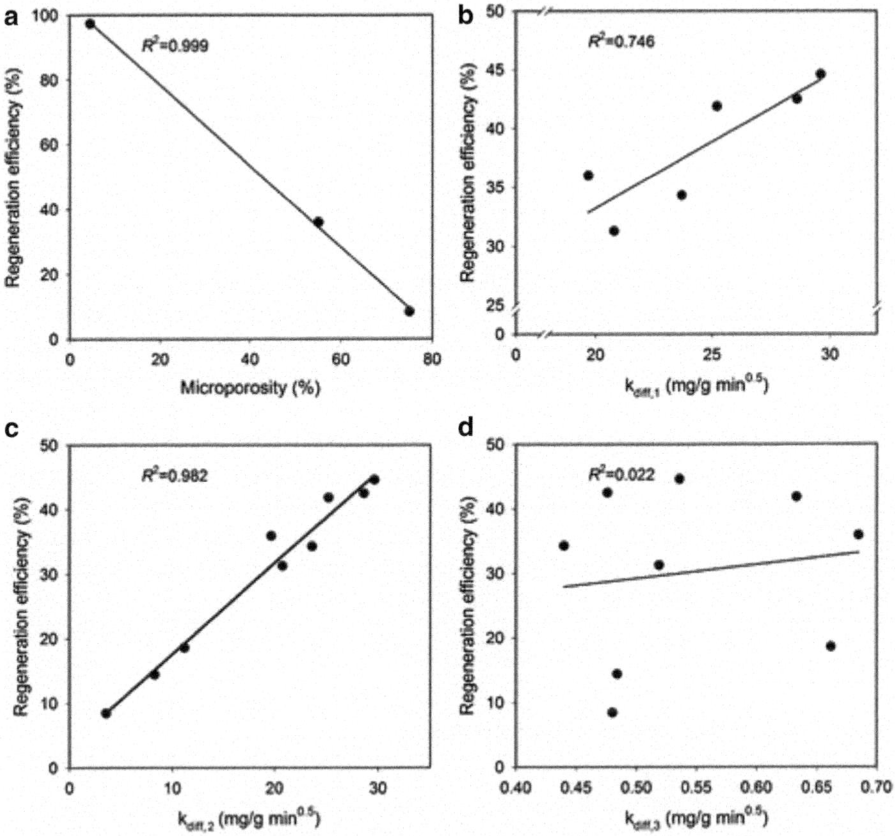

Fig. 9.8 Relationship between regeneration efficiency and pore structure: (**a**) microporosity; (**b–d**) intraparticle diffusion rate constants. (Xiao and Hill (2017). Reprinted (adapted) with permission from (Xiao, Y. and J. M. Hill (2017). "Impact of Pore Size on Fenton Oxidation of Methyl Orange Adsorbed on Magnetic Carbon Materials: Trade-Off between Capacity and Regenerability." Environmental Science & Technology 51(8): 4567–4575). Copyright (2017) American Chemical Society)

$$Fe^{3+} + e^- \rightarrow Fe^{2+} \tag{9.19}$$

Two types of electro-Fenton reactors have been applied for the treatment of water—namely, divided and undivided electrochemical cells (Brillas et al. 2009). The benefit of the divided cells is that the oxidation of H_2O_2 to O_2 at the anode through reactions (9.20) and (9.21) was avoided (Brillas et al. 1995).

$$H_2O_2 \rightarrow HO_2^{\bullet} + H^+ + e^- \tag{9.20}$$

$$HO_2^{\bullet} \rightarrow O_{2(g)} + H^+ + e^- \tag{9.21}$$

These two reactions reduce the concentration of H_2O_2 in the system. In addition, the influence of the anode on the concentration of Fe^{2+} through reaction (9.22) is avoided (Brillas et al. 2009).

$$Fe^{2+} \rightarrow Fe^{3+} + e^- \tag{9.22}$$

However, the overall mineralization of the organic pollutants was difficult to achieve because of the generation of Fe(III)-carboxylate complexes similar in regular Fenton oxidation process (Brillas et al. 2009). In the undivided electro-Fenton cells, it is possible to decompose the Fe(III)-carboxylate complexes by oxidant produced from the anode, especially in the cases when boron doped diamond (BDD) is used as the anode (Brillas et al. 2007). The heterogeneous BDD (•OH) has a higher oxidation ability than the homogeneous •OH formed by Fenton's reaction in bulk solution (Liu et al. 2015a). The Fe(III)-carboxylate complexes, such as Fe^{3+}-oxalato complexes, could be efficiently decomposed by UV light in photoelectron-Fenton process with the combination of photolysis and electro-Fenton oxidation (Brillas et al. 2000; Boye et al. 2002).

9.8.4 Electro-Fenton Regeneration

As indicated above, electro-Fenton oxidation has the advantage of in situ generation of H_2O_2. Bañuelos et al. introduced the electro-Fenton method for regeneration of adsorbents saturated with organic pollutants (Bañuelos et al. 2013). In this process, the H_2O_2 generated at the cathode from the reduction of oxygen interacted with the iron ion in the electrolyte or Fe-loaded resins in the cathode to generate hydroxide radicals, which can decompose molecules adsorbed on the activated carbon (Bañuelos et al. 2013; Bañuelos et al. 2015). Since then, electro-Fenton oxidation has been applied for the regeneration of adsorbents saturated with organic pollutants (Roth et al. 2016; Trellu et al. 2016; Trellu et al. 2018). Only one group, however, has applied this process for the regeneration of magnetic adsorbents (Xiao and Hill 2019), and in this study, the application of electrochemistry improved the regeneration process. The influence of adsorbent microporosity on the regeneration efficiency was reduced when a cathodic potential of −3.0 V (vs Ag/AgCl electrode) was applied. The regeneration efficiency for a γ-Fe_2O_3/activated carbon using electro-Fenton oxidation was 64% compared with ~10% for Fenton oxidation. The electro-Fenton oxidation regeneration was demonstrated to be controlled by two processes, electro-desorption and oxidation as indicated in Fig. 9.9 for anionic pollutants. As shown in Fig. 9.9, the desorption process, however, dominated the regeneration process for the anionic pollutants because this process directly regenerated the

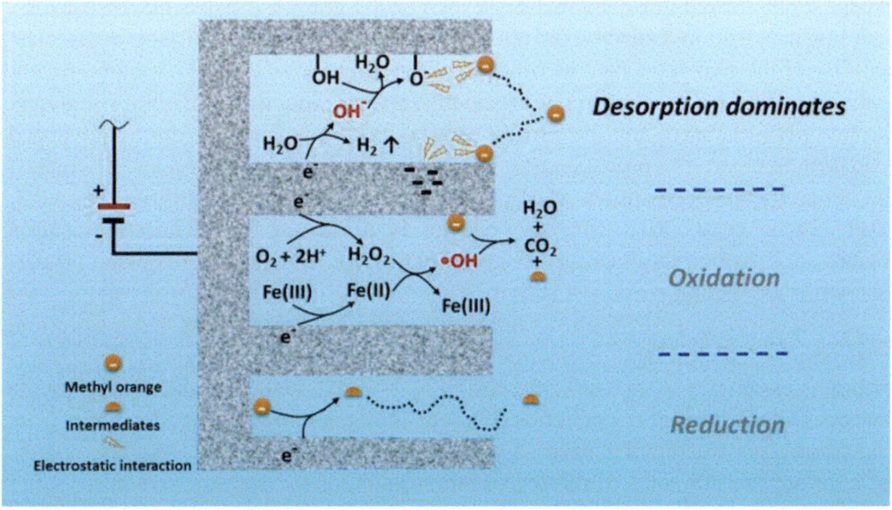

Fig. 9.9 Illustration of electro-Fenton regeneration mechanism. (Xiao and Hill (2019). Reprinted from Journal of Hazardous Materials, 367, Xiao, Y. and Hill, J. M. Mechanistic insights for the electro-Fenton regeneration of carbon materials saturated with methyl orange: Dominance of electrodesorption, 59–67, Copyright (2019), with permission from Elsevier)

adsorbent surface. The subsequent oxidation processes contributed to the decomposition of the desorbed organic pollutants.

9.9 Future Outlook

More and more studies have focused on the application of magnetic adsorbents in water treatment, which evoked the research on the regeneration of saturated adsorbents. Desorption-based regeneration methods were efficient to remove adsorbates from the adsorbent through pH variation, solvent extraction, or salt addition, and are beneficial for the recovery of valuable adsorbates. Decomposition-based regeneration technologies, especially advanced oxidation methods, however, can regenerate the adsorbent and degrade the pollutants simultaneously under ambient conditions. Magnetic adsorbents are more easily separated from the solution phase after regeneration and are generally stable under the regeneration conditions used (there are stability issues with highly acidic ($pH < 3$) regeneration solutions, as addressed below). In addition, the presence of magnetic species on the adsorbents assisted the advanced oxidation process acting as catalysts.

Despite these advantages, further research on the regeneration of magnetic adsorbents by advanced oxidation technologies is required for practical applications. Specifically, the pore structure of the adsorbents requires tailoring to balance adsorption capacity with regeneration efficiency. Most of the mesoporous materials, except sophisticated designed materials such as 3D graphene-based macrostructure, had low adsorption capacity (<100 mg/g) because of lower surface areas. The stability of the magnetic species was affected during the advanced oxidation regeneration process. Magnetic species, especially ferrites, were dissolved into solution at the low pH (~3) used, and the stirring used to improve efficiency during advanced oxidation regeneration process hastened the loss of the magnetic species. Coating of the materials has been shown to be effective in improving the stability (Pang et al. 2011; Zheng et al. 2018) but has been rarely discussed for advanced oxidation regeneration, which requires further investigation for maintaining magnetic properties while preventing oxidation activity loss.

The development of in situ techniques for the investigation of the adsorption, desorption, oxidation, and/or electrochemical mechanisms are required. This information is essential for appropriately designing both the materials and scaling-up the process. Application of regeneration methods in regenerating spent adsorbents from real wastewater treatment is necessary to analyze the stability of these technologies. In addition, the influence of the regeneration methods on the water chemistry after reusing the regenerated adsorbents is of great importance to evaluate the environmental impact of different technologies.

References

Adewuyi YG (2005) Sonochemistry in environmental remediation. 1. Combinative and hybrid sonophotochemical oxidation processes for the treatment of pollutants in water. Environ Sci Technol 39(10):3409–3420. https://doi.org/10.1021/es049138y

Ahad JME, Pakdel H, Savard MM, Calderhead AI, Gammon PR, Rivera A, Peru KM, Headley JV (2013) Characterization and quantification of mining-related "naphthenic acids" in groundwater near a major oil sands tailings pond. Environ Sci Technol 47(10):5023–5030. https://doi.org/10.1021/es3051313

Alizadeh Fard M, Barkdoll B (2018) Using recyclable magnetic carbon nanotube to remove micropollutants from aqueous solutions. J Mol Liq 249:193–202. https://doi.org/10.1016/j.molliq.2017.11.039

Alizadeh Fard M, Barkdoll B (2019) Magnetic activated carbon as a sustainable solution for removal of micropollutants from water. Int J Environ Sci Technol 16(3):1625–1636. https://doi.org/10.1007/s13762-018-1809-5

Andreozzi R, Caprio V, Insola A, Marotta R (1999) Advanced oxidation processes (AOP) for water purification and recovery. Catal Today 53(1):51–59. https://doi.org/10.1016/S0920-5861(99)00102-9.

Anslyn EV, Dougherty DA (2006) Modern physical organic chemistry. University Science Books, Sausalito

Aslibeiki B, Kameli P, Ehsani MH (2016) $MnFe_2O_4$ bulk, nanoparticles and film: a comparative study of structural and magnetic properties. Ceram Int 42(11):12789–12795. https://doi.org/10.1016/j.ceramint.2016.05.041.

Azari A, Noorisepehr M, Dehganifard E, Karimyan K, Hashemi SY, Kalhori EM, Norouzi R, Agarwal S, Gupta VK (2019) Experimental design, modeling and mechanism of cationic dyes biosorption on to magnetic chitosan-glutaraldehyde composite. Int J Biol Macromol 131:633–645. https://doi.org/10.1016/j.ijbiomac.2019.03.058

Bai B, Xu X, Li C, Xing J, Wang H, Suo Y (2018) Magnetic Fe_3O_4@ chitosan carbon microbeads: removal of doxycycline from aqueous solutions through a fixed bed via sequential adsorption and heterogeneous Fenton-like regeneration. J Nanomater 2018. https://doi.org/10.1155/2018/5296410

Bañuelos JA, Rodríguez FJ, Manríquez Rocha J, Bustos E, Rodríguez A, Cruz JC, Arriaga LG, Godínez LA (2013) Novel electro-Fenton approach for regeneration of activated carbon. Environ Sci Technol 47(14):7927–7933. https://doi.org/10.1021/es401320e

Bañuelos J, García-Rodríguez O, Rodríguez-Valadez F, Manríquez J, Bustos E, Rodríguez A, Godínez L (2015) Cathodic polarization effect on the electro-Fenton regeneration of activated carbon. J Appl Electrochem 45(5):523–531. https://doi.org/10.1007/s10800-015-0815-2

Bazhin N (2012) The born formula describes enthalpy of ions solvation. ISRN Thermodynamics 2012:3. https://doi.org/10.5402/2012/204104

Beverskog B, Puigdomenech I (1996) Revised pourbaix diagrams for iron at 25–300 °C. Corros Sci 38(12):2121–2135. https://doi.org/10.1016/S0010-938X(96)00067-4

Bhatia D, Datta D (2019) Removal of Bisphenol-a using amine-modified magnetic multiwalled carbon nanotubes: batch and column studies. J Chem Eng Data 64(6):2877–2887. https://doi.org/10.1021/acs.jced.9b00240

Bhatia D, Datta D, Joshi A, Gupta S, Gote Y (2019) Adsorption of isonicotinic acid from aqueous solution using multi-walled carbon nanotubes/Fe_3O_4. J Mol Liq 276:163–169. https://doi.org/10.1016/j.molliq.2018.11.127

Boukhalfa N, Boutahala M, Djebri N, Idris A (2019) Maghemite/alginate/functionalized multiwalled carbon nanotubes beads for methylene blue removal: adsorption and desorption studies. J Mol Liq 275:431–440. https://doi.org/10.1016/j.molliq.2018.11.064

Boye B, Dieng MM, Brillas E (2002) Degradation of herbicide 4-chlorophenoxyacetic acid by advanced electrochemical oxidation methods. Environ Sci Technol 36(13):3030–3035. https://doi.org/10.1021/es0103391

Brillas E, Martínez-Huitle CA (2015) Decontamination of wastewaters containing synthetic organic dyes by electrochemical methods. An updated review. Appl Catal B Environ 166-167:603–643. https://doi.org/10.1016/j.apcatb.2014.11.016

Brillas E, Bastida RM, Llosa E, Casado J (1995) Electrochemical destruction of aniline and 4-chloroaniline for wastewater treatment using a carbon-PTFE O_2 - fed cathode. J Electrochem Soc 142(6):1733–1741. https://doi.org/10.1149/1.2044186

Brillas E, Calpe JC, Casado J (2000) Mineralization of 2,4-D by advanced electrochemical oxidation processes. Water Res 34(8):2253–2262. https://doi.org/10.1016/S0043-1354(99)00396-6

Brillas E, Baños MÁ, Skoumal M, Cabot PL, Garrido JA, Rodríguez RM (2007) Degradation of the herbicide 2,4-DP by anodic oxidation, electro-Fenton and photoelectro-Fenton using platinum and boron-doped diamond anodes. Chemosphere 68(2):199–209. https://doi.org/10.1016/j.chemosphere.2007.01.038

Brillas E, Sirés I, Oturan MA (2009) Electro-Fenton process and related electrochemical technologies based on Fenton's reaction chemistry. Chem Rev 109(12):6570–6631. https://doi.org/10.1021/cr900136g

Brown NW, Roberts EPL, Chasiotis A, Cherdron T, Sanghrajka N (2004) Atrazine removal using adsorption and electrochemical regeneration. Water Res 38(13):3067–3074. https://doi.org/10.1016/j.watres.2004.04.043

Cao S-W, Zhu Y-J, Chang J (2008) Fe_3O_4 polyhedral nanoparticles with a high magnetization synthesized in mixed solvent ethylene glycol–water system. New J Chem 32(9):1526–1530. https://doi.org/10.1039/B719436F

Cao D, Li H, Pan L, Li J, Wang X, Jing P, Cheng X, Wang W, Wang J, Liu Q (2016) High saturation magnetization of γ-Fe_2O_3 nano-particles by a facile one-step synthesis approach. Sci Rep 6(1):32360. https://doi.org/10.1038/srep32360

Çavuşoğlu FC, Akan S, Arı EA, Çetinkaya E, Çolak E, Daştan GN, Deniz S, Erdem D, Köksal M, Korkmaz S, Onsekiz N, Oruçoğlu B, Özkaya D, Uslu HB, Ünal Ç, Yıldız O, Özkara-Aydınoğlu

Ş, Bayazit ŞS (2019) Preparation of magnetic activated carbon-chitosan nanocomposite for crystal violet adsorption. Korean J Chem Eng 36(11):1915–1921. https://doi.org/10.1007/s11814-019-0377-9

Centi G, Perathoner S, Torre T, Verduna MG (2000) Catalytic wet oxidation with H_2O_2 of carboxylic acids on homogeneous and heterogeneous Fenton-type catalysts. Catal Today 55 (1):61–69. https://doi.org/10.1016/S0920-5861(99)00226-6

Chang Q, Jiang G-D, Hu M-X, Huang J, Tang H-Q (2014) Adsorption of methylene blue from aqueous solution onto magnetic Fe_3O_4/ graphene oxide nanoparticles. Huan jing ke xue= Huanjing kexue 35(5):1804–1809

Cheah W, Hosseini S, Khan MA, Chuah TG, Choong TSY (2013) Acid modified carbon coated monolith for methyl orange adsorption. Chem Eng J 215–216:747–754. https://doi.org/10.1016/j.cej.2012.07.004

Chen M, Ohman K, Metcalfe C, Ikonomou MG, Amatya PL, Wilson J (2006) Pharmaceuticals and endocrine disruptors in wastewater treatment effluents and in the water supply system of Calgary, Alberta, Canada. Water Qual Res J Can 41(4):351–364

Chen Q, Liu H, Yang Z, Tan D (2017) Regeneration performance of spent granular activated carbon for tertiary treatment of dyeing wastewater by Fenton reagent and hydrogen peroxide. J Mat Cycles Waste Manage 19(1):256–264. https://doi.org/10.1007/s10163-015-0410-y

Chen B, Chen S, Zhao H, Liu Y, Long F, Pan X (2019a) A versatile B-cyclodextrin and polyethyleneimine bi-functionalized magnetic nanoadsorbent for simultaneous capture of methyl orange and Pb(II) from complex wastewater. Chemosphere 216:605–616. https://doi.org/10.1016/j.chemosphere.2018.10.157

Chen S, Bai B, He Y, Hu N, Wang H, Suo Y (2019b) Controllable conversion of Prussian blue@yeast bio-template into 3D cage-like magnetic Fe_3O_4@N-doped carbon absorbent and its cohesive regeneration by persulfate activation. RSC Adv 9(2):1151–1164. https://doi.org/10.1039/c8ra08886a

Chong MN, Jin B, Chow CWK, Saint C (2010) Recent developments in photocatalytic water treatment technology: a review. Water Res 44(10):2997–3027. https://doi.org/10.1016/j.watres.2010.02.039

Clemente JS, Fedorak PM (2005) A review of the occurrence, analyses, toxicity, and biodegradation of naphthenic acids. Chemosphere 60(5):585–600. https://doi.org/10.1016/j.chemosphere.2005.02.065

Crangle J, Goodman GM (1971) The magnetization of pure Iron and nickel. Proc R Soc Lond A Math Phys Sci 321(1547):477–491

Cui C, Zheng Q, Han Y, Xin Y (2015) Rapid microwave-assisted regeneration of magnetic carbon nanotubes loaded with p-nitrophenol. Appl Surf Sci 346:99–106. https://doi.org/10.1016/j.apsusc.2015.03.212

Deng L, Shi Z, Peng X, Zhou S (2016) Magnetic calcinated cobalt ferrite/magnesium aluminum hydrotalcite composite for enhanced adsorption of methyl orange. J Alloys Compd 688:101–112. https://doi.org/10.1016/j.jallcom.2016.06.227

Dhoble RM, Lunge S, Bhole AG, Rayalu S (2011) Magnetic binary oxide particles (MBOP): a promising adsorbent for removal of As (III) in water. Water Res 45(16):4769–4781. https://doi.org/10.1016/j.watres.2011.06.016

Dinesen AR, Pedersen CT, Bender Koch C (2001) The thermal conversion of Lepidocrocite (γ-FeOOH) revisited. J Therm Anal Calorim 64(3):1303–1310. https://doi.org/10.1023/A:1011582004634

Do MH, Phan NH, Nguyen TD, Pham TTS, Nguyen VK, Vu TTT, Nguyen TKP (2011) Activated carbon/Fe_3O_4 nanoparticle composite: fabrication, methyl orange removal and regeneration by hydrogen peroxide. Chemosphere 85(8):1269–1276. https://doi.org/10.1016/j.chemosphere.2011.07.023

Eskandarloo H, Arshadi M, Azizi M, Enayati M, Abbaspourrad A (2019) High-throughput, green, low-cost, and efficient recovery of heparin from a biological mixture using bio-originated magnetic nanofibers. ACS Sustain Chem Eng 7(4):3895–3908. https://doi.org/10.1021/acssuschemeng.8b04945

Fallou H, Cimetière N, Giraudet S, Wolbert D, Le Cloirec P (2016) Adsorption of pharmaceuticals onto activated carbon fiber cloths – modeling and extrapolation of adsorption isotherms at very low concentrations. J Environ Manag 166:544–555. https://doi.org/10.1016/j.jenvman.2015.10.056

Fan S, Huang Z, Zhang Y, Hu H, Liang X, Gong S, Zhou J, Tu R (2019) Magnetic chitosan-hydroxyapatite composite microspheres: preparation, characterization, and application for the adsorption of phenolic substances. Bioresour Technol 274:48–55. https://doi.org/10.1016/j.biortech.2018.11.078

Fang Q, Lin J, Zhan Y, Yang M, Zheng W (2014) Synthesis of hydroxyapatite/magnetite/zeolite composite for Congo red removal from aqueous solution. Huan jing ke xue= Huanjing kexue 35 (8):2992–3001

Farzin Nejad N, Shams E, Amini MK, Bennett JC (2013) Synthesis of magnetic mesoporous carbon and its application for adsorption of dibenzothiophene. Fuel Process Technol 106:376–384. https://doi.org/10.1016/j.fuproc.2012.09.002

Feng L, Cao M, Ma X, Zhu Y, Hu C (2012) Superparamagnetic high-surface-area Fe_3O_4 nanoparticles as adsorbents for arsenic removal. J Hazard Mater 217-218:439–446. https://doi.org/10.1016/j.jhazmat.2012.03.073

Foo KY, Hameed BH (2010) Insights into the modeling of adsorption isotherm systems. Chem Eng J 156(1):2–10. https://doi.org/10.1016/j.cej.2009.09.013

Gao Y, Zhong D, Zhang D, Pu X, Shao X, Su C, Yao X, Li S (2014) Thermal regeneration of recyclable reduced graphene oxide/Fe_3O_4 composites with improved adsorption properties. J Chem Technol Biotechnol 89(12):1859–1865. https://doi.org/10.1002/jctb.4268

Ghaemi M, Absalan G, Sheikhian L (2014) Adsorption characteristics of Titan yellow and Congo red on $CoFe_2O_4$ magnetic nanoparticles. J Iran Chem Soc 11(6):1759–1766. https://doi.org/10.1007/s13738-014-0448-0

Guo Y, Du E (2012) The effects of thermal regeneration conditions and inorganic compounds on the characteristics of activated carbon used in power plant. Energy Procedia 17(Part A):444–449. https://doi.org/10.1016/j.egypro.2012.02.118

Guo S, Huang L, Li W, Wang Q, Wang W, Yang Y (2019) Willow tree-like functional groups modified magnetic nanoparticles for ultra-high capacity adsorption of dye. J Taiwan Inst Chem Eng 101:99–104. https://doi.org/10.1016/j.jtice.2019.04.041

Gupta VK, Suhas (2009) Application of low-cost adsorbents for dye removal – a review. J Environ Manage 90(8):2313–2342. https://doi.org/10.1016/j.jenvman.2008.11.017

Hassanzadeh-Tabrizi SA, Motlagh MM, Salahshour S (2016) Synthesis of ZnO/CuO nanocomposite immobilized on γ-Al_2O_3 and application for removal of methyl orange. Appl Surf Sci 384:237–243. https://doi.org/10.1016/j.apsusc.2016.04.165

Hritcu D, Humelnicu D, Dodi G, Popa MI (2012) Magnetic chitosan composite particles: evaluation of thorium and uranyl ion adsorption from aqueous solutions. Carbohydr Polym 87 (2):1185–1191. https://doi.org/10.1016/j.carbpol.2011.08.095

Hu Y, Zhu Y, Zhang Y, Lin T, Zeng G, Zhang S, Wang Y, He W, Zhang M, Long H (2019) An efficient adsorbent: simultaneous activated and magnetic ZnO doped biochar derived from camphor leaves for ciprofloxacin adsorption. Bioresour Technol 288:121511. https://doi.org/10.1016/j.biortech.2019.121511

Huang Y, Keller AA (2013) Magnetic nanoparticle adsorbents for emerging organic contaminants. ACS Sustain Chem Eng 1(7):731–736. https://doi.org/10.1021/sc400047q

Huling SG, Jones PK, Lee TR (2007) Iron optimization for Fenton-driven oxidation of MTBE-spent granular activated carbon. Environ Sci Technol 41(11):4090–4096. https://doi.org/10.1021/es062666k

Huling SG, Kan E, Wingo C (2009) Fenton-driven regeneration of MTBE-spent granular activated carbon—effects of particle size and iron amendment procedures. Appl Catal B Environ 89 (3–4):651–658. https://doi.org/10.1016/j.apcatb.2009.02.002

Hussain SN, Asghar HMA, Campen AK, Brown NW, Roberts EPL (2013) Breakdown products formed due to oxidation of adsorbed phenol by electrochemical regeneration of a graphite adsorbent. Electrochim Acta 110:550–559. https://doi.org/10.1016/j.electacta.2013.03.017

I Schwerdt J, F Goya G, Pilar Calatayud M, B Herenu C, C Reggiani P, G Goya R (2012) Magnetic field-assisted gene delivery: achievements and therapeutic potential. Curr Gene Ther 12 (2):116–126

Inyang M, Dickenson E (2015) The potential role of biochar in the removal of organic and microbial contaminants from potable and reuse water: a review. Chemosphere 134:232–240. https://doi.org/10.1016/j.chemosphere.2015.03.072

Israelachvili JN (2011) Intermolecular and surface forces: revised third edition. Academic Press

Kalska-Szostko B, Wykowska U, Piekut K, Satuła D (2014) Stability of Fe_3O_4 nanoparticles in various model solutions. Colloids Surf A Physicochem Eng Asp 450:15–24. https://doi.org/10.1016/j.colsurfa.2014.03.002

Kan E, Huling SG (2009) Effects of temperature and acidic pre-treatment on Fenton-driven oxidation of MTBE-spent granular activated carbon. Environ Sci Technol 43(5):1493–1499. https://doi.org/10.1021/es802360f

Kannan K, Koistinen J, Beckmen K, Evans T, Gorzelany JF, Hansen KJ, Jones PD, Helle E, Nyman M, Giesy JP (2001) Accumulation of perfluorooctane sulfonate in marine mammals. Environ Sci Technol 35(8):1593–1598

Kazemi J, Javanbakht V (2019) Alginate beads impregnated with magnetic Chitosan@Zeolite nanocomposite for cationic methylene blue dye removal from aqueous solution. Int J Biol Macromol. https://doi.org/10.1016/j.ijbiomac.2019.11.024

Khan MA, Otero M, Kazi M, Alqadami AA, Wabaidur SM, Siddiqui MR, Alothman ZA, Sumbul S (2019) Unary and binary adsorption studies of lead and malachite green onto a nanomagnetic copper ferrite/drumstick pod biomass composite. J Hazard Mater 365:759–770. https://doi.org/10.1016/j.jhazmat.2018.11.072

Kolpin DW, Furlong ET, Meyer MT, Thurman EM, Zaugg SD, Barber LB, Buxton HT (2002) Pharmaceuticals, hormones, and other organic wastewater contaminants in US streams, 1999–2000: a national reconnaissance. Environ Sci Technol 36(6):1202–1211

Kondo K, Jin T, Miura O (2010) Removal of less biodegradable dissolved organic matters in water by superconducting magnetic separation with magnetic mesoporous carbon. Phys C 470 (20):1808–1811. https://doi.org/10.1016/j.physc.2010.05.212

Kumar M, Dosanjh HS, Singh H (2018) Magnetic zinc ferrite–chitosan bio-composite: synthesis, characterization and adsorption behavior studies for cationic dyes in single and binary systems. J Inorg Organomet Polym Mater 28(3):880–898. https://doi.org/10.1007/s10904-017-0752-0

Kumar M, Dosanjh HS, Singh H (2019) Surface modification of spinel ferrite with biopolymer for adsorption of cationic and anionic dyes in single and ternary dye system. Fibers Polym 20 (4):739–751. https://doi.org/10.1007/s12221-019-8462-6

Kumarasamy G, Nachimuthu P (2018) Biodegradable glucose and glucosamine grafted polyacrylamide/graphite composites for the removal of acid violet 17 from an aqueous solution. E-Polymers 18(4):297–311. https://doi.org/10.1515/epoly-2017-0187

Kwan WP, Voelker BM (2003) Rates of hydroxyl radical generation and organic compound oxidation in mineral-catalyzed Fenton-like systems. Environ Sci Technol 37(6):1150–1158. https://doi.org/10.1021/es020874g

Li S, Gong Y, Yang Y, He C, Hu L, Zhu L, Sun L, Shu D (2015) Recyclable CNTs/Fe_3O_4 magnetic nanocomposites as adsorbents to remove bisphenol A from water and their regeneration. Chem Eng J 260:231–239. https://doi.org/10.1016/j.cej.2014.09.032

Li B, Ma J, Zhou L, Qiu Y (2017) Magnetic microsphere to remove tetracycline from water: adsorption, H_2O_2 oxidation and regeneration. Chem Eng J 330(Supplement C):191–201. https://doi.org/10.1016/j.cej.2017.07.054

Li Z, Tang X, Liu K, Huang J, Peng Q, Ao M, Huang Z (2018) Fabrication of novel sandwich nanocomposite as an efficient and regenerable adsorbent for methylene blue and Pb (II) ion removal. J Environ Manag 218:363–373. https://doi.org/10.1016/j.jenvman.2018.04.082

Li T, Liu L, Zhang Z, Han Z (2020) Preparation of nanofibrous metal-organic framework filter for rapid adsorption and selective separation of cationic dye from aqueous solution. Sep Purif Technol 237:116360. https://doi.org/10.1016/j.seppur.2019.116360

Liang K, Liu Q, ding Y (2014) Removal of methyl violet and cationic gold yellow from aqueous with porous magnetic polymer microspheres and its adsorption kinetics. Polym Polym Compos 22(9):809–816. https://doi.org/10.1177/096739111402200908

Lim T-T, Yap P-S, Srinivasan M, Fane AG (2011) TiO_2/AC composites for synergistic adsorption-photocatalysis processes: present challenges and further developments for water treatment and reclamation. Crit Rev Environ Sci Technol 41(13):1173–1230. https://doi.org/10.1080/10643380903488664

Lin S-S, Gurol MD (1998) Catalytic decomposition of hydrogen peroxide on Iron oxide: kinetics, mechanism, and implications. Environ Sci Technol 32(10):1417–1423. https://doi.org/10.1021/es970648k

Lin S-H, Juang R-S (2009) Adsorption of phenol and its derivatives from water using synthetic resins and low-cost natural adsorbents: a review. J Environ Manag 90(3):1336–1349. https://doi.org/10.1016/j.jenvman.2008.09.003

Liu Y, Chen S, Quan X, Yu H, Zhao H, Zhang Y (2015a) Efficient mineralization of Perfluorooctanoate by electro-Fenton with H_2O_2 electro-generated on hierarchically porous carbon. Environ Sci Technol 49(22):13528–13533. https://doi.org/10.1021/acs.est.5b03147

Liu Y, Tian Y, Luo C, Cui G, Yan S (2015b) One-pot preparation of a MnO_2-graphene-carbon nanotube hybrid material for the removal of methyl orange from aqueous solutions. New J Chem 39(7):5484–5492. https://doi.org/10.1039/C5NJ00697J

Liu Y, Xie J, Ong CN, Vecitis CD, Zhou Z (2015c) Electrochemical wastewater treatment with carbon nanotube filters coupled with in situ generated H_2O_2. Environ Sci Water Res Technol. https://doi.org/10.1039/C5EW00128E

Liu T, Gu Y, Xing DY, Dong W, Wu X (2018) Rapid and high-capacity adsorption of PFOS and PFOA by regenerable ammoniated magnetic particle. Environ Sci Pollut Res 25 (14):13813–13822. https://doi.org/10.1007/s11356-018-1578-1

Lou Z, Zhang W, Hu X, Zhang H (2017) Synthesis of a novel functional group-bridged magnetized bentonite adsorbent: characterization, kinetics, isotherm, thermodynamics and regeneration. Chin J Chem Eng 25(5):587–594. https://doi.org/10.1016/j.cjche.2016.10.010

Lu Z-Y, Muir DM (1988) Dissolution of metal ferrites and iron oxides by HCl under oxidising and reducing conditions. Hydrometallurgy 21(1):9–21. https://doi.org/10.1016/0304-386X(88)90013-8

Luo W, Zhu L, Wang N, Tang H, Cao M, She Y (2010) Efficient removal of organic pollutants with magnetic Nanoscaled $BiFeO_3$ as a reusable heterogeneous Fenton-like catalyst. Environ Sci Technol 44(5):1786–1791. https://doi.org/10.1021/es903390g

Ma J, Yu F, Zhou L, Jin L, Yang M, Luan J, Tang Y, Fan H, Yuan Z, Chen J (2012) Enhanced adsorptive removal of methyl Orange and methylene blue from aqueous solution by alkali-activated multiwalled carbon nanotubes. ACS Appl Mater Interfaces 4(11):5749–5760. https://doi.org/10.1021/am301053m

Mahdavinia GR, Mosallanezhad A (2016) Facile and green rout to prepare magnetic and chitosan-crosslinked κ-carrageenan bionanocomposites for removal of methylene blue. J Water Proc Eng 10:143–155. https://doi.org/10.1016/j.jwpe.2016.02.010

Malato S, Fernández-Ibáñez P, Maldonado MI, Blanco J, Gernjak W (2009) Decontamination and disinfection of water by solar photocatalysis: recent overview and trends. Catal Today 147 (1):1–59. https://doi.org/10.1016/j.cattod.2009.06.018

Martínez-Huitle CA, Brillas E (2009) Decontamination of wastewaters containing synthetic organic dyes by electrochemical methods: a general review. Appl Catal B Environ 87(3):105–145. https://doi.org/10.1016/j.apcatb.2008.09.017

Minh TD, Lee BK, Tri NLM (2018) A novel framework of ternary Fe_3O_4@γ-APTES@rGO Nanohybrid by [CH_3OH]-soluble distribution for synergistic removal of 1,2,3-Benzotriazole. Macromol Res 26(8):763–773. https://doi.org/10.1007/s13233-018-6098-9

Mittal H, Alhassan SM, Ray SS (2018) Efficient organic dye removal from wastewater by magnetic carbonaceous adsorbent prepared from corn starch. J Environ Chem Eng 6(6):7119–7131. https://doi.org/10.1016/j.jece.2018.11.010

Mokhtari P, Ghaedi M, Dashtian K, Rahimi MR, Purkait MK (2016) Removal of methyl orange by copper sulfide nanoparticles loaded activated carbon: kinetic and isotherm investigation. J Mol Liq 219:299–305. https://doi.org/10.1016/j.molliq.2016.03.022

Moreno-Castilla C (2004) Adsorption of organic molecules from aqueous solutions on carbon materials. Carbon 42(1):83–94. https://doi.org/10.1016/j.carbon.2003.09.022

Munoz M, de Pedro ZM, Casas JA, Rodriguez JJ (2015) Preparation of magnetite-based catalysts and their application in heterogeneous Fenton oxidation – a review. Appl Catal B Environ 176–177(Supplement C):249–265. https://doi.org/10.1016/j.apcatb.2015.04.003

Muranaka CT, Julcour C, Wilhelm A-M, Delmas H, Nascimento CAO (2010) Regeneration of activated carbon by (photo)-Fenton oxidation. Ind Eng Chem Res 49(3):989–995. https://doi.org/10.1021/ie900675d

Narbaitz RM, Cen J (1994) Electrochemical regeneration of granular activated carbon. Water Res 28(8):1771–1778. https://doi.org/10.1016/0043-1354(94)90250-X

Narbaitz RM, Cen J (1997) Alternative methods for determining the percentage regeneration of activated carbon. Water Res 31(10):2532–2542. https://doi.org/10.1016/S0043-1354(97)00085-7

Ni R, Wang Y, Wei X, Chen J, Xu P, Xu W, Meng J, Zhou Y (2019) Ionic liquid modified molybdenum disulfide and reduced graphene oxide magnetic nanocomposite for the magnetic separation of dye from aqueous solution. Anal Chim Acta 1054:47–58. https://doi.org/10.1016/j.aca.2018.12.037

Opdyke MD, Channell JE (1996) Magnetic stratigraphy. Academic press, New York

Pan B, Xing B (2008) Adsorption mechanisms of organic chemicals on carbon nanotubes. Environ Sci Technol 42(24):9005–9013. https://doi.org/10.1021/es801777n

Pang Y, Zeng G, Tang L, Zhang Y, Liu Y, Lei X, Li Z, Zhang J, Liu Z, Xiong Y (2011) Preparation and application of stability enhanced magnetic nanoparticles for rapid removal of Cr(VI). Chem Eng J 175:222–227. https://doi.org/10.1016/j.cej.2011.09.098

Peng X, Luan Z, Zhang H (2006) Montmorillonite–Cu(II)/Fe(III) oxides magnetic material as adsorbent for removal of humic acid and its thermal regeneration. Chemosphere 63(2):300–306. https://doi.org/10.1016/j.chemosphere.2005.07.019

Pera-Titus M, García-Molina V, Baños MA, Giménez J, Esplugas S (2004) Degradation of chlorophenols by means of advanced oxidation processes: a general review. Appl Catal B Environ 47(4):219–256. https://doi.org/10.1016/j.apcatb.2003.09.010

Pignatello JJ, Oliveros E, MacKay A (2006) Advanced oxidation processes for organic contaminant destruction based on the Fenton reaction and related chemistry. Crit Rev Environ Sci Technol 36(1):1–84. https://doi.org/10.1080/10643380500326564

Pliego G, Zazo JA, Garcia-Muñoz P, Munoz M, Casas JA, Rodriguez JJ (2015) Trends in the intensification of the Fenton process for wastewater treatment: an overview. Crit Rev Environ Sci Technol 45(24):2611–2692. https://doi.org/10.1080/10643389.2015.1025646

Qin Y, Long M, Tan B, Zhou B (2014) RhB adsorption performance of magnetic adsorbent Fe_3O_4/RGO composite and its regeneration through a Fenton-like reaction. Nano-Micro Lett 6(2):125–135. https://doi.org/10.1007/bf03353776

Qurrat Ul A, Khatoon J, Shah MR, Malik MI, Khan IAT, Khurshid S, Naz R (2019) Convenient pH-responsive removal of Acid Black 1 by green l-histidine/iron oxide magnetic nanoadsorbent from water: performance and mechanistic studies. RSC Adv 9(6):2978–2996. https://doi.org/10.1039/c8ra09279f

Reguyal F, Sarmah AK, Gao W (2017) Synthesis of magnetic biochar from pine sawdust via oxidative hydrolysis of $FeCl_2$ for the removal sulfamethoxazole from aqueous solution. J Hazard Mater 321:868–878. https://doi.org/10.1016/j.jhazmat.2016.10.006

Ribeiro RS, Silva AMT, Figueiredo JL, Faria JL, Gomes HT (2016) Catalytic wet peroxide oxidation: a route towards the application of hybrid magnetic carbon nanocomposites for the degradation of organic pollutants. A review. Appl Catal B Environ 187(Supplement C):428–460. https://doi.org/10.1016/j.apcatb.2016.01.033

Rong Y, Han R (2019) Adsorption of p-chlorophenol and p-nitrophenol in single and binary systems from solution using magnetic activated carbon. Korean J Chem Eng 36(6):942–953. https://doi.org/10.1007/s11814-019-0267-1

Rongcheng W, Jiuhui Q (2004) Removal of azo dye from water by magnetite adsorption-Fenton oxidation. Water Environ Res 76(7):2637–2642. https://doi.org/10.2175/106143004X141861

Roth H, Gendel Y, Buzatu P, David O, Wessling M (2016) Tubular carbon nanotube-based gas diffusion electrode removes persistent organic pollutants by a cyclic adsorption – electro-Fenton process. J Hazard Mater 307:1–6. https://doi.org/10.1016/j.jhazmat.2015.12.066

Rusevova K, Kopinke F-D, Georgi A (2012) Nano-sized magnetic iron oxides as catalysts for heterogeneous Fenton-like reactions—influence of Fe(II)/Fe(III) ratio on catalytic performance. J Hazard Mater 241-242:433–440. https://doi.org/10.1016/j.jhazmat.2012.09.068

Salvador F, Martin-Sanchez N, Sanchez-Hernandez R, Sanchez-Montero MJ, Izquierdo C (2015a) Regeneration of carbonaceous adsorbents. Part I: thermal regeneration. Microporous Mesoporous Mater 202:259–276. https://doi.org/10.1016/j.micromeso.2014.02.045

Salvador F, Martin-Sanchez N, Sanchez-Hernandez R, Sanchez-Montero MJ, Izquierdo C (2015b) Regeneration of carbonaceous adsorbents. Part II: chemical, microbiological and vacuum regeneration. Microporous Mesoporous Mater 202:277–296. https://doi.org/10.1016/j.micromeso.2014.08.019

Sandeep Kumar TK, Viswanathan NN, Ahmed H, Dahlin A, Andersson C, Bjorkman B (2019) Investigation of magnetite oxidation kinetics at the particle scale. Metall Mater Trans B 50 (1):150–161. https://doi.org/10.1007/s11663-018-1459-5

Saroyan HS, Giannakoudakis DA, Sarafidis CS, Lazaridis NK, Deliyanni EA (2017) Effective impregnation for the preparation of magnetic mesoporous carbon: application to dye adsorption. J Chem Technol Biotechnol 92(8):1899–1911. https://doi.org/10.1002/jctb.5210

Sharif F, Gagnon LR, Mulmi S, Roberts EPL (2017) Electrochemical regeneration of a reduced graphene oxide/magnetite composite adsorbent loaded with methylene blue. Water Res 114:237–245. https://doi.org/10.1016/j.watres.2017.02.042

Shuang C, Li P, Li A, Zhou Q, Zhang M, Zhou Y (2012) Quaternized magnetic microspheres for the efficient removal of reactive dyes. Water Res 46(14):4417–4426. https://doi.org/10.1016/j.watres.2012.05.052

Singhal S, Namgyal T, Bansal S, Chandra K (2010) Effect of Zn substitution on the magnetic properties of cobalt ferrite nano particles prepared via sol-gel route. J Electromagn Anal Appl 2 (6):6. https://doi.org/10.4236/jemaa.2010.26049

Sirés I, Brillas E, Oturan M, Rodrigo M, Panizza M (2014) Electrochemical advanced oxidation processes: today and tomorrow. A review. Environ Sci Pollut Res 21(14):8336–8367. https://doi.org/10.1007/s11356-014-2783-1

Song R, Bai B, Puma GL, Wang H, Suo Y (2015) Biosorption of azo dyes by raspberry-like Fe_3O_4@yeast magnetic microspheres and their efficient regeneration using heterogeneous Fenton-like catalytic processes over an up-flow packed reactor. React Kinet Mech Catal 115 (2):547–562. https://doi.org/10.1007/s11144-015-0854-z

Sun ZH, Wang LF, Liu PP, Wang SC, Sun B, Jiang DZ, Xiao F-S (2006) Magnetically motive porous sphere composite and its excellent properties for the removal of pollutants in water by adsorption and desorption cycles. Adv Mater 18(15):1968–1971. https://doi.org/10.1002/adma.200600337

Tajizadegan H, Torabi O, Heidary A, Golabgir MH, Jamshidi A (2016) Study of methyl orange adsorption properties on ZnO–Al_2O_3 nanocomposite adsorbent particles. Desalin Water Treat 57(26):12324–12334. https://doi.org/10.1080/19443994.2015.1049558

Terzaghi E, Zanardini E, Morosini C, Raspa G, Borin S, Mapelli F, Vergani L, Di Guardo A (2018) Rhizoremediation half-lives of PCBs: role of congener composition, organic carbon forms, bioavailability, microbial activity, plant species and soil conditions, on the prediction of fate and persistence in soil. Sci Total Environ 612:544–560. https://doi.org/10.1016/j.scitotenv.2017.08.189

Toledo LC, Silva ACB, Augusti R, Lago RM (2003) Application of Fenton's reagent to regenerate activated carbon saturated with organochloro compounds. Chemosphere 50(8):1049–1054. https://doi.org/10.1016/S0045-6535(02)00633-1

Trellu C, Péchaud Y, Oturan N, Mousset E, Huguenot D, van Hullebusch ED, Esposito G, Oturan MA (2016) Comparative study on the removal of humic acids from drinking water by anodic oxidation and electro-Fenton processes: mineralization efficiency and modelling. Appl Catal B Environ 194:32–41. https://doi.org/10.1016/j.apcatb.2016.04.039

Trellu C, Oturan N, Keita FK, Fourdrin C, Pechaud Y, Oturan MA (2018) Regeneration of activated carbon fiber by the electro-Fenton process. Environ Sci Technol 52(13):7450–7457. https://doi.org/10.1021/acs.est.8b01554

USEPA (2000) Wastewater technology fact sheet: granular activated carbon adsorption and regeneration. U.S. EPA, Washington, D.C.

USEPA. Persistent organic pollutants: a global issue, a global response. Retrieved October 27, 2017, from https://www.epa.gov/international-cooperation/persistent-organic-pollutants-global-issue-global-response

Wan J, Deng H, Shi J, Zhou L, Su T (2014) Synthesized magnetic manganese ferrite nanoparticles on activated carbon for sulfamethoxazole removal. Clean (Weinh) 42(9):1199–1207. https://doi.org/10.1002/clen.201300432

Wang Y, Cheng R, Wen Z, Zhao L (2012) Investigation on the room-temperature preparation and application of chain-like iron flower and its ramifications in wastewater purification. Chem Eng J 203:277–284. https://doi.org/10.1016/j.cej.2012.06.063

Wang X, Cai W, Wang G, Wu Z, Zhao H (2013) One-step fabrication of high performance micro/nanostructured Fe_3S_4–C magnetic adsorbent with easy recovery and regeneration properties. CrystEngComm 15(15):2956–2965. https://doi.org/10.1039/C3CE26856J

Wang Y, Gao Q, You Q, Liao G, Xia H, Wang D (2016) Porous polyimide framework: a novel versatile adsorbent for highly efficient removals of azo dye and antibiotic. React Funct Polym 103:9–16. https://doi.org/10.1016/j.reactfunctpolym.2016.04.004

Wang Q, Tang A, Zhong L, Wen X, Yan P, Wang J (2018a) Amino-modified γ-Fe_2O_3/sepiolite composite with rod-like morphology for magnetic separation removal of Congo red dye from aqueous solution. Powder Technol 339:872–881. https://doi.org/10.1016/j.powtec.2018.08.055

Wang X, Liu Z, Ying Z, Huo M, Yang W (2018b) Adsorption of trace estrogens in ultrapure and wastewater treatment plant effluent by magnetic graphene oxide. Int J Environ Res Public Health 15(7). https://doi.org/10.3390/ijerph15071454

Wei Y, Latour RA (2010) Correlation between desorption force measured by atomic force microscopy and adsorption free energy measured by surface Plasmon resonance spectroscopy for peptide−surface interactions. Langmuir 26(24):18852–18861. https://doi.org/10.1021/la103685d

Worch E (2012) Adsorption technology in water treatment: fundamentals, processes, and modeling. Walter de Gruyter.

Wu R, Qu J (2005) Removal of water-soluble azo dye by the magnetic material $MnFe_2O_4$. J Chem Technol Biotechnol 80(1):20–27. https://doi.org/10.1002/jctb.1142

Wu R, Qu J, He H, Yu Y (2003) Adsorption and catalytic combustion of ARB on CuO-Fe_2O_3. Chin Sci Bull 48(21):2311–2316. https://doi.org/10.1360/03wb0083

Wu R, Qu J, Chen Y (2005) Magnetic powder MnO–Fe_2O_3 composite—a novel material for the removal of azo-dye from water. Water Res 39(4):630–638. https://doi.org/10.1016/j.watres.2004.11.005

Xiao Y (2018) Regeneration of carbon materials saturated with organic pollutants by Fenton oxidation based technologies (University of Calgary, Calgary). http://dx.doi.org/10.11575/PRISM/31999

Xiao Y, Hill JM (2017) Impact of pore size on Fenton oxidation of methyl orange adsorbed on magnetic carbon materials: trade-off between capacity and Regenerability. Environ Sci Technol 51(8):4567–4575. https://doi.org/10.1021/acs.est.7b00089

Xiao Y, Hill JM (2019) Mechanistic insights for the electro-Fenton regeneration of carbon materials saturated with methyl orange: dominance of electrodesorption. J Hazard Mater 367:59–67. https://doi.org/10.1016/j.jhazmat.2018.12.066

Xu B, Zheng H, Zhou H, Wang Y, Luo K, Zhao C, Peng Y, Zheng X (2018) Adsorptive removal of anionic dyes by chitosan-based magnetic microspheres with pH-responsive properties. J Mol Liq 256:424–432. https://doi.org/10.1016/j.molliq.2018.02.061

Xu P, Zheng M, Chen N, Wu Z, Xu N, Tang J, Teng Z (2019) Uniform magnetic chitosan microspheres with radially oriented channels by electrostatic droplets method for efficient removal of Acid Blue. J Taiwan Inst Chem Eng 104:210–218. https://doi.org/10.1016/j.jtice.2019.09.016

Yadav S, Asthana A, Chakraborty R, Jain B, Singh AK, Carabineiro SA, Susan M, Hassan AB (2020) Cationic dye removal using novel magnetic/activated charcoal/β-cyclodextrin/alginate polymer nanocomposite. Nano 10(1):170

Yamashita N, Kannan K, Taniyasu S, Horii Y, Petrick G, Gamo T (2005) A global survey of perfluorinated acids in oceans. Mar Pollut Bull 51(8):658–668. https://doi.org/10.1016/j.marpolbul.2005.04.026

Yan B, Niu CH, Wang J (2017) Kinetics, electron-donor-acceptor interactions, and site energy distribution analyses of norfloxacin adsorption on pretreated barley straw. Chem Eng J 330:1211–1221. https://doi.org/10.1016/j.cej.2017.08.056

Yang X, Wang Z, Jing M, Liu R, Jin L, Shen X (2013) Efficient removal of dyes from aqueous solution by mesoporous nanocomposite $Al_2O_3/Ni_{0.5}Zn_{0.5}Fe_2O_4$ Microfibers. Water Air Soil Pollut 225(1):1819. https://doi.org/10.1007/s11270-013-1819-3

Yimin D, Jiaqi Z, Danyang L, Lanli N, Liling Z, Yi Z, Xiaohong Z (2018) Preparation of Congo red functionalized Fe_3O_4@SiO_2 nanoparticle and its application for the removal of methylene blue. Colloids Surf A Physicochem Eng Asp 550:90–98. https://doi.org/10.1016/j.colsurfa.2018.04.033

Yu H, Wang T, Dai W, Li X, Hu X, Ma N (2015) Single and bicomponent anionic dyes adsorption equilibrium studies on magnolia-leaf-based porous carbons. RSC Adv 5(79):63970–63977. https://doi.org/10.1039/C5RA11568J

Zanella O, Tessaro IC, Féris LA (2014) Desorption- and decomposition-based techniques for the regeneration of activated carbon. Chem Eng Technol 37(9):1447–1459. https://doi.org/10.1002/ceat.201300808

Zeng H, Qiao T, Zhai L, Zhang J, Li D (2019) Fe_3O_4@C particles synthesized with iron-containing water treatment residuals and its potential for methylene blue removal. J Chem Technol Biotechnol 94(12):3970–3980. https://doi.org/10.1002/jctb.6202

Zhan J, Wang H, Pan X, Wang J, Yu G, Deng S, Huang J, Wang B, Wang Y (2016) Simultaneous regeneration of p-nitrophenol-saturated activated carbon fiber and mineralization of desorbed pollutants by electro-peroxone process. Carbon 101:399–408. https://doi.org/10.1016/j.carbon.2016.02.023

Zhang X, Ding Y, Tang H, Han X, Zhu L, Wang N (2014) Degradation of bisphenol A by hydrogen peroxide activated with $CuFeO_2$ microparticles as a heterogeneous Fenton-like catalyst: efficiency, stability and mechanism. Chem Eng J 236:251–262. https://doi.org/10.1016/j.cej.2013.09.051

Zhang M, Mao Y, Wang W, Yang S, Song Z, Zhao X (2016) Coal fly ash/$CoFe_2O_4$ composites: a magnetic adsorbent for the removal of malachite green from aqueous solution. RSC Adv 6(96):93564–93574. https://doi.org/10.1039/C6RA08939A

Zhang B, Li Y, Wu T, Sun D, Chen W, Zhou X (2018) Magnetic iron oxide/graphene oxide nanocomposites: formation and interaction mechanism for efficient removal of methylene blue and p-tert-butylphenol from aqueous solution. Mater Chem Phys 205:240–252. https://doi.org/10.1016/j.matchemphys.2017.11.015

Zhao H, Lang Y (2018) Adsorption behaviors and mechanisms of florfenicol by magnetic functionalized biochar and reed biochar. J Taiwan Inst Chem Eng 88:152–160. https://doi.org/10.1016/j.jtice.2018.03.049

Zheng X, Zheng H, Zhao R, Sun Y, Sun Q, Zhang S, Liu Y (2018) Polymer-functionalized magnetic nanoparticles: synthesis, characterization, and methylene blue adsorption. Materials 11(8). https://doi.org/10.3390/ma11081312

Zhu H, Jia S, Wan T, Jia Y, Yang H, Li J, Yan L, Zhong C (2011) Biosynthesis of spherical Fe_3O_4/bacterial cellulose nanocomposites as adsorb ents for heavy metal ions. Carbohydr Polym 86 (4):1558–1564. https://doi.org/10.1016/j.carbpol.2011.06.061

Chapter 10
Magnetic Nanofibers for Contaminants' Removal from Water

Andrei V. Igansi, Bruna S. Farias, Luiz A. A. Pinto, and Tito R. S. Cadaval Jr

Contents

Abstract The increase in the population and, consequently, in the industrial and agricultural production has led the planet to the depletion of natural resources and, in particular, water resources. Seeking to minimize this problem, some operations can be performed using new hybrid materials, being an attractive way to remove contaminants, in low concentration, from aqueous effluents. The development of nanomaterials has advanced and the use of nanostructures has been receiving the attention of researchers for its application in wastewater treatment. This chapter presents information regarding the applicability of magnetic nanofibers to remove contaminants from the water medium. Three-dimensional networks formed by nanofibers are nanoarchitectures with controlled characteristics and tunable physicochemical features, with particular interest for the development of novel and functional materials. Polymeric membranes containing high porosity, with

A. V. Igansi · B. S. Farias · L. A. A. Pinto · T. R. S. Cadaval Jr (✉)
School of Chemistry and Food, Federal University of Rio Grande–FURG, Rio Grande, Brazil
e-mail: dqmpinto@furg.br

L. Meili, G. L. Dotto (eds.), *Advanced Magnetic Adsorbents for Water Treatment*, Environmental Chemistry for a Sustainable World 61,
https://doi.org/10.1007/978-3-030-64092-7_10

nanochannels, have been presented as suitable templates for obtaining 3D magnetic nanofiber networks that present magnetic and magneto-transport characteristics. Firstly, some concepts about the electrospinning process to obtain general nanofibers, the aspects involved in the synthesis, the polymers used to produce nanofibers, and some applications are presented. The magnetic nanofiber aspects and different methodologies to produce this hybrid material are also presented. The main necessary analysis to characterize a magnetic nanofiber with regard to textural, morphological, thermal stability, and magnetic aspects are elucidated. In order to illustrate the characterization essays, some examples of the literature are shown. Finally, some applications of the magnetic nanofibers in the removal of contaminants from the water medium and the characteristics of the different processes, their operating conditions, and the results of the treatment are identified.

Keywords Magnetic · Nanofiber · Contaminant · Water treatment

10.1 Introduction

Industrial and agricultural production, nowadays, has led the planet to the depletion of natural resources and, in particular, water resources. Thus, the scarcity of clean and safe drinking water has become a reality in several countries around the world. Inadequate discharges of domestic and industrial effluents not treated or treated inappropriately lead to the aquatic sources daily various types of contaminants such as toxic metals, pesticides, drugs, dyes, detergents, and organic matter contaminated with microorganisms. The presence of these harmful substances can cause a series of diseases to the population, limiting the use of water sources for consumption, in addition to impairing the development of local flora and fauna (Gouider et al. 2009; Hamza et al. 2018; Marques Jr. et al. 2018; Alves et al. 2019; Affonso et al. 2020).

In this context, treatments are needed that allow the removal of these contaminants from the effluents before their disposal in the springs. Currently, many techniques have been studied for the removal of these substances in order to reach the appropriate parameters for the return of this water to nature. The studied methods for water purification include chemical precipitation, filtration, ozonation, ion exchange, oxidation, electrolysis, reverse osmosis, adsorption, coagulation/flocculation, and biological procedures. Some of these operations can have their performance enhanced using hybrid materials. This can be an attractive way to remove contaminants, in low concentrations, from aqueous effluents (Barbosa et al. 2016; Zhang et al. 2017; Silva et al. 2017; Marques et al. 2018; Guo et al. 2019).

The development of nanomaterials has advanced and the use of nanostructures has been receiving important attention from the scientific community over application in wastewater treatment. Among the different forms of nanostructures, we can highlight the nanofibers that are defined as solid that have diameters from 1 to 1000 nm. With the development of nanotechnology, these materials have attracted widespread attention in research and new technological possibilities. Nanofibers have a wide range of advantages, such as large specific surface area and significant shape anisotropy. They provide possibilities to be precisely controlled in bulk density, surface properties, diameter, and connectivity. Figure 10.1 shows different nanofiber structures developed based on distinct materials (Feng et al. 2013; Rolandi and Rolandi 2014).

Three-dimensional nanofiber networks are nanoarchitectures with controlled topologies and tunable physical and chemical properties. They are of high interest for the development and applicability of novel and functional devices. These kinds of materials are also widely studied because of their interesting magneto-transport aspects, which can be used for the development of systems with controlled magnetic anisotropy (Wang et al. 2014; Gomes et al. 2017; Erfan et al. 2019). Thus, this chapter presents the applicability of magnetic nanofibers for contaminants' removal from water, showing how to obtain, characterize, and apply these materials in different operations in order to remove contaminants from an aqueous medium.

10.2 Aspects Involved in Synthesis of Nanofibers

Nanofibers can be defined as particles with a dimension in the range of 1–1000 nm. Many distinct methods such as thermal oxidation, chemical vapor deposition, sol-gel, and electrospinning have been used to prepare nanofiber structures. Among these, the electrospinning method is the most widely adopted to prepare nanofibers, being recognized as a very efficient process for the fabrication of polymeric nanofibers with a diameter of between 5 and 500 nm, presenting values from 10^2 to 10^4 times smaller than nanofibers obtained by other methods. The industrial applicability started around 1990, and since then, the study of electrospinning has shown that a large variety of polymers can be used to produce nanofibers (Zhao et al. 2011).

Solution and polymer characteristics, such as conductivity, molecular weight, viscosity, and surface tension, are important parameters in electrospinning. For example, viscosity is a key parameter since it is related to the extent that the macromolecule chain is entangled in the solution. The polymer solution characteristics, the tip to collector distance, the electric voltage, and the feeding rate are very important and can directly influence the production and characteristics of the material obtained (Zhao et al. 2011; Elsabee et al. 2012; Sarbatly et al. 2016).

The difference in electrical potential is the most relevant parameter, since it establishes the forces of electrostatic interaction inducing the formation of the polymeric jet. There is a critical concentration below which the chain tangles are insufficient to stabilize the charge repulsion in the jet, leading to the formation of beads instead of fibers. When the applied electric field exceeds the surface tension of the polymeric solution, uniform and homogeneous fibers are molded. In some cases, an increase in fiber diameter can be observed with increased voltage. The formation of droplets, beads, and fibers can be driven by the surface tension of the polymeric

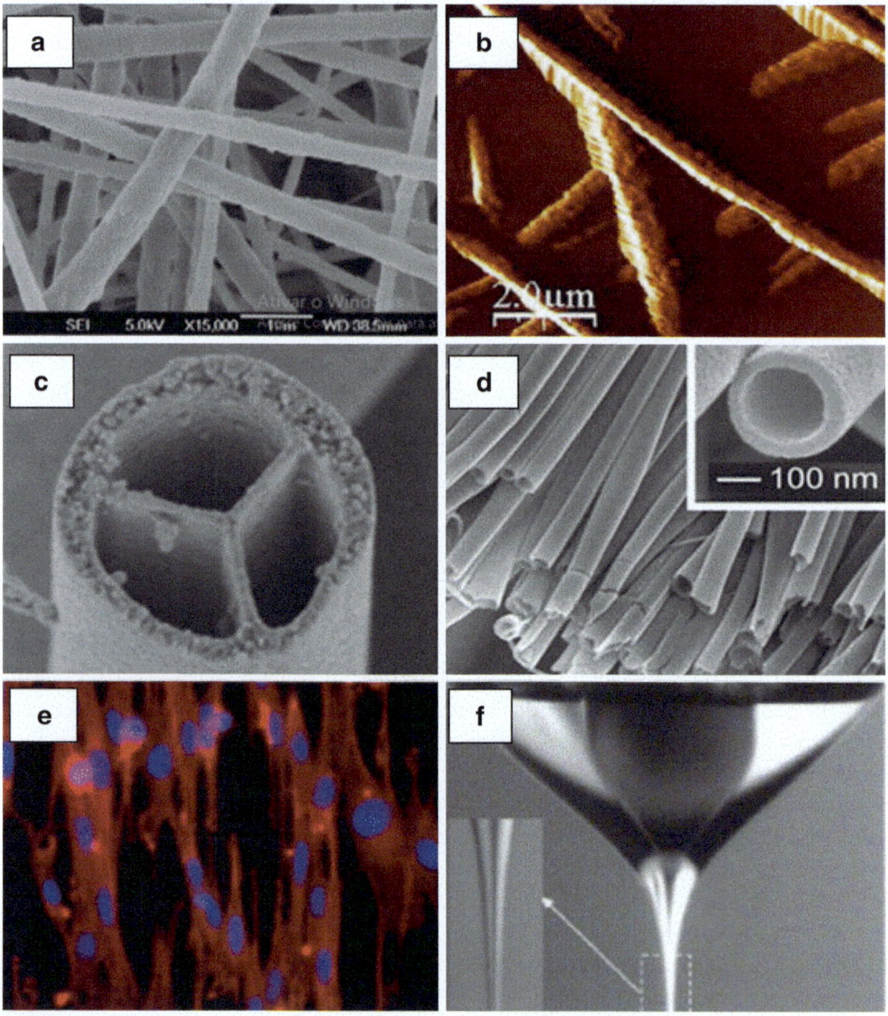

Fig. 10.1 Nanofiber structures developed based on distinct materials. (**a**) SEM images of electrospun PVDF (with permission of Elsevier, Eur. Polym. J., 46, 2010, 1957–1965). (**b**) 2D AFM images of PVDF nanofiber membrane (with permission of Elsevier, J. Membr, Sci., 311, 2008, 1–6). (**c**) The cross-sectional SEM images of hollow fibers with three channels (with permission of Elsevier, Sep. Purif. Technol., 102, 2013, 118–135). (**d**) SEM images of a uniaxially aligned array of nanotubes made of anatase (with permission of Elsevier, Sep. Purif. Technol., 102, 2013, 118–135). (**e**) Fluorescence images of a cytoskeleton (with permission of Elsevier, Adv. Colloid Interface Sci., 207, 2014, 216–222). (**f**) Taylor cone image (with permission of Elsevier, Sep. Purif. Technol., 102, 2013, 118–135)

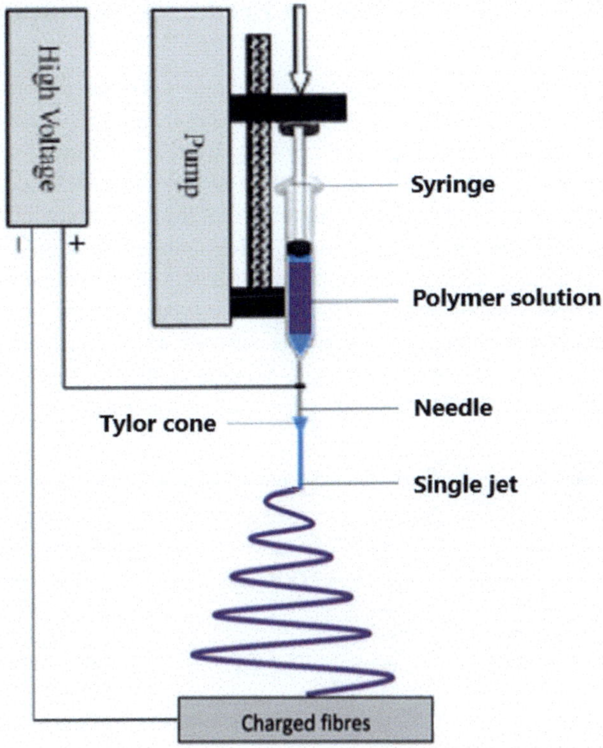

Fig. 10.2 Schematic setup of electrospinning

solution. This variable defines the lower and upper limits of the electrospinning operating range if the other parameters are kept unchanged. Usually, lower surface tension leads to electrospinning to occur in lower electric fields. Figure 10.2 shows a schematic setup of electrospinning (Zhao et al. 2011; Elsabee et al. 2012).

The process of manufacturing nanofibers by electrospinning is based on the unidirectional elongation of a viscoelastic suspension. There are basically four components in an electrospun setup: a high voltage power supply, a pump, a capillary with a needle, and a metal collector. The high voltage electric field is applied between the needle capillary and the collector. Thus, the surface load of the suspension is deformed into a conical shape. When the electric field surpasses a limit value where the electrostatic repulsion force of the surface charges overcome surface tension, the charged fluid jet is ejected from the tip of the conical protrusion (Taylor cone). The main advantage of electrospinning nano-manufacturing is that it has a low cost when compared to most methods. The nanofibers prepared using electrospinning are often continuous and uniform and most of the time do not require purification (Zhao et al. 2011; Elsabee et al. 2012).

10.3 Polymers Used to Produce Electrospun Nanofibers

The electrospinning process can obtain nanofibers from a huge distinct of macromolecules presenting high flexibility regarding raw material. Many polymeric matrices have already been successfully used in obtaining nanofibers. Most operations focus on using synthetic polymers as raw material since biopolymers are considerably more complex due to the fact that they present a large distribution of molecular weights besides ramified and functionalized structures. Much research has been developed trying to obtain nanofibers using biomacromolecules matrices. Any macromolecule can be electrospun into nanofibers. It is just to have an adequate molecular weight and that the solvent evaporated during the jet transit interval (Hemamalini and Dev 2018; Miguel et al. 2018). Table 10.1 shows nanofibers provide from different polymers and its applications.

Table 10.1 Nanofibers obtained from different polymers and their applications

Polymers	Polymers Source	Solvent	Application	Reference
Carboxyethyl chitosan/PVA	Natural/ synthetic blend	Deionized water	Wound dressing	Zhou et al. (2007)
Chitosan/arginine-chitosan	Natural blend	TFA/DCM	Wound dressing	Antunes et al. (2015)
Hexanoyl/ Chitosan	Natural blend	Acetic acid/water solution	Biomedical applications	Jia et al. (2007)
Polyvinylidene	Synthetic	N,N-dimethylacetamide (DMAC)/acetone mixture	Pretreatment of water prior to RO or UF	Gopal et al. (2006)
Polysulfone	Synthetic	N,N-dimethylformamide	Pre-filters for particulate removal	Gopal et al. (2007)
Polyethersulfone	Synthetic	N,N-dimethylformamide	Water and other liquid separation	Homaeigohar et al. (2010)
Amaranth protein isolate/ pullulan	Natural blend	Formic acid solution	Folic acid encapsulation	Aceituno–Medina et al. (2015)
Type B gelatin	Natural polymer	Acetic aqueous solution	Curcumin encapsulation	Deng et al. (2017)
Type B gelatin/ zein protein	Natural blend	Acetic aqueous solution	Bioactive delivery	Deng et al. (2018)
Pullulan	Natural polymer	Distilled water	Fish oil encapsulation	García-Moreno et al. (2017)
Chitosan/poly (ethylene oxide)	Natural/ synthetic blend	Acetic aqueous solution	Biosorption of glycerol impurities from biodiesel production	Farias et al. (2019)
Cyclodextrins	Natural polymer	Water	Linalool encapsulation	Aytac et al. (2017)

10.4 Magnetic Nanofibers

Regarding the development and challenges of nanoscience and nanotechnology, nanofibers stood out as a form of materials to be explored in different technological applications. Nanofibers own many applications due to their large specific surface area and shape anisotropy. Besides that, during the process of production, nanofibers can be precisely controlled in surface properties, connectivity, density, and diameter. Magnetic nanomaterials have been showing very interesting electromagnetic properties and with high potential in the nanoscience applicability like quantum size, the small size effect of magnetic ordered particles, macroscopic quantum tunneling, and superparamagnetic effects (Long et al. 2011; Chen et al. 2018).

Additionally, the nanomagnetic structures normally can present characteristics such as paramagnetism, ferromagnetism, ferrimagnetism, diamagnetism, single-domain structure, and coercivity. Due to the characteristics presented above, charged nanofibers with magnetic nanoparticles can incorporate the advantages of both nanofibers and nanoparticles into only one magnetic nanofiber structure. Magnetic nanofibers can be obtained by different techniques such as electrospinning, template, phase separation, and magnetic field-assisted. The following will present the basic principles of each of the techniques mentioned (Long et al. 2011; Chen et al. 2018; Erfan et al. 2019).

10.4.1 Electrospinning

Electrospinning is the most effective nanofiber manufacturing form, providing the possibility to prepare continuous magnetic nanofibers. Polymer solution containing magnetic nanoparticles is easy to be made into magnetic nanofibers using electrospinning, by which different polymers (such as biodegradable, conductive, and biocompatible polymers) and magnetic nanoparticles can be incorporated to obtain various functionalities. Nanoparticles can be stabilized in the polymer solution to enable the nanofibers to integrate magnetic, antibacterial, photocatalytic, and anti-ultraviolet properties. Polymers provide different roles in magnetic nanofibers such as (a) a template to control the shape, size, and structure; (b) as protective separators, insulators, or molding; (c) the mechanical properties; and (d) electrical, optical, and chemical properties (Yuan et al. 2011; Chen et al. 2018).

There are three different methods to obtain magnetic nanomaterials using electrospinning as follows. Figure 10.3 shows a schematic illustration of the different processes to prepare magnetic nanofibers by electrospinning. In the first step, magnetic nanofibers are prepared through the solubilization of the polymer and dissolution of the metal salts in the same solution. After, the metal salt is converted into magnetic nanoparticles by a gas-phase chemical reaction. The addition of metal salts is usual in the preparation of inorganic nanofibers. The properties of inorganic components can be developed after the nanofibers calcination at high temperatures.

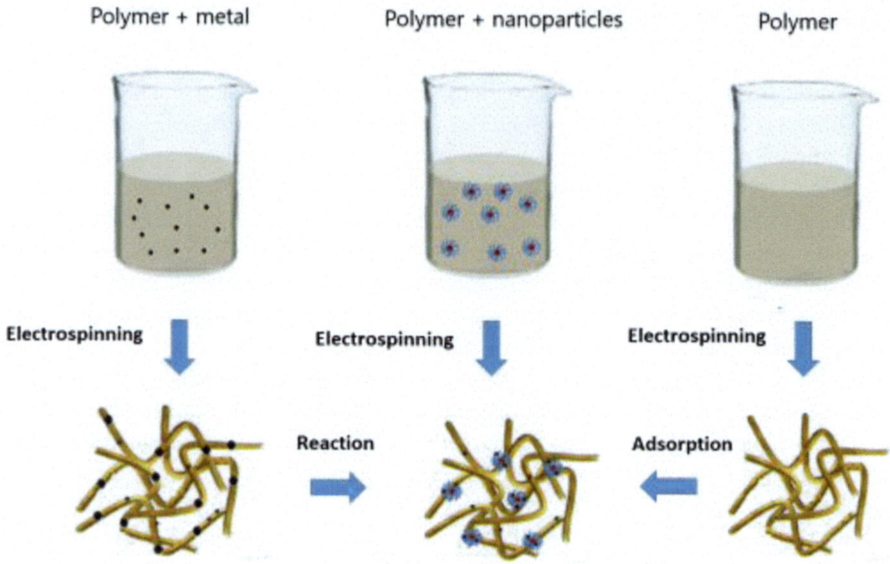

Fig. 10.3 Schematic illustration of the different methods to prepare hybrid magnetic nanofibers by electrospinning

In the second process, the nanoparticles are synthesized and subsequently added in the polymeric solution at an appropriate proportion to obtain magnetic nanofibers. The third methodology proposes that the first step is to prepare pure polymer nanofibers, and then the magnetic functionalization is performed by adsorption of the magnetic nanoparticles on the surface of nanofibers (Yuan et al. 2011; Chen et al. 2018).

10.4.2 Template Method

The template method is a widely used approach for magnetic nanofiber production. The general preparation of this method is to select a template condition, and then the template must be filled with semiconductor, metal, or nonmetal, by electrochemical deposition, liquid phase deposition, or vapor deposition. The templates are removed and nanofibers are obtained. In this way, the diameter of the fibers can be adjusted by changing the shape of the template. The nanofibers can be oriented in certain directions to achieve controllable architectures and ordered arrangements (Keller et al. 2004; Liu et al. 2009).

Some template examples are alumina, macroporous glass, molecular hoof, macromolecule, carbon nanotube, and porous polymer (Keller et al. 2004; Liu et al. 2009).

10.4.3 Phase Separation Method

In this methodology, first, the mixture is separated in different phases, one phase rich in polymer and another phase rich in solvent, which are then freeze-dried to remove solvent and moisture. After this procedure, the porous nanofibers are obtained. The main processing variables of this methodology include oil-water ratio, polymer solution energy, and molecular weight of the polymer. According to different types of solvent, phase separation methods can be classified into three freeze-drying methods: emulsion, solution, and hydrogel. The main advantage of this preparation of nanofibers is do not be necessary to use calcination in elevated temperatures. This process becomes interested when the magnetic nanofiber be used for the controlled delivery of bioactive substances. The type of sample that can be prepared using this procedure is limited, and it can only be applied to a few kinds of polymers (Whang et al. 1995; Tarrés et al. 2017).

10.4.4 Magnetic Field-Assisted Method

The magnetic field-assisted method can improve the performance and controls the morphology of the nanofibers by applying an external magnetic field during the process. The ordinary electromagnetic processes use electromagnetic stirring or an electromagnetic brake. The magnetic field-assisted method inserts high-intensity magnetization into the sample, which then changes its thermodynamic state, resulting in the rearrangement of atoms and molecules.

Magnetic nanofibers can be assembled in a superstructure with the influence of a strong magnetic field. The fibers obtained in this process, using the magnetic field, have an intense magnetoelectric coupling. Thus, the magnetic field-induced structures of the multiferroic nanofibers usually are applied as magnetic sensors and high-frequency devices (Hu et al. 2008; Rahmani et al. 2014). Figure 10.4 shows the images of different hybrid magnetic nanofibers.

10.5 Magnetic Nanofibers for Contaminants' Removal

Magnetic nanofibers have unparalleled advantages in industrial wastewater treatment due to the properties that are not available in a conventional material, such as ease of recycling and good adsorption performance. It is a new kind of material with small dimensions, surface effect, large surface area, and magnetic response, which can be used and recovered been very interesting for water treatment. The application of these materials in the field of wastewater treatment mainly includes filtration, adsorption, and photodegradation of heavy metals, pesticides, dyes, and drugs, for example.

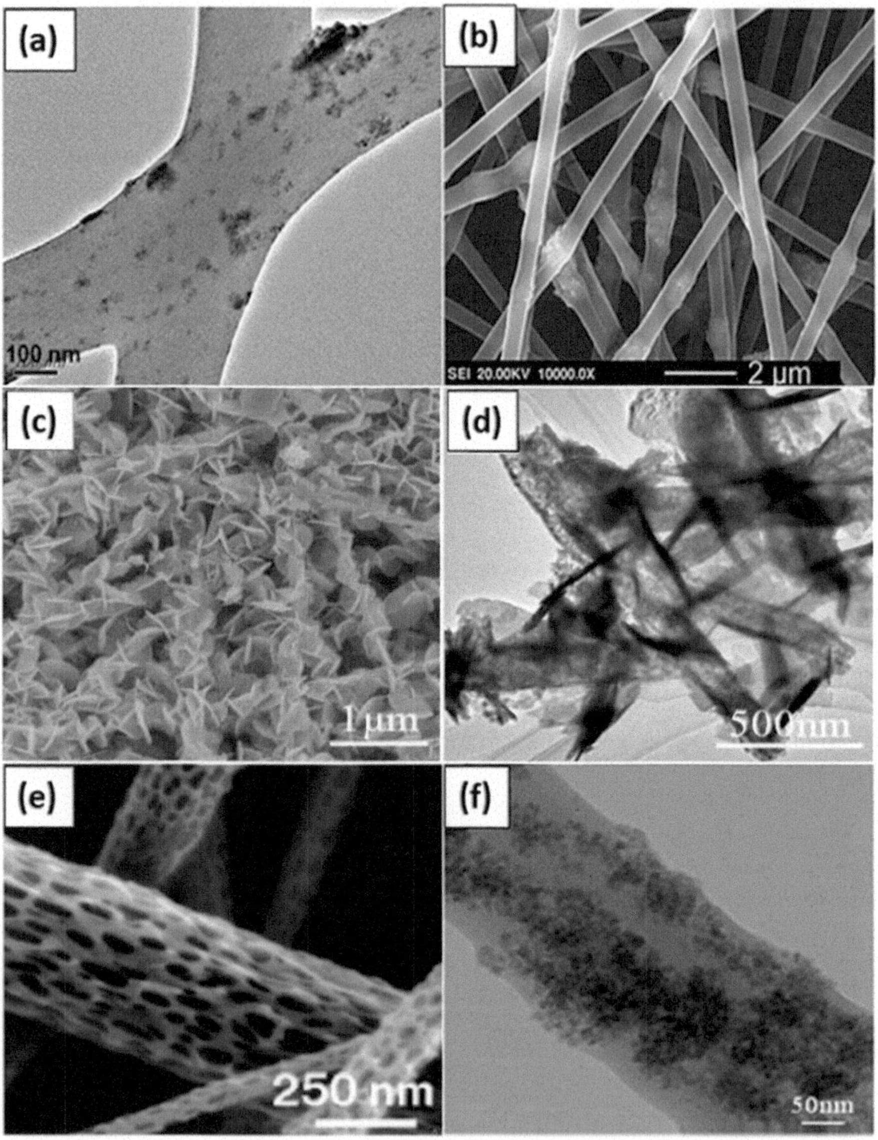

Fig. 10.4 Images of different hybrid magnetic nanofibers. (**a**) Poly-D, L-lactide nanofiber and iron oxide nanoparticles (with permission of Elsevier, Mater. Sci. Eng., C, 34, 2014, 252–261). (**b**) Scanning electron microscopy images of Fe_3O_4/gelatin composite nanofibers (with permission of Elsevier, Mater. Sci. Eng., B, 190, 2014, 126–132). (**c**) SEM image of CFO/BiOI-30 (with permission of Elsevier, Colloids Surf., A, 562, 2019, 127–135). (**d**) TEM image of CFO/BiOI-30 (with permission of Elsevier, Colloids Surf., A, 562, 2019, 127–135). (**e**) Poly(lactide-co-glycolide) nanofiber (with permission of Elsevier, Prog. Polym. Sci., 36, 2011, 1415–1442). (**f**) Scanning transmission electron microscopy of magnetic nanofiber (with permission of Elsevier, Appl. Surf. Sci., 258, 2012, 7530–7535)

10.5.1 Characterization of Magnetic Nanofibers for Contaminants' Removal

After synthesis and before use, nanofibers need to be analyzed for their morphological, textural, thermal stability, crystallinity, infrared spectrum, and magnetic characteristics. Zou et al. (2015) studied the dye removal capacity using magnetic $CuFe_2O_4$@CeO_2 nanofibers as adsorbent. In this search, magnetic adsorbents using CeO_2 nanoparticles coated on $CuFe_2O_4$ nanofibers were obtained using the combining electrospinning technique and chemical precipitation. The phases of nanofibers were determined by X-ray diffractometry. The morphologies of magnetic nanofibers were studied by field emission scanning electron microscopy; the textural characteristics were obtained by the Brunauer-Emmett-Teller (BET). Pore size distribution was analyzed from the adsorption of the isotherm based on the BJH method. The Fourier transform infrared (FT-IR) measurement of the sample was recorded in order to identify the functional groups of the material. Magnetization was measured using a magnetometer, and the point of zero charges (PZC) of nanofibers were determined using the mass titration method.

XRD patterns of the magnetic nanofibers, (a) $CuFe_2O_4$ and (b) $CuFe_2O_4$@CeO_2, show that the reflection peaks occurred due to the cubic CuFe2O4, with a little impurity phase of orthorhombic $CuFe_2O_4$. After coated with the nanoparticles of CeO_2, the peaks of orthorhombic $CuFe_2O_4$ disappear. The disappearance of the orthorhombic $CuFe_2O_4$ phase was attributed to further calcination at high temperatures. Using SEM images, the nanofibers showed a coarse and porous surface with a diameter of about 200 nm. The CeO_2 nanoparticles were evenly distributed on the fiber surface and presented a close contact with nanofibers. The nanofibers prevented the aggregation of nanoparticles, and the composite nanofibers band together to form meso- and macropores. These structures were useful for adsorption, with an improvement of contaminants adsorption capacity.

For the specific surface area and pore structure measurement of the $CuFe_2O_4$@CeO_2 composite nanofibers, the nitrogen adsorption-desorption isotherm was used. The isotherm showed a curve with an H3 hysteresis loop that belongs to type V and was attributed to the accumulation of nitrogen indicating the presence of pore structures in the material. The calculated specific surface area, pore volume, and average pore diameter were 64.12 m^2/g, 0.15 cm^3/g, and 4.74 nm, respectively. The pore size distribution was from 50 to 150 A, which confirmed that the $CuFe_2O_4$@CeO_2 composite nanofibers have a mesoporous structure containing a very high pore volume. Figure 10.5a shows the adsorption-desorption isotherm and the pore size distribution of the magnetic nanofiber.

Magnetic nanofibers are characterized by magnetic parameters such as the remanent magnetism (Mr), specific saturation magnetization (Ms), and coercivity (Hc). The hysteresis loops of both $CuFe_2O_4$ and $CuFe_2O_4$@CeO_2 nanofibers exhibited typical soft-magnetic materials behavior with a Ms. of 28.32 Am^2/kg. The $CuFeO_4$ nanofibers showed a Mr. of 12.85 Am^2/kg and a Hc of 25.11 kA/m, while the $CuFe_2O_4$@CeO_2 composite nanofibers presented a Ms. of 20.51 Am^2/kg, a Mr. of

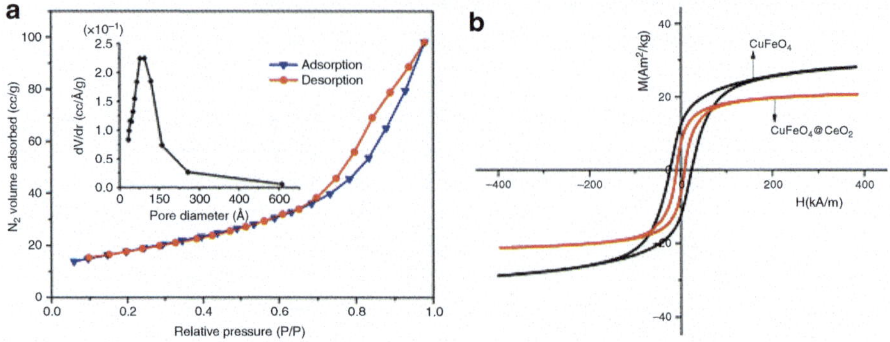

Fig. 10.5 (**a**) Adsorption-desorption isotherm and the pore size distribution of the magnetic nanofiber. (**b**) Hysteresis loops of $CuFe_2O_4$ and $CuFe_2O_4@CeO_2$ nanofibers (with permission of Elsevier, Appl. Surf. Sci., 332, 2015, 674–681)

7.24 Am^2/kg, and a Hc of 8.18 kA/m, respectively. After having coated CeO_2 nanoparticles, the Hc value of the $CuFe_2O_4$ nanofibers decreases by about 67.4%, while the Ms. value shows a little reduction of about 27.6%. Figure 10.5b shows the hysteresis loops of $CuFe_2O_4$ and $CuFe_2O_4@CeO_2$ nanofibers.

Li et al. (2019) studied the immobilization of horseradish peroxidase in magnetic nanofibers and applied this material for phenol removal. Fe_3O_4/polyacrylonitrile magnetic nanofibers were obtained through the electrospinning method in order to immobilize the horseradish peroxidase, making a platform for phenol removal. Transmission electron microscopy and scanning electron microscopy were used to observe the morphology of the hybrid nanostructure. Thermogravimetric and magnetometric techniques were applied in order to measure thermal stability and magnetization. The functional groups on the magnetic nanofibers were identified by attenuated total reflections. The ultraviolet-visible spectrum was used to obtain the phenol concentration and enzyme activity.

Fe_3O_4 magnetic nanoparticles were generated by the chemical coprecipitation method and characterized by transmission electron microscopy. The Fe_3O_4 magnetic nanoparticles showed 10 nm, and the remanent magnetization (Mr) indicated that the nanoparticles had superparamagnetism due to the small coercive force and remanence. The saturation magnetization (Ms) value was 81.389 emu/g.

The morphology of PAN nanofibers and Fe_3O_4/PAN magnetic nanofibers were obtained through scanning electron microscopy and transmission electron microscopy. The average diameter of both PAN nanofibers and magnetic nanofibers was about 200–400 nm. The pure PAN nanofibers showed a smooth morphology. The surface of the material became coarser and the average diameter tended to increase with the use of nanoparticles. A good dispersion of Fe_3O_4 nanoparticles in the nanofiber was observed when the magnetic particle loading capacity was higher than 20%. In the thermogravimetric analysis, the residual amount of blank nanofibers after burning at 900 °C reached 2%, while the residual amount of magnetic nanofibers changed from 10% to 25%, when the Fe_3O_4 nanoparticles amount increased from 10% to 40%. According to the vibrating sample magnetometer results, all the hysteresis curves were uniformly symmetrical and crossing through the origin. This result showed that Fe_3O_4 nanoparticles show superparamagnetism characteristics. This characteristic remained after the nanoparticles were dispersed on the nanofibers. The Ms. increased evenly during the increase of nanoparticles loading capacity and reached a maximum of 19.03 emu/g when the loading amount of Fe_3O_4 nanoparticles was 40%.

10.5.2 Application of Magnetic Nanofibers in the Contaminants' Removal

Due to their versatility, magnetic nanofibers can be used in different operations in the process of removing contaminants from wastewater. Thus, its application can be both as an adsorbent, where its textural characteristics of high surface area can assist, as well as in the formation of beds molded by nanofibers, and also as a structure capable of degrading molecules with the aid of magnetic nanoparticles stabilized on its surface.

Zou et al. (2015) studied the application of magnetic $CuFe_2O_4@CeO_2$ nanofibers in dye removal from water. The dye adsorption capacity of these magnetic $CuFe_2O_4@CeO_2$ nanofibers was used in column bed, and the methyl orange solution was chosen as the wastewater. In this work, the experimental procedures were performed in a continuum column bed, and the parameters such as the flow rate of dye solution, pH of the solution, and initial dye concentration were evaluated. The breakthrough curves showed the typical S-shape, indicating that the electrostatic interaction between the dye molecules and adsorbents is the main adsorption mechanism during the operation. When the pH was increased from 4.0 to 10.0, the breakpoint decreases from 262 to 11 min, $t_{0.5}$ reduced from 357 to 20 min, respectively, showing that in acid pH the adsorption was more effective.

The breakthrough time and the time to reach about 5% of the initial dye concentration at the column top decreased from 262 to 49 min and from 357 to 118 min, respectively, as the flow rate increases from 2.0 to 5.0 mL/min. This was attributed to the reduction of the contact time between the adsorbent and the adsorbate.

Regarding the initial dye concentration effect, the values of 0.025, 0.05, and 0.1 mg/mL were evaluated. The breakpoint occurred faster with increasing initial concentration and the highest adsorption capacity was reached around 100 mg/g in pH 4, initial concentration of 0.05 mg/ml, and flow rate of 2.0 mL/min. Kinetic analysis showed that the Thomas model was the most adequate to represent the adsorption operation. The regeneration tests were performed using 0.5 M NaCl, ethyl alcohol, and deionized water, with a flow rate of 3.5 mg/mL and a time of 12 h. The NaCl solution showed the more appropriate ability for desorption among these three solvents. The order of desorption activities was NaCl > CH_3CH_2OH > H_2O.

EL–Rafei et al. (2017) evaluated the production of electrospun magnetically separable calcium ferrite nanofibers produced using the electrospinning method for photocatalytic water purification. Nanofibers with three-dimensional random calcium ferrite were prepared. The material showed active hydroxyl radicals under simulated solar light irradiation showing its applicability for photocatalysis in water purification. In addition, these magnetic nanofibers were easily separated from the aqueous medium by applying an external magnetic field. Photocatalytic activities of the nanofibers heat-treated at 500 °C, 700 °C, and 800 °C were evaluated based on their capacity for the production of the active •OH. Terephthalic acid was used as a probe molecule for •OH. These radicals have a high ability for the oxidation of organic pollutants in water and wastewater, as they are strong oxidizing agents.

Based on HPLC chromatograms of the irradiated Terephthalic acid (TPA) with magnetic nanofibers using solar light for 4 h was recorded peaks attributed to the fluorescent 2-hydroxyterephthalic acid (HTPA) formed by the reaction between •OH and nonfluorescent TPA. This result supports using the materials as a separable photocatalyst for the treatment of contaminated water. The amount of the •OH produced using the nanofibers calcined at 500 °C was higher than that formed by nanofibers calcined at 700 °C. This behavior can be attributed to the smaller grain size of the sample calcined at 500 °C where more active sites are available for the •OH production. However, nanofibers calcined at 700 °C showed 20% higher magnetization saturation than that calcined at 500 °C. Thus, from the practical point of view, nanofibers calcined at 700 °C were considered as the optimal sample for water purification as it can be separated from the effluent easily with reasonable photocatalytic activity.

In other studies with regard to photodegradation, Tang et al. (2018) evaluated the fabrication of magnetic Fe_3O_4/silica nanofiber composites with enhanced Fenton-like catalytic performance for the degradation of rhodamine B. The catalytic performances were observed through degradation kinetic profile evaluation. The experimental assays on adsorption and degradation were carried out to evaluate the Fenton-like catalytic activity. Compared to the naked Fe_3O_4 nanoparticles, the Fe_3O_4/porous silica nanofibers show higher adsorption capacity and Fenton-like catalytic efficiency for dye degradation. The Fe_3O_4/porous silica nanofibers show high catalytic performance under low catalyst dosage and low H_2O_2 concentration. The authors proposed that the high catalytic activity was attributed to the formation of the highly dispersed Fe_3O_4 nanoparticles due to the presence of porous silica nanofibers. Moreover, the Fe_3O_4/porous silica nanofibers exhibit high stability, which can be easily recovered by magnetic separation and reused in other reactions. This research observed that the use of porous silica nanofibers as support was an effective manner to improve the Fenton-like catalytic performance of Fe_3O_4 nanoparticles.

10.6 Conclusions

Three-dimensional nanofiber networks are very attractive nanoarchitectures, with controlled topologies and adjustable physicochemical properties, which are of particular interest for the development of new and functional devices. Porous polymeric nanofibers can have nanochannels, being excellent models for the synthesis of 3D interconnected magnetic nanofiber networks that promote interesting magnetic and magnet transport properties. This chapter presented the applicability of magnetic nanofibers for removing contaminants from water, showing how this material can be obtained, characterized, and applied in different operations to remove contaminants from the aqueous medium. The electrospinning method is the most widely adopted for the preparation of nanofibers, being recognized as an efficient technique for the manufacture of polymeric nanofibers. Many polymeric matrices have already been used successfully in obtaining nanofibers, and most processes focus on using synthetic polymers as raw material, since biopolymers are considerably more complex, due to the fact that they present a large molecular weight distribution in addition to branched and functionalized structures. Regarding the development and challenges of nanoscience and nanotechnology, nanofibers stand out as a form of materials to be explored in different technological applications. This material has a broad spectrum of applicability due to its large specific surface area and the significant anisotropy. In addition, during the production process, nanofibers can be precisely controlled on surface properties, connectivity, density, and diameter. Magnetic nanofibers have unique advantages in the water purification due to the properties that are not available in conventional materials, such as good superparamagnetic adsorption performance and ease of recycling. It is a new type of material with small size, surface effect, and magnetic response. The application of this product in the field of wastewater treatment mainly includes adsorption, filtration, and photodegradation of heavy metals, dyes, pesticides, and drugs.

References

Aceituno–Medina M, Mendoza S, Lagaron JM, López–Rubio A (2015) Photoprotection of folic acid upon encapsulation in food-grade amaranth (*Amaranthus hypochondriacus* L.) protein isolate – Pullulan electrospun fibers. Lebensm Wiss Technol 62(2):970–975. https://doi.org/10.1016/j.lwt.2015.02.025

Affonso LN, Marques JL Jr, Lima VVC, Gonçalves JO, Barbosa SC, Primel EG, Burgo TAL, Dotto GL, Pinto LAA, Cadaval TRS Jr (2020) Removal of fluoride from fertilizer industry effluent using carbon nanotubes stabilized in chitosan sponge. J Hazard Mater 388:122042. https://doi.org/10.1016/j.jhazmat.2020.122042

Alves DCS, Gonçalves JO, Coseglio BB, Burgo TAL, Dotto GL, Pinto LAA, Cadaval TRS Jr (2019) Adsorption of phenol onto chitosan hydrogel scaffold modified with carbon nanotubes. J Env Chem Eng 7:103460. https://doi.org/10.1016/j.jece.2019.103460

Antunes BP, Moreira AF, Gaspar VM, Correia IJ (2015) Chitosan/arginine–chitosan polymer blends for assembly of nanofibrous membranes for wound regeneration. Carbohydr Polym 130:104–112. https://doi.org/10.1016/j.carbpol.2015.04.072
Aytac Z, Yildiz ZI, Kayaci–Senirmak F, Tekinay T, Uyar T (2017) Electrospinning of cyclodextrin/linalool–inclusion complex nanofibers: Fast–dissolving nanofibrous web with prolonged release and antibacterial activity. Food Chem 231:192–201. https://doi.org/10.1016/j.foodchem.2017.03.113
Barbosa GP, Debone HS, Severino P, Souto EB, Da Silva CF (2016) Design and characterization of chitosan/zeolite composite films - effect of zeolite type and zeolite dose on the film properties. Mater Sci Eng C 60:246–254. https://doi.org/10.1016/j.msec.2015.11.034
Chen X, Cheng L, Li H, Barhoum A, Zhang Y, He X, Yang W, Bubakir MM, Chen H (2018) Magnetic nanofibers: unique properties, fabrication techniques, and emerging applications. Electro Phys Theor Chem 3:9127–9143. https://doi.org/10.1002/slct.201702480
Deng L, Kang X, Liu Y, Feng F, Zhang H (2017) Effects of surfactants on the formation of gelatin nanofibres for controlled release of curcumin. Food Chem 231:70–77. https://doi.org/10.1016/j.foodchem.2017.03.027
Deng L, Kang X, Liu Y, Feng F, Zhang H (2018) Characterization of gelatin/zein films fabricated by electrospinning vs solvent casting. Food Hydrocolloids 74:324–332. https://doi.org/10.1016/j.foodhyd.2017.08.023
EL–Rafei AM, El–Kalliny AS, Gad–Allah TA (2017) Electrospun magnetically separable calcium ferrite nanofibers for photocatalytic water purification. J Magn Magn Mater 428:92–98. https://doi.org/10.1016/j.jmmm.2016.12.020
Elsabee MZ, Naguib HF, Morsi RE (2012) Chitosan based nanofibers, review. Mater Sci Eng C 32:1711–1726. https://doi.org/10.1016/j.msec.2012.05.009
Erfan NA, Barakat NAM, Muller–Borer BJ (2019) Preparation and characterization of ß-lactoglobulin/poly(ethylene oxide) magnetic nanofibers for biomedical applications. Colloids Surf A 576:63–72. https://doi.org/10.1016/j.colsurfa.2019.05.035
Farias BS, Gründmann DDR, Strieder MM, Silveira Jr. N, Cadaval Jr. TRS, Pinto LAA (2019) Biosorption of glycerol impurities from biodiesel production onto electrospun chitosan-based nanofibers: equilibrium and thermodynamic evaluations. Environ Sci Pollut Res Int 26:28436–28443. 10.1007/s11356-019-04525-3
Feng C, Khulbe KC, Matsuura T, Tabe S, Ismail AF (2013) Preparation and characterization of electro-spun nanofiber membranes and their possible applications in water treatment. Sep Purif Technol 102:118–135. https://doi.org/10.1016/j.seppur.2012.09.037
García-Moreno PJ, Özdemir N, Stephansen K, Mateiu RV, Echegoyen Y, Lagaron JM, Jacobsen C (2017) Development of carbohydrate–based nano–microstructures loaded with fish oil by using electrohydrodynamic processing. Food Hydrocolloids 69:273–285. https://doi.org/10.1016/j.foodhyd.2017.02.013
Gomes TCSC, Medina JDLT, Velázquez–Galván YG, Martínez–Huerta JM, Encinas A, Piraux L (2017) 3D interconnected magnetic nanofiber networks with multifunctional properties. IEEE Trans Magn 53:2301006. https://doi.org/10.1109/TMAG.2017.2719658
Gopal R, Kaur S, Feng CY, Chan C, Ramakrishna S, Tabe S, Matsuura T (2007) Electrospun nanofibrous polysulfone membranes as pre-filters: particulate removal. J Membr Sci 289:210–219. https://doi.org/10.1016/j.memsci.2006.11.056
Gopal R, Kaur S, Ma Z, Chan C, Ramakrishna S, Matsuura T (2006) Electrospun nanofibrous filtration membrane. J Membr Sci 281:581–586. https://doi.org/10.1016/j.memsci.2006.04.026
Gouider M, Feki M, Sayadi S (2009) Separative recovery with lime of phosphate and fluoride from an acidic effluent containing H_3PO_4, HF and/or H_2SiF_6. J Hazard Mater 170:962–968. https://doi.org/10.1016/j.jhazmat.2009.05.067
Guo Y, Jia Z, Wang S, Su Y, Ma H, Wang L, Meng W (2019) Sandwich membranes based on PVDF–g–4VP and surface modified graphene oxide for Cu(II) adsorption. J Hazard Mater 377:17–23. https://doi.org/10.1016/j.jhazmat.2019.05.044

Hamza MF, Roux JC, Guibal E (2018) Uranium and europium sorption on amidoxime-functionalized magnetic chitosan micro-particles. Chem Eng J 344:124–137. https://doi.org/10.1016/j.cej.2018.03.029

Hemamalini T, Dev VRG (2018) Comprehensive review on electrospinning of starch polymer for biomedical application. Int J Biol Macromol 106:712–718. https://doi.org/10.1016/j.ijbiomac.2017.08.079

Homaeigohar SS, Buhr K, Ebert K (2010) Polyethersulfone electron nanofibrous composite membrane for liquid filtration. J Membr Sci 365:68–77. https://doi.org/10.1016/j.memsci.2010.08.041

Hu MJ, Lin B, Yu SH (2008) Magnetic field-induced solvothermal synthesis of one–dimensional assemblies of Ni–Co alloy microstructures. Nano Res 1:303–313. https://doi.org/10.1007/s12274-008-8031-6

Jia Y, Gong J, Gu X, Kim H, Dong J, Shen X (2007) Fabrication and characterization of poly (vinyl alcohol)/chitosan blend nanofibers produced by electrospinning method. Carbohydr Polym 67:403–409. https://doi.org/10.1016/j.carbpol.2006.06.010

Keller N, Pham-Huu C, Shiga T, Estournès C, Grenèche JM, Ledoux MJ (2004) Mild synthesis of $CoFe_2O_4$ nanowires using carbon nanotube template: a high–coercivity material at room temperature. J Magn Magn Mater 272:1642–1644. https://doi.org/10.1016/j.jmmm.2003.12.228

Li J, Chen X, Xu D, Pan K (2019) Immobilization of horseradish peroxidase on electrospun magnetic nanofibers for phenol removal. Ecotoxicol Environ Saf 170:716–721. https://doi.org/10.1016/j.ecoenv.2018.12.043

Liu L, Li H, Fan S, Gu J, Li Y, Sun H (2009) Fabrication and magnetic properties of Ni–Zn nanowire arrays. J Magn Magn Mater 321:3511–3514. https://doi.org/10.1016/j.jmmm.2009.06.062

Long Y, Li M, Gu C, Wan M, Duvail J, Liu Z, Fan Z (2011) Recent advances in synthesis, physical properties and applications of conducting polymer nanotubes and nanofibers. Prog Polym Sci 36:1415–1442. https://doi.org/10.1016/j.progpolymsci.2011.04.001

Miguel SP, Figueira DR, Simões D, Ribeiro MP, Coutinho P, Ferreira P, Correia IJ (2018) Electrospun polymeric nanofibres as wound dressings: a review. Colloids Surf B 169:60–71. https://doi.org/10.1016/j.colsurfb.2018.05.011

Marques BS, Frantz TS, Cadaval TRS Jr, Pinto LAA, Dotto GL (2018) Adsorption of a textile dye onto piaçava fibers: kinetic, equilibrium, thermodynamics, and application in simulated effluents. Environ Sci Pollut Res Int:1–9. https://doi.org/10.1007/s11356-018-3587-5

Marques JL Jr, Lütke SF, Frantz TS, Espinelli JBS Jr, Carapelli R, Pinto LAA, Cadaval TRS Jr (2018) Removal of Al (III) and Fe (III) from binary system and industrial effluent using chitosan films. Int J Biol Macromol 120:1667–1673. https://doi.org/10.1016/j.ijbiomac.2018.09.13

Rahmani S, Rafizadeh M, Taromi FA (2014) Statistical analysis of nanofibers alignment in magnetic-field-assisted electrospinning including an alignment percentage formula. J Appl Polym Sci 131:41179. https://doi.org/10.1002/app.41179

Rolandi M, Rolandi R (2014) Self–assembled chitin nanofibers and applications. Adv Colloid Interface Sci 207:216–222. https://doi.org/10.1016/j.cis.2014.01.019

Sarbatly R, Krishnaiah D, Kamin Z (2016) A review of polymer nanofibres by electrospinning and their application in oil–water separation for cleaning up marine oil spills. Mar Pollut Bull 106:8–16. https://doi.org/10.1016/j.marpolbul.2016.03.037

Silva JM, Farias BS, Gründmann DDR, Cadaval TRS, Moura JM, Dotto GL, Pinto LAA (2017) Development of chitosan/spirulina bio–blend films and its biosorption potential for dyes. J Appl Polym Sci 134:1–8. https://doi.org/10.1002/app.44580

Tang X, Feng Q, Liu K, Li Z, Wang H (2018) Fabrication of magnetic Fe_3O_4/silica nanofiber composites with enhanced Fenton-like catalytic performance for rhodamine B degradation. J Mater Sci 53:369–384. https://doi.org/10.1007/s10853-017-1490-y

Tarrés Q, Deltell A, Espinach FX, Pèlach MA, Delgado-Aguilar M, Mutjé P (2017) Magnetic bionanocomposites from cellulose nanofibers: fast, simple and effective production method. Int J Biol Macromol 99:29–36. https://doi.org/10.1016/j.ijbiomac.2017.02.072

Wang S, Sun Z, Yan E, Yuan J, Gao Y, Bai Y, Chen Y, Wang C, Zheng Y, Jing T (2014) Magnetic composite nanofibers fabricated by electrospinning of Fe_3O_4/gelatin aqueous solutions. Mater Sci Eng B 190:126–132. https://doi.org/10.1016/j.mseb.2014.10.001

Whang K, Thomas CH, Healy KE, Nuber G (1995) A novel method to fabricate bioabsorbable scaffolds. Polymer 36:837–842. https://doi.org/10.1016/0032-3861(95)93115-3

Yuan J, Xu Y, Muller AHE (2011) One-dimensional magnetic inorganic–organic hybrid nanomaterials. Chem Soc Rev 40:640–655. https://doi.org/10.1039/c0cs00087f

Zhang J, Chen N, Li M, Su P, Feng C (2017) Fluoride removal from aqueous solution by Zirconium-Chitosan/Graphene Oxide Membrane. React Funct Polym 114:127–135. https://doi.org/10.1016/j.reactfunctpolym.2017.03.008

Zhao LM, Shi LE, Zhang ZL, Chen JM, Shi DD, Yang J, Tang ZX (2011) Preparation and application of chitosan nanoparticles and nanofibers. Braz J Chem Eng 28:353–362. https://doi.org/10.1590/S0104-66322011000300001

Zhou Y, Yang D, Chen X, Xu Q, Lu F, Nie J (2007) Electrospun water–soluble carboxyethyl chitosan/poly(vinyl alcohol) nanofibrous membrane as potential wound dressing for skin regeneration biomacromolecules. 9:349–354. https://doi.org/10.1021/bm7009015

Zou L, Wang Q, Shen X, Wang Z, Jing M, Luo Z (2015) Fabrication and dye removal performance of magnetic $CuFe_2O_4$@CeO_2 nanofibers. Appl Surf Sci 332:674–681. https://doi.org/10.1016/j.apsusc.2015.01.176

Chapter 11
Magnetic Solid-Phase-Based Sorbents for Isolation/Preconcentration and Removal of Pesticides

Wan Aini Wan Ibrahim and Hamid Rashidi Nodeh

Contents

Abstract Recently, extractions, preconcentrations, and removals of pesticides from various samples are frequently performed with magnetic materials prior to instrument analysis. Trace-level pesticide studies require potential material to trap pesticides from huge interference of complex matrices of foods, biological media, and environmental samples. In this regard, scientists have gained specific strategies to improve the efficiency, chemical stability, and selectivity of the magnetic materials with different types of substances, such as carbon-based material, silica-based organic–inorganic hybrid composites, metal organic framework (MOF)-based sorbent, polymer-based sorbents, and green solvents. Combination of magnetic nanoparticles and efficient substances allows the formation of magnetic platform

W. A. W. Ibrahim (✉)
Department of Chemistry, Faculty of Science, Universiti Teknologi Malaysia, Johor Bahru, Johor, Malaysia
e-mail: waini@utm.my

H. Rashidi Nodeh
Food Technology and Agricultural Products Research Centre, Standard Research Institute, Karaj, Iran

L. Meili, G. L. Dotto (eds.), *Advanced Magnetic Adsorbents for Water Treatment*, Environmental Chemistry for a Sustainable World 61,
https://doi.org/10.1007/978-3-030-64092-7_11

for easy handling with the aid of external magnetic field without further centrifugation or filtration. In this chapter, the use of magnetic solid-phase-based sorbents such as multiwalled carbon nanotubes, graphene, activated carbon, silica sol–gel, MOFs, polymers/biopolymers, and green solvents as ionic liquids or deep eutectic solvents as potential sorbents in different sample preparation or removal techniques for pesticide extraction and remediation is described comprehensively. The advantages of each magnetic materials and proposed interactions between the sorbents and pesticides are described adequately.

Keywords Sorbents · Magnetic · Adsorption · Pesticides · Carbon based · Silica based · Graphene based · Metal organic framework (MOF) · Polymer based · Ionic liquids · Deep eutectic solvent · Interaction mechanism

11.1 Introduction

Pesticides include a wide range of toxic organic compounds and approximately 2.5 million tons per year are used worldwide (45% Europe, 25% the United States, and 25% rest of the world) to combat plant control, destroy unwanted living organism/crops/pests, and followed by improving the quality of food, fruits, vegetables, and agricultural products (De et al. 2014; John et al. 2018; Rajan et al. 2020). Wide usage of pesticides has led to increase in their residues in surface water and food chain that result in dangerous risk to mammals, environment, and human health (Shrestha et al. 2018; Kumar and Kumar 2019; de Souza et al. 2020). Arsenic, ethylene oxide, and lindane are among pesticides rated as Group 1 carcinogens (Deadman 2017; Khanjani et al. 2017). Some pesticides are known to cause birth defects and affect the central nervous, respiratory, and endocrine systems (Rashidi Nodeh 2015; Kim et al. 2017). Pesticides are present as a large variety of subclasses, such as insecticides, fungicides, herbicides, rodenticides, nematicides, and molluscicides. Pesticides can also be divided as subclasses of organophosphorus (OPP), organochlorine (OCP), carbamates, triazine herbicides, pyrrole-based, pyrethroid pesticides (Gou et al. 2000), and benzoylurea pesticides (Fan et al. 2020). The structures of some of these pesticides are shown in Fig. 11.1. Due to the toxicity of these pesticides, European Union (EU), World Health Organization (WHO), and Environmental Protection Agency (EPA) have set a maximum permissible level (MPL) of pesticides in surface water at 0.5 ng mL^{-1}, drinking water at 0.1 ng mL^{-1}, and food, fruits, and vegetables at 0.01–0.5 mg kg^{-1} (EU 2005; WHO 2011). Thus, due to wide variety of toxic pesticide contamination, trace analysis with low LOD and simultaneous identification is often necessary (Bolygo and Atreya 1991; Amiri et al. 2020).

Quantitative analysis of pesticides in matrices such as water, food, fruits, vegetables, and biological media (urine and blood) is a critical step and requires preconcentration and/or clean-up to remove the complex matrices and interferences

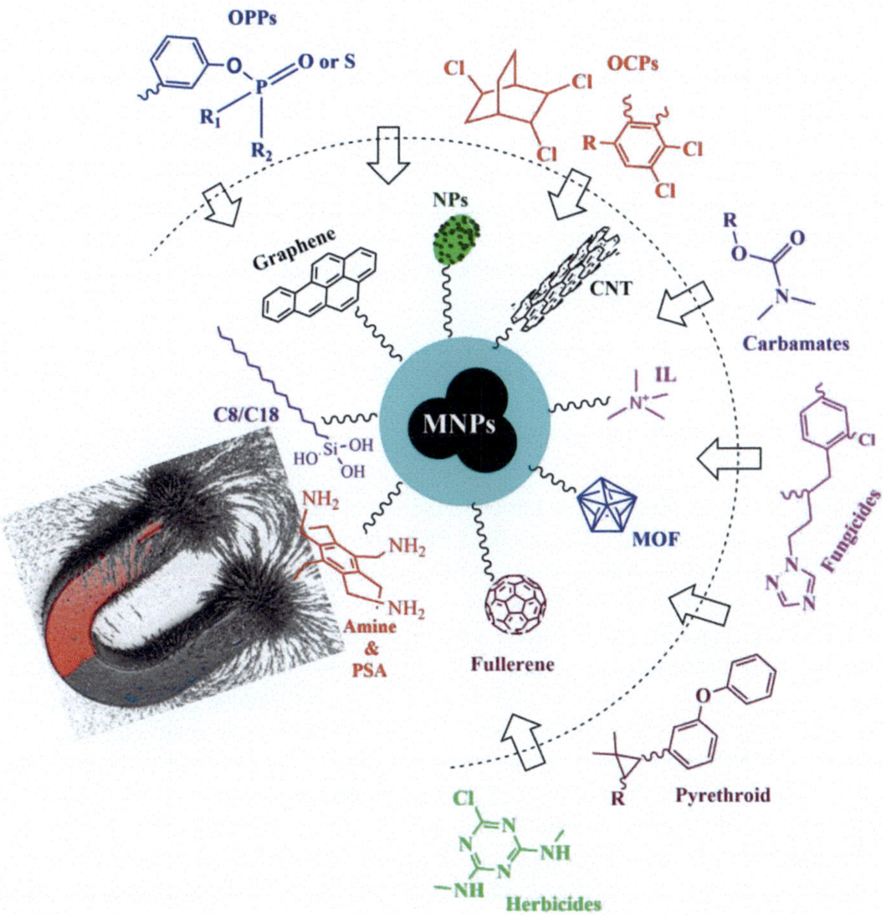

Fig. 11.1 Schematic illustration for functionalized magnetic nanoparticles (MNPs) with different types of sorbent materials (graphene, fullerene, molecular organic framework (MOF), ionic liquid (IL), carbon nanotube (CNT), C18/C8, amine, and nanoparticles) to interact with various pesticides: OPPs, OCPs, carbamates, fungicides, herbicides, and pyrethroids

to enhance the analytes with high efficiency and sensitivity (Wan Ibrahim et al. 2015a, b; Rashidi Nodeh et al. 2016; Yu et al. 2017; Wen 2020). Solid-phase extraction (SPE) is a standard method and most popular technique widely used to isolate, clean-up, preconcentrate, and extract or remove pesticides from various media (Covaci and Schepens 2001; Huang et al. 2011; Andrade-Eiroa et al. 2016; Atapattu and Johnson 2020). The advantages of SPE include flexibility for various stationary phases, significant recovery, high enrichment factor, great efficiency, good recovery, low cost, high adsorption capacity, high clean-up for trace targets, and consume less organic solvents (Cai et al. 2003; Shahrebabak et al. 2019; Wen 2020). Although the conventional SPE provided potential benefits, still researchers are reporting some drawbacks of SPE, including back pressure, sorbent channeling, tediousness, time consuming, and low reusability (Wan Ibrahim et al. 2015b; Andrade-Eiroa et al. 2016). Scientists have produced various kinds of sorbents for use in SPE and other modes of solid-phase-based extraction materials in an effort to overcome the shortcomings. Magnetic sorbents based on various kinds of materials have been seen in the literatures in the quest for alternative or potential sorbents for various applications. This review was compiled in an effort to assess the general extent of scientific knowledge regarding the use and advances of magnetic sorbents for pesticide analysis (preconcentration and removal) with focus on carbon, silica, polymer, metal organic framework, and ionic liquid-based ones. Over 100 references were reviewed and analyzed in this study.

11.2 Magnetic Sorbents

Recently, research interests have turned to the use of magnetic nanoparticles (MNPs) as sorbent to speed up the conventional SPE sorbents and limitations (Wanjeri et al. 2018; Zhou et al. 2019; Marsin et al. 2020). Iron oxide of γ-Fe_2O_3 and Fe_3O_4 are the most common forms of MNPs widely used as an sorbent due to their super paramagnetic properties, high surface area, relatively small size, high breakthrough volume, good adsorption properties, high enrichment factor, superior high surface area, small particle size, and ease of dispersion in water (Maddah And Shamsi 2012; Ibarra et al. 2014; Rashidi Nodeh et al. 2017a, b). Hence, the main characteristic rewards of MNPs can be described as easy synthesis via coprecipitation method, fine functionalization with various synthetic organic/inorganic compounds (ionic liquids, metal oxides, silica, metal organic framework (MOF), polymer, graphene, carbonaceous material, C18, amine (-NH_2), etc.) (Fig. 11.1), biomolecules (protein, enzyme, N-halamine) (Dong et al. 2011; Maddah and Shamsi 2012; Lakshmanan et al. 2013; López et al. 2013; Ibarra et al. 2014; Zhang et al. 2014a, b; Mahpishanian et al. 2015; Qin et al. 2016; Vaghari et al. 2016; Rashidi Nodeh et al. 2017a, b; Shrivas et al. 2019; Niu et al. 2020), and easy engineering of quite small particles in the range of cells (10–100 nm), viruses (20–450 nm), proteins (5–50 nm), and gen dimension (2 nm wide and 10–100 nm long) that led to a rise in MNP application in biomedical and drug delivery (Pankhurst et al. 2003).

Researchers tend to functionalize or modify MNPs with different substances (Fig. 11.2) to improve the compatibility, sensitivity, stability, and selectively and to exhibit various interactive forces (Heidari and Razmi 2012; Tong and Chen 2013; Wang et al. 2014; Wan Ibrahim et al. 2015b; Ma et al. 2016; Xu et al. 2017; Lemasson et al. 2017; Wang et al. 2019a, b; Turan et al. 2020). As illustrated in Fig. 11.2, multiple interactive force material simultaneously shows van der Waals forces, dipole force, hydrophobic interaction, hydrophilic interaction, electrostatic interaction, and hydrogen-bonding enthalpy and entropy effect (basis of size exclusion chromatography) (He et al. 2014; Asadi et al. 2020; Jost and Habedank 2020).

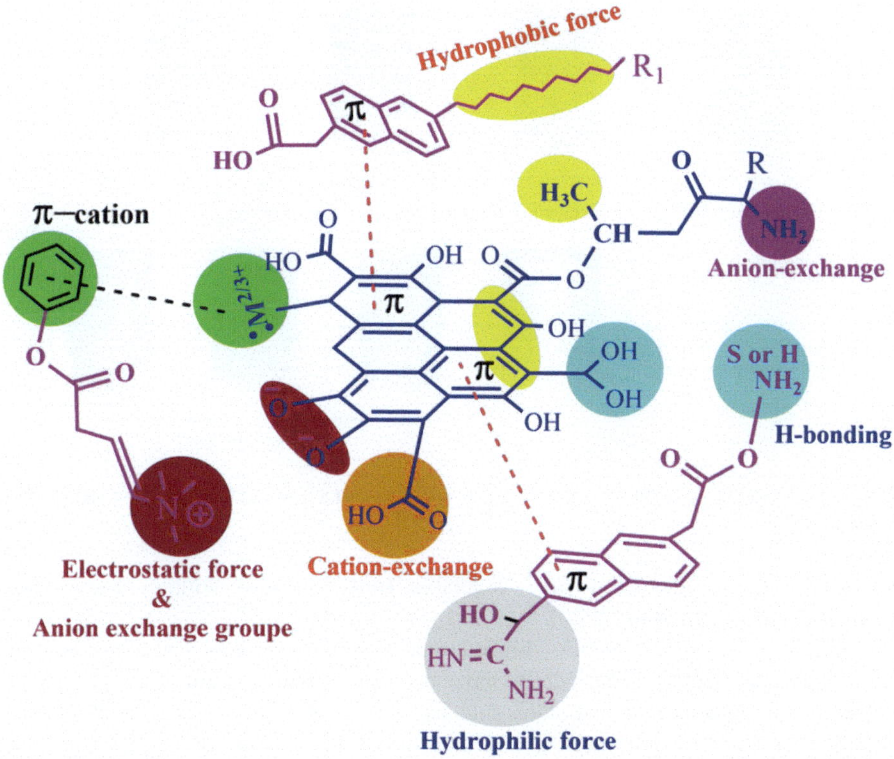

Fig. 11.2 Schematic of sorbent material and analytes with multifunctional groups and multiforce interactions

11.3 Magnetic Solid-Phase-Based Extractions

Since the introduction of magnetic separation technology (MST) in the early twentieth century (Gunther 1909), it has been established itself as a powerful separation approach in industrial processes, bio-separation, and environmental and material science. MNPs are used in extraction methodology to overcome the conventional SPE drawbacks in a method called magnetic solid-phase extraction (MSPE) (Šafaříková and Šafařík 1999). Introducing magnetic Fe_3O_4 nanoparticles in the sorbent can enhance the extraction process separation from solution by using an external magnet without the need for filtration and centrifugation process (Mehdinia et al. 2011; Li et al. 2013a, b, c; Wang et al. 2018a, b; Wang et al. 2019a, b; Turan et al. 2020) (Fig. 11.3). Magnetic material SPE sorbent advantages include low cost, easily extracted with an external magnet from liquid samples, quick, short extraction time, water dispersive, high adsorption capacity, and sensitive to polar and nonpolar pesticides (Wan Ibrahim et al. 2015b). Hence, MSPE is good candidate for

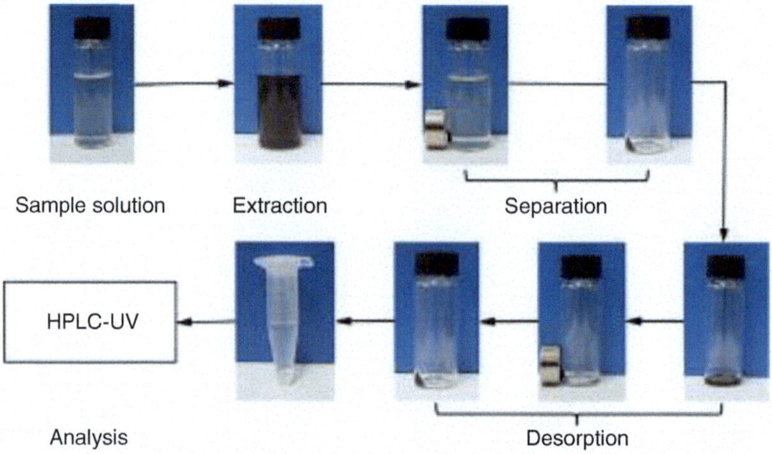

Fig. 11.3 Magnetic solid-phase extraction procedure to extract pesticides from aqueous media. (Li et al. 2015)

extraction of pesticides from water, vegetable, fruits, and milk samples (Farajzadeh et al. 2020; do Lago et al. 2020).

Literature survey on MNPs showed that it is widely used for extraction of pesticides since 2007–2020. MNPs have been modified with different groups of materials for extraction of pesticides, such as silica, ionic liquids (ILs), metal organic frameworks (MOFs), polymers, and carbon-based material in order to improve the extraction efficiency, selectivity, and compatibility. Pesticides were successfully preconcentrated/extracted from various matrices such as aqueous solution, foods, fruits, and vegetables using the modified MNP-based MSPE.

11.3.1 Carbon-Based Sorbents

The large group of modified magnetic sorbents derived from carbon-based material includes graphene oxide, three-dimensional graphene, multiwalled carbon nanotubes (MWCNTs), porous carbon, biochar, activated carbon (AC), and graphitized black carbon (GCB). These materials are predominantly used for pesticide isolation (Zheng et al. 2013; Mandal et al. 2017; Kermani et al. 2019; Shahrebabak et al. 2019; Gao et al. 2020; Sereshti et al. 2020a, b). The proposed materials possess high surface area and large delocalized π-electrons that demand proper π–π interaction with benzene ring of pesticides and hydrophobic interaction with nonpolar pesticides ($log\ K_{O/W} > 3$) (Rashidi Nodeh et al. 2015).

Graphene-magnetite NPs was successfully applied for extraction of different kinds of pesticides (fungicides, herbicides, carbamates, pyrethroid pesticides and OPPs) from various kinds of matrices. Example of applications are for extraction of fungicides from water and juice samples achieving a low LOD at part per trillion level and high enrichment factor (1495-1849 times) (Li et al. 2013c) and in an earlier study by Zhao et al. (2011), four triazine herbicides were successfully extracted from water samples with LOD 6-25 times higher than the one achieved by Li et al. (2013c) using graphene-coated MNPs. A study by Wu et al. (2011) for five carbamate pesticides from water samples found a similar level of LOD with the study of Zhao et al. (2011) but with a lower enrichment factor. Hou et al. (2013) achieved a LOD 2 times lower than the one achieved by Wu et al. (2011) for pyrethroids. Carbon-coated MNPs provided much lower LOD (4.3 ng L^{-1}) for nonpolar OPPs and higher LOD (47 ng L^{-1}) for mid-polar OPPs (Heidari and Razmi 2012).

Magnetic carbon nanospheres were used for six chiral pesticide extraction from food using MSPE and analysis using chiral liquid chromatography with Chiralpak IG column coupled with tandem mass spectrometry with limits of quantification between 1 and 4 ng g^{-1} (Zhao et al. 2019). Magnetic bio char (MBC) (surface area 404 m^2/g) was successfully used for atrazine removal with adsorption capacity of 298 mg g^{-1} but a much lower adsorption capacity for 2,4-dichlorophenol (102.17 mg/g). However, the regeneration capability of the MBC sorbent was low; sorption capacity dropped to 2/3–3/4 of its original adsorption capacity (Dai et al. 2019).

Magnetic graphene provided high efficiency for very nonpolar bifenthrin pyrethroid (log $K_{ow} > 6.0$) at medium acidic pH. The low efficiency at high pH probably is due to the hydrolysis of pyrethroids pesticides (Hou et al. 2013). MWCNTs applied for OPPs and OCPs gave LOD < 80 ng mL^{-1} and less interference effects (González-Curbelo et al. 2013; Deng et al. 2014). These trends can be concluded that the hydrophobic nature of graphene and carbon-based material improved the extraction of nonpolar compounds.

Further functionalization of MWCNTs, graphene, and carbon based with hydrophilic groups, that is, amine ($-NH_2$), cyano ($-C \equiv N$), pyrrole, hybrid inorganic–organic, amide, polyethylene diamine, and glucamine, increases the affinity toward polar pesticides (*log* $K_{O/W} < 3$) and mid-polar pesticides (*log* $K_{O/W}$ *1.7–2.8*) (Wu et al. 2011; Deng et al. 2014; Kamboh et al. 2016; Soutoudehnia Korrani et al. 2016; Rashidi Nodeh et al. 2016; Rashidi Nodeh et al. 2017b; Oellig and Schmid 2019; Marsin et al. 2020).

The report by Shahrebabak et al. described the simultaneous extraction of acidic and basic pesticides (Fig. 11.4) from food samples using magnetic graphene triazine polymer at pH 7.4 (Shahrebabak et al. 2019). The synergic affinity of proposed magnetic composite toward imidacloprid and dichlorophenoxyacetic acid could be attributed to the three interactive forces of hydrogen bonding, π–π interaction, and electrostatic interactions (Shahrebabak et al. 2019).

The latest application of magnetite–graphene is represented as stir bar sorptive-dispersive microextraction (SBSDμE) for extraction of multiclass of seven pesticides from aqueous solution (Madej et al. 2019). Comparison was made between MSPE and SBSDμE for these seven pesticides using the material magnetite–graphene. Extraction was found to faster for SBSDμE (35 min) compared to MSPE (50 min). The recoveries, R, and relative standard deviations (RSDs) of the seven pesticides by

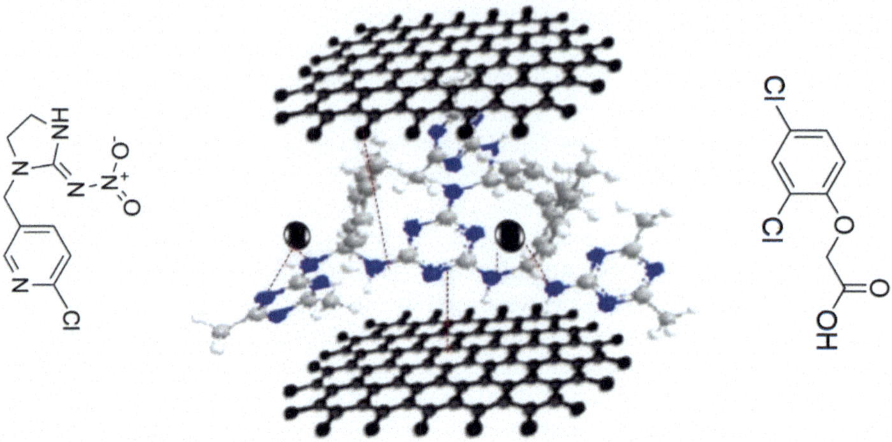

Fig. 11.4 Magnetic graphene triazine polymer nanocomposite and imidacloprid (basic pesticides) and dichlorophenoxyacetic acid (acidic pesticides) (Shahrebabak et al. 2019)

both methods were similar (%R 20–82% and RSDs 1–13%). However, the SBSDμE approach is better than MSPE in terms of faster extraction time (30% faster).

A magnetic sorbent based on magnetic silica core shells modified with graphene oxide and phenylethyl amine exhibits good reusability up to 30 times showing only a 5% decrease in performance for MSPE of six organophosphorus pesticides from fruits, vegetables, and water samples. They attributed the good reusability to the covalent bonding formed between GO and the magnetic silica core shell (Mahpishanian et al. 2015).

Nickel ferrite/nickel oxide nanoparticles co-loaded three-dimensional reduced graphene oxide (3DRGO-$NiFe_2O_4$/NiO NPs) was used as a potential paper sensor for determination of dichlorvos, an organophosphorus pesticide in vegetables and fruits achieving a limit of detection of 10μg mL^{-1}. They claimed that a simple and inexpensive assay for rapid detection of dichlorvos was achieved using this new graphene oxide-based sensor (Wei et al. 2020).

11.3.2 Organic–Inorganic Hybrid Silica-Based Sorbents

Silica-based materials as a cheap and potential material are widely used in sample preparation process for extraction of pesticides and clean-up complex matrices. The mesoporous silica and commercial long-chain silica gel of C18 and C8 material have merged with magnetic nanoparticles to offer fast extraction of pesticides from different media (Maddah and Shamsi 2012; Jiang et al. 2013; Moliner-Martinez et al. 2014; Farajzadeh et al. 2018; Yu et al. 2018; Zhang et al. 2019). MNPs-C18 isolated OPPs from aqueous media via hydrophobic interaction with LOD of 0.014 ng mL^{-1}. This study showed low efficiency at acidic and basic pHs; this is probably due to protonation of NH and OH groups at low pH and hydrolysis of OPPs at high pH (Maddah and Shamsi 2012; Kamboh et al. 2016; Rashidi Nodeh et al. 2017a, b).

Magnetic core shell silica nanoparticles ($Fe_3O_4@SiO_2$) modified with phenyl were successfully used as an extractant after QuEChERS (Quick, Easy, Cheap, Effective, Rugged, Safe) method to provide efficient and sensitive preconcentration of pesticides (Farajzadeh et al. 2019). SiO_2–MNPs provided LOD at parts per trillion levels (50–100 ng L^{-1}) for chlorpyrifos and chlorfenvinphos in water samples prior to capillary liquid chromatography-diode array detector (LC-DAD) analysis (Moliner-Martinez et al. 2014). Magnetic–silica–titania–NH_2 as hybrid nanocomposite provided appropriate LOD (1.4 ng g^{-1}) for selected pesticides in coffee bean (Asadi and Sereshti 2020). The results obtained by the latter produced a LOD 71 times higher than the one obtained with the hybrid organic-inorganic silica nanocomposite, methyltrimethoxysilane–tetraethoxysilane for chlorpyrifos (Wan Ibrahim et al. 2012).

An amphiphilic (containing hydrophilic-lipophilic moiety) silica-based magnetic nanocomposite based on divinyl benzene (DVB) and N-vinyl pyrrolidone (NVP) ($Fe_3O_4@SiO_2@$DVB-NVP) simultaneously extracted and preconcentrated 96 pesticides from sea water with satisfactory LOQs in the range of 1–10 ng L^{-1} using liquid chromatography-tandem mass spectrometry (LCMS/MS) analysis (Liu et al. 2020). The method is claimed to be able to meet the monitoring requirements for multi-pesticide residue analysis in seawater.

11.3.3 Polymer-Based Sorbents

Recently, plain polymer or imprinted polymer material, cross-linked polymers, polymeric microbeads, and biopolymers have gained interest to modify MNPs since polymer network can enhance the covalent bonding, electrostatic, and π–π interaction (Faraji 2016; Badawy et al. 2018; Mohd Hassan et al. 2020; Kermani et al. 2020; Özer et al. 2020). Similarly, the tunability, stability, and hydrophilic and hydrophobic nature of most polymers made them appropriate sorbent to extract both polar and nonpolar analytes (He et al. 2014; Jiao et al. 2016; Yang et al. 2016).

Magnetic cross-linked polystyrene/divinylbenzene provided 750 times enrichment for atrazine pesticides in soil samples (Kermani et al. 2020). Polyaniline–MNPs successfully enhanced the dispersive SPE of five pyrethroid pesticides from tea drink (Wang et al. 2014). LOD at ppb level was achieved (0.025–0.032 ng mL^{-1}) with good recoveries (72.1–118.4%), and the sorbent was reusable up to 15 times without showing signs of significant loss in the recoveries.

Polydopamine–MNPs were used for magnetic dispersive solid-phase extraction (MDSPE) of four benzoylurea insecticides in environmental water with low limit of detection ($S/N = 3{:}1$) of 0.75μg L^{-1} and limit of quantification ($S/N = 10{:}1$) of 2.5μg L^{-1} (Fig. 11.5a) (Huang et al. 2019). Precision of the polydopamine–MNPs MDSPE was acceptably good for real water samples spiked at 100μg L^{-1}

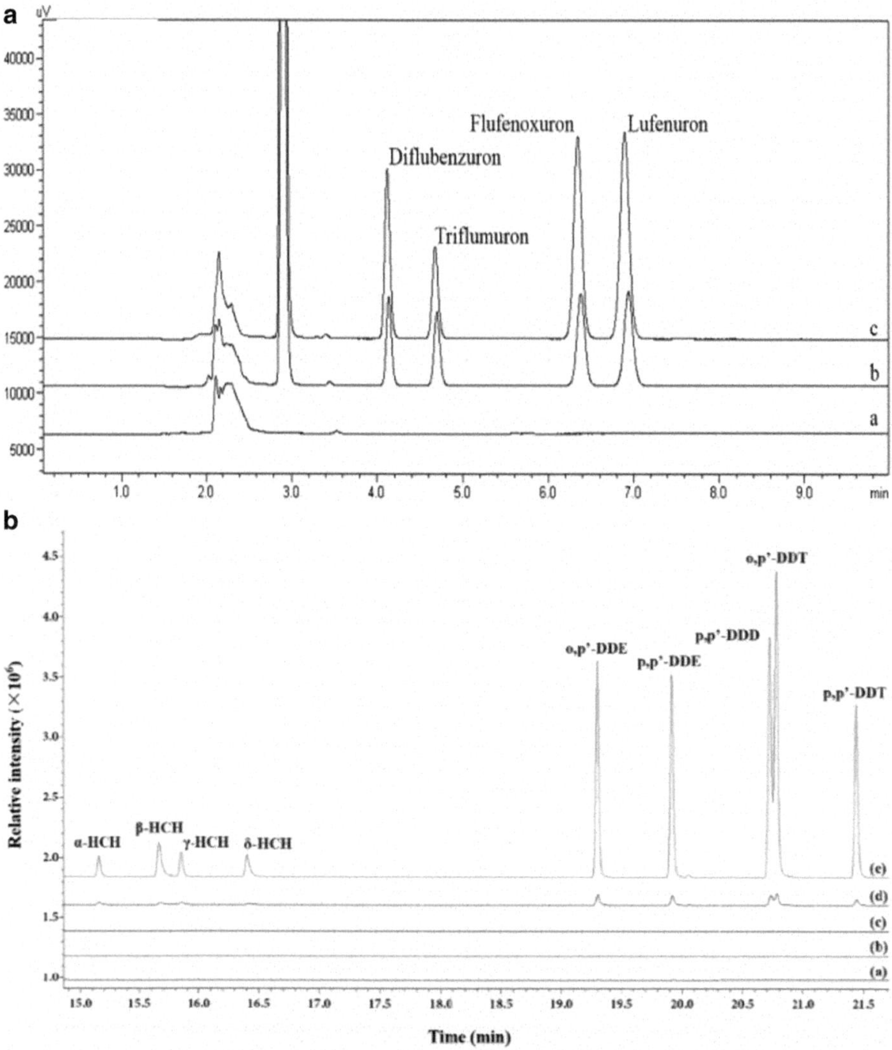

Fig. 11.5 (**a**) Chromatogram of polydopamine–MNPs for benzoylurea insecticides (Huang et al. 2019). (**b**) GC-MS for OCPs extraction from water (Huang et al. 2018). (**c**) GC-FPD chromatogram of OPPs extracted with magnetic MOF (Li et al. 2019). (**d**) HPLC-DAD result for pyrrole pesticides using magnetic MIL-101-MOF (Ma et al. 2016)

(71.8–91.6%). The extraction was fast (30 s) and uses only 10 mg of the sorbent. The sorbent was reusable up to five times where the recovery efficiency remained satisfactory.

Magnetic polypyrrole nanowires were successfully used for MSPE of 11 pesticides (organochlorine, organophosphorus, and pyrethroids) from teas, juices, and environmental water samples using GC-MS analysis (Zhao et al. 2013) achieving the

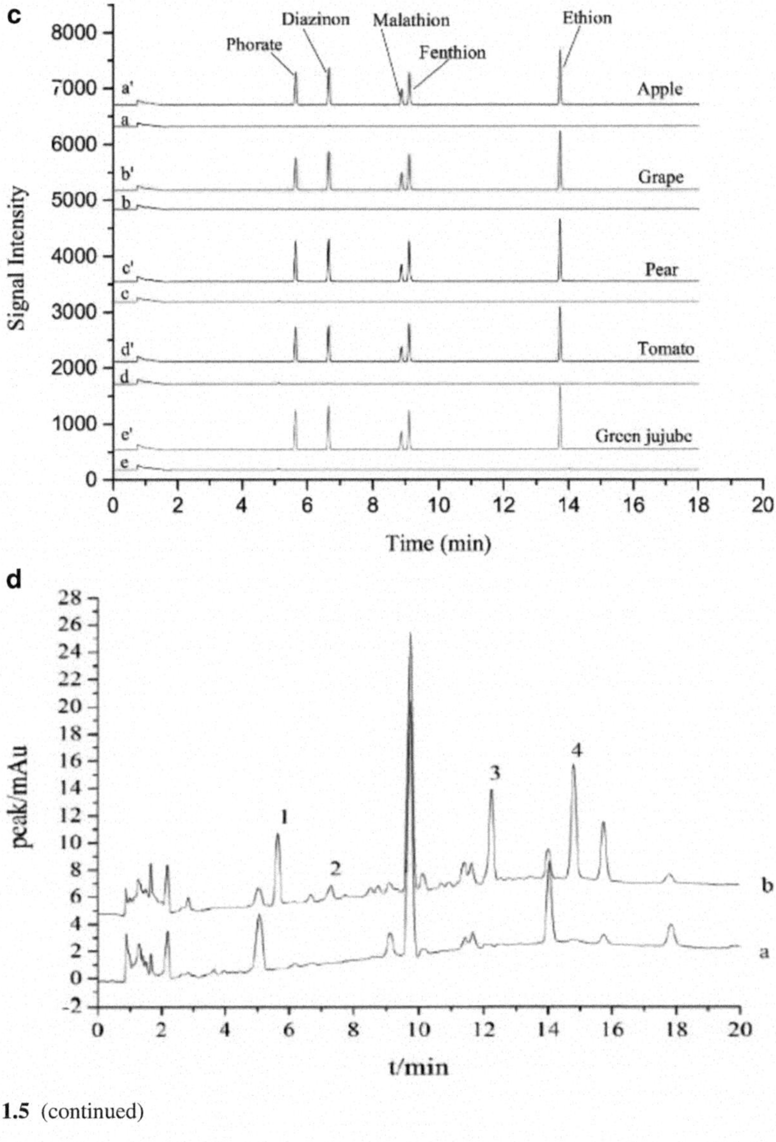

Fig. 11.5 (continued)

limit of quantification at parts per billion level (0.09–0.29μg L^{-1}). Recoveries of the 11 pesticides from real samples ranged from 63% to 129% with RSDs <14% using only 2 mg of the sorbent and 5-min extraction time.

Chitosan as biopolymers modified with siloxane MNPs coupled with HPLC analysis attained an LOD of 0.002 and 0.046μg mL^{-1} for fenamiphos and diazinon, respectively, with a recovery of 60.8–95.3%. The sorbent amount used in this work

(200 mg) (Badawy et al. 2018) is 100 times higher than the amount used by Zhao et al. (2013). The behavior displayed by the polymers to pesticides is because of the large π-electron stacking and electrostatic interaction (Wang et al. 2014; Huang et al. 2019; Zhao et al. 2013).

Feizbakhsh and Ehteshami (2016) produced a magnetic nanocomposite based on polythiophene and chitosan for DMSPE of triazine herbicides in aquatic media attaining excellent relative recovery percentages (96–102%). Limit of detection and quantitation obtained (ppg level) is sufficient for real sample analysis of the triazine herbicides.

A novel stir bar sportive kit was developed based on MNPs and polyether sulfone polymer for organophosphorus pesticides and hexythiazox pesticide with 40 times theoretical enrichment factor prior to chemometrics modeling (Gorji et al. 2019).

Triazine-imine magnetic core–shell organic polymer with excellent stability and high adsorption capacity provided π-conjugated system, high enrichment, and fast separation of eight pesticides from fruits. LOD of 0.4 ng L^{-1} was attained using the sorbent coupled with ultra HPLC (UHPLC) and tandem mass spectrometry (MS/MS) (Liang et al. 2019). Assisted with the developed method using triazine-imine magnetic core–shell organic polymer with UHPLC-MS/MS, the researchers found four pesticides (fenhexamid, carbendazim, diflufenican, and quinoxyfen) in strawberries and two pesticides (fenhexamid and quinoxyfen) in grapes.

Clay polymers such as magnetic attapulgite and gum montmorillonite have also been used successfully as sorbents for pesticide studies (Yang et al. 2016). Magnetic attapulgite polypyrrole sorbent efficiently isolated pyrethroids (0.3μg L^{-1}) from honey sample (Yang et al. 2020b) and gum montmorillonite biopolymer modified MNPs efficiently removed diazinon from water with sorption capacity of 47.17 mg g^{-1} (Nikzad et al. 2019).

11.3.4 Metal Organic Framework Sorbents

Metal organic framework (MOF) also known as porous coordination polymers is a class of material made by the self-assembly of organic ligands and metal containing nodes. Introduction of magnetism into the MOF produces magnetic MOF (MMOF), and these materials have gained increased attention from researchers and have been harnessed in several studies for various applications including pesticide analysis (Ma et al. 2013, 2018; Zhang et al. 2014b; Cao et al. 2017; Xu et al. 2017; Zhou et al. 2017; Huang et al. 2018; Liu et al. 2017; Liu et al. 2018b, c; Zheng et al. 2018).

The first use of MMOF (MIL-101(Cr)) was for MSPE of polycyclic aromatic hydrocarbon (PAH) analysis from water samples (Huo and Yan 2012). The developed method achieved detection and quantification limits at parts per trillion levels for the PAHs studied.

A magnetic MWCNT–MOF based on zeolitic imidazolate framework (ZIF) termed as M-M-ZIF-67 appropriately isolated nine OCPs from agriculture water (Fig. 11.5b) with appropriate resolution, efficiency, and less interference effects with low LOD (S/N = 3:1) of 0.07–1.03μg L^{-1} (Huang et al. 2018). The analysis of OCPs was completed within 20 min. Good analytes recoveries (74.9–116.3% and 75.1–112.7%) resulted from the developed method at spiked levels of 10 and 100μg L^{-1}, respectively. In another study (Liu et al. 2018c), a magnetic multiwalled (M-M) based on ZIF (M-M-ZIF-8) successfully studied the sorbent for 8 OPPs from soil and environmental water.

Magnetic carbon Zn/Co-MOF provided high surface area and high extraction efficiency for OPPs with low LOD of 0.01 ng mL^{-1}. The proposed MOF obviously extracted OPPs from fruits samples with less interferences effects as shown in the GC-FPD chromatogram (Fig. 11.5c) (Li et al. 2019). This could be attributed to the high affinity of the metal cations (Zr, Zn, Co) to phosphate moiety of OPPs (P=S or P=O) (Jiang et al. 2016; Li et al. 2019). Magnetic core–shell MOF nanocomposite enhanced recoveries for OPPs in fruit samples (72–111%) (Lin et al. 2020).

Magnetic MIL-101 MOF applied for pyrazole/pyrrole pesticides in river water with LOD of 0.3 ng mL^{-1} followed by HPLC-DAD determination (Fig. 11.5d) (Ma et al. 2016). MIL-101 MOF provided high extraction efficiency when NaCl added into extraction vessel the salt addition, since NaCl would weaken the electric interaction between adsorbent and analytes (Ma et al. 2016).

High affinity of the MOF-199 and MOF-100(Fe) toward triazole and OCPs is attributed to the interactions of bimetallic and heterocycle member of MOF (Fig. 11.6) with benzene rings of pesticides. Hydrophobic interaction, hydrogen bonding, and electrostatic interactions are possible as well (Liu et al. 2017, 2018b; Li et al. 2019; Lin et al. 2020).

Magnetic graphene oxide-zinc MOF coupled with HPLC and tandem MS used for extraction of heterocyclic fungicides from vegetables produced low LOD

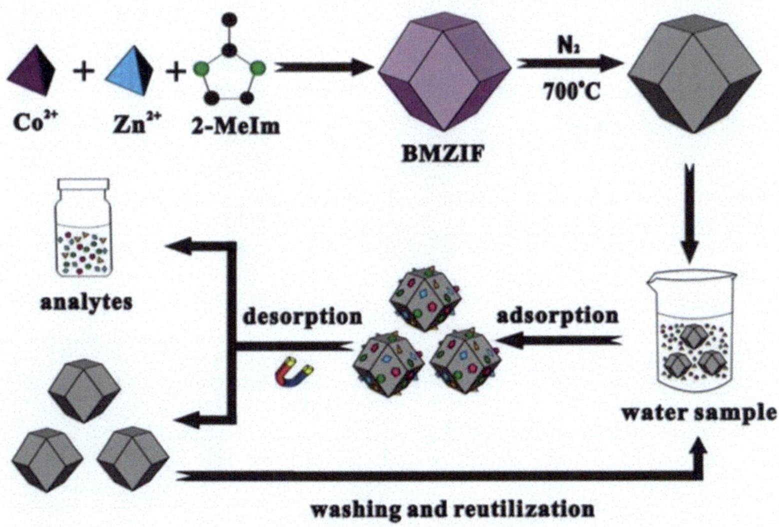

Fig. 11.6 Schematic structure, synthesis, and application of magnetic MOF to extract pesticides. (Liu et al. 2017)

(0.25–1.0 ng mL^{-1}) and good recoveries (74.82–99.52%) and RSDs <7.3% using 10 mg sorbent mass (Liu et al. 2019b).

11.3.5 *Ionic Liquids and Deep Eutectic Solvent-Based Sorbents*

Sample preparation using magnetic ionic liquids (MILs) is gaining interest of many researchers (Wang et al. 2015; Liu et al. 2018a; Chatzimitakos et al. 2018a, b). Ionic liquid as eco-friendly material was applied to modify MNPs in order to achieve rapid, sensitive, and accurate extraction and analysis of pesticides (Zhang et al. 2014a; Zheng et al. 2014; Fan et al. 2017; Amiri et al. 2018; Chatzimitakos et al. 2018a, b; Huang et al. 2019).

Ionic liquid functionalized magnetic polyoxometalate effectively extracted OPPs from fruit juice with low LOD of 0.02 ng mL^{-1} (Amiri et al. 2018). Magnetic polyionic liquid with large effective contact area and high viscosity was synthesized for repeatable extraction of pesticides from aqueous media (Zheng et al. 2014). Pyrethroid pesticides were successfully extracted using anion exchange magnetic ionic liquids (Zheng et al. 2014; Fan et al. 2017) (Fig. 11.7).

Magnetic solid-phase dispersion using rare earth-based magnetic ionic liquids (MIL) was for the first time demonstrated by Chatzimitakos et al. (2018b) for several classes of pesticide (organophosphate, organochloride, and triazine pesticides)

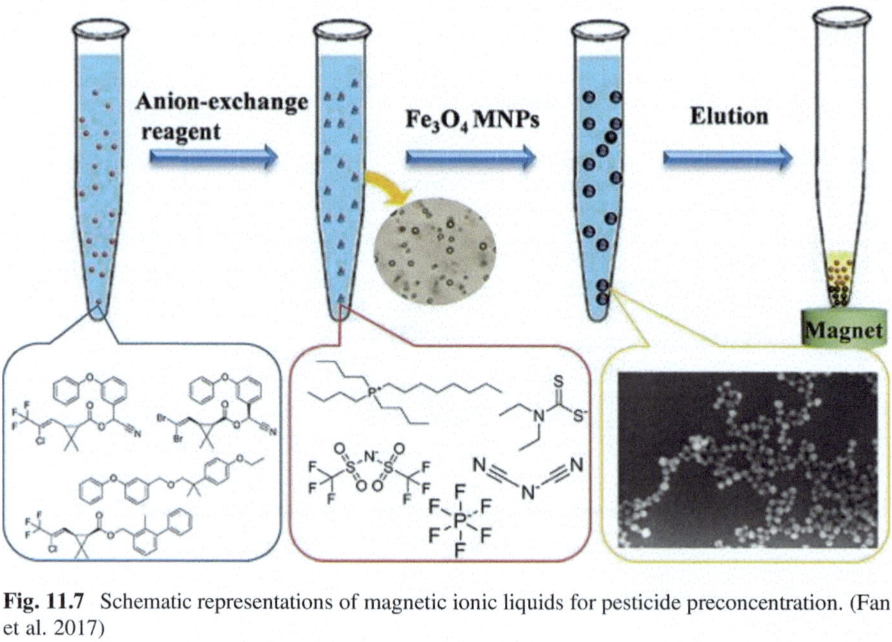

Fig. 11.7 Schematic representations of magnetic ionic liquids for pesticide preconcentration. (Fan et al. 2017)

analysis from four different vegetables. LOQ achieved was sufficient to meet the maximum residue limits (MRLs) established for pesticides (0.01 mg kg^{-1}) (Commission regulation (EU) 2018).

The water solubility of the ionic liquid is the main limit to apply them as extraction sorbent. In order to overcome this issue, imidazolium ionic liquid was modified with MNPs for pesticide isolation from aqueous media (Nacham et al. 2016). The main forces between ionic liquids and pesticides are anion exchange ability and π–π electron–donor/acceptor interaction (Faraji 2016).

Recently, the new type of green solvent, deep eutectic solvent (DES), was successfully developed as sorbent for pesticide extraction (Sereshti et al. 2020a; Yang et al. 2020a). DES is a new kind of ionic liquid and is a combination of ammonium salts (choline chloride) and hydrogen bond donors (polyols).

Polydopamine-based magnetic DES combined with HPLC analysis provided high efficiency for four sulfonylurea herbicides in water samples (Wang et al. 2019a, b) attaining an enrichment factor of between 495 and 630 and LOD at parts per billion level. The best DES combination consists of three parts of propylene glycol and one part of proline. DES modified with magnetic MWCNTs with ultrasound assistance recovered 76–97% of pesticide residues from food product (Zhao et al. 2020).

Table 11.1 gives a summary of some applications of MSPE sorbents from 2007 until early 2020.

11.4 Interaction Mechanism

Various forces drive uptake of pesticides onto modified MNPs including hydrophobic, hydrophilic, dipole–dipole, electrostatic, π–π donor/acceptor, anion exchange, and hydrogen bonding (Kamboh et al. 2016; Rashidi Nodeh et al. 2017a, b; Wang et al. 2018b; Wang et al. 2019b; Marsin et al. 2020).

Kamboh and coworkers studied the proposed mechanism for extraction of pesticides using magnetic calixarene ethylenediamine at different pH (Fig. 11.8a) (Kamboh et al. 2016). As can be seen, at high acidic pH, the surface association of the adsorbent (OH_2^+) and protonation of chlorpyrifos resulted in repulsion as the main force that causes decrease in extraction efficiency. At neutral pHs, amine and hydroxyl groups show binding abilities to anionic sites (O, S, N, and Cl) via hydrogen bonding or dipole–dipole interaction, which can increase the extraction efficiency. At high pH, following deprotonation process, repulsion occurs between adsorbent and adsorbate that can decrease the extraction percentage due to variety of benzene ring in calixarene structure containing double bond that make proper π–π interaction with chlorpyrifos and diazinon.

In another proposed mechanism (Fig. 11.8b), mixed hemimicelles and admicelles improve the affinity of the magnetic materials toward analytes driven by two hydrophobic, hydrophilic, and electrostatic attractions simultaneously (Li et al. 2013a; Ranjbar Bandforuzi and Hadjmohammadi 2019).

Table 11.1 Some applications of MSPE sorbents based on MNPs for the extraction of pesticides from different matrices from 2007 until early 2020

MSPE sorbent	Pesticides	Sample matrix	LOD	% R	EF	Detector	Reference
Palm AC–polypyrrole–MNPs	Endosulfan and dieldrin	Water	7.3–8.6 ng L^{-1}	86–103	–	GC-μ ECD	Marsin et al. (2020)
MOF–MNPs	Benzoylurea insecticides	Tea	0.7–3.2μg L^{-1}	78–114	72	GC-MS	Niu et al. (2020)
Lanthanum phosphate GO@MNPs	Chlorpyrifos and hexaconazole	Water Fruit	0.67–0.89μg L^{-1}	78–120	–	GC-μ ECD	Asadi et al. (2020)
Hollow TiO_2–SiO_2–MNPs	OPPs and hexaconazole	Coffee bean	1.33–1.42μg g^{-1}	74–113	–	GC-μ ECD	Asadi and Sereshti (2020)
Polypyrrole–MNPs	Pyrethroids	Honey	0.2–0.34μg L^{-1}	81–106	–	HPLC-UV	Yang et al. (2020b)
Hydroxyapatite–MNPs	OPPs	Fruit juice	0.03–0.22 ng mL^{-1}	89–99.7	–	GC-FID	Chahkandi et al. (2019)
Triazine polymer MNPS–GO	Imidacloprid, dichlorophenoxyacetic	Water and food	1.7μg L^{-1}	90	–	HPLC-UV	Shahrebabak et al. (2019)
Mesoporous silica MNPs	Pyrethroids	Water	1–6μg L^{-1}	86.5–98.8	>500	HPLC-UV	Zhang et al. (2019)
MNPs–rGO–CNT	OPPs	Vegetable and fruit	6.5μg L^{-1}	84–106	–	GC-MS	Yuan et al. (2019)
Polyamidoamine–MNPs	Benzoylurea insecticides	Water	0.39 ng mL^{-1}	75–111	74.4	HPLC-DAD	Liu et al. (2019c)
MOF–MNPs–MIL-100	Organochlorine pest	Plants	0.62 ng mL^{-1}	81–113	–	GC-MS	Zhou et al. (2019)
Hyperbranch polyester–MNPs	Benzoylurea insecticides (hexaflumuron, flufenoxuron, lufenuron, and chlorfluazuron)	Tea	0.15μg L^{-1}	90–98.4	–		Liu et al. (2019a)

MOF–MNPs	OPPs (phorate, diazinon, ethion, malathion, fenthion)	Fruit	0.018–1μg L^{-1}	84–116		GC-FPD	Li et al. (2019)
Polyphenyl diamine–MNPs	OPPs	Fruit juice	0.1–0.3 ng mL^{-1}	88–99.2	163	GC-FID	Targhoo et al. (2018)
BWO–IL–MNPs	OPPs	Water and fruit juice	0.02–0.06 ng mL^{-1}	70–89.2		GC-FID	Amiri et al. (2018)
Hybrid silica–MNPs	OPPs	Water	0.02–0.06 ng mL^{-1}	72–103	>100	GC-μ ECD	Malek et al. (2018)
Polystyrene–MNPs	Pyrethroids	Fruit	0.1–0.5 ng mL^{-1}	73–123		HPLC-DAD	Yu et al. (2018)
MNPs–SiO_2–C8	Pesticides	Fruit and vegetable	0.03–0.1μg kg^{-1}	64–82	2133		Farajzadeh et al. (2018)
Chitosan–siloxane–MNPs	Abamectin, diazinon, fenamiphos, imidacloprid, λ-cyhalothrin, methomyl, and thiophanate-methyl	Water	0.002–0.004μg L^{-1}	73–99	11	HPLC-VWD	Badawy et al. (2018)
Graphene@SiO_2@MNPs	OPPs (malathion, chlorpyrifos, isocarbophos, fenamiphos, profenophos)	Water	0.1–5 mg L^{-1}	90–102		GC-FPD	Wang et al. (2018a)
MOF–MIL101–GO–MNPs	Thiazine herbicides	Rice	0.01–0.08μg kg^{-1}	83–103		HPLC-UV	Liang et al. (2018)
AgO–3DG–MNPs	Fenitrothion, chlorpyrifos, hexaconazole	Fruit	0.07–0.13 ng g^{-1}	72–109		GC-μ ECD	Sereshti et al. (2018)
IL–MNPs–anion exchange	Pyrethroids	Water	0.16–0.21μg L^{-1}	80–117	213		Fan et al. (2017)
MGO@SiO_2-TMESPED	Fenitrothion, chlorpyrifos, hexaconazole	Fruit	0.2–0.3μg kg^{-1}	82–113		GC-μ ECD	Rashidi Nodeh et al. (2017a)
Polymethacrylate–MNPs	OPPs	Water	0.01–0.25μg L^{-1}	71–98		HPLC-DAD	Meseguer-Lloret et al. (2017)

(continued)

Table 11.1 (continued)

MSPE sorbent	Pesticides	Sample matrix	LOD	% R	EF	Detector	Reference
MNPs@G-TEOS-MTMOS	OPPs	Water	1.4–19.8 pg mL^{-1}	83–105	>100	GC-μ ECD	Rashidi Nodeh et al. (2017b)
DES-bucky gel – MWCNT–MNPs	Organochlorine pesticides	Water	0.4–0.27μg L^{-1}		340	GC-ECD	Yousefi et al. (2017)
MNPs–QuEChERS	Multi-pesticides	Blood	0.1–0.16μg L^{-1}	70–111		GC-MS	Yu et al. (2017)
Bimetallic MOF–MNPs	Organochlorine pesticides	Water	0.3–0.7 ng L^{-1}				Liu et al. (2017)
β-cyclodextrin–rGO–MNPs	Organochlorine pesticides	Honey	0.52–3.21μg L^{-1}	78–116	370	GC-μ ECD	Mahpishanian and Sereshti (2017)
3DG–aerogel–MNPs	OPPs	Water	1.2–5.6 ng L^{-1}	86–107		GC-NPD	Mahpishanian and Sereshti (2016)
MOF–MNPs	Pyrrole pesticides	Water	0.3–1.5μg L^{-1}	81–107		HPLC-DAD	Ma et al. (2016)
rGO–MNPs–Au NPs	Organochlorine pesticides	Water	0.4–4.1 ng L^{-1}	69–114		GC-ECD	Mehdinia et al. (2016)
ATP–MNPs–PANI	Benzoylurea insecticides	Water	0.02–0.43μg L^{-1}	77–103		HPLC-UV	Yang et al. (2016)
MNPs–EDTA–Zr	OPPs	Water	0.1–10 ng L^{-1}	86–112		GC-MS	Jiang et al. (2016)
Oleate–MNPs	Polychlorinated biphenyls	Fruit juice	3–6 ng L^{-1}	70		GC-MS	Pérez et al. (2016)
Graphene–MNPs–SiO_2	Chlorinated pesticides	Water	0.12–0.28 pg mL^{-1}	80–106	1000	GC-μ ECD	Rashidi Nodeh et al. (2015)

Graphene–MNPs	Carbamate pesticides (metolcarb and methiocarb)	Tomatoes	0.58–2.6 ng g^{-1}	90–101	434	HPLC-UV	Li et al. (2015)
Octylate–MNPs	OPPS (ethoprophos, fenchlorphos, methyl parathion, chlorpyrifos, prothiofos, and methyl azinphos)	Water	0.02–0.1μg L^{-1}	>82	143	GC-MS	Soon and Tay (2015)
QuEChERS–MNPs	11 pesticides	Juice of fruit and vegetable	2–49.9 ng g^{-1}	70–114		GC-MS	Zheng et al. (2015)
Poly IL–MNPs	OPPs (parathion, fenthion, phoxim, and temephos)	Tea drink	0.01μg L^{-1}	81–112	161	HPLC	Zheng et al. (2014)
QuEChERS–MNPs	101 multi-pesticides	Fruits		71–111		GC-MS	Li et al. (2014)
Fatty acid–MNPs	OPPs, organochlorine, fungicides	Water	20–30 ng L^{-}		1000	GC-MS	Tavakoli et al. (2014)
SiO_2–MNPs	OPPs (chlorpyrifos, chlorfenvinphos)	Water	50–100 ng L^{-}	94–97		Capillary LC-DAD	Moliner-Martinez et al. (2014)
Chitosan–MNPs	OPPs	Fruit	0.3–3.59 ng g^{-1}	79–98	364	GC	Tang et al. (2014)
MWCNTs–MNPs	Pesticides	Tea	0.02–0.08 mg kg^{-1}	72–109		GC-MS	Deng et al. (2014)
MNPs–chitosan	Pyrethroids	Water	5.5 pg mL^{-1}			HPLC	Tong and Chen (2013)
Graphene–MNPs	Chloroacetanilide herbicides	Water	0.02–0.05 ng mL^{-1}	80.7–105.3	649–1078	GC-ECD	Li et al. (2013b)
Graphene–MNPs	Herbicides	Green tea and water	0.01–0.03 ng mL^{-1}	80.2–105.3	3399–4002	GC-FID	Bai et al. (2013)

(continued)

Table 11.1 (continued)

MSPE sorbent	Pesticides	Sample matrix	LOD	% R	EF	Detector	Reference
MNPs @SiO_2–graphene	Neonicotinoid pesticides	Pear and tomato	0.08–0.15 ng g^{-1}	93.1–107.4	160–195	HPLC-DAD	Ma et al. (2013)
C18–MNPs @SiO_2	OPPs	Water	1.8–5.0 ng g^{-1}	–	–	GC-MS	Xie et al. (2013)
MNPs @DDAC@SiO_2	Herbicides	Water	0.078–0.10 ng mL^{-1}	80.4–107.1	1200–1410	HPLC-UV	He et al. (2012)
MNPs/PSDVB	Fenitrothion	Biology	0.5 ng mL^{-1}	97.2–100.0	–	UV-Vis	Eskandari and Naderi-Darehshori (2012)
MNPs @poly styrene	Pyrethroid pesticides	Water	0.01–0.02 ng mL^{-1}	78.0–96.5	500	UFLC-UV	Yu et al. (2012)
Graphene–MNPs	Neonicotinoid insecticides	Water	0.004–0.01 ng mL^{-1}	86–110	3325	UFLC-UV	Wang et al. (2012)
MNPs@SiO_2–C18	Carbamates and OPPs	Water	1–8 ng mL^{-1}	70.2–110.2	1015	GC-MS	Xiong et al. (2012)
MNPs–graphene	Atrazine, prometon, prometryn, and propazine	Water	0.02 ng mL^{-1}			HPLC-UV	Zhao et al. (2011)
MNPs–graphene	Carbamates	Water	0.04 ng mL^{-1}		474–868	HPLC-DAD	Wu et al. (2011)
MNPs–C18	OPPs	Water	–	84–92%	–	GC-NDP	Shen et al. (2007)

MOF metal organic framework, *BWO* $BeW_{12}O_{40}$, *3DG* three-dimensional graphene, *TMESPED* N-[3-(trimethoxysilyl)propyl]ethylenediamine, *TEOS-MTMOS* tetraethoxysilane-methyl trimethoxysilane, *DES* deep eutectic solvent, *PANI* polyaniline, *EDTA* ethylene diamine tetraacetic acid, *QuEChERS* Quick, Easy, Cheap, Effective, Rugged and Safe, *DDAC* dioctadecyl dimethyl ammonium chloride, *PSDVB* polystyrene divinyl benzene

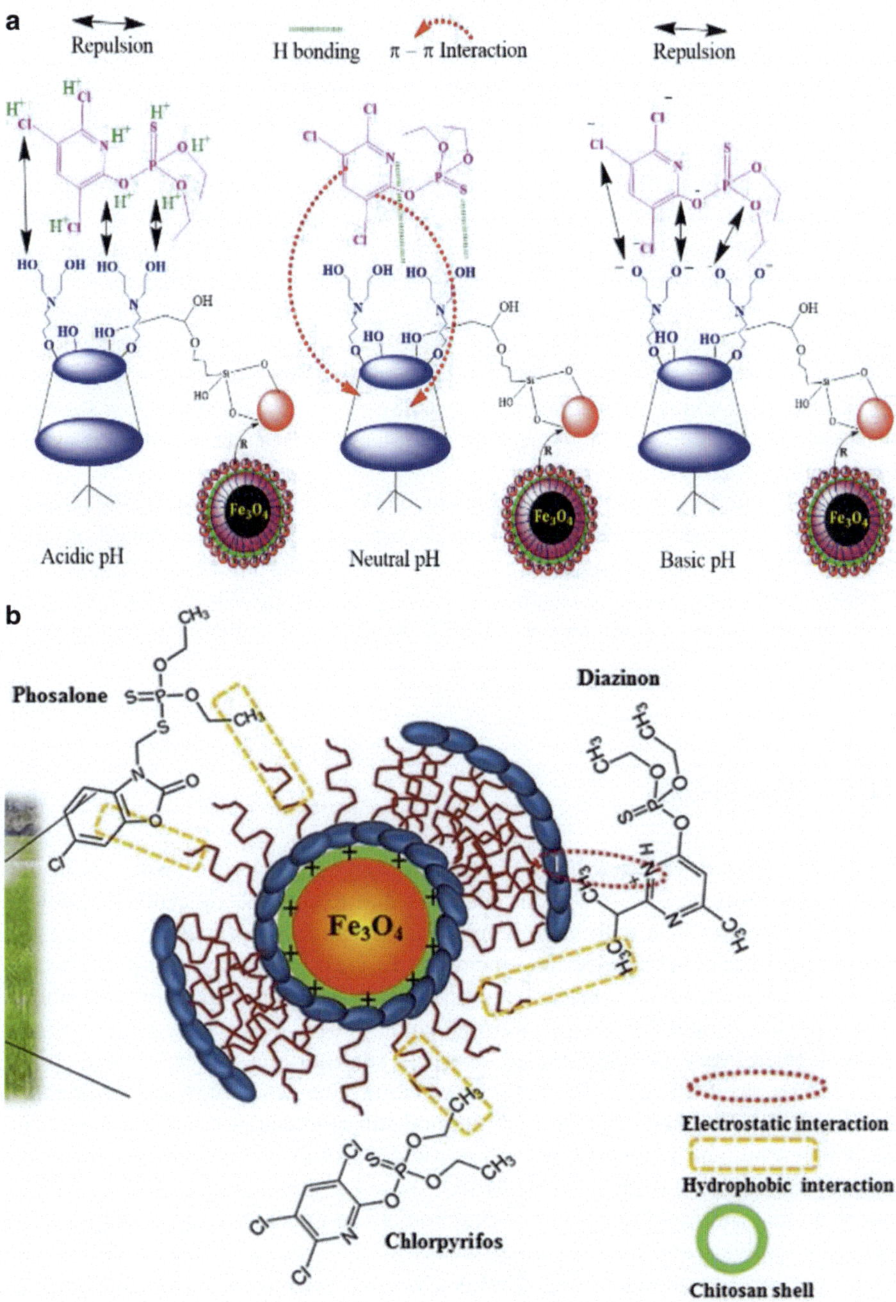

Fig. 11.8 (**a**) Magnetic calixarene ethylenediamine (Kamboh et al. 2016) and (**b**) OPP interaction with magnetic chitosan hemimicelle. (Ranjbar Bandforuzi and Hadjmohammadi 2019)

Magnetic C18 and C8 gained hydrophobic interaction with nonpolar pesticides, that is, chlorpyrifos (Log $K_{O/W}$ 4.7) and hexaconazole (Log $K_{O/W}$ 3.6). Anion exchange ability and π–π electron–donor/acceptor interaction mechanism between the pesticides and ionic liquids were identified (Zhang et al. 2014a, b; Faraji 2016; Fan et al. 2017; Amiri et al. 2018; Huang et al. 2019).

Molecular level studies are required for insight into the interactions between the different sorbents and the different pesticides. This is an interesting area of research to explore. Molecular docking study on the adsorption mechanisms of ketoconazole as guest analyte on magnetic graphene composite found out that π–π stacking should be the most important interaction (Wang et al. 2018b). In a similar study on triazole fungicides, Wang et al. (2019b) studied the adsorption mechanism interaction between tebuconazole as the analyte with magnetic graphene using experimental and molecular docking study. Molecular docking results indicated that π–π stacking between the benzene ring and the bulk π systems of graphene remained as the predominant interaction.

An understanding of the role of graphene oxide as sorbent for pesticide adsorption was studied using molecular docking (Wang et al. 2020). They found out that the pesticides kept on moving on top of the GO surface during the whole molecular docking simulation process through π–π stacking interaction. More docking studies with different kinds of pesticides and MMOF sorbents are an interesting research to be explored.

11.5 Conclusions

Solid-phase-based MNPs will continue to be one of the favorite sorbents in the analysis of pesticides from various matrices due to its ease of synthesis, tailor-made functionalization with various organic/inorganic compounds to achieve the desired qualities of selectivity and sensitivity. Endless magnetic solid-phase-based sorbents can be produced for analysis, adsorption, and removal of organic pollutions such as pesticides from the environment and food. Magnetic graphene-based sorbents were very popular due to its large surface area and π–π interactions afforded by the sorbents with analytes containing benzene rings and conjugated systems. MMOF-based sorbents exhibit good supramolecular recognition, enrichment potential, and recoveries toward pesticides from various matrices. Inexpensive, eco-friendly, and more stable MMOF-based sorbents in aqueous media and one that can offer rapid removal or extraction are an interesting area to explore for pesticide analysis and also for other analytes from various matrices. Studies at molecular level to elucidate important interactions between analyte/s and sorbent are an interesting exploration as there are not many studies focusing on this aspect. Recently introduced DES-based sorbents are also an area worth expanding for pesticide analysis.

References

Amiri A, Saadati-Moshtaghin HR, Zonoz FM (2018) A hybrid material composed of a polyoxometalate of type $BeW_{12}O_{40}$ and an ionic liquid immobilized onto magnetic nanoparticles as a sorbent for the extraction of organophosphorus pesticides prior to their determination by gas chromatography. Microchim Acta 185:176. https://doi.org/10.1007/s00604-018-2713-x

Amiri A, Baghayeri M, Vahdati-Nasab N (2020) Effective extraction of organophosphorus pesticides using sol–gel based coated stainless steel mesh as novel solid-phase extraction sorbent. J Chromatogr A 7:461020. https://doi.org/10.1016/j.chroma.2020.461020

Andrade-Eiroa A, Canle M, Leroy-Cancellieri V, Cerdà V (2016) Solid-phase extraction of organic compounds: a critical review. Part II. TrAC Trends Anal Chem 80:655–667. https://doi.org/10.1016/j.trac.2015.08.014

Asadi M, Sereshti H (2020) Magnetic amino-functionalized hollow silica-titania microsphere as an efficient sorbent for extraction of pesticides in green and roasted coffee beans. J Sep Sci 43:2115. https://doi.org/10.1002/jssc.201901135

Asadi M, Sereshti H, Rashidi Nodeh H (2020) Development of magnetic dispersive microsolid-phase extraction using lanthanum phosphate nanoparticles doped on magnetic graphene oxide as a highly selective adsorbent for pesticide residues analysis in water and fruit samples. Res Chem Intermed 14:1–5. https://doi.org/10.1007/s11164-020-04121-y

Atapattu SN, Johnson KR (2020) Pesticide analysis in cannabis products. J Chromatogr A 1612:460656. https://doi.org/10.1016/j.chroma.2019.460656

Badawy MEI, Marei AEM, El-Nouby MAM (2018) Preparation and characterization of chitosan-siloxane magnetic nanoparticles for the extraction of pesticides from water and determination by HPLC. Sep Sci Plus 1:506–519. https://doi.org/10.1002/sscp.201800084

Bai SS, Li Z, Zang XH, Wang C, Wang Z (2013) Graphene-based magnetic solid phase extraction-dispersive liquid-liquid microextraction combined with gas chromatographic method for determination of five acetanilide herbicides in water and green tea samples. Fenxi Huaxue/Chinese J Anal Chem 41:1177–1182. https://doi.org/10.1016/S1872-2040(13)60672-6

Bolygo E, Atreya NC (1991) Solid-phase extraction for multi-residue analysis of some triazole and pyrimidine pesticides in water. Fresenius J Anal Chem 339:423–430. https://doi.org/10.1007/Bf00322364

Cai Y, Jiang G, Liu J, Zhou Q (2003) Multiwalled carbon nanotubes as a solid-phase extraction adsorbent for the determination of bisphenol A, 4-n-nonylphenol, and 4-tert-octylphenol. Anal Chem 75:2517–2521. https://doi.org/10.1021/ac0263566

Cao X, Jiang Z, Wang S, Hong S, Li H, Shao Y, She Y, Wang J, Jin F, Jin M (2017) One-pot synthesis of magnetic zeolitic imidazolate framework/graphene oxide composites for the extraction of neonicotinoid insecticides from environmental water samples. J Sep Sci 40:4747–4756. https://doi.org/10.1002/jssc.201700674

Chahkandi M, Amiri A, Arami SRS (2019) Extraction and preconcentration of organophosphorus pesticides from water samples and fruit juices utilizing hydroxyapatite/Fe3O4 nanocomposite. Microchem J 144:261–269. https://doi.org/10.1016/j.microc.2018.09.018

Chatzimitakos TG, Pierson SA, Anderson JL, Stalikas CD (2018a) Enhanced magnetic ionic liquid-based dispersive liquid-liquid microextraction of triazines and sulfonamides through a one-pot, pH-modulated approach. J Chromatogr A 1571:47–54. https://doi.org/10.1016/j.chroma.2018.08.013

Chatzimitakos TG, Anderson JA, Stalikas CD (2018b) Matrix solid-phase dispersion based on magnetic ionic liquids: an alternative sample preparation approach for the extraction of pesticides from vegetables. J Chromatogr A 1581–1582:168–172. https://doi.org/10.1016/j.chroma.2018.11.008

Commission Regulation (EU) 2018/62 of 17 January 2018, replacing Annex I to Regulation (EC) No 396/2005 of the European Parliament and of the Council, UK

Covaci A, Schepens P (2001) Simplified method for determination of organochlorine pollutants in human serum by solid-phase disk extraction and gas chromatography. Chemosphere 43:439–447. https://doi.org/10.1016/S0045-6535(00)00392-1

Dai S, Zhao Y, Niu D, Li Q, Chen Y (2019) Preparation and reactivation of magnetic biochar by molten salt method: relevant performance for chlorine-containing pesticides abatement. J Air Waste Manage Assoc 69:58–70. https://doi.org/10.1080/10962247.2018.1510441

De Souza RM, Seibert D, Quesada HB, de Jesus Bassetti F, Fagundes-Klen MR, Bergamasco R (2020) Occurrence, impacts and general aspects of pesticides in surface water: a review. Process Saf Environ 135:22–37. https://doi.org/10.1016/j.psep.2019.12.035

De A, Bose R, Kumar A, Mozumdar S (2014) Worldwide pesticide use. In: Targeted delivery of pesticides using biodegradable polymeric nanoparticles. Springer India, New Delhi, pp 5–6

Deadman ML (2017) Sources of pesticide residues in food: toxicity, exposure, and risk associated with use at the farm level. In: Khan MS, Rahman MS (eds) Pesticide residue in foods. Springer International Publishing AG, Cham, pp 7–35

Deng X, Guo Q, Chen X, Xue T, Wang H, Yao P (2014) Rapid and effective sample clean-up based on magnetic multiwalled carbon nanotubes for the determination of pesticide residues in tea by gas chromatography–mass spectrometry. Food Chem 145:853–858. https://doi.org/10.1016/j.foodchem.2013.08.137

Do Lago AC, da Silva Cavalcanti MH, Rosa MA, Silveira AT, Tarley CR, Figueiredo EC (2020) Magnetic restricted-access carbon nanotubes for dispersive solid phase extraction of organophosphates pesticides from bovine milk samples. Anal Chim Acta 1102:11–23. https://doi.org/10.1016/j.aca.2019.12.039

Dong A, Lan S, Huang JF, Wang T, Zhao TY, Xiao LH, Wang WW, Zheng X, Liu FQ, Gao G, Chen YX (2011) Modifying Fe3O4-functionalized nanoparticles with N-halamine and their magnetic/antibacterial properties. ACS Appl Mater Interfaces 3:4228–4235. https://doi.org/10.1021/Am200864p

Eskandari H, Naderi-Darehshori A (2012) Preparation of magnetite/poly(styrene-divinylbenzene) nanoparticles for selective enrichment-determination of fenitrothion in environmental and biological samples. Anal Chim Acta 743:137–144. https://doi.org/10.1016/j.aca.2012.07.012

EU (2005) Regulation (EC) No 396/2005 of the European Parliament and of the Council of 23 February 2005 on maximum residue levels of pesticides in or on food and feed of plant and animal origin and amending Council Directive 91/414/EEC Text with EEA relevance. Off J Eur Union 70:1–16

Fan C, Liang Y, Dong H, Ding G, Zhang W, Tang G, Yang J, Kong D, Wang D, Cao Y (2017) In-situ ionic liquid dispersive liquid-liquid microextraction using a new anion-exchange reagent combined Fe_3O_4 magnetic nanoparticles for determination of pyrethroid pesticides in water samples. Anal Chim Acta 975:20–29. https://doi.org/10.1016/j.aca.2017.04.036

Fan J, Liu Z, Li J, Zhou W, Gao H, Zhang S, Lu R (2020) PEG-modified magnetic Schiff base network-1 materials for the magnetic solid phase extraction of benzoylurea pesticides from environmental water samples. J Chromatogr A 5:460950. https://doi.org/10.1016/j.chroma.2020.460950

Faraji M (2016) Recent analytical applications of magnetic nanoparticles. Nanochem Res 1:264–290. https://doi.org/10.7508/ncr.2016.02.014

Farajzadeh MA, Yadeghari A, Khoshmaram L (2018) Magnetic solid phase extraction using Fe_3O_4@SiO_2@C8 nanoparticles performed in a narrow-bore tube followed by dispersive liquid–liquid microextraction for extraction and preconcentration of nine pesticides. New J Chem 42:6215–6224. https://doi.org/10.1039/C8NJ00501J

Farajzadeh MA, Safi R, Yadeghari A (2019) Combination of QuEChERS extraction with magnetic solid phase extraction followed by dispersive liquid–liquid microextraction as an efficient procedure for the extraction of pesticides from vegetable, fruit, and nectar samples having high content of solids. Microchem J 147:571–581. https://doi.org/10.1016/j.microc.2019.03.074

Farajzadeh MA, Safi R, Yadeghari A (2020) Magnetic solid-phase extraction method for extraction of some pesticides in vegetable and fruit juices. J Sep Sci 43:1523. https://doi.org/10.1002/jssc.201900790

Feizbakhsh A, Ehteshami S (2016) Polythiophene chitosan magnetic nanocomposite as a novel sorbent for disperse magnetic solid phase extraction of triazine herbicides in aquatic media. Chromatographia 79:2921706. https://doi.org/10.1007/s10337-018-3588-z

Gao T, Wang J, Wu Q, Wang C, Wang ZA (2020) Graphene oxide–based composite for solid-phase extraction of carbamate pesticides from vegetables. Food Anal Method 13:690–698. https://doi.org/10.1007/s12161-019-01685-3

González-Curbelo MÁ, Herrera-Herrera AV, Hernández-Borges J, Rodríguez-Delgado MÁ (2013) Analysis of pesticides residues in environmental water samples using multiwalled carbon nanotubes dispersive solid-phase extraction. J Sep Sci 36:556–563. https://doi.org/10.1002/jssc.201200782

Gorji S, Biparva P, Bahram M, Nematzadeh G (2019) Stir bar sorptive extraction kit for determination of pesticides in water samples with chemometric data processing. Microchem J 148:313–321. https://doi.org/10.1016/j.microc.2019.04.056

Gou Y, Eisert R, Pawliszyn J (2000) Automated in-tube solid-phase microextraction–high-performance liquid chromatography for carbamate pesticide analysis. J Chromatogr A 873:137–147. https://doi.org/10.1016/s0021-9673(99)01125-5

Gunther CG (1909) Electro-magnetic ore separation. Hill Publishing Company, New York. Available (April 2020) at https://www.yumpu.com/en/document/read/6583830/electro-magnetic-ore-separation

He Z, Liu D, Li R, Zhou Z, Wang P (2012) Magnetic solid-phase extraction of sulfonylurea herbicides in environmental water samples by Fe_3O_4@dioctadecyl dimethyl ammonium chloride@silica magnetic particles. Anal Chim Acta 747:29–35. https://doi.org/10.1016/j.aca.2012.08.015

He Z, Wang P, Liu D, Zhou Z (2014) Hydrophilic–lipophilic balanced magnetic nanoparticles: preparation and application in magnetic solid-phase extraction of organochlorine pesticides and triazine herbicides in environmental water samples. Talanta 127:1–8. https://doi.org/10.1016/j.talanta.2014.03.074

Heidari H, Razmi H (2012) Multi-response optimization of magnetic solid phase extraction based on carbon coated Fe3O4 nanoparticles using desirability function approach for the determination of the organophosphorus pesticides in aquatic samples by HPLC–UV. Talanta 99:13–21. https://doi.org/10.1016/j.talanta.2012.04.023

Hou M, Zang X, Wang C, Wang Z (2013) The use of silica-coated magnetic graphene microspheres as the adsorbent for the extraction of pyrethroid pesticides from orange and lettuce samples followed by GC–MS analysis. J Sep Sci 36:3242–3248. https://doi.org/10.1002/jssc.201300656

Huang C, Xie W, Li X, Zhang J (2011) Speciation of inorganic arsenic in environmental waters using magnetic solid phase extraction and preconcentration followed by ICP-MS. Microchim Acta 173:165–172. https://doi.org/10.1016/j.chroma.2012.01.022

Huang X, Liu G, Xu D, Xu X, Li L, Zheng S, Lin H, Gao H (2018) Novel zeolitic imidazolate frameworks based on magnetic multiwalled carbon nanotubes for magnetic solid-phase extraction of organochlorine pesticides from agricultural irrigation water samples. Appl Sci 8:959. https://doi.org/10.3390/app8060959

Huang X, Qiao K, Li L, Liu G, Xu X, Lu R, Gao H, Xu D (2019) Preparation of a magnetic graphene/polydopamine nanocomposite for magnetic dispersive solid-phase extraction of benzoylurea insecticides in environmental water samples. Sci Rep 9:8919. https://doi.org/10.1038/s41598-019-45186-z

Huo S, Yan X (2012) Facile magnetization of metal–organic framework MIL-101 for magnetic solid-phase extraction of polycyclic aromatic hydrocarbons in environmental water samples. Analyst 137:3445–3451. https://doi.org/10.1039/C2AN35429B

Ibarra IS, Miranda JM, Rodriguez JA, Nebot C, Cepeda A (2014) Magnetic solid phase extraction followed by high-performance liquid chromatography for the determination of sulphonamides in milk samples. Food Chem 157:511–517. https://doi.org/10.1016/j.foodchem.2014.02.069

Jiang C, Sun Y, Yu X, Gao Y, Zhang L, Wang Y, Zhang H, Song D (2013) Liquid–solid extraction coupled with magnetic solid-phase extraction for determination of pyrethroid residues in vegetable samples by ultra fast liquid chromatography. Talanta 114:167–175. https://doi.org/10.1016/j.talanta.2013.04.004

Jiang L, Huang T, Feng S, Wang J (2016) Zirconium(IV) functionalized magnetic nanocomposites for extraction of organophosphorus pesticides from environmental water samples. J Chromatogr A 1456:49–57. https://doi.org/10.1016/j.chroma.2016.06.005

Jiao Z, Zhang Y, Fan H (2016) Ultrasonic-microwave method in preparation of polypyrrole-coated magnetic particles for vitamin D extraction in milk. J Chromatogr A 1457:7–13. https://doi.org/10.1016/j.chroma.2016.06.041

John EM, Varghese EM, Krishnasree N, Jisha MS (2018) In situ bioremediation of chlorpyrifos by Klebsiella sp. isolated from pesticide contaminated agricultural soil. Int J Curr Microbiol Appl Sci 7:1418–1429. https://doi.org/10.20546/ijcmas.2018.703.170

Jost U, Habedank F (2020) Two-dimensional hydrophilic interaction and reversed phase liquid chromatography easily extracted pesticides and polar pesticides multi-residue method –a concept. J Chromatogr A 13:461040. https://doi.org/10.1016/j.chroma.2020.461040

Kamboh MA, Wan Ibrahim WA, Rashidi Nodeh H, Sanagi MM, Sherazi STH (2016) The removal of organophosphorus pesticides from water using a new amino-substituted calixarene-based magnetic sporopollenin. New J Chem 40:3130–3138. https://doi.org/10.1039/C5NJ02284C

Kermani M, Jafari MT, Saraji M (2019) Porous magnetized carbon sheet nanocomposites for dispersive solid-phase microextraction of organophosphorus pesticides prior to analysis by gas chromatography-ion mobility spectrometry. Microchim Acta 186:88–98. https://doi.org/10.1007/s00604-018-3215-6

Kermani M, Sereshti H, Nikfarjam N (2020) Application of magnetic nanocomposite of cross-linked poly(styrene/divinylbenzene) as adsorbent for magnetic dispersive solid phase extraction-dispersive liquid-liquid microextraction of atrazine in soil and aqueous samples. Anal Method 12:1834. https://doi.org/10.1039/D0AY00374C

Khanjani N, Jafarnejad A-B, Tavakkoli L (2017) Arsenic and breast cancer: a systematic review of epidemiologic studies. Rev Environ Health 32:267–277. https://doi.org/10.1515/reveh-2016-0068

Kim K-H, Kabir E, Jahan SA (2017) Exposure to pesticides and the associated human health effects. Sci Total Environ 575:525–535. https://doi.org/10.1016/j.scitotenv.2016.09.009

Kumar V, Kumar P (2019) Pesticides in agriculture and environment: impacts on human health. In: Contaminants in agriculture and environment: health risks and remediation. Agro Environ Media – Agriculture and Environmental Science Academy, Haridwar, pp 76–95

Lakshmanan R, Okoli C, Boutonnet M, Järås S, Rajarao GK (2013) Effect of magnetic iron oxide nanoparticles in surface water treatment: trace minerals and microbes. Bioresour Technol 129:612–615. https://doi.org/10.1016/j.biortech.2012.12.138

Lemasson E, Hennig P, Bertin S, Lesellier E, Caroline W (2017) Mixed-mode chromatography-a review. LC-GC 30:22–33

Li C, Chen L, Li W (2013a) Magnetic titanium oxide nanoparticles for hemimicelle extraction and HPLC determination of organophosphorus pesticides in environmental water. Microchim Acta 180:1–8. https://doi.org/10.1007/s00604-013-1029-0

Li Z, Bai S, Hou M, Wang C, Wang Z (2013b) Magnetic graphene nanoparticles for the preconcentration of chloroacetanilide herbicides from water samples prior to determination by GC-ECD. Anal Lett 46:1012–1024. https://doi.org/10.1080/0032719.2012.745086

Li Z, Hou M, Bai S, Wang C, Wang Z (2013c) Extraction of imide fungicides in water and juice samples using magnetic graphene nanoparticles as adsorbent followed by their determination with gas chromatography and electron capture detection. Anal Sci 29:325–331. https://doi.org/10.2116/analsci.29.325

Li Y, Qiao L, Li F, Ding Y, Yang Z, Wang M (2014) Determination of multiple pesticides in fruits and vegetables using a modified quick, easy, cheap, effective, rugged and safe method with magnetic nanoparticles and gas chromatography tandem mass spectrometry. J Chromatogr A 1361:77–87. https://doi.org/10.1016/j.chroma.2014.08.011

Li N, Chen J, Shi Y (2015) Magnetic graphene solid-phase extraction for the determination of carbamate pesticides in tomatoes coupled with high performance liquid chromatography. Talanta 141:212–219. https://doi.org/10.1016/j.talanta.2015.04.018

Li D, He M, Chen B, Hu B (2019) Metal organic frameworks-derived magnetic nanoporous carbon for preconcentration of organophosphorus pesticides from fruit samples followed by gas chromatography-flame photometric detection. J Chromatogr A 1583:19–27. https://doi.org/10.1016/j.chroma.2018.11.012

Liang L, Wang X, Sun Y, Ma P, Li X, Piao H, Jiang Y, Song D (2018) Magnetic solid-phase extraction of triazine herbicides from rice using metal-organic framework MIL-101 (Cr) functionalized magnetic particles. Talanta 179:512–519. https://doi.org/10.1016/j.talanta.2017.11.017

Liang R, Peng Y, Hu Y, Li G (2019) A hybrid triazine-imine core-shell magnetic covalent organic polymer for analysis of pesticides in fruit samples by ultra high performance liquid chromatography with tandem mass spectrometry. J Sep Sci 42:1432–1439. https://doi.org/10.1002/jssc.201801299

Lin X, Wang X, Wang J, Yuan Y, Di S, Wang Z, Xu H, Zhao H, Qi P, Ding W (2020) Facile synthesis of a core-shell structured magnetic covalent organic framework for enrichment of organophosphorus pesticides in fruits. Anal Chim Acta 1101:65–73. https://doi.org/10.1016/j.aca.2019.12.012

Liu Y, Gao Z, Wu R, Wang Z, Chen X, Chan T-WD (2017) Magnetic porous carbon derived from a bimetallic metal–organic framework for magnetic solid-phase extraction of organochlorine pesticides from drinking and environmental water samples. J Chromatogr A 1479:55–61. https://doi.org/10.1016/j.chroma.2016.12.014

Liu F, Yang X, Wu X, Xi X, Gao H, Zhang S, Zhou W, Lu R (2018a) A dispersive magnetic solid phase microextraction based on ionic liquid coated and cyclodextrin-functionalized magnetic core dendrimer nanocomposites for the determination of pyrethroids in juice samples. Food Chem 268:485–491. https://doi.org/10.1016/j.foodchem.2018.06.105

Liu G, Li L, Huang X, Zheng S, Xu D, Xu X, Zhang Y, Lin H (2018b) Determination of triazole pesticides in aqueous solution based on magnetic graphene oxide functionalized MOF-199 as solid phase extraction sorbents. Microporous Mesoporous Mater 270:258–264. https://doi.org/10.1016/j.micromeso.2018.05.023

Liu G, Li L, Huang X, Zheng S, Xu X, Liu Z, Zhang Y, Wang J, Lin H, Xu D (2018c) Adsorption and removal of organophosphorus pesticides from environmental water and soil samples by using magnetic multi-walled carbon nanotubes@organic framework ZIF-8. J Mater Sci 53:10772–10783. https://doi.org/10.1007/s10853-018-2352-y

Liu C, Liu X, Marriott PJ, Qian H, Meng Z, Yang Z, Lu R, Gao H, Zhou W (2019a) Magnetic solid-phase extraction of benzoylurea insecticides in tea samples with Fe_3O_4 -hyperbranched polyester magnetic composite as sorbent. J Sep Sci 42:1610–1619. https://doi.org/10.1002/jssc.201801159

Liu G, Huang X, Lu M, Li L, Li T, Xu D (2019b) Facile synthesis of magnetic zinc metal-organic framework for extraction of nitrogen-containing heterocyclic fungicides from lettuce vegetable samples. J Sep Sci 42:1451–1458. https://doi.org/10.1002/jssc.201801169

Liu Z, Lin X, Liu X, Li J, Zhou W, Gao H, Zhang S, Lu R (2019c) Magnetic nanoparticles modified with hyperbranched polyamidoamine for the extraction of benzoylurea insecticides prior to their quantitation by HPLC. Microchim Acta 186:351–361. https://doi.org/10.1007/s00604-019-3450-5

Liu Z, Qi P, Wang J, Wang Z, Di S, Xu H, Zhao H, Wang Q, Wang X, Wang X (2020) Development, validation, comparison, and implementation of a highly efficient and effective method using magnetic solid-phase extraction with hydrophilic-lipophilic-balanced materials for LC-MS/MS analysis of pesticides in seawater. Sci Total Environ 708:135221. https://doi.org/10.1016/j.scitotenv.2019.135221

López KA, Piña MN, Morey J (2013) Squaramide-coated Fe_3O_4 nanoparticles and their selective complexation with carboxylate anions in water. Sensors Actuators B Chem 181:267–273. https://doi.org/10.1016/j.snb.2013.01.069

Ma X, Wang J, Sun M, Wang W, Wu Q, Wang C, Wang Z (2013) Magnetic solid-phase extraction of neonicotinoid pesticides from pear and tomato samples using graphene grafted silica-coated Fe_3O_4 as the magnetic adsorbent. Anal Methods 5:2809–2815. https://doi.org/10.1039/C3AY40207J

Ma J, Yao Z, Hou L, Lu W, Yang Q, Li J, Chen L (2016) Metal organic frameworks (MOFs) for magnetic solid-phase extraction of pyrazole/pyrrole pesticides in environmental water samples followed by HPLC-DAD determination. Talanta 161:686–692. https://doi.org/10.1016/j.talanta.2016.09.035

Ma J, Wu G, Li S, Tan W, Wang X, Li J, Chen L (2018) Magnetic solid-phase extraction of heterocyclic pesticides in environmental water samples using metal-organic frameworks coupled to high performance liquid chromatography determination. J Chromatogr A 1553:57–66. https://doi.org/10.1016/j.chroma.2018.04.034

Maddah B, Shamsi J (2012) Extraction and preconcentration of trace amounts of diazinon and fenitrothion from environmental water by magnetite octadecylsilane nanoparticles. J Chromatogr A 1256:40–45. https://doi.org/10.1016/j.chroma.2012.07.085

Madej K, Jonda A, Borcuch A, Piekoszewski W, Chmielarz L, Gil B (2019) A novel stir bar sorptive-dispersive microextraction in combination with magnetically modified graphene for isolation of seven pesticides from water samples. Microchem J 147:962–971. https://doi.org/10.1016/j.microc.2019.04.002

Mahpishanian S, Sereshti H (2016) Three-dimensional graphene aerogel-supported iron oxide nanoparticles as an efficient adsorbent for magnetic solid phase extraction of organophosphorus pesticide residues in fruit juices followed by gas chromatographic determination. J Chromatogr A 1443:43–53. https://doi.org/10.1016/j.chroma.2016.03.046

Mahpishanian S, Sereshti H (2017) One-step green synthesis of β-cyclodextrin/iron oxide-reduced graphene oxide nanocomposite with high supramolecular recognition capability: application for vortex-assisted magnetic solid phase extraction of organochlorine pesticides residue from honey samples. J Chromatogr A 1485:32–43. https://doi.org/10.1016/j.chroma.2017.01.035

Mahpishanian S, Sereshti H, Baghdadi M (2015) Superparamagnetic core-shells anchored onto graphene oxide grafted with phenylethyl amine as a nano-adsorbent for extraction and enrichment of organophosphorus pesticides from fruit, vegetable and water samples. J Chromatogr A 1406:48–58. https://doi.org/10.1016/j.chroma.2015.06.025

Malek SK, Rashidi Nodeh H, Akbari-Adergani B (2018) Silica-based magnetic hybrid nanocomposite for the extraction and preconcentration of some organophosphorus pesticides before gas chromatography. J Sep Sci 41:2934–2941. https://doi.org/10.1002/jssc.201800090

Mandal A, Singh N, Purakayastha TJ (2017) Characterization of pesticide sorption behaviour of slow pyrolysis biochars as low cost adsorbent for atrazine and imidacloprid removal. Sci Total Environ 577:376–385. https://doi.org/10.1016/j.scitotenv.2016.10.204

Marsin FM, Wan Ibrahim WA, Rashidi Nodeh H, Sanagi MM (2020) New magnetic oil palm fiber activated carbon-reinforced polypyrrole solid phase extraction combined with gas chromatography-electron capture detection for determination of organochlorine pesticides in water samples. J Chromatogr A 1612:460638. https://doi.org/10.1016/j.chroma.2019.460638

Mehdinia A, Roohi F, Jabbari A (2011) Rapid magnetic solid phase extraction with in situ derivatization of methylmercury in seawater by Fe_3O_4/polyaniline nanoparticle. J Chromatogr A 1218:4269–4274. https://doi.org/10.1016/j.chroma.2011.04.070

Mehdinia A, Rouhani S, Mozaffari S (2016) Microwave-assisted synthesis of reduced graphene oxide decorated with magnetite and gold nanoparticles, and its application to solid-phase extraction of organochlorine pesticides. Microchim Acta 183:1177–1185. https://doi.org/10.1007/s00604-015-1691-5

Meseguer-Lloret S, Torres-Cartas S, Catalá-Icardo M, Simó-Alfonso EF, Herrero-Martínez JM (2017) Extraction and preconcentration of organophosphorus pesticides in water by using a polymethacrylate-based sorbent modified with magnetic nanoparticles. Anal Bioanal Chem 409:3561–3571. https://doi.org/10.1007/s00216-017-0294-x

Mohd Hassan A, Wan Ibrahim WA, Sanagi MM, Sutirman ZA, Rashidi Nodeh H, Mokhter MA (2020) New effective 3-aminopropyltrimethoxysilane functionalized magnetic sporopollenin-based silica coated graphene oxide adsorbent for removal of Pb(II) from aqueous environment. J Environ Manag 253:109658. https://doi.org/10.1016/j.jenvman.2019.109658

Moliner-Martinez Y, Vitta Y, Prima-Garcia H, González-Fuenzalida RA, Ribera A, Campíns-Falcó P, Coronado E (2014) Silica supported Fe_3O_4 magnetic nanoparticles for magnetic solid-phase extraction and magnetic in-tube solid-phase microextraction: application to organophosphorous compounds. Anal Bioanal Chem 406:2211–2215. https://doi.org/10.1007/s00216-013-7379-y

Nacham O, Clark KD, Anderson JL (2016) Synthesis and characterization of the physicochemical and magnetic properties for perfluoroalkyl ester and Fe(III) carboxylate-based hydrophobic magnetic ionic liquids. RSC Adv 6:11109–11117. https://doi.org/10.1039/C5RA25002A

Nikzad S, Amooey AA, Alinejad-Mir A (2019) Adsorption of diazinon from aqueous solutions by magnetic guar gum-montmorillonite. Chem Data Collect 20:100187. https://doi.org/10.1016/j.cdc.2019.100187

Niu M, Li Z, He W, Zhou W, Lu R, Li J, Gao H, Zhang S, Pan C (2020) Attapulgite modified magnetic metal-organic frameworks for magnetic solid phase extraction and determinations of benzoylurea insecticides in tea infusions. Food Chem 317:126425. https://doi.org/10.1016/j.foodchem.2020.126425

Oellig C, Schmid S (2019) Polyethyleneimine as weak anionic exchanger adsorbent for clean-up in pesticide residue analysis of fruits and vegetables. J Chromatogr A 1597:9–17. https://doi.org/10.1016/j.chroma.2019.03.020

Özer ET, Osman B, Parlak B (2020) An experimental design approach for the solid phase extraction of some organophosphorus pesticides from water samples with polymeric microbeads. Microchem J 154:104537–1045cc. https://doi.org/10.1016/j.microc.2019.104537

Pankhurst QA, Connolly J, Jones SK, Dobson J (2003) Applications of magnetic nanoparticles in biomedicine. J Phys D Appl Phys 36:R167–R181. https://doi.org/10.1088/0022-3727/36/13/201

Pérez RA, Albero B, Tadeo JL, Sánchez-Brunete C (2016) Oleate functionalized magnetic nanoparticles as sorbent for the analysis of polychlorinated biphenyls in juices. Microchim Acta 183:157–165. https://doi.org/10.1007/s00604-015-1617-2

Qin S, Yin H, Yang C, Dou Y, Liu Z, Zhang P, Yu H, Huang Y, Feng J, Hao J, Hao J, Deng L, Yan X, Dong X, Zhao Z, Jiang T, Wang H-W, Luo S-J, Xie C (2016) A magnetic protein biocompass. Nat Mater 15:217–226. https://doi.org/10.1038/nmat4484

Rajan M, Chandran V, Shahena S, Mathew L (2020) Controlled release pesticides as a route to sustainable crop production. In: Rakhimol KR, Thomas S, Volova T, Jayachandran K (eds) Controlled release of pesticides for sustainable agriculture. Springer, Cham, pp 111–125. https://doi.org/10.1007/978-3-030-23396-9_4

Ranjbar Bandforuzi S, Hadjmohammadi MR (2019) Modified magnetic chitosan nanoparticles based on mixed hemimicelle of sodium dodecyl sulfate for enhanced removal and trace determination of three organophosphorus pesticides from natural waters. Anal Chim Acta 1078:90–100. https://doi.org/10.1016/j.aca.2019.06.026

Rashidi Nodeh H (2015) Synthesis and applications of functionalized graphene-based magnetic nanoparticles as adsorbent for pesticides and arsenic preconcentration. Universiti Teknologi Malaysia (Thesis)

Rashidi Nodeh H, Wan Ibrahim WA, Kamboh MA, Sanagi MM (2015) Dispersive graphene-based silica coated magnetic nanoparticles as a new adsorbent for preconcentration of chlorinated pesticides from environmental water. RSC Adv 5:76424–76434. https://doi.org/10.1039/C5RA13450A

Rashidi Nodeh H, Wan Ibrahim WA, Sanagi MM, Aboul-Enein HY (2016) Magnetic graphene-based cyanopropyltriethoxysilane as an adsorbent for simultaneous determination of polar and non-polar organophosphorus pesticides in cow's milk. RSC Adv 6:24853–24864. https://doi.org/10.1039/c5ra26742k

Rashidi Nodeh H, Sereshti H, Gaikani H, Kamboh MA, Afsharsaveh Z (2017a) Magnetic graphene coated inorganic-organic hybrid nanocomposite for enhanced preconcentration of selected pesticides in tomato and grape. J Chromatogr A 1509:26–34. https://doi.org/10.1016/j.chroma.2017.06.032

Rashidi Nodeh H, Wan Ibrahim WA, Kamboh MA, Sanagi MM (2017b) New magnetic graphene-based inorganic–organic sol-gel hybrid nanocomposite for simultaneous analysis of polar and non-polar organophosphorus pesticides from water samples using solid-phase extraction. Chemosphere 166:21–30. https://doi.org/10.1016/j.chemosphere.2016.09.054

Šafaříková M, Šafařík I (1999) Magnetic solid-phase extraction. J Magn Magn Mater 194:108–112. https://doi.org/10.1016/S0304-8853(98)00566-6

Sereshti H, Afsharsaveh Z, Gaikani H, Rashidi Nodeh H (2018) Electroless-coated magnetic three-dimensional graphene with silver nanoparticles used for the determination of pesticides in fruit samples. J Sep Sci 41:1567. https://doi.org/10.1002/jssc.201700956

Sereshti H, Jamshidi F, Nouri N, Rashidi Nodeh H (2020a) Hyphenated dispersive solid-and liquid-phase microextraction technique based on a hydrophobic deep eutectic solvent: application for trace analysis of pesticides in fruit juices. J Sci Food Agr 100:2534–2543. https://doi.org/10.1002/jsfa.10279

Sereshti H, Toloutehrani A, Rashidi Nodeh H (2020b) Determination of cholecalciferol (vitamin D3) in bovine milk by dispersive micro-solid phase extraction based on the magnetic three-dimensional graphene-sporopollenin sorbent. J Chromatogr B 1136:121907. https://doi.org/10.1016/j.jchromb.2019.121907

Shahrebabak SM, Saber-Tehrani M, Faraji M, Shabanian M, Aberoomand-Azar P (2019) Simultaneous magnetic solid phase extraction of acidic and basic pesticides using triazine-based polymeric network modified magnetic nanoparticles/graphene oxide nanocomposite in water and food samples. Microchem J 146:630–639. https://doi.org/10.1016/j.microc.2019.01.047

Shen H, Zhu Y, Wen X, Zhuang Y (2007) Preparation of Fe_3O_4-C18 nano-magnetic composite materials and their cleanup properties for organophosphorous pesticides. Anal Bioanal Chem 387:2227–2237. https://doi.org/10.1007/s00216-006-1082-1

Shrestha S, Parks CG, Goldner WS, Kamel F, Umbach DM, Ward MH, Lerro CC, Koutros S, Hofmann JN, Beane Freeman LE, Sandler DP (2018) Pesticide use and incident hypothyroidism in pesticide applicators in the agricultural health study. Environ Health Perspect 126:097008. https://doi.org/10.1289/EHP3194

Shrivas K, Patel S, Maji P, Sinha D (2019) Application of magnetic nanoparticles for removal of pesticides from environmental samples prior to instrumental analysis. In: Magnetic nanostructures. Springer, Berlin, pp 247–260

Soon YX, Tay KS (2015) N-octylated magnetic nanoparticle-based microextraction for the determination of organophosphorus pesticides in water. Anal Lett 48:1604–1618. https://doi.org/10.1080/00032719.2014.991964

Soutoudehnia Korrani Z, Wan Ibrahim WA, Rashidi Nodeh H, Aboul-Enein HY, Sanagi MM (2016) Simultaneous preconcentration of polar and non-polar organophosphorus pesticides from water samples by using a new sorbent based on mesoporous silica. J Sep Sci 39:1144–1151. https://doi.org/10.1002/jssc.201500896

Tang Q, Wang X, Yu F, Qiao X, Xu Z (2014) Simultaneous determination of ten organophosphate pesticide residues in fruits by gas chromatography coupled with magnetic separation. J Sep Sci 37:820–827. https://doi.org/10.1002/jssc.201301161

Targhoo A, Amiri A, Baghayeri M (2018) Magnetic nanoparticles coated with poly (p-phenylenediamine-co-thiophene) as a sorbent for preconcentration of organophosphorus pesticides. Microchim Acta 185:15–20. https://doi.org/10.1007/s00604-017-2560-1

Tavakoli M, Hajimahmoodi M, Shemirani F (2014) Trace level monitoring of pesticides in water samples using fatty acid coated magnetic nanoparticles prior to GC-MS. Anal Methods 6:2988–2997. https://doi.org/10.1039/c3ay41915k

Tong J, Chen L (2013) Determination of pyrethroids in environmental waters using magnetic chitosan extraction coupled with high performance liquid chromatography detection. Anal Lett 46:1183–1197. https://doi.org/10.1080/00032719.2012.755687

Turan NB, TuğbaZaman B, Bakırdere S (2020) Application of oleic acid functionalized magnetic nanoparticles for a highly sensitive and efficient dispersive magnetic solid phase extraction of fenazaquin in almond samples for determination by gas chromatography mass spectrometry. Microchem J 153:104329. https://doi.org/10.1016/j.microc.2019.104329

Vaghari H, Jafarizadeh-Malmiri H, Mohammadlou M, Berenjian A, Anarjan N, Jafari N, Nasiri S (2016) Application of magnetic nanoparticles in smart enzyme immobilization. Biotechnol Lett 38:223–233. https://doi.org/10.1007/s10529-015-1977-z

Wan Ibrahim WA, Veloo KV, Sanagi MM (2012) Novel sol–gel hybrid methyltrimethoxysilane–tetraethoxysilane as solid phase extraction sorbent for organophosphorus pesticides. J Chromatogr A 1229:55–62. https://doi.org/10.1016/j.chroma.2012.01.022

Wan Ibrahim WA, Rashidi Nodeh R, Aboul-Enein HY, Sanagi MM (2015a) Magnetic solid-phase extraction based on modified ferum oxides for enrichment, preconcentration, and isolation of pesticides and selected pollutants. Crit Rev Anal Chem 45:270–287. https://doi.org/10.1080/10408347.2014.938148

Wan Ibrahim WA, Rashidi Nodeh H, Sanagi MM (2015b) Graphene-based materials as solid phase extraction sorbent for trace metal ions, organic compounds and biological sample preparation. Crit Rev Anal Chem 46:267–283. https://doi.org/10.1080/10408347.2015.1034354

Wang W, Li Y, Wu Q, Wang C, Zang X, Wang Z (2012) Extraction of neonicotinoid insecticides from environmental water samples with magnetic graphene nanoparticles as adsorbent followed by determination with HPLC. Anal Methods 4:766–772. https://doi.org/10.1039/C2AY05734D

Wang Y, Sun Y, Gao Y, Xu B, Wu Q, Zhang H, Song D (2014) Determination of five pyrethroids in tea drinks by dispersive solid phase extraction with polyaniline-coated magnetic particles. Talanta 119:268–275. https://doi.org/10.1016/j.talanta.2013.11.007

Wang Y, Sun Y, Xu B, Li X, Wang X, Zhang H, Song D (2015) Matrix solid-phase dispersion coupled with magnetic ionic liquid dispersive liquid–liquid microextraction for the determination of triazine herbicides in oilseeds. Anal Chim Acta 888:67–74. https://doi.org/10.1016/j.aca.2015.07.028

Wang P, Luo M, Liu D, Zhan J, Liu X, Wang F, Zhou Z, Wang P (2018a) Application of a magnetic graphene nanocomposite for organophosphorus pesticide extraction in environmental water samples. J Chromatogr A 1535:9–16. https://doi.org/10.1016/j.chroma.2018.01.003

Wang Z, Zhao P, Yu J, Jiang Z, Guo X (2018b) Experimental and molecular docking study on graphene/Fe_3O_4 composites as a sorbent for magnetic solid-phase extraction of seven imidazole antifungals in environmental water samples prior to LC-MS/MS for enantiomeric analysis. Microchem J 140:222–231. https://doi.org/10.1016/j.microc.2018.04.027

Wang D, Zhao Y, Ou yang M, Guo H, Yang Z (2019a) Magnetic polydopamine modified with deep eutectic solvent for the magnetic solid-phase extraction of sulfonylurea herbicides in water samples. J Chromatogr A 1601:53–59. https://doi.org/10.1016/j.chroma.2019.05.011

Wang Z, Zhang J, Hu B, Yu B, Wang J, Gui X (2019b) Graphene/Fe_3O_4 nanocomposite for effective removal of ten triazole fungicides from water solution: tebuconazole as an example for investigation of the adsorption mechanism by experimental and molecular docking study. J Taiwan Inst Chem Eng 95:635–642. https://doi.org/10.1016/j.jtice.2018.09.026

Wang H, Hua B, Gao Z, Zhang Z, Wang J (2020) Emerging role of graphene oxide as sorbent for pesticides adsorption: experimental observations analyzed by molecular modeling. J Mater Sci Technol. https://doi.org/10.1016/j.jmst.2020.02.033

Wanjeri VWO, Sheppard CJ, Prinsloo ARE, Ngila JC, Ndungu PG (2018) Isotherm and kinetic investigations on the adsorption of organophosphorus pesticides on graphene oxide based silica coated magnetic nanoparticles functionalized with 2-phenyl-ethylamine. J Environ Chem Eng 6:1333–1346. https://doi.org/10.1016/j.ece.2018.01.064

Wei Z, Li H, Wu J, Dong Y, Zhang H, Chen H, Ren C (2020) 3DRGO-$NiFe_2O_4$/NiO nanoparticles for fast and simple detection of organophosphorus pesticides. Chinese Chem Lett 31:177–180. https://doi.org/10.1016/j.cclet.2019.05.031

Wen Y (2020) Recent advances in solid-phase extraction techniques with nanomaterials. In: Hussain CM (ed) Handbook of nanomaterials in analytical chemistry: modern trends in analysis. Elsevier, Amsterdam, pp 57–73. https://doi.org/10.1016/B978-0-12-816699-4.00004-9

WHO (2011) Guidelines for drinking-water quality, 4th edn. WHO Press, Geneva

Wu Q, Zhao G, Feng C, Wang C, Wang Z (2011) Preparation of a graphene-based magnetic nanocomposite for the extraction of carbamate pesticides from environmental water samples. J Chromatogr A 1218:7936–7942. https://doi.org/10.1016/j.chroma.2011.09.027

Xie J, Liu T, Song G, Hu Y, Deng C (2013) Simultaneous analysis of organophosphorus pesticides in water by magnetic solid-phase extraction coupled with GC–MS. Chromatographia 76:535–540. https://doi.org/10.1007/s10337-013-2408-8

Xiong Z, Zhang L, Zhang R, Zhang Y, Chen J, Zhang W (2012) Solid-phase extraction based on magnetic core–shell silica nanoparticles coupled with gas chromatography-mass spectrometry for the determination of low concentration pesticides in aqueous samples. J Sep Sci 35:2430–2437. https://doi.org/10.1002/jssc.201200260

Xu M, Chen K, Luo C, Song G, Hu Y, Cheng H (2017) Synthesis of Fe_3O_4@m-SiO_2/ PSA@Zr-MOF nanocomposites for bifenthrin determination in water samples. Chromatographia 80:463–471. https://doi.org/10.1007/s10337-017-3253-y

Yang X, Qiao K, Ye Y, Yang M, Li J, Gao H, Zhang S, Zhou W, Lu R (2016) Facile synthesis of multifunctional attapulgite/Fe3O4/polyaniline nanocomposites for magnetic dispersive solid phase extraction of benzoylurea insecticides in environmental water samples. Anal Chim Acta 934:114–121. https://doi.org/10.1016/j.aca.2016.06.027

Yang D, Wang Y, Li H, Yang Y (2020a) Acid-base-governed deep eutectic solvent-based microextraction combined with magnetic solid-phase extraction for determination of phenolic compounds. Microchim Acta 187(2):1–9. https://doi.org/10.1007/s00604-020-4109-y

Yang X, Mi Y, Liu F, Li J, Gao H, Zhang S, Zhou W, Lu R (2020b) Preparation of magnetic attapulgite/polypyrrole nanocomposites for magnetic effervescence-assisted dispersive solid-phase extraction of pyrethroids from honey samples. J Sep Sci 43:2419. https://doi.org/10.1002/jssc.202000049

Yousefi SM, Shemirani F, Ghorbanian SA (2017) Deep eutectic solvent magnetic bucky gels in developing dispersive solid phase extraction: application for ultra trace analysis of organochlorine pesticides by GC-micro ECD using a large-volume injection technique. Talanta 168:73–81. https://doi.org/10.1016/j.talanta.2017.03.020

Yu X, Sun Y, Jiang C, Sun X, Gao Y, Wang Y, Zhang H, Song D (2012) Magnetic solid-phase extraction of five pyrethroids from environmental water samples followed by ultrafast liquid chromatography analysis. Talanta 98:257–264. https://doi.org/10.1016/j.talanta.2012.07.022

Yu T, Wang T, Huang Z, Huang N, Zhang H, Luo Z, Li H, Ding S, Feng W (2017) Determination of multiple pesticides in human blood using modified QuEChERS method with Fe_3O_4 magnetic nanoparticles and GC–MS. Chromatographia 80:165–170. https://doi.org/10.1007/s10337-016-3206-x

Yu X, Li Y, Ng M, Yang H, Wang S (2018) Comparative study of pyrethroids residue in fruit peels and fleshes using polystyrene-coated magnetic nanoparticles based clean-up techniques. Food Control 85:300–307. https://doi.org/10.1016/j.foodcont.2017.10.016

Yuan C, Qian Y, Hong X, Wang H, He H, Liu H, Chai Z, Zhang Y, Zhao P, Wang Y (2019) A new reversed-dispersive micro-solid-phase extraction of organophosphorus pesticides based on three-dimensional magnetic nanoparticles supported by graphene-carbon nanotubes nanocomposite. J Biobaased Mater Bioenergy 13:170–174. https://doi.org/10.1166/jbmb.2019.1840

Zhang R, Su P, Yang L, Yang Y (2014a) Microwave-assisted preparation of poly(ionic liquids)-modified magnetic nanoparticles for pesticide extraction. J Sep Sci 37:1503–1510. https://doi.org/10.1002/jssc.201400125

Zhang S, Jiao Z, Yao W (2014b) A simple solvothermal process for fabrication of a metal-organic framework with an iron oxide enclosure for the determination of organophosphorus pesticides in biological samples. J Chromatogr A 1371:74–81. https://doi.org/10.1016/j.chroma.2014.10.088

Zhang M, Yang J, Geng X, Li Y, Zha Z, Cui S, Yang J (2019) Magnetic adsorbent based on mesoporous silica nanoparticles for magnetic solid phase extraction of pyrethroid pesticides in water samples. J Chromatogr A 1598:20–29. https://doi.org/10.1016/j.chroma.2019.03.048

Zhao G, Song S, Wang C, Wu Q, Wang Z (2011) Determination of triazine herbicides in environmental water samples by high-performance liquid chromatography using graphene-coated magnetic nanoparticles as adsorbent. Anal Chim Acta 708:155–159. https://doi.org/10.1016/j.aca.2011.10.006

Zhao Q, Lu Q, Feng YQ (2013) Dispersive microextraction based on magnetic polypyrrole nanowires for the fast determination of pesticide residues in beverage and environmental water samples. Anal Bioanal Chem 405:1–12. https://doi.org/10.1007/s00216-013-6866-5

Zhao P, Li S, Chen X, Guo X, Zhao L (2019) Simultaneous enantiomeric analysis of six chiral pesticides in functional foods using magnetic solid-phase extraction based on carbon nanospheres as adsorbent and chiral liquid chromatography coupled with tandem mass spectrometry. J Pharm Biomed Anal 175:112784. https://doi.org/10.1016/j.jpba.2019.112784

Zhao J, Meng Z, Zhao Z, Zhao L (2020) Ultrasound-assisted deep eutectic solvent as green and efficient media combined with functionalized magnetic multi-walled carbon nanotubes as solid-phase extraction to determine pesticide residues in food products. Food Chem 310:125863. https://doi.org/10.1016/j.foodchem.2019.125863

Zheng H, Zhao Q, Mo J, Huang Y, Luo Y, Yu Q, Feng Y (2013) Quick, easy, cheap, effective, rugged and safe method with magnetic graphitized carbon black and primary secondary amine as adsorbent and its application in pesticide residue analysis. J Chromatogr A 1300:127–133. https://doi.org/10.1016/j.chroma.2013.04.040

Zheng X, He L, Duan Y, Jiang X, Xiang G, Zhao W, Zhang S (2014) Poly(ionic liquid) immobilized magnetic nanoparticles as new adsorbent for extraction and enrichment of organophosphorus pesticides from tea drinks. J Chromatogr A 1358:39–45. https://doi.org/10.1016/j.chroma.2014.06.078

Zheng H, Ding J, Zheng S, Yu Q, Yuan B, Feng Y (2015) Magnetic "one-step" quick, easy, cheap, effective, rugged and safe method for the fast determination of pesticide residues in freshly squeezed juice. J Chromatogr A 1398:1–10. https://doi.org/10.1016/j.chroma.2015.04.021

Zheng X, Wang J, Xue X, Liu W, Kong Y, Cheng R, Yuan D (2018) Facile synthesis of Fe_3O_4@MOF-100(Fe) magnetic microspheres for the adsorption of diclofenac sodium in aqueous solution. Environ Sci Pollut Res 25:31705–31717. https://doi.org/10.1007/s11356-018-3134-4

Zhou L, Su P, Deng Y, Yang Y (2017) Self-assembled magnetic nanoparticle supported zeolitic imidazolate framework-8: an efficient adsorbent for the enrichment of triazine herbicides from fruit, vegetables, and water. J Sep Sci 40:909–918. https://doi.org/10.1002/jssc.201601089

Zhou Y, Zhu J, Yang J, Lv Y, Zhu Y, Bi W, Yang X, Chen DDY (2019) Magnetic nanoparticles speed up mechanochemical solid phase extraction with enhanced enrichment capability for organochlorines in plants. Anal Chim Acta 1066:49–57. https://doi.org/10.1016/j.aca.2019.03.049

Chapter 12
Characterization and Application of Fe-Magnetic Materials and Nanomaterials for Application in the Aqueous Matrices Decontamination

Pamela Sepúlveda, Ricardo Salazar, L. Carolina Espinoza, and Alejandra García García

Contents

P. Sepúlveda (✉)
Facultad de Química y Biología, Universidad de Santiago de Chile, Santiago, Chile

Departamento de Física, Universidad de Santiago de Chile, Santiago, Chile

Centro para el Desarrollo de la Nanociencia y la Nanotecnología, CEDENNA, Santiago, Chile
e-mail: pamela.sepulvedaor@usach.cl

R. Salazar
Laboratorio de Electroquímica del Medio Ambiente (LEQMA), Universidad de Santiago de Chile, Santiago, Chile
e-mail: ricardo.salazar@usach.cl

L. C. Espinoza
Laboratorio de Electroquímica del Medio Ambiente (LEQMA), Universidad de Santiago de Chile, Santiago, Chile

Centro de Investigación de Procesos Redox (CIPREx), Facultad de Ciencias Químicas y Farmacéuticas, Universidad de Chile, Independencia, Chile
e-mail: carolina.espinoza@ciq.uchile.cl

A. G. García
Laboratorio de síntesis y modificación de nanoestructuras y materiales bidimensionales, Centro de Investigación en Materiales Avanzados S.C. (CIMAV), Apodaca, NL, Mexico
e-mail: alejandra.garcia@cimav.edu.mx

L. Meili, G. L. Dotto (eds.), *Advanced Magnetic Adsorbents for Water Treatment*, Environmental Chemistry for a Sustainable World 61,
https://doi.org/10.1007/978-3-030-64092-7_12

Abstract The high level of pollutants presents in water have generated important problems on the environmental and human, considering that some of these inorganic and organic contaminants are toxic to low levels of concentration. An alternative to this problem is the use of Fe-magnetic materials and nanomaterials in the adsorption process and advanced oxidation processes because they have high pollutant removal capacities and can be extracted using an external magnetic field. The efficiency of the removal of pollutants depend on the characteristics of the materials. For this, the morphology, structure, and physical and chemical properties of these materials are of great interest. Microscopy, spectroscopy, and structural micro-analysis are the main tools used for characterization.

Herein, we present a review about Fe-magnetic materials and nanomaterials, including (1) general aspects of Fe-magnetic materials and nanomaterial as excellent adsorbent substrate to remove inorganic and organic pollutant, considering its different removal mechanisms and reuse, and (2) conventional and unconventional characterization techniques to evidence the morphology, structure, size, oxidation state, and crystallinity of the materials, associating each of the characteristics found (as the case may be) with its ability to eliminate pollutants. Specifically, in this section, we review examples and applications of transmission electron microscopy (TEM), scanning electron microscopy (SEM), X-ray diffraction (XRD), X-ray photoelectronic spectroscopy (XPS), and Raman and Mössbauer spectroscopy. Finally, we show two excellent removal techniques that use magnetic materials: (3) adsorption process to eliminate metal contaminants and metalloids, such as arsenic, lead, and chromium, among others, and (4) advanced oxidation processes (AOP) used for degradation of organic pollutants, for example, dyes, antibiotics, evidencing the high efficiency and removal capacity of the Fe-magnetic materials and nanomaterials, both of which are used in the removal process.

Keywords Fe- Magnetic materials · Adsorption process · Advanced oxidation processes (AOPs) · Organic and inorganic pollutant and characterization techniques

12.1 Introduction

At present, water pollution is a global and important environmental problem; therefore, the application of different water treatment technologies is necessary (Mahdi et al. 2019). However, the contaminants can be classified as inorganic, organic, and biological (Mahdi et al. 2019; Ntombenhle et al. 2019), and in this context, the selected technology should be appropriate for efficient removal of each type of pollutant. A variety of removal processes are used for decontamination, such as adsorption, coagulation, ion exchange (Khan and Malik 2019), and advanced drinking water treatment (ADWT) processes, for example, advanced oxidation processes (AOPs), membrane processes, adsorption on different materials or nanomaterials, among others (Teodosiu et al. 2018). Specifically, the adsorption process is an excellent alternative for pollutant removal because of its low cost, easy operation, and a small number of by-products generated (Zhang et al. 2010). Likewise, AOPs are other technology used for water decontamination; in this process, hydroxyl radicals (•OH) with high oxidation capability are produced to degrade organic substances present in aqueous matrices (Quiroz et al. 2011). To improve current technologies, the use of materials and nanomaterials has been proposed (Khan and Malik 2019). For example, nano-scale zero-valent iron (nZVI) or iron nanoparticles are used for water purification, producing reactions of reduction, oxidation, adsorption, and other processes that involve the physical or chemical interaction between the surface of nanomaterial and pollutants. This nanostructure also exhibits antimicrobial, antiparasitic, and magnetic properties, where the latter characteristic allows the nanomaterial to be separated by an external magnetic field (Li et al. 2006). Other magnetic materials have been used for the adsorption and AOPs process. Nano-Fe_3O_4 were applied in the heterogeneous electro-Fenton process for amoxicillin removal, obtaining over 98% of degradation (Kalantary et al. 2018). However, the physical and chemical characteristics of these materials are still being studied to determine how the structural properties such as morphology, structure, size, and others, such as magnetic ones, affect their removal capacity (Ferrando et al. 2008). In this context, different characterization techniques have been applied, such as transmission and scanning electron microscopy (TEM and SEM, respectively) to evaluate morphology and size (Inkson 2016), X-ray diffraction (XRD) to determine crystalline phases and structure in polycrystalline materials (Elici 2012; Rius 2015), X-ray photoelectron spectroscopy (XPS) to evaluate the surface composition, oxidation state, and quantitative analysis, (Olefjord 2007), Raman spectroscopy to phase identification and transitions, analysis of amorphous nanodomains, size, and mechanical characterization (Gouadec and Colomban 2007), and finally, the determination of the magnetic properties by Mössbauer spectroscopy in addition to providing complementary information about chemical and structural properties of the materials (Peters et al. 2019).

The aim of this chapter is to highlight the scientific evidence about the use of Fe-based magnetic materials and nanomaterials in the removal of pollutants (organic and inorganic) present in water. The adsorption technique for the removal of inorganic pollutants and AOPs for organic compounds degradation are presented. Besides, the characterization techniques used for the proposed materials and nanomaterials are described, such as TEM, SEM, XRD, XPS, Raman, and Mössbauer spectroscopy to evaluate important aspects of the physical, chemical, and magnetic properties involved in the effective removal of pollutants.

12.2 General Aspect of Fe-Magnetic Materials and Nanomaterials

In the last decade, the use of metallic materials and nanomaterials for the elimination of inorganic and organic pollutants has been widely reported. The nanostructures allow the adsorption and transformation of pollutants, particularly, the nanoparticles exhibit a high removal capacity with respect to bulk material, associated with surface energy and chemical reactivity (Quinn et al. 2009). Iron (Fe) nanoparticles are the main material for environmental applications, generating the transformation of different kinds of pollutants, such as organochlorine pesticides, organic dyes, metal and metalloid (Sun et al. 2006). Besides, owing to its magnetic properties, following the removal process, this nanostructure can be separated from the solution easily, making it an easy and low-cost technology (Reddy et al. 2016). The nanoscale zero-valent iron (nZVI) has a "core-shell" structure, with metallic iron in the core and the amorphous shell composed of iron oxi-hydroxide and oxides (Crane and Scott 2012; Taylor et al. 2006; Yan et al. 2012; Yousef et al. 2011). Fig. 12.1 shows different mechanisms of removal of pollutants using iron nanoparticles, demonstrating that this core-shell structure promotes diverse processes with the contaminant, such as surface processes (blue arrow) like adsorption and precipitation mechanism, intraparticle processes (yellow arrow) as reduction, oxidation and coprecipitation mechanisms and reduction process on the core, generating that these nanomaterials have high level of removal (O'Carroll et al. 2013; Sun et al. 2006; Zhang et al. 2015).

However, Fe nanoparticles have a disadvantage that is associated with their high surface energy when they are in an aqueous medium and in systems with high humidity, such as soil, and they tend to agglomerate and oxidize, diminishing their ability to remove pollutants (Hosseini and Tosco 2013; Kim et al. 2013; O'Carroll et al. 2013). To solve this problem and improve the stability, efficiency, and performance of iron nanoparticle, the incorporation of a second metal (bimetallic nanoparticles) and functionalization process on these monometallic nanoparticles

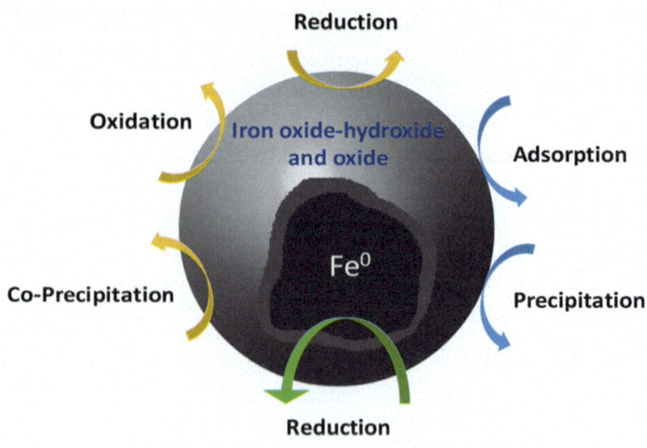

Fig. 12.1 Removal mechanisms on the core-shell structure of Fe nanoparticles. (Li et al. 2006; O'Carroll et al. 2013)

have been reported (Liu et al. 2014; Suazo-Hernández et al. 2019). For example, Sepúlveda et al. (2018) informed that As (V) can be removed by Fe-Cu bimetallic nanoparticles (FeCu BMNPs), where a mass ratio of 0.9:0.1 of Fe:Cu (core-shell structure) sorbed more arsenic than FeCu BMNPs with a mass ratio of 0.5:0.5 of Fe: Cu (Janus like structure) (Sepúlveda et al. 2018). Suazo-Hernández et al. (2019) reported that the use of composites of zeolite (Z) and iron nanoscale zero-valent iron (nZVI) allows that the sorption capacity of As (V), Se (VI), and As-Se was 1.5 times greater than with nZVI as adsorbent material, making this nanocomposite (ZnZVI) an excellent alternative to remove different pollutants present in water (Suazo-Hernández et al. 2019). Likewise, magnetic materials of iron oxide have been applied to eliminate organic water contaminants. Particularly, the heterogeneous electro-Fenton process using iron materials (Fe_3O_4 and Fe_2O_3) is an excellent catalyst for organic pollutant degradation because it enhances catalysis of H_2O_2 and production of •OH, making this process more efficient (Kalantary et al. 2018). Wang et al. (2018) informed that the material's structure limits the different removal process, where graphitized carbon structure is beneficial to organic dyes adsorption and the amorphous iron material is good for catalyzing photo Fenton-like reaction of Rhodamine B (Wang et al. 2018), similar to that reported by Sepúlveda et al. (2018). Therefore, in the case of bimetallic nanoparticles, iron oxides, and composites or nanocomposites, it is necessary to determine the morphology, structure, oxidation state, and other parameters to evaluate the reactivity of these materials by means characterization techniques (Czaplinska et al. 2014; Lai et al. 2014; Wanjala et al. 2011; Ye and Crooks 2007). These analyses are shown in the next section.

With the aforementioned background, the use of Fe- magnetic materials and nanomaterials (monometallic, bimetallic, composites, and other) is an excellent alternative to remove inorganic and organic pollutants, considering its different removal mechanisms and reuse.

12.3 Conventional and Nonconventional Characterization Techniques of Fe-Magnetic Materials

Nowadays, the study of the physicochemical properties of a material is the most outstanding research activity on nanotechnology and materials science. The characterization techniques have been a crucial tool; however, for the last 9 decades, instrumental development has enabled us to obtain fundamental properties of materials and the relationship with their performance in different application areas. The electronic microscope is one of the most revolutionary and important tools in material science because the physicochemical properties of these are directly involved with its shape and size. Thanks to the progress of this tool, we can know the periodic organization of atoms and its relationship with a determined crystalline structure, which allows classifying materials with respect to shape and structural properties and then correlate it with its removal capacity. To determined the relationship between structural properties and chemical composition, X-ray diffraction is one of the support tools with which we can determine diverse phases present on the material and its structural characteristics. However, to the exact understanding of the chemistry of materials, the use of nonconventional characterization techniques has been explored. For example, as previously mentioned, X-ray photoelectronic spectroscopy (XPS) is a technique that provides information on the materials about its elemental chemistry and oxidation states when the analysis has to be carried out on the surface. Besides, this technique informs about the binding energy and the types of atoms present when the XPS analysis is performed to a different level of depth or thickness. On the other hand, vibrational spectroscopy is an excellent tool for the characterization of organic and inorganic materials, making it possible for the study of adsorbent and catalytic materials post-removal process and therefore to determine the changes of these materials generated by the inorganic contaminant adsorption or organic pollutant degradation. One type of vibrational spectroscopy is Raman spectroscopy, which provides information on the molecules through an induced change in their dipole moment, resulting in specific vibrational or rotational characteristics of studied materials.

The following details the conventional and nonconventional techniques used for the physical and chemical characterization of micro- and nanomaterials applied in the removal processes.

12.3.1 Electron Microscopy

The electron microscope emerged in the decade of 1920 in quantum physics, where the fundamental principle of its design is based on the wave–particle duality principle. The microscope and its design were improved with time to reach high definition; however, it was necessary for the development of lenses to obtained atomic resolution. By the 1940s appeared the scanning electron microscope, which was built by German researcher Von Ardenne in 1938, but distributed 26 years later by the Cambridge Instruments company. This microscope arose for the study of surface of materials, obtaining information about its topography and morphology. The image taken from electron microscope will depend on the type of sample (size, density, and elemental composition); for example, in the case of TEM, the electron beam will be transmitted across of the sample (see Fig. 12.2a), giving an image with different contrasts, only if the sample contains areas with different atomic composition or arrangement. Namely, the electrons will be dispersed differently because the sample has distinct zones and this phenomenon will be directly reflected in the image. In the case of a scanning electron microscope, the image will depend on the capacity of the sample to emit secondary electrons (see Fig. 12.2b). Therefore, if a zone emits more electron than another zone, the resulting image will appear with different contrast with respect to the zone with distinct emission. In this way, the built image will be the addition of different contrast (from light to dark) of the diverse zone present in the sample. The high-resolution electron microscopy, as well

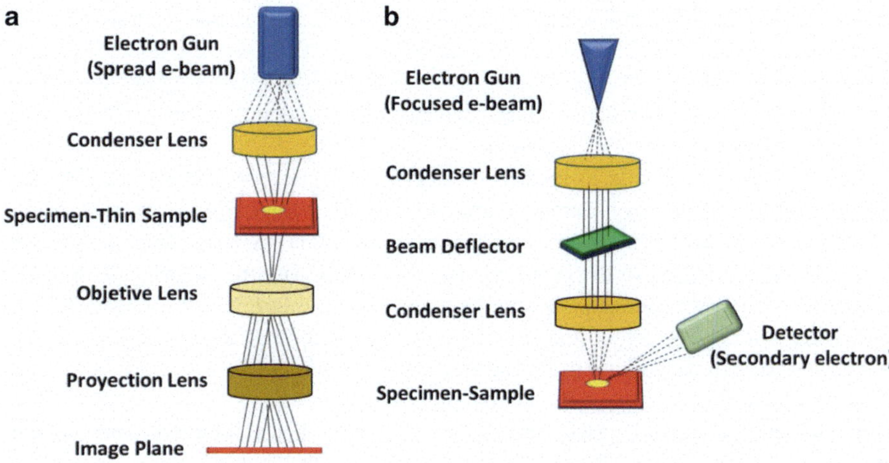

Fig. 12.2 Representative diagrams of the basic components of (**a**) TEM and (**b**) SEM

as SEM, is a fundamental tool for the study of materials and nanostructure applied to diverse scientific fields. Specifically, in this section, we highlight the use of electron microscopy in the characterization of nanoparticles and nanostructures applied to the pollutant removal present in aqueous matrices.

Transmission Electron Microscopy (TEM)

This technique is very important to nanostructure characterization because it allows determining its size and morphology. These characteristics are related to physics, magnetics, and electronic properties of materials. Knowing these characteristics helps us to answer questions like the following: How do the size and morphology of particles influence the removal capacity of water contaminants? Is the homogeneous distribution of nanoparticles important to improve its optic and electronics properties? Do the morphology and size of materials change after applications in pollutants removal processes? On the other hand, in the case of functionalized materials and bimetallic nanoparticles, the questions are: Do the structure and morphology of these new materials change with respect to the individual materials or nanoparticles? What is the distribution of the metals in the bimetallic nanoparticles? How is the physical or chemical interaction of the contaminant with the surface of the bimetallic nanoparticles? Therefore, to respond to these questions, TEM is an excellent technique for the characterization of nanostructure applied to pollutant removal.

Liu and Zhang (2014) carried out extensive characterization of iron nanostructure by electron microscopy, and informed the different operating modes that can be used to obtain diverse structural characteristics of materials (Liu and Zhang 2014). In the case of nZVI and its core-shell structure, to study the chemical reaction in the surface, interface, and the core of the nanoparticle, the scanning transmission electron microscopy (STEM) operating mode is one of the recommended because it allows physical and chemical identification at high resolution. The images in STEM mode can be obtained in dark- and bright-field, which allow us to identify structural differences by phase contrast.

Zhang's group was able to identify the aggregation degree of nanoparticles, the characteristic morphology (chains formation), as well as thickness of oxide–hydroxides shell (2–3 nm), being identified by different contrasts (the shell has greater transparency compared to the core). Using X-ray spectroscopy (EDS), the compositional evolution was detected, while the nanoscale structure was visualized. In this sense, the authors performed the linear analysis or mapping, which gave a map of colors assigned to each element found. In the case of bimetallic nanoparticles, core-shell or alloys, this tool is useful to visually identify the areas of the sample where each element is located. Also, the behavior of nZVI was studied with greater precision from the chemical point of view by means of electron energy loss spectroscopy (EELS), studying the bonds present locally and not in all the material as normally done by XPS (a technique that will be described later). This study allowed them to see along with a localized particle, the chemical changes and elements present, as well as its elementary valences. In the case of nZVI, it was possible to locate the shell and core composition, observing at shell a mixture of Fe (II) and Fe (III) while, in the core, the presence of Fe (0) was confirmed.

In a subsequent article, Liu et al. in 2017 studied the evolution of nZVI in water by TEM, observing the structural and morphological changes that the material has in the presence and absence of oxygen. Under conditions without oxygen, the core-shell type structure remains stable until 72 hand under conditions in the presence of oxygen, there is a morphological change from spherical to acicular, attributed to the corrosive conditions of medium, completely losing the shape of the initial material (Liu et al. 2017).

To study the morphological changes that occur after the sorption processes, electron microscopy is also useful, as it shows the structure of the material after the contaminant sorption. Hao et al. reported in their studies on absorbent materials used for the removal of As, the inhibition in the growth of γ-Fe_2O_3 nanoparticles after the sorption of As (V) ions on the surface, resulting in average sizes lower than that of the nanoparticles (Hao et al. 2018).

High-resolution transmission microscopy (HRTEM) makes it possible to observe characteristic fringes of the crystalline network present in different areas of the particles used as an absorbent. Wen et al. conducted a study on bimetallic oxides of Fe-Ce for the sorption of As and dyes, and showed by analysis of HRTEM micrographs, a long-range arrangement where the interplanar spacing had a separation of 0.254 nm corresponding to the plane (311) of Fe_3O_4, while the cerium oxides phase was assigned as amorphous by not finding reticular fringes (Wen et al. 2019).

Scanning Electron Microscopy (SEM)

The importance of scanning electron microscopy is that it helps to define surface characteristics of the materials, which allow understanding the importance of the surface in the sorption processes. Liu and et al. reported in their 2019 work, a study on obtaining composite materials for As sorption, where surface characteristics such as porosity perform an important role. Through SEM, they determined the conditions to generate surface changes in composite materials to increase the sorption efficiency of pollutants. The materials with higher porosity were those with the highest As sorption, stabilizing in cluster form on the composite material surface (H. Liu et al. 2019a). The aggregation of nanoparticles is another characteristic that is easily detectable with SEM, since unlike TEM, with SEM, we have a general outlook of the state of the sample. Murtaza and coworkers showed in their research work the aggregation of monometallic and bimetallic nanoparticles. In the case of monometallic iron nanoparticles (Fe^0), these were aggregated in a chain form due to the magnetic interaction between them, while in the bimetallic nanoparticles (Bi/Fe^0), the presence of a second metal stabilizes the structure, avoiding further aggregation. Through SEM, the final morphology of the Bi/Fe^0 bimetallic material was obtained, observing a combination of spherical particles corresponding to Fe and flower-like structures composed of Bi sheets, resulting in a good coupling of both metals (Murtaza et al. 2019). On the other hand, Sepúlveda et al. (2018) reported results on bimetallic Fe/Cu particles used for the sorption of As, wherefrom the results of SEM showed the importance of the synthesis method and the concentration of Fe with respect to Cu for obtaining nanoparticles. It was observed that at 1:1 concentration of both metals, the morphology of the bimetallic nanoparticles are alloy type, while at concentrations of 9:1 iron with respect to copper, the morphology obtained is core-shell type, where the core is formed by iron and the shell is formed by copper oxides (Sepúlveda et al. 2018).

12.3.2 X-Ray Diffraction (XRD)

XRD technique is one of the most exploited in the field of materials science and nanotechnology because it is a nondestructive technique, and it gives extensive information about the crystalline structure, prevailing phases, network parameters, crystallite size, and grain size. X-ray diffraction is one of the oldest techniques with more than 100 years since its discovery. It is based on the principle of constructive interference of monochromatic X-rays and a crystalline sample. In crystalline samples, the atoms are ordered periodically in the 3 dimensions, generating the X-ray diffraction only in those that comply with the Bragg's law. Whereas amorphous solids that lack periodicity maintain only short-range order, the result of this is an X-ray dispersion curve characterized by wide peaks. Particularly, powders diffraction is used when it is not possible to obtain monocrystalline sample. The

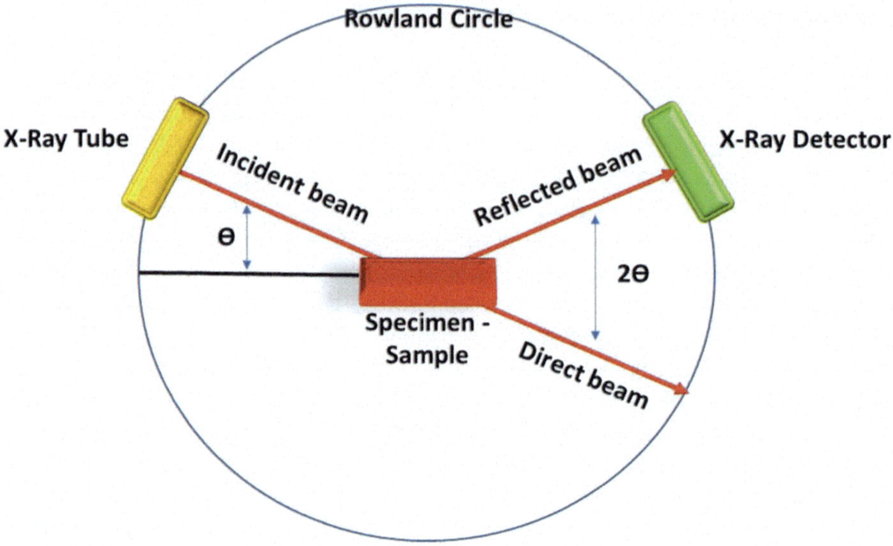

Fig. 12.3 Schematic representation of X-ray diffractometer

identification of crystalline phases in these samples is by means of a comparative study between obtained diffraction patrons and databases, where the concordance between them will suggest the presence of a determinate crystalline phase (Rius 2015). In the case of XRD analysis of absorbent materials such as nZVI nanoparticles, a characteristic peak located at 44.66° 2 theta will always be observed in the diffraction spectrum as reported by several authors, including Sepúlveda et al. (Sepúlveda et al. 2018). XRD is a sensitive technique to phase changes due to oxidation or reduction of the surface species. Therefore, this technique can give us information before and after the sorption processes, indicating the structural changes that may occur in contact with the pollutants. When systems are more complex with composite materials such as nano iron-activated carbon and iron oxides (nZVI/GAC), XRD is useful for determining the presence of these materials within the main matrix of the compound and monitor the phases of iron oxides and the transformation of nano iron to other species such as oxyhydroxides (Wu et al. 2019) (Fig. 12.3).

Figure 12.3 shows the schematic representation of the instrumentation of X-ray diffraction.

12.3.3 X-Ray Photoelectron Spectroscopy (XPS)

XPS is one of the most used quantitative analytical techniques for chemical surface analysis in materials; its physical principle is the photoelectric effect observed by Robinson and Rawlinson in 1914, and it is commonly used to elucidate the electronic structure, elemental composition, and oxidation states of the elements that make up a material.

This technique was employed at the end of the 1960s, 10 years after Siegbahn and his coworkers started to develop the technique, based on the principle that (Olefjord 2007), an X-ray-irradiated surface releases characteristic photoelectron both from the element from which each electron is emitted and from the bond between that atom and neighboring atoms. The emitted photoelectrons can be generated from a depth between 1 and 10 nm of the surface, and these are not modified by collision with intermediate atoms and are suitable for analysis (Ferrando et al. 2008). The energy of released photoelectrons from the depths below the surface is severely diminished by the atomic collision. These photoelectrons do not reach the detector or have energy so distinctively lower that they can be easily discriminated against. The XPS technique provides information on the elemental composition and chemical bonding, which is an advantage. For example, it has been reported by several authors, including Liu et al., that the surface of the nZVI is composed of a mixture of Fe (II) and Fe (III) and the core is formed by Fe (0) (Liu and Liu and Zhang 2014). This analysis was achieved by doing a depth profile by XPS, concluding that the Fe (III) oxide is the one that predominates at the surface. This observation was got by deconvolution of each of the bands in the high-resolution spectrum of Fe 2p and O 1s. This characterization tool also helped in the study of nZVI to know the thickness of the shell which was less than 5 nm.

Another type of core-shell magnetic nanoparticles are Mn−Fe/MnO_2, reported by Chen et al., where Mn–Fe bond was detected at core of the nanoparticles, showing a shift toward higher binding energies after coating with MnO_2, where, the oxidation state of Mn corresponds to the Mn (II) at core and Mn (IV) located at shell. The results of XPS for the analysis after the reaction of Pb (II) on the surface of Mn-Fe/MnO_2 provided an idea of the absorption processes present. The sorption of Pb (II) was confirmed with the appearance of a new peak in the spectrum corresponding to the initial material (Mn-Fe/MnO_2), where it was determined that the valence state of the Pb absorbed on the surface was divalent. The authors concluded that the absorption of lead in this core-shell material is mainly achieved by the formation of a complex on the surface between the groups -OH and Pb (Chen et al. 2014).

Bimetallic nanoparticles also are used for the simultaneous pollutant removal, such as the case reported by Wen et al. (2019), who presented XPS results that supported the proposed absorption mechanism from the changes in the surface state of the nanoparticles of Fe-Ce used for this purpose (Wen et al. 2019).

On the other hand, in studies where the occurrence of reactions after the sorption of contaminants is important, XPS is a technique that helps to monitor and confirm them. In the work of Xia and collaborators, chloramphenicol was removed from aqueous solutions, where XPS confirmed the reduction of organic compounds in water from superficial oxidation of Fe (0) (Xia et al. 2014).

12.3.4 Raman Spectroscopy

The origin of Raman spectroscopy dates back to the mid-nineteenth century when physicist Smekal postulated the phenomenon of inelastic light scattering and was subsequently observed experimentally by physicist Chandrasekhara Venkata Raman, which led him to be awarded the Nobel Prize of Physics in 1930.

The phenomenon of inelastic light scattering allowed the rotational and vibrational study of molecules. This phenomenon occurs when light interacts with matter, that is, the photons that make up the light can be absorbed or dispersed by matter, or they may not interact with the material and pass through it. If the energy of the incident photon corresponds to the difference between the basal state of a molecule and its excited state, then said photon can be absorbed and promote the molecule to a higher energy state. This change is measured in absorption spectroscopy through detection of loss of that radiation energy from the incident light in the material. On the other hand, for Raman spectroscopy, a monochromatic beam of light with a known wavelength is required to influence the material and analyze the light scattered throughout the sample. Most of the scattered light will have the same wavelength as the incident beam, and this is an elastic dispersion or, a Rayleigh dispersion, which is a dominant dispersion process. However, a small part will have a frequency change as a result of the interaction of the photons with the electron cloud around the nucleus; this small part of incident photons that changed their frequency in a unit will have exchanged energy, either from the molecule to the incident photon or vice versa. This process is known as Raman dispersion.

As can be seen in Fig. 12.4, the creation of "virtual states" is due to the interaction of laser that polarizes electrons around the nucleus, and its energy will be determined by the frequency of incident light. During the Raman dispersion process, the fundamental state "m" leads to the absorption of energy by the molecule to a higher

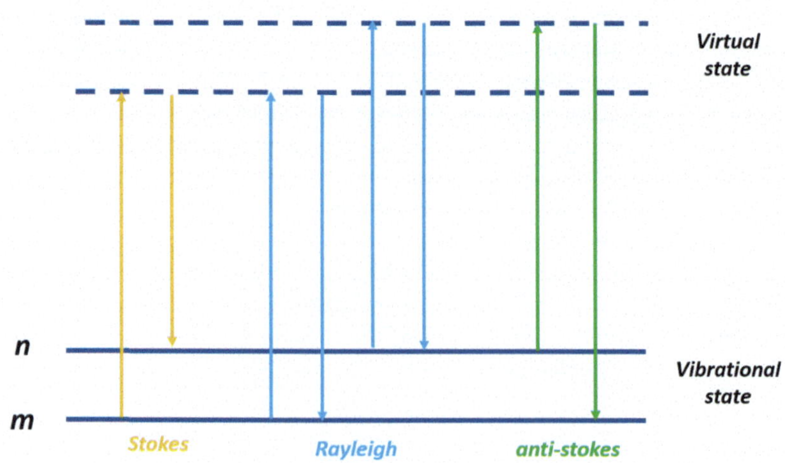

Fig. 12.4 Vibrational excitation modes in Raman scattering

energy state "n." This process is known as Stokes dispersion. However, due to thermal energy, some molecules will have an excitation state "n"; therefore, the dispersion from this state to fundamental state "m" is called anti-Stokes dispersion and involves the transfer of energy from molecule to the incident photon. Each material will have a set of characteristic values of its polyatomic structure and nature of chemical bonds that form it. Since most of the molecules at room temperature are in a vibrational state of lower energy, the probability of energy transfers that give rise to anti-Stokes dispersion is very low; therefore, measurements are usually made using Stokes dispersion.

Strictly speaking, the Raman dispersion should be expressed as a change in the incident energy value and should be expressed as Δcm^{-1} but for simplicity, only cm^{-1} is used.

Raman is currently a technique used for the characterization of nanocatalysts based on zero-valent iron since it allows the researcher to determine the molecular structure of different species present and evaluate the catalytic efficiency after use. The analyzed compounds after sorption processes have well-defined peaks in Raman spectroscopy, which can be consulted in the work published by Li et al. (2019), who, in their work, summarized the different peaks found for iron species used in the sorption treatment of water pollutants (Li et al. 2019).

X. Song et al. (2019) used Raman spectroscopy to characterize magnetic biocomposites obtained by combining calcination and hydrothermal methods, which were applied for water remediation. The characteristic peaks of this material were reported at 1330 and 1590 cm^{-1}, which are described as D and G bands, respectively, of carbonaceous materials, and are attributed to the photonic vibration of the sp^3 atoms and the vibration in the plane of the sp^2 atoms, respectively (Song et al. 2019). These results indicate that high pyrolysis temperatures of the raw materials result in the formation of aromatic and amorphous carbon that was partially graphitized. Indirectly, the Raman technique was used to determine the amount of magnetic iron nanoparticles in the biocomposite, calculating the intensity ratio of the D and G bands (I_D/I_G), where an increase in the ratio (>1) indicates the presence of defects in carbon network, and therefore, the growth of nanoparticles on the surface takes place.

On the other hand, Raman is also an identification tool in iron species, when its indexation is difficult due to overlapping peaks in XRD diffraction spectra. According to Ahmed et al. (2017), the zero-valent iron particle (ZVI) shows characteristic bands at 215.7, 278, 392, 483, 590 cm^{-1}, and a maximum peak at 1282.5 cm^{-1}. While the peaks at 215.7, 278, and 483 cm^{-1} were assigned to hematite, the peak at ~394.7 cm^{-1} indicates the presence of α-FeOOH and γ-FeOOH oxyhydroxides. The way in which Raman distinguishes one phase from another is from peaks intensity at a certain frequency, which is linked to the vibration of each molecule, for example, phases of oxyhydroxides α-FeOOH, β-FeOOH, γ-FeOOH, and δ-FeOOH have characteristic resonance frequencies for each one, vibrating more or less at a certain frequency (Ahmed et al. 2017).

12.3.5 Mössbauer Spectroscopy

Mössbauer spectroscopy is a versatile technique used to study nuclear structure by absorption and remission of gamma rays, which are part of the electromagnetic spectrum. The technique uses a combination of the Mössbauer effect and Doppler change to test the transition between the excited state and the basal state of an atomic nucleus. Mössbauer spectroscopy requires the use of a solid or crystal, which has the probability of absorbing the photon without the possibility of photon recoil. Many isotopes exhibit Mössbauer characteristics, but the most commonly studied is the ^{57}Fe isotope.

Rudolf L. Mössbauer was a physics student at the technical university of Munich and an assistant professor at the mathematics institute in 1955, during which time he began writing his thesis. In 1958, at the age of 28, Mössbauer graduated and showed experimental evidence for resonant absorption without recoil in the nucleus, called the Mössbauer Effect later. This demonstration of the effect led him to win the Nobel Prize in Physics in 1961.

The Mössbauer effect is the recoil energy associated with the absorption or resonant emission of a photon and can be described by the conservation of momentum. For this resonant absorption to occur, it is necessary that during the emission or absorption of radiation, the corresponding nuclei do not recede, since this recoil takes part of the radiation energy that, in the case of free nuclei, produces a shift in the emission line (and absorption) that is generally greater than the natural width of the line corresponding to nuclear transition, and which causes the "tuning" of the resonant process to be lost. Therefore, we find that the recoil energy depends inversely on the mass of the system. Thus, for gas, the mass of a single core is small compared to a solid. The solid or crystal absorbs the energy as phonons, quantizes vibrational states, but there is a possibility that no phonon is created and that the whole crystal acts as a mass, resulting in an emission without gamma-ray recoil. The new radiation has adequate energy to excite (non-phonic absorption) the next nucleus from its fundamental state.

In applications for pollutant removal, specifically with iron nanoparticles, the Mössbauer spectroscopy is a technique that can probe small changes in the energy levels of Fe nucleus, when it comes to nZVI in response to its environment or, it can give us information on phases present when it comes to a combination of iron materials such as oxides and/or oxyhydroxides. The Mössbauer absorption spectrum is obtained from the exposure of the iron sample to a beam of gamma radiation, measuring the intensity of the beam transmitted through it.

Li and coworkers made known in their research work a complete table with the existing oxyhydroxides as a passivating layer on nZVI, of which data for 5 oxides and 4 oxyhydroxides are summarized, as well as for zero-valence iron. Among the summary results on nuclear interactions are isomeric shift, electric quadrupole interaction, and magnetic interaction for different oxyhydroxides (Li et al. 2019). The isomeric shift provides important information about the nuclear structure and chemical environment of atoms. This shift is due to the combined effect of the

Mössbauer transition free of recoil between two nuclear isomeric states and the transition between two atomic states in those two environments. That is, for a given nuclear isomer in two different physical or chemical environments (different physical phases or different chemical combinations), the electron wave functions are also different. Therefore, in addition to the isomeric shift in the atomic spectral lines, which is due to the difference between the two states of nuclear isomers, there will be a change between the two environments (source reported by Li was ^{57}Co in a platinum matrix, steel or ruthenium, and absorbent was the nZVI sample). The quadripolar electric interaction represents a small disturbance in the energy of the nucleus, dividing the energy levels into doublets (in the case of ^{57}Fe the resonance line is divided into two energy levels); therefore, this parameter allows us to deduce the difference of energy between the two new levels, drawing conclusions about the shape of the nucleus. On the other hand, in the reported table, the magnetic interaction is also presented, which is generated between the magnetic moment of the core and a magnetic field, which can be internal or external. The result is a splitting of the line into six lines in the case of ^{57}Fe, and this parameter helps us to determine the magnetic field in the vicinity of the nucleus, in addition to its spin state and the orientation of the nuclear magnetic moments with respect to the field.

In the case of bimetallic systems such as Fe-Cr, Shen et al. (2005) published their results on the synthesis by grinding method as a time function in different atmospheres (Shen et al. 2005). To the obtained materials, Mössbauer's characterization was carried out to monitor the structural and magnetic changes in the material. At a time of 45 h, the Mössbauer characterization showed that time was insufficient to obtain the desired alloy, and instead, it was a material of a magnetic and paramagnetic character, attributing this behavior to disorder present in the crystalline network. At a time of 85 h, the paramagnetic contribution was almost 100% in the Mössbauer spectrum associated mainly with the increase in crystalline order. This technique also helped to determine if the obtaining atmosphere was better in argon or in a vacuum. The results in vacuum showed the formation of a solid solution with a magnetic BCC crystalline structure at a time greater than 45 h. While, in argon, the formation of the alloy was not achieved. Therefore, in this work, it is concluded that the hyperfine distributions for the Fe60Cr40 material in Argon are formed by crystalline grains and messy grain boundaries. However, for the material obtained under vacuum, hyperfine distributions indicate the formation of disordered alloys with heterogeneous nanoscale structures, as well as homogenization continues for more than 45 h of grinding.

In summary, the use of different characterization techniques is crucial to determine specific information about size, shape, morphology, crystallinity, and magnetic properties of materials and nanomaterials for a better interpretation of removal mechanisms of pollutants.

12.4 Adsorption Process for the Removal of Inorganic Pollutants from Wastewater

The high levels of inorganic pollutants in water, such as lead (Pb), copper (Cu), chromium (Cr), arsenic (As) and others, have generated an important problem worldwide, affecting millions of people (ALOthman et al. 2016; McGrory et al. 2017; Sherman M Ponder et al. 2000). Different human diseases attributable to the presence of theses pollutants in water have been reported. For example, Alzheimer's and Parkinson's diseases (Santurtún et al. 2016; Zhou et al. 2018) are associated with Pb exposure, and cancer, hyperkeratosis, and neurological disorders are related to As (Camacho et al. 2011; Huq et al. 2019; Mohan and Pittman 2007), and alimentary tract cancer, stomach cancer, and pneumonia in the case of Cr exposure (Biswas et al. 2017).

The metal (Cr, Pb, Cu, and others) and metalloid (arsenic) are introduced in the environment by natural and anthropogenic sources and can be present in water with different inorganic forms depending on the environmental conditions, such as pH, redox potential, and oxic and anoxic conditions (Luther et al. 2012; Wang et al. 2014). In the case of lead, this ion is present as Pb^{2+}, $Pb(OH)^{+}$, $Pb(OH)_2{}^{0}$, and $Pb(OH)_3{}^{-}$ depending on the pH value; where pH is >6, the predominant specie is Pb^{2+} (Li et al. 2011). The inorganic arsenic exists in different oxidation states: −3, 0, +3, and +5, where two forms are in natural waters as $AsO_3{}^{3-}$ (arsenite) and $AsO_4{}^{3-}$(arsenate) (Manning et al. 2002; Mohan and Pittman 2007), but only As (V) ion is predominant in oxygenated water and As (III) present in anaerobic matrices (Gomes et al. 2007).

To solve this environmental problem generated with the presence of toxic pollutants in water, different removal processes and adsorbent materials have been applied to remove these analytes. Specifically, a variety of technologies have been used to remove this contaminant, for example, precipitation, electrocoagulation, biological process, and adsorption process (Islam et al. 2011; Katsoyiannis and Zouboulis 2004; Kumar et al. 2004). The adsorption process is a promising method for inorganic pollutants removal (Lata and Samadder 2016). The adsorption is associated with the mass transfer process where the analyte present in the aqueous matrix is transferred to the surface of the substrate by means of chemical and/or physical interaction (Fig. 12.5).

Different adsorbent materials have been used for pollutant removal. Activated carbon is an excellent material for water decontamination because it is a low-cost adsorbent and has a high surface area and adsorbents pores (Mohan and Pittman 2006). Similarly, the use of clays for inorganic pollutant present in water has been used too, where physicochemical parameters, such as degree of ionization at the surface, functional groups, and physical size, affect the surface of these adsorbents and therefore conditioning its applicability in the removal process (Sen Gupta and Bhattacharyya 2014). Particularly, McKay et al. (1985) reported that the capacity removal using carbon activated was 35 and 138 mg/g of chromium (III) and mercury (II), respectively (McKay et al. 1985), and in the case of clays, Zamzow et al.

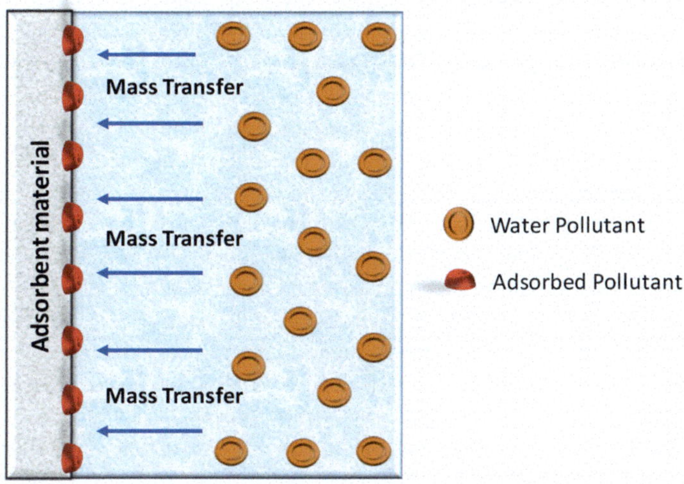

Fig. 12.5 Schematic representation of adsorption process

Table 12.1 Magnetic nanomaterials and their removal capacity of inorganic and organic pollutants

Magnetic nanomaterial	Removal capacity	References
$CoFe_2O_4@SiO_2@m\text{-}SiO_2\text{-}SH/NH_2$	504.34 mg Hg (II)/g	Zhang et al. (2019)
Mn–Fe/MnO_2	261.10 mg Pb (II)/g	Chen et al. (2014)
Spinel $ZnFe_2O_4$	45.45 mg Evans Blue/g	Vergis et al. (2019)
Fe-Cu bimetallic nanoparticles	60.00 mg As (V)/g	Sepúlveda et al. (2018)
Magnetic graphene oxide	252.00 mg tetracycline/g 234.00 mg Cd(II)/g 14.00 mg As (V)/g	Huang et al. (2019)
$MnFe_2O_4$	94.00 mg As (III)/g 90.00 mg As (V)/g	Zhang et al. (2010)
$CoFe_2O_4$	100.00 mg As (III)/g 74.00 mg As (V)/g	Zhang et al. (2010)
γ-Fe_2O_3-NPs-AC	95.55 mg methylene blue/g 207.04 mg malachite green/g	Asfaram et al. (2016)
Fe_3O_4 magnetic polypyrrole–graphene	400.00 mg Hg (II)/g	Zhou et al. (2017)
rGO/Ni/MMO hybrid nanomaterial	210.80 mg Methyl Orange/g	Yang et al. (2013)

informed that clinoptilolite has high affinity for different inorganic pollutant, obtaining the following adsorption sequence: Pb(II) ~ Cd(II) > Cs (I) ~ Cu (II) > Co(II) ~ Cr(III) > Zn > Ni(II) ~ Hg(II) (Dionisiou and Matsi 2016; Zamzow et al. 1990).

However, to improve the application and efficiency of the removal process, other materials have been used. Magnetic nanomaterial has been an excellent alternative as adsorbent material for remediation by its high removal capacity and its magnetic characteristic that allows recover it from aqueous matrixes using and external magnetic force (for magnetic nanoparticles) (Khin et al. 2012). In Table 12.1, magnetic nanomaterials and their removal capacity are shown.

With respect to the adsorption studies, different analyses can be applied to determine maximum removal capacity, saturation time, and information about the adsorbent–adsorbate interaction. Adsorption kinetics provides an analysis of the adsorption mechanism or rates of a chemical process (Azizian 2004). The adsorption of the analyte on the solid materials can occur through 3 steps: (1) external diffusion: transport of the pollutant from the bulk to the external surface of the adsorbent material; (2) internal diffusion: diffusion to the internal active center present in the solid; and (3) adsorption: adsorbate–adsorbent interaction by chemisorption or physisorption. In this sense, diverse equations or kinetic models had been used to explain the rate and adsorption mechanism (Sen Gupta and Bhattacharyya 2011), for example, pseudo first order (PFO), pseudo second order (PSO), Elovich model, intraparticle diffusion, and others (Boroumand Jazi et al. 2014; Guo and Wang 2019; Shahwan 2014). Besides, isotherm studies can be applied to correlate the kinetic and thermodynamic information about the adsorption process associated with interface interaction between the adsorbate and the adsorbent (Ghosal and Gupta 2017). Two typical models are used to explain the adsorption process—Langmuir and Freundlich models. The Langmuir model assumes that the adsorption of the analyte occurs by monolayer adsorption with a homogeneous surface and equivalent active site, and on the other hand, the Freundlich model says that the surface is heterogeneous and the adsorption process is by multi-layer (Naseem et al. 2019). Table 12.2 shows the values of the removal capacity of inorganic pollutants obtained by kinetic and isotherm adsorption studies.

As mentioned earlier, kinetic and isotherm studies allow obtained information about of removal mechanism of analytes. In this context, nanoscale zero-valent iron (nZVI) has been applied to remove different inorganic pollutants. Li et al. reported that the nZVI has a "core-shell" structure, where the amorphous shell is of iron oxides and the core of metallic or zero-valent iron (Li et al. 2008). The core-shell structure and the electron donor properties of this nanoparticles make it excellent adsorbent materials for remediation of water pollutant, and consider that precipitation, adsorption, reduction, coprecipitation or transformation processes can occur in the nZVI surface (Li et al. 2008; Yan et al. 2010). O-O'Carrol postulated that the removal mechanism of heavy metals using nanomaterials depends on standard redox potential (E°). Pollutants (metals) with E°, similar to or more negative than metallic iron, are principally eliminated from the water matrix by adsorption mechanism on the shell of the nZVI. Conversely, if the metals have E° more positive than metallic iron, precipitation and reduction processes are preferentially mechanism for the elimination of these pollutants, but, if the E° of the metal is slightly more positive than Fe^0, reduction and adsorption process can occur. Finally, in the case of oxidation and coprecipitation reactions where these are subject to pH, speciation and initial concentration of pollutants, redox condition and other (O'Carroll et al. 2013), Ponder et al. (2000) reported by means characterization techniques that Cr (VI) and Pb (II) were removed by reduction process of the analyte to Cr (III) and Pb (0), respectively, and other insoluble phases, as well as oxidation of Fe^0 to α-FeOOH. Besides, by kinetic studies, the authors informed that the removal mechanism of both pollutants is physical (i.e., that it involves occlusion of the

Table 12.2 The best kinetic and isotherm model to inorganic pollutant removal

Inorganic pollutant	Adsorbent nanomaterial	Kinetic adsorption parameter (25 °C)	Isotherm adsorption parameter (25 °C)	References
Pb (II) Cu (II) Cd (II) Cr (III)	Amino-functionalized $Fe_3O_4@SiO_2$	*Pseudo second-order* q_e (mg/g) = 476.2 for Pb (II) 454.6 for Cu (II) 434.8 for Cd (II). 357.1 for Cr (III)	*Langmuir* q_m (mg/g) = 555.6 for Pb (II) 526.3 for Cu (II) 476.2 for Cd (II). 434.8 for Cr (III)	Wang et al. (2010)
Pb (II) Cd (II)	Magnetic graphene oxide composites	*Pseudo second-order* q_e (mg/g) = 172.0 for Pb (II) 76.0 for Cd (II).	*Langmuir* q_m (mg/g) = 417.9 for Pb (II) 104.6 for Cd (II).	Bao et al. (2020)
As (III)	$MnFe_2O_4$ $CoFe_2O_4$ Fe_3O_4	*Pseudo second-order* q_e (mg/g) = Not reported	*Freundlich* K_F ($mg^{1-1/n}$ $L^{1/n}$ g^{-1})= 29.6 for $MnFe_2O_4$ 36.9 for $CoFe_2O_4$ 15.2 for Fe_3O_4	Zhang et al. (2010)
As (V)		*Pseudo second-order* q_e (mg/g) = *Not reported*	*Langmuir* q_m (mg/g) = 90.4 for $MnFe_2O_4$ 73.8 for $CoFe_2O_4$ 44.1 for Fe_3O_4	
Cd (II)	CNSR functionalized Fe_3O_4 magnetic nanoparticles	*Pseudo second-order* q_e (mg/g) = 49.3	$K_F(mg/g(L/mg)^{1/n})$= 0.0028	Devi et al. (2017)

zero-valent iron) (Ponder et al. 2000). Similarly, Yan et al. (2012) studied the solid-phase redox transformations of As (III)-iron nanoparticles using X-ray photoelectron spectroscopy, where the authors informed that the As (III) oxidation takes place at the shell of the nZVI, that is, at the iron oxides, while the reduction from As(III) to As(0) is a slower reaction, which is generated at the subsurface layer near the Fe (0) core, concluding that the nZVI is an excellent sequestration agent for arsenic removal (Yan et al. 2012). Other magnetics nanomaterials have been applied to pollutants removal. For example, Zhang et al. (2010) studied the adsorption of As (III) and As (V) on $MnFe_2O_4$ and $CoFe_2O_4$, observing by characterization studies that a little As(III) was oxidized to As (V) and that the arsenic can be adsorbed by inner-sphere complexes between As (V) or As (III) and magnetic nanomaterials (Zhang et al. 2010).

This section summarizes the effect that mono and bimetallic nanoparticles have on the sorption of aqueous pollutants, with bimetallic ones having the highest level of adsorption of inorganic pollutants and toxic ions such as $As^{3+,5+}$, Pb^{2+}, $Cr^{3+,6+}$, Cd^{2+}, and other. The high absorption capacity is due to the increase in surface area of nanomaterials in addition to surface activation through active groups such as oxygen, redox process, and other removal mechanisms.

12.5 Use of Fe-Based Nanomaterials in AOPs for the Removal of Organic Pollutants

The importance of water for life, for the environment, for human beings, and for industries is indisputable. Among the different aspects that can limit the development of a country, freshwater scarcity is identified as a central and prominent point. Nevertheless, the quality and availability of this resource are affected by numerous variables, including increases in consumption and climate change (Moursi et al. 2017). In the context of scarcity, wastewater can be considered as a resource that should be tapped, and the need to developing technology for its recovery is very important (Capocelli et al. 2019). However, some of this wastewater has recalcitrant pollutants that are toxic, and its treatment is complex (Klauck et al. 2017; Rivas et al. 2019; Souza et al. 2013). This pollution causes depletion of dissolved oxygen, which has an adverse effect on the marine ecological system. The vast majority of organic pollutants present in different kinds of wastewater contain persistent organics pollutants (POPs) or emergent contaminants (EC), with high stability to sunlight irradiation and resistance to microbial attack and temperature (Petrie et al. 2015; Sorensen et al. 2015). As a result, many POPs and EC have been detected in low amounts, usually micrograms per liter, in rivers, lakes, oceans, and even drinking water all over the world (Huang et al. 2015; Smarr et al. 2016). Although the carcinogenic, mutagenic, and bactericidal properties of most POPs and/or EC remain unknown, there exists great concern for their removal from the aquatic environment to avoid the toxic consequences and potential hazardous health effects on living organisms, including human beings (Echeveste et al. 2016; Lignell et al. 2016; Pal et al. 2013; Smarr et al. 2016).

The most extended wastewater treatment around the world is the activated sludge biological process. This type of treatment has been applied since the beginning of 1900 and its importance is related to its simplicity, efficiency, and competitive cost. Nevertheless, the activated sludge process presents some limitations and does not always provide good results. Toxic and non-biodegradable wastewater cannot be treated by biological processes or, at the most, can be just partially treated (Lapertot et al. 2006). A typical example of this type of water is the industrial wastewater, the characteristics of which are highly dependent on the sort of origin (industrial, municipal, underwater, among others), but it usually presents ecotoxic compounds, high dissolved organic carbon, and salts concentrations. In this sense, advanced oxidation processes, known as AOPs, have emerged as especially efficient alternatives (Oller et al. 2011).

These processes are physicochemical treatments capable of producing high reactive transitory species (mainly hydroxyl radicals) that are very effective in the oxidation of organic matter which can be converted, in some cases, into H_2O, CO_2 and innocuous mineral salts when optimal operating conditions are applied. There are many and varied AOPs facilitating compliance with the specific treatment requirements. Although their efficiency has been widely recognized, their cost is significantly high (Ballesteros Martín et al. 2008); therefore, a special interest has been raised on those AOPs that can use Fe-based nanomaterials for the removal of organic contaminants. Fe-based nanomaterials act not only as adsorbents of contaminants but also due to its high reactivity for the transformation of organic contaminants in compounds less toxic. Among them, the scientific community agrees that heterogeneous photo-catalysis and photo-Fenton processes are the most promising water treatments due to their high conversion efficiency and versatility. On the other hand, one of the most prominent processes involving Fe nanomaterials (called zero-valent nanoparticles of iron, ZVI NPs) is its use in permeable reactive barriers (PRB) promoting the dehalogenation of chlorinated hydrocarbons. This reaction proceeds according to the process summarized in eq. 12.1 (Li et al. 2017; San Román et al. 2013).

$$Fe^0 + RX + H^+ \rightarrow Fe^{2+} + RH + X^- \tag{12.1}$$

The reaction has been successfully evaluated for many different chlorinated hydrocarbons including atrazine derivatives, clopyralid, lindane, perchloroethylene, and chlorophenols, among others (Chang and Cheng 2006; Choi et al. 2008; Dombek et al. 2004; Joo et al. 2004; Kim et al. 2010; Liu et al. 2011; Shih et al. 2011; Truex et al. 2011; Vidal et al. 2018; Xu and Wang 2011).

As mentioned earlier, in the Fenton process, the redox capacity of iron is used, where Fe^0 (2) or Fe^{2+} (3) reduces hydrogen peroxide and produces $^{\bullet}OH$ radicals, which are highly reactive and manage to degrade various contaminants present in the water (see Table 12.3) in some cases until its complete transformation to CO_2.

$$Fe^0 + H_2O_2 \rightarrow Fe^{2+} + 2OH^- \tag{12.2}$$

$$Fe^{2+} + H_2O_2 \rightarrow \cdot OH + Fe^{3+} + OH^- \tag{12.3}$$

An example of the ZVI NPs use is the work of Ortiz de la Plata et al., where they studied the 2-chlorophenol (2-CP) degradation using these NPs as Fenton catalyst through the heterogeneous photo Fenton reaction (Sanganyado and Gwenzi 2019). The most important results are related to the mineralization or transformation into CO_2 because the total organic carbon (TOC) decay does not exceed 5%, which indicates the formation of intermediates and/or reaction byproducts that have not been degraded.

This type of nanomaterial has also been applied in the simultaneous removal of aniline, antimony (Sb), and chromium (Cr) coupled with H_2O_2 (Gang et al. 2019). Experiment with iso-propanol (as the radical scavenger) confirmed that hydroxyl

Table 12.3 Use of zero-valent iron nanoparticles in different chemical and electrochemical processes to degrade organic compounds

Process	Pollutant	Experimental conditions	Main results	References
Heterogeneous Fenton	4-phenylazophenol (4-PAP)	1×10^{-4} mol/L of 4-PAP, 0.2 g/L of catalyst, 2×10^{-3} mol/L of H_2O_2, pH 3.0	Complete elimination was reached at 8 min of treatment while in the absence of H_2O_2, the same effect was achieved at 60 min	Donadelli et al. (2020)
Heterogeneous Fenton	Ametryn	4.54 mg/L of Ametryn, 17 mg/L H_2O_2, 2.83 mg/L of iron NPs, pH of 3.5, stirring at 210 rpm and room temperature	A 100% of herbicide removal at 135 min of treatment was achieved	Sangami and Manu (2017)
Heterogeneous Fenton	Methylene Blue (MB)	MB concentration of 30 mg/L, 0.1 Mol/L of H_2O_2, 1 g/L of the as-prepared catalyst powders, pH 7.0, 45 °C	100% of MB elimination and a 63% of mineralization at 40 min of treatment Nanomaterial reuse: 5 times	Li et al. (2020)
Adsorption and electro-Fenton	Sulfathiazole (STZ)	50 mg/L of STZ, 0.05 mol/L of Na_2SO_4 as the supporting electrolyte, 1 g/L of catalyst, air flow of 1.5 L/min, pH of the solution 5.0, 15 °C $j = 7.5$ mA cm^{-2}. Porous graphite was used as cathode and Ni as anode	Almost 100% STZ was removed within 5 min and a complete mineralization was achieved after 160 min	Chen et al. (2019)
Heterogeneous Fenton	Ibuprofen in real wastewater	5×10^{-6} mol/L of Ibuprofen, 0.08 g/L of catalyst and 1×10^{-4} mol/L of H_2O_2, pH 6	A complete elimination after 12 min of treatment	Minella et al. (2019)
Heterogeneous Fenton	Amoxicillin (AMX) and	Initial antibiotic concentration of 20 mg/L, 25 g/L of	A greater degradation at lower treatment	Ghauch et al. (2009)

(continued)

Table 12.3 (continued)

Process	Pollutant	Experimental conditions	Main results	References
	Ampicillin (AMP)	catalyst, pH of 6.60, stirring at 750 rpm	times was achieved with the ZVI supported on polyethylene glycol	
Heterogeneous photo electro-Fenton	Phenol	Initial phenol concentration of 200 mg/L, pH of 6.2, 0.5 g/L of catalyst and 500 mg/L of H_2O_2, mercury lamp of 8 W $j = 12$ mA/cm^2 Two stainless steel were used as electrodes	A complete removal of phenol was observed within 30 min	Babuponnusami and Muthukumar (2012)

radicals derived from Fenton reaction account for aniline degradation, but not for Sb and Cr. The elimination of Sb and Cr was due to the action of Fe^0 and iron (hydro)oxides, the last formed by the corrosion of Fe^0 with H_2O_2.

The magnetic properties of ZVI nanomaterials are very attractive because they can be easily separated from the aqueous solution with the application of a magnetic field. However, they have a high reactivity in the presence of water, which is associated with the high surface energy and oxidation/passivation processes. One way to solve this problem is to modify them with a noble metal such as Pd or Au, called bimetallic nanoparticles. The addition of a noble metal protects the iron core from rapid oxidation and also serves as a catalyst for water remediation. It is believed that noble metal improves iron oxidation and can act as a catalyst for electron transfer and hydrogenation reactions, achieving significantly higher degradation rates, and prevent the formation of toxic by-products.

In Table 12.4, different bimetallic NPs were used to eliminate organic compounds like dyes and antibiotics. Lien et al. used Fe/Al bimetallic NPs to eliminate formic acid (FA). In this bimetallic system, iron is responsible for the •OH formation, while aluminum acts as an electron donor to maintain the electron supply and preserve the outer layer of iron at the zero-valence state with enhanced surface areas. The authors propose a reaction mechanism that involves both a surface-mediated reaction and an aqueous phase reaction. The first step occurring at the bimetallic Fe/Al surface, ZVI was oxidized to Fe (II) upon corrosion and produced hydrogen under reducing conditions, while a surface-mediated process of oxygen reduction took place to generate superoxide (O•-). In the aqueous phase, these O•– radicals further reacted with hydrogen ions, thereby leading to H_2O_2 production. In the last step, a typical Fenton reaction occurred, where Fe (II) and H_2O_2 reacted to form Fe (III) and •OH.

Table 12.4 Bimetallic nanoparticles used in different chemical and electrochemical processes to degrade organic compounds

Nanomaterial	Process	Pollutant	Experimental conditions	Main results	References
Magnetic carbon nanomaterials coated with iron NPs	Heterogeneous photo Fenton	Rhodamine B (RhB)	30 mL/L of RhB, 0.67 g/L Fe/C, 0.43 mol/L H_2O_2, irradiation of 300 W with a xenon lamp	95.10% of degradation at 45 min with the material of Fe/C-300	Wang et al. (2018)
Fe/Cu allophane nanoclays	Heterogeneous electro-Fenton	Phenol	Initial phenol concentration of 5×10^{-4} mol/L, 0.05 mol/L of Na_2SO_4 as supporting electrolyte, pH of 5.5. A glassy carbon electrode modified with the bimetallic Fe-Cu NPs was used as working electrode, platinum wire as counter electrode, and $Ag/AgCl_{sat}$ as reference electrode	100% of phenol conversion and 80% of COD removal after 2 h of treatment	Garrido-Ramírez et al. (2016)
Fe/Ni	Adsorption	Profenofos (PFF)	1 mg/L of PFF, 0.8 g/L of catalyst, pH 7	96% of adsorption after 16 min of contact time. Max. Adsorption capacity of 0.2020 mg/g.	Mansouriieh et al. (2016)
Fe/Cu	Ozonation	Indigo carmine (IC)	500 mg/L of IC, 1 g/l of Fe/Cu particles, pH 3.0	97% of indigo carmine removal and 92% of TOC removal	Torres-Blancas et al. (2017)
Fe/Cu supported on bentonite	Adsorption	Tetracycline (TC)	Initial TC concentration of 10 mg/L, 0.15 g/L at pH 7.0	95% of TC removal Nanomaterial reuse: 5 times	Gopal et al. (2020)

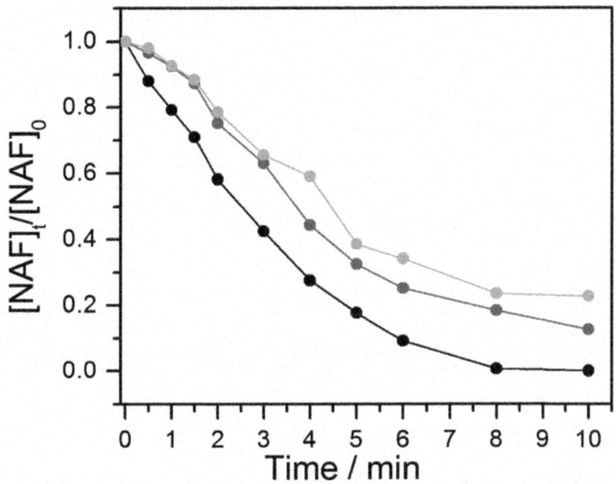

Fig. 12.6 Reuse of Fe/Cu NPs as catalysts in HEF. (•) 1 st cycle, (●) second cycle, (●) third cycle. Experimental conditions: 5 mA cm^{-2}, Na_2SO_4 0.05 M, pH 7, [NAF] = 36 mg/L. (Campos et al. 2020)

Recently, Campos et al. (2020) used Fe/Cu bimetallic nanoparticles (Fe/Cu NPs) for the degradation of Nafcillin (NAF) by the heterogeneous electro-Fenton process (Campos et al. 2020). NAF was completely eliminated after 7 min of electrolysis (see Fig. 12.6). In a second experiment with recovery bimetallic nanoparticles, the degradation reached 88% of NAF decay at the same time. Finally, in the last reuse cycle, at the beginning of treatment, elimination resembles the last cycle, with de exception that, after 10 min of electrolysis, NAF elimination is lower, allowing 77% of degradation. The reduction in the effectiveness of the HEF process is due to changes in the surface of the NP, because new oxides were formed. Nevertheless, NAF elimination through this process and using Fe/Cu bimetallic NPs is still significant.

The Fenton-like reaction has been used as an alternative to the conventional Fenton reaction since it allows another source of oxidants production different from Fe^{2+} or Fe^0. Diverse technologies combine the Fenton-like reaction with the application of UV or solar radiation to make the process more efficient. Among these, it is important to mention the heterogeneous photocatalysis with Fe oxide-based nanomaterials or the mixtures of iron oxides with other materials for the elimination of organic compounds (see Table 12.5). It has been recognized as "green" and inexpensive because it can be carried out under mild experimental conditions (ambient temperature and pressure with cheap and nontoxic semiconductors as photocatalysts), often using water as solvent, O_2 as oxidizing agent, and solar or artificial light with low-energy consumption as the irradiation sources (Bardhan et al. 2015). Furthermore, photocatalytic reactions do not generate toxic by-products (Corma Canos et al. 2007; Granone et al. 2018; Sheldon 2014). With this in environmental conditions, hematite is the most stable phase of iron oxide

Table 12.5 Iron oxide and mixtures of iron oxides used in photocatalysis to degrade organic compounds

Nanomaterial	Pollutant	Experimental conditions	Main results	References
Hematite	Phenol	1×10^{-4} mol/L of phenol, 1.0×10^{-3} mol/L of H_2O_2, and 200 mg/L of catalyst. Irradiance of 18 W/m^2 and solution pH of 4.0	100% of elimination after 240 min of irradiation	Demarchis et al. (2015)
Hematite	Methylene Blue (MB)	7×10^{-3} g/L of MB, 0.05 mol/of H_2O_2, catalyst dosage of 0.16 g/L. A 150 W xenon lamp was used	A 99.6% of photocatalytic efficiency was achieved	J. Liu et al., (2019b)
Magnetite	Diphenhydramine (DP)	100 mg/L of DP, 0.02 g/L of catalyst, H_2O_2 with a concentration of 0.16 mol/L, pH 2.8, irradiance of 0.033 W/cm^2 and room temperature	Complete degradation of DP with a 78% of mineralization was achieved	Pastrana-Martínez et al. (2015)
Magnetite	Sunset Yellow (SY) dye	Initial SY dye concentration of 100 mg/L, pH of 3.0, UVA irradiation (300–400 nm), 0.1 mol/L of Na^2SO^4 or 0.13 mol/L of NaCl at 20 °C three electrode system: Gas diffusion electrode (GDE) used as working electrode, Ag/AgCl as reference and platinum (Pt) screen as counter electrode	Concentration decay of SY dye was 80 and 99% In the presence of Na_2SO_4 and NaCl respectively. A mineralization ~50% was achieved at 90 min Nanomaterial reuse: 8 times	Pinheiro et al. (2020)
Hematite combine with graphitic carbon nitride (α-Fe_2O_3/g-C_3N_4)	Rhodamine B (RhB) and tetracycline (TC)	10 mg/L of RhB or TC, 1 g/L of catalyst and neutral pH Stirring rate of 500 rpm	96% of degradation after 90 min of treatment for RhB and 95% of elimination for TC after 150 min	Wang et al. (2020)
Ammonia-modified graphene	Mixture of phenol, 2-nitrophenol (2-NP) and	Concentration of phenolic compounds 4×10^{-4}	93.56% phenol, 98.76% 2-NP and 98.06% of 2-CP	Boruah et al. (2017)

(continued)

Table 12.5 (continued)

Nanomaterial	Pollutant	Experimental conditions	Main results	References
(AG) sheets decorated with Fe_3O_4 (AG/Fe_3O_4)	2-chlorophenol (2-CP)	mol/L, catalyst loading 0.5 g/L and pH 7. Intensity of the sunlight between 600–700 w/m^2	degradation were achieved within 50–80 min using AG/Fe_3O_4 nanocomposite under sunlight irradiation	

(α-Fe_2O_3) (Mishra and Chun 2015). This n-type semiconductor is abundant, nontoxic, stable in photocatalytic reactions and in aqueous solvents and is one of the materials cheaper among the available semiconductors (Maji et al. 2012). On the other hand, hematite has a high absorption of visible light (500–600 nm), so it can collect a greater amount of energy from the sun. The use of Fe_2O_3 on a nanometric scale considerably increases its photocatalytic properties for the degradation of organic pollutants.

Due to the advantages of this material, Domacena et al. (2020) studied the effects of type of additives on the morphology and corresponding photocatalytic property of various hematite nanostructures (α-Fe_2O_3 croissant-like structures, urchin-like structures, and textured microspheres) in the methyl orange (MO) degradation under UV-C lamp (Domacena et al. 2020). The urchin-like α-Fe_2O_3 hierarchical structures exhibited the best photocatalytic behavior with a 76.5% removal of MO after a 2-h irradiation. This is attributed to the high surface area of the urchin-like morphology which provides more active sites for the degradation.

Another important photocatalysis work is the one reported by Soares Emídio et al. (2020) in which a catalyst based on a mixture of copper oxide and magnetite is synthesized ($Fe_{3-x}Cu_xO_4$) to degrade the anticancer drugs 5-fluorouracil and cyclophosphamide (Emídio et al. 2020). Among the results stands out the higher catalytic activity for H_2O_2 conversion to $^{\bullet}OH$, compared to pristine magnetite (Fe_3O_4) and the low leaching of Cu and Fe, which demonstrated the stability of the catalyst, with high catalytic activity (>90%) maintained after use in four consecutive cycles.

Due to the versatility of Fe-based materials, they are a great alternative for the degradation of organic pollutants present in water through advanced oxidation processes as photocatalysis or processes based on the Fenton reaction. It also highlights the magnetic properties of Fe-based materials, which allow the material to be recovered and reused in the degradation of contaminants.

12.6 Conclusions

The use of Fe-based materials and nanomaterials with magnetic characteristic are an excellent alternative to remove different types of pollutant present in water, considering the high removal capacity associated with its diverse removal mechanisms (adsorption, redox, precipitation, coprecipitation, and others) and reusing due to its magnetic characteristic. In this context, we evidence the versatility of these materials, and nanomaterials can be used in two excellent removal processes: adsorption and AOPs. The first is extensively applied to remove inorganic pollutants, for example, Pb^{2+}, $As^{3+,5+}$, $Cr^{3+,6+}$, and Cd^{2+}, where the use of adsorbent substrates based on bimetallic nanoparticles and/or functionalized Fe-magnetic nanomaterials can improved the removal capacity of these pollutant ions. On the other hand, the versatility of Fe-based materials allows its use in advanced oxidation processes, achieving the elimination of persistent pollutants present in water, producing the de-halogenation of toxic organochlorine compounds or even the total mineralization of the initial pollutant and its intermediates. The main AOPs that are involved in Fe-based materials are the zero-valent iron method, heterogeneous photocatalysis, and the Fenton process, where the material participates as a catalyst and also allows its reuse in several degradation cycles.

Finally, in this chapter, different characterization techniques were summarized too, where TEM, SEM, XRD, XPS, Raman, and Mössbauer spectroscopy are an important techniques to determine specific information about size, shape, morphology, crystallinity, and magnetic and chemical properties of materials and nanomaterials and from these characteristics to determine the possible removal mechanisms of pollutants associated with changes of the material.

Acknowledgments Pamela Sepúlveda acknowledges ANID-FONDECYT/Post-Doctoral Grant N° 3200063 and Centers of Excellence with Basal/ANID Financing, GRANT AFB 180001, CEDENNA; L. Carolina Espinoza acknowledges ANID-FONDECYT/Post-Doctoral Grant N° 3200062. Alejandra García acknowledges Alejandro Arizpe Zapata from CIMAV for technical support. Finally, Pamela Sepúlveda, Ricardo Salazar, and L. Carolina Espinoza thank Dirección de Investigación Científica y Tecnológica, *DICYT-USACH.*

References

Ahmed MB, Zhou JL, Ngo HH, Guo W, Johir MAH, Sornalingam K, Belhaj D, Kallel M (2017) Nano-Fe0 immobilized onto functionalized biochar gaining excellent stability during sorption and reduction of chloramphenicol via transforming to reusable magnetic composite. Chem Eng J 322:571–581. https://doi.org/10.1016/j.cej.2017.04.063

ALOthman ZA, Habila MA, Al-Shalan NH, Alfadul SM, Ali R, Rashed IGA a, Alfarhan B (2016) Adsorptive removal of Cu(II) and Pb(II) onto mixed-waste activated carbon: kinetic, thermodynamic, and competitive studies and application to real wastewater samples. Arab J Geosci 9. https://doi.org/10.1007/s12517-016-2350-9

Asfaram A, Ghaedi M, Hajati S, Goudarzi A (2016) Synthesis of magnetic γ-Fe2O3-based nanomaterial for ultrasonic assisted dyes adsorption: modeling and optimization. Ultrason Sonochem 32:418–431. https://doi.org/10.1016/j.ultsonch.2016.04.011

Azizian S (2004) Kinetic models of sorption: a theoretical analysis. J Colloid Interface Sci 276:47–52. https://doi.org/10.1016/j.jcis.2004.03.048

Babuponnusami A, Muthukumar K (2012) Removal of phenol by heterogenous photo electro Fenton-like process using nano-zero valent iron. Sep Purif Technol 98:130–135. https://doi.org/10.1016/j.seppur.2012.04.034

Ballesteros Martín MM, Sánchez Pérez JA, Acién Fernández FG, Casas López JL, García-Ripoll AM, Arques A, Oller I, Malato Rodríguez S (2008) Combined photo-Fenton and biological oxidation for pesticide degradation: effect of photo-treated intermediates on biodegradation kinetics. Chemosphere 70:1476–1483. https://doi.org/10.1016/j.chemosphere.2007.08.027

Bao S, Yang W, Wang Y, Yu Y, Sun Y (2020) One-pot synthesis of magnetic graphene oxide composites as an efficient and recoverable adsorbent for Cd(II) and Pb(II) removal from aqueous solution. J Hazard Mater 381:120914. https://doi.org/10.1016/j.jhazmat.2019.120914

Bardhan SK, Gupta S, Gorman ME, Haider MA (2015) Biorenewable chemicals: Feedstocks, technologies and the conflict with food production. Renew Sust Energ Rev 51:506–520. https://doi.org/10.1016/j.rser.2015.06.013

Biswas P, Karn AK, Balasubramanian P, Kale PG (2017) Biosensor for detection of dissolved chromium in potable water: a review. Biosens Bioelectron 94:589–604. https://doi.org/10.1016/j.bios.2017.03.043

Boroumand Jazi M, Arshadi M, Amiri MJ, Gil A (2014) Kinetic and thermodynamic investigations of Pb(II) and Cd(II) adsorption on nanoscale organo-functionalized SiO2Al2O3. J Colloid Interface Sci 422:16–24. https://doi.org/10.1016/j.jcis.2014.01.032

Boruah PK, Sharma B, Karbhal I, Shelke MV, Das MR (2017) Ammonia-modified graphene sheets decorated with magnetic Fe3O4 nanoparticles for the photocatalytic and photo-Fenton degradation of phenolic compounds under sunlight irradiation. J Hazard Mater 325:90–100. https://doi.org/10.1016/j.jhazmat.2016.11.023

Camacho LM, Parra RR, Deng S (2011) Arsenic removal from groundwater by MnO2-modified natural clinoptilolite zeolite: effects of pH and initial feed concentration. J Hazard Mater 189:286–293. https://doi.org/10.1016/j.jhazmat.2011.02.035

Campos S, Salazar R, Arancibia-Miranda N, Rubio MA, Aranda M, García A, Sepúlveda P, Espinoza LC (2020) Nafcillin degradation by heterogeneous electro-Fenton process using Fe, Cu and Fe/Cu nanoparticles. Chemosphere 247. https://doi.org/10.1016/j.chemosphere.2020.125813

Capocelli M, Prisciandaro M, Piemonte V, Barba D (2019) A technical-economical approach to promote the water treatment & reuse processes. J Clean Prod 207:85–96. https://doi.org/10.1016/j.jclepro.2018.09.135

Chang JH, Cheng SF (2006) The remediation performance of a specific electrokinetics integrated with zero-valent metals for perchloroethylene contaminated soils. J Hazard Mater 131:153–162. https://doi.org/10.1016/j.jhazmat.2005.09.026

Chen J, He F, Zhang H, Zhang X, Zhang G, Yuan G (2014) Novel core-shell structured Mn-Fe/MnO2 magnetic nanoparticles for enhanced Pb(II) removal from aqueous solution. Ind Eng Chem Res 53:18481–18488. https://doi.org/10.1021/ie502967a

Chen YP, Yang LM, Paul Chen J, Zheng YM (2019) Electrospun spongy zero-valent iron as excellent electro-Fenton catalyst for enhanced sulfathiazole removal by a combination of adsorption and electro-catalytic oxidation. J Hazard Mater 371:576–585. https://doi.org/10.1016/j.jhazmat.2019.03.043

Choi JH, Choi SJ, Kim YH (2008) Hydrodechlorination of 2,4,6-trichlorophenol for a permeable reactive barrier using zero-valent iron and catalyzed iron. Korean J Chem Eng 25:493–500. https://doi.org/10.1007/s11814-008-0083-5

Corma Canos A, Iborra S, Velty A (2007) Chemical routes for the transformation of biomass into chemicals. Chem Rev 107:2411–2502. https://doi.org/10.1021/cr050989d

Crane RA, Scott TB (2012) Nanoscale zero-valent iron: future prospects for an emerging water treatment technology. J Hazard Mater 211–212:112–125. https://doi.org/10.1016/j.jhazmat.2011.11.073

Czaplinska J, Sobczak I, Ziolek M (2014) Bimetallic AgCu/SBA-15 system: the effect of metal loading and treatment of catalyst on surface properties. J Phys Chem C 118:12796–12810. https://doi.org/10.1021/jp5011764

Demarchis L, Minella M, Nisticò R, Maurino V, Minero C, Vione D (2015) Photo-Fenton reaction in the presence of morphologically controlled hematite as iron source. J Photochem Photobiol A Chem 307–308:99–107. https://doi.org/10.1016/j.jphotochem.2015.04.009

Devi V, Selvaraj M, Selvam P, Kumar AA, Sankar S, Dinakaran K (2017) Preparation and characterization of CNSR functionalized Fe3O4 magnetic nanoparticles: an efficient adsorbent for the removal of cadmium ion from water. J Environ Chem Eng 5:4539–4546. https://doi.org/10.1016/j.jece.2017.08.036

Dionisiou NS, Matsi T (2016) Natural and surfactant-modified zeolite for the removal of pollutants (mainly inorganic) from natural waters and wastewaters, environmental materials and waste: resource recovery and pollution prevention. Elsevier Inc. https://doi.org/10.1016/B978-0-12-803837-6.00023-8

Domacena AMG, Aquino CLE, Balela MDL (2020) Photo-Fenton degradation of methyl orange using hematite (α-Fe 2 O 3) of various morphologies. Mater Today Proc 22:248–254. https://doi.org/10.1016/j.matpr.2019.08.095

Dombek T, Davis D, Stine J, Klarup D (2004) Degradation of terbutylazine (2-chloro-4-ethylamino-6-terbutylamino-1,3,5- triazine), deisopropyl atrazine (2-amino-4-chloro-6-ethylamino-1,3,5-triazine), and chlorinated dimethoxy triazine (2-chloro-4,6-dimethoxy-1,3,5-triazine) by zero valent iron and e. Environ Pollut 129:267–275. https://doi.org/10.1016/j.envpol.2003.10.008

Donadelli JA, Caram B, Kalaboka M, Kapsi M, Sakkas VA, Carlos L, García Einschlag FS (2020) Mechanisms of 4-phenylazophenol elimination in micro- and nano-ZVI assisted-Fenton systems. J Environ Chem Eng 8:103624. https://doi.org/10.1016/j.jece.2019.103624

Echeveste P, Galbán-Malagón C, Dachs J, Berrojalbiz N, Agustí S (2016) Toxicity of natural mixtures of organic pollutants in temperate and polar marine phytoplankton. Sci Total Environ 571:34–41. https://doi.org/10.1016/j.scitotenv.2016.07.111

Elici ROF (2012) Surface X-ray diffraction 1424–1443

Emídio ES, Hammer P, Nogueira RFP (2020) Simultaneous degradation of the anticancer drugs 5-fluorouracil and cyclophosphamide using a heterogeneous photo-Fenton process based on copper-containing magnetites (Fe3-xCuxO4). Chemosphere 241:124990. https://doi.org/10.1016/j.chemosphere.2019.124990

Ferrando R, Jellinek J, Johnston RL (2008) Nanoalloys: from theory to applications of alloy clusters and nanoparticles. 108

Gang X, Wang Q, Qian Y, Gao P, Su Y, Liu Z, Chen H, Li X, Chen J (2019) Simultaneous removal of aniline, antimony and chromium by ZVI coupled with H2O2: implication for textile wastewater treatment. J Hazard Mater 368:840–848. https://doi.org/10.1016/j.jhazmat.2019.02.009

Garrido-Ramírez EG, Marco JF, Escalona N, Ureta-Zañartu MS (2016) Preparation and characterization of bimetallic Fe-Cu allophane nanoclays and their activity in the phenol oxidation by heterogeneous electro-Fenton reaction. Microporous Mesoporous Mater 225:303–311. https://doi.org/10.1016/j.micromeso.2016.01.013

Ghauch A, Tuqan A, Assi HA (2009) Antibiotic removal from water: elimination of amoxicillin and ampicillin by microscale and nanoscale iron particles. Environ Pollut 157:1626–1635. https://doi.org/10.1016/j.envpol.2008.12.024

Ghosal PS, Gupta AK (2017) Development of a generalized adsorption isotherm model at solid-liquid interface: a novel approach. J Mol Liq 240:21–24. https://doi.org/10.1016/j.molliq.2017.05.042

Gomes JAG, Daida P, Kesmez M, Weir M, Moreno H, Parga JR, Irwin G, McWhinney H, Grady T, Peterson E, Cocke DL (2007) Arsenic removal by electrocoagulation using combined Al-Fe electrode system and characterization of products. J Hazard Mater 139:220–231. https://doi.org/10.1016/j.jhazmat.2005.11.108

Gopal G, Sankar H, Natarajan C, Mukherjee A (2020) Tetracycline removal using green synthesized bimetallic nZVI-Cu and bentonite supported green nZVI-Cu nanocomposite: a comparative study. J Environ Manag 254:109812. https://doi.org/10.1016/j.jenvman.2019.109812

Gouadec G, Colomban P (2007) Raman spectroscopy of nanomaterials: how spectra relate to disorder, particle size and mechanical properties. Prog Cryst Growth Charact Mater 53:1–56. https://doi.org/10.1016/j.pcrysgrow.2007.01.001

Granone LI, Sieland F, Zheng N, Dillert R, Bahnemann DW (2018) Photocatalytic conversion of biomass into valuable products: a meaningful approach? Green Chem 20:1169–1192. https://doi.org/10.1039/c7gc03522e

Guo X, Wang J (2019) A general kinetic model for adsorption: theoretical analysis and modeling. J Mol Liq 288:111100. https://doi.org/10.1016/j.molliq.2019.111100

Hao L, Liu M, Wang N, Li G (2018) A critical review on arsenic removal from water using iron-based adsorbents. RSC Adv 8:39545–39560. https://doi.org/10.1039/c8ra08512a

Hosseini SM, Tosco T (2013) Transport and retention of high concentrated nano-Fe/Cu particles through highly flow-rated packed sand column. Water Res 47:326–338. https://doi.org/10.1016/j.watres.2012.10.002

Huang S, He S, Xu H, Wu P, Jiang R, Zhu F, Luan T, Ouyang G (2015) Monitoring of persistent organic pollutants in seawater of the Pearl River Estuary with rapid on-site active SPME sampling technique. Environ Pollut 200:149–158. https://doi.org/10.1016/j.envpol.2015.02.016

Huang D, Wu J, Wang L, Liu X, Meng J, Tang X, Tang C, Xu J (2019) Novel insight into adsorption and co-adsorption of heavy metal ions and an organic pollutant by magnetic graphene nanomaterials in water. Chem Eng J 358:1399–1409. https://doi.org/10.1016/j.cej.2018.10.138

Huq ME, Fahad S, Shao Z, Sarven MS, Al-Huqail AA, Siddiqui MH, Habib ur Rahman M, Khan IA, Alam M, Saeed M, Rauf A, Basir A, Jamal Y, Khan SU (2019) High arsenic contamination and presence of other trace metals in drinking water of Kushtia district, Bangladesh. J Environ Manag 242:199–209. https://doi.org/10.1016/j.jenvman.2019.04.086

Inkson BJ (2016) Scanning Electron Microscopy (SEM) and Transmission Electron microscopy (TEM) for materials characterization, materials characterization using nondestructive evaluation (NDE) methods. Elsevier Ltd. https://doi.org/10.1016/B978-0-08-100040-3.00002-X

Islam M, Mishra PC, Patel R (2011) Arsenate removal from aqueous solution by cellulose-carbonated hydroxyapatite nanocomposites. J Hazard Mater 189:755–763. https://doi.org/10.1016/j.jhazmat.2011.03.051

Joo SH, Feitz AJ, Waite TD (2004) Oxidative degradation of the carbothioate herbicide, molinate, using nanoscale zero-valent iron. Environ Sci Technol 38:2242–2247. https://doi.org/10.1021/es035157g

Kalantary RR, Farzadkia M, Kermani M, Rahmatinia M (2018) Heterogeneous electro-Fenton process by Nano-Fe3O4 for catalytic degradation of amoxicillin: process optimization using response surface methodology. J Environ Chem Eng 6:4644–4652. https://doi.org/10.1016/j.jece.2018.06.043

Katsoyiannis IA, Zouboulis AI (2004) Application of biological processes for the removal of arsenic from groundwaters. Water Res 38:17–26. https://doi.org/10.1016/j.watres.2003.09.011

Khan ST, Malik A (2019) Engineered nanomaterials for water decontamination and purification: from lab to products. J Hazard Mater 363:295–308. https://doi.org/10.1016/j.jhazmat.2018.09.091

Khin MM, Nair AS, Babu VJ, Murugan R, Ramakrishna S (2012) A review on nanomaterials for environmental remediation. Energy Environ Sci 5:8075–8109. https://doi.org/10.1039/c2ee21818f

Kim H, Hong HJ, Jung J, Kim SH, Yang JW (2010) Degradation of trichloroethylene (TCE) by nanoscale zero-valent iron (nZVI) immobilized in alginate bead. J Hazard Mater 176:1038–1043. https://doi.org/10.1016/j.jhazmat.2009.11.145

Kim SA, Kamala-Kannan S, Lee KJ, Park YJ, Shea PJ, Lee WH, Kim HM, Oh BT (2013) Removal of Pb(II) from aqueous solution by a zeolite-nanoscale zero-valent iron composite. Chem Eng J 217:54–60. https://doi.org/10.1016/j.cej.2012.11.097

Klauck CR, Giacobbo A, Altenhofen CG, Silva LB, Meneguzzi A, Bernardes AM, Rodrigues MAS (2017) Toxicity elimination of landfill leachate by hybrid processing of advanced oxidation process and adsorption. Environ Technol Innov 8:246–255. https://doi.org/10.1016/j.eti.2017.07.006

Kumar PR, Chaudhari S, Khilar KC, Mahajan SP (2004) Removal of arsenic from water by electrocoagulation. Chemosphere 55:1245–1252. https://doi.org/10.1016/j.chemosphere.2003.12.025

Lai B, Zhang YH, Yuan Y, Chen ZY, Yang P (2014) Influence of preparation conditions on characteristics, reactivity, and operational life of microsized FE/CU bimetallic particles. Ind Eng Chem Res 53:12295–12304. https://doi.org/10.1021/ie501756m

Lapertot M, Pulgarín C, Fernández-Ibáñez P, Maldonado MI, Pérez-Estrada L, Oller I, Gernjak W, Malato S (2006) Enhancing biodegradability of priority substances (pesticides) by solar photo-Fenton. Water Res 40:1086–1094. https://doi.org/10.1016/j.watres.2006.01.002

Lata S, Samadder SR (2016) Removal of arsenic from water using nano adsorbents and challenges: a review. J Environ Manag 166:387–406. https://doi.org/10.1016/j.jenvman.2015.10.039

Li X, Elliott DW, Zhang W (2006) Zero-valent iron nanoparticles for abatement of environmental pollutants: zero-valent iron nanoparticles for abatement of environmental pollutants: materials and engineering aspects. https://doi.org/10.1080/10408430601057611

Li XQ, Elliott DW, Zhang WX (2008) Zero-valent iron nanoparticles for abatement of environmental pollutants: materials and engineering aspects. Part Syst Nano-Biotechnol 8436:309–329. https://doi.org/10.1080/10408430601057611

Li J, Chen S, Sheng G, Hu J, Tan X, Wang X (2011) Effect of surfactants on Pb(II) adsorption from aqueous solutions using oxidized multiwall carbon nanotubes. Chem Eng J 166:551–558. https://doi.org/10.1016/j.cej.2010.11.018

Li X, Zhou M, Pan Y, Xu L (2017) Pre-magnetized Fe0/persulfate for notably enhanced degradation and dechlorination of 2,4-dichlorophenol. Chem Eng J 307:1092–1104. https://doi.org/10.1016/j.cej.2016.08.140

Li J, Dou X, Qin H, Sun Y, Yin D, Guan X (2019) Characterization methods of zerovalent iron for water treatment and remediation. Water Res 148:70–85. https://doi.org/10.1016/j.watres.2018.10.025

Li D, Yang T, Li Y, Liu Z, Jiao W (2020) Facile and green synthesis of highly dispersed tar-based heterogeneous Fenton catalytic nanoparticles for the degradation of methylene blue. J Clean Prod 246. https://doi.org/10.1016/j.jclepro.2019.119033

Lignell S, Winkvist A, Bertz F, Rasmussen KM, Glynn A, Aune M, Brekke HK (2016) Environmental organic pollutants in human milk before and after weight loss. Chemosphere 159:96–102. https://doi.org/10.1016/j.chemosphere.2016.05.077

Liu A, Zhang W (2014) Fine structural features of nanoscale zero-valent iron characterized by spherical aberration corrected scanning transmission electron microscopy (Cs-STEM). Analyst 139:4512–4518

Liu X, Zhao W, Sun K, Zhang G, Zhao Y (2011) Dechlorination of PCBs in the simulative transformer oil by microwave-hydrothermal reaction with zero-valent iron involved. Chemosphere 82:773–777. https://doi.org/10.1016/j.chemosphere.2010.10.074

Liu WJ, Qian TT, Jiang H (2014) Bimetallic Fe nanoparticles: recent advances in synthesis and application in catalytic elimination of environmental pollutants. Chem Eng J 236:448–463. https://doi.org/10.1016/j.cej.2013.10.062

Liu A, Liu J, Han J, Zhang WX (2017) Evolution of nanoscale zero-valent iron (nZVI) in water: microscopic and spectroscopic evidence on the formation of nano- and micro-structured iron oxides. J Hazard Mater 322:129–135. https://doi.org/10.1016/j.jhazmat.2015.12.070

Liu H, Li P, Yu H, Zhang T, Qiu F (2019a) Controlled fabrication of functionalized nanoscale zero-valent iron/celluloses composite with silicon as protective layer for arsenic removal. Chem Eng Res Des 151:242–251. https://doi.org/10.1016/j.cherd.2019.09.020

Liu J, Wang B, Li Z, Wu Z, Zhu K, Zhuang J, Xi Q, Hou Y, Chen J, Cong M, Li J, Qian G, Lin Z (2019b) Photo-Fenton reaction and H2O2 enhanced photocatalytic activity of α-Fe2O3 nanoparticles obtained by a simple decomposition route. J Alloys Compd 771:398–405. https://doi.org/10.1016/j.jallcom.2018.08.305

Luther S, Borgfeld N, Kim J, Parsons JG (2012) Removal of arsenic from aqueous solution: a study of the effects of pH and interfering ions using iron oxide nanomaterials. Microchem J 101:30–36. https://doi.org/10.1016/j.microc.2011.10.001

Mahdi S, Sahebi H, Zandavar H, Mirsadeghi S (2019) Fabrication of Fe 3 O 4 nanoparticles coated by extracted shrimp peels chitosan as sustainable adsorbents for removal of chromium contaminates from wastewater: the design of experiment. Compos Part B 175:107130. https://doi.org/10.1016/j.compositesb.2019.107130

Maji SK, Mukherjee N, Mondal A, Adhikary B (2012) Synthesis, characterization and photocatalytic activity of α-Fe 2O 3 nanoparticles. Polyhedron 33:145–149. https://doi.org/10.1016/j.poly.2011.11.017

Manning BA, Hunt ML, Amrhein C, Yarmoff JA (2002) Arsenic (III) and arsenic (V) reactions with zerovalent iron corrosion products. Environ Sci Technol 36:5455–5461. https://doi.org/10.1021/es0206846

Mansouriieh N, Sohrabi MR, Khosravi M (2016) Adsorption kinetics and thermodynamics of organophosphorus profenofos pesticide onto Fe/Ni bimetallic nanoparticles. Int J Environ Sci Technol 13:1393–1404. https://doi.org/10.1007/s13762-016-0960-0

McGrory ER, Brown C, Bargary N, Williams NH, Mannix A, Zhang C, Henry T, Daly E, Nicholas S, Petrunic BM, Lee M, Morrison L (2017) Arsenic contamination of drinking water in Ireland: a spatial analysis of occurrence and potential risk. Sci Total Environ 579:1863–1875. https://doi.org/10.1016/j.scitotenv.2016.11.171

McKay G, Bino MJ, Altamemi AR (1985) The adsorption of various pollutants from aqueous solutions on to activated carbon. Water Res 19:491–495. https://doi.org/10.1016/0043-1354(85)90041-7

Minella M, Bertinetti S, Hanna K, Minero C, Vione D (2019) Degradation of ibuprofen and phenol with a Fenton-like process triggered by zero-valent iron (ZVI-Fenton). Environ Res 179:108750. https://doi.org/10.1016/j.envres.2019.108750

Mishra M, Chun DM (2015) α-Fe2O3 as a photocatalytic material: a review. Appl Catal A Gen 498:126–141. https://doi.org/10.1016/j.apcata.2015.03.023

Mohan D, Pittman CU (2006) Activated carbons and low cost adsorbents for remediation of tri- and hexavalent chromium from water. J Hazard Mater 137:762–811. https://doi.org/10.1016/j.jhazmat.2006.06.060

Mohan D, Pittman CU (2007) Arsenic removal from water/wastewater using adsorbents-a critical review. J Hazard Mater 142:1–53. https://doi.org/10.1016/j.jhazmat.2007.01.006

Moursi H, Kim D, Kaluarachchi JJ (2017) A probabilistic assessment of agricultural water scarcity in a semi-arid and snowmelt-dominated river basin under climate change. Agric Water Manag 193:142–152. https://doi.org/10.1016/j.agwat.2017.08.010

Murtaza B, Shah NS, Sayed M, Khan JA, Imran M, Shahid M, Khan ZUH, Ghani A, Murtaza G, Muhammad N, Khalid MS, Niazi NK (2019) Synergistic effects of bismuth coupling on the reactivity and reusability of zerovalent iron nanoparticles for the removal of cadmium from aqueous solution. Sci Total Environ 669:333–341. https://doi.org/10.1016/j.scitotenv.2019.03.062

Naseem K, Begum R, Wu W, Usman M, Irfan A, Al-Sehemi AG, Farooqi ZH (2019) Adsorptive removal of heavy metal ions using polystyrene-poly (N-isopropylmethacrylamide-acrylic acid) core/shell gel particles: adsorption isotherms and kinetic study. J Mol Liq 277:522–531. https://doi.org/10.1016/j.molliq.2018.12.054

Ntombenhle G, Thabang P, Meyyappan M, Moothi K (2019) Simultaneous removal of pollutants from water using nanoparticles: a shift from single pollutant control to multiple pollutant control. Sci Total Environ 656:808–833. https://doi.org/10.1016/j.scitotenv.2018.11.257

O'Carroll D, Sleep B, Krol M, Boparai H, Kocur C (2013) Nanoscale zero valent iron and bimetallic particles for contaminated site remediation. Adv Water Resour 51:104–122. https://doi.org/10.1016/j.advwatres.2012.02.005

Olefjord I (2007) X-ray photoelectron spectroscopy. Surf Charact A User's Sourceb:291–319

Oller I, Malato S, Sánchez-Pérez JA (2011) Combination of advanced oxidation processes and biological treatments for wastewater decontamination-a review. Sci Total Environ 409:4141–4166. https://doi.org/10.1016/j.scitotenv.2010.08.061

Pal S, Blais JM, Robidoux MA, Haman F, Krümmel E, Seabert TA, Imbeault P (2013) The association of type 2 diabetes and insulin resistance/secretion with persistent organic pollutants in two first nations communities in northern Ontario. Diabetes Metab 39:497–504. https://doi.org/10.1016/j.diabet.2013.01.006

Pastrana-Martínez LM, Pereira N, Lima R, Faria JL, Gomes HT, Silva AMT (2015) Degradation of diphenhydramine by photo-Fenton using magnetically recoverable iron oxide nanoparticles as catalyst. Chem Eng J 261:45–52. https://doi.org/10.1016/j.cej.2014.04.117

Peters G, Naidoo D, Masenda H, Genga RM, Freemantle CS, Cornish LA (2019) A Mössbauer spectroscopy study of Fe based cemented carbides. Int J Refract Met Hard Mater:105127. https://doi.org/10.1016/j.ijrmhm.2019.105127

Petrie B, Barden R, Kasprzyk-Hordern B (2015) A review on emerging contaminants in wastewaters and the environment: current knowledge, understudied areas and recommendations for future monitoring. Water Res 72:3–27. https://doi.org/10.1016/j.watres.2014.08.053

Pinheiro ACN, Bernardino TS, Junior FEB, Lanza MRV, Barros WRP (2020) Enhanced electrodegradation of the sunset yellow dye in acid media by heterogeneous Photoelectro-Fenton process using Fe3O4 nanoparticles as a catalyst. J Environ Chem Eng 8:103621. https://doi.org/10.1016/j.jece.2019.103621

Ponder SM, Darab JG, Mallouk TE (2000) Remediation of Cr (VI) and Pb (II) aqueous solutions using supported, nanoscale zero-valent iron. Environ Sci Technol 34:2564–2569. https://doi.org/10.1021/es9911420

Quinn J, Elliott D, O'Hara S, Billow A (2009) Use of nanoscale iron and bimetallic particles for environmental remediation: a review of field-scale applications. ACS Symp Ser 1027:263–283. https://doi.org/10.1021/bk-2009-1027.ch015

Quiroz MA, Bandala ER, Martinez-Huitle CA (2011) Advanced oxidation processes (AOPs) for removal of pesticides from aqueous media. In: Pesticides-formulations, effects, fate. InTech, Rijeka

Reddy AVB, Yusop Z, Jaafar J, Reddy YVM, Aris AB, Majid ZA, Talib J, Madhavi G (2016) Recent progress on Fe-based nanoparticles: synthesis, properties, characterization and environmental applications. J Environ Chem Eng 4:3537–3553. https://doi.org/10.1016/j.jece.2016.07.035

Rius J (2015) Structure from diffraction methods. Crystallogr Rev 21:213–216. https://doi.org/10.1080/0889311X.2014.991393

Rivas FJ, Solís RR, Beltrán FJ, Gimeno O (2019) Sunlight driven photolytic ozonation as an advanced oxidation process in the oxidation of bezafibrate, cotinine and iopamidol. Water Res 151:226–242. https://doi.org/10.1016/j.watres.2018.12.013

San Román I, Alonso ML, Bartolomé L, Galdames A, Goiti E, Ocejo M, Moragues M, Alonso RM, Vilas JL (2013) Relevance study of bare and coated zero valent iron nanoparticles for lindane degradation from its by-product monitorization. Chemosphere 93:1324–1332. https://doi.org/10.1016/j.chemosphere.2013.07.050

Sangami S, Manu B (2017) Synthesis of Green Iron Nanoparticles using Laterite and their application as a Fenton-like catalyst for the degradation of herbicide Ametryn in water. Environ Technol Innov 8:150–163. https://doi.org/10.1016/j.eti.2017.06.003

Sanganyado E, Gwenzi W (2019) Antibiotic resistance in drinking water systems: occurrence, removal, and human health risks. Sci Total Environ 669:785–797. https://doi.org/10.1016/j.scitotenv.2019.03.162

Santurtún A, Delgado-Alvarado M, Villar A, Riancho J (2016) Geographical distribution of mortality by Parkinson's disease and its association with air lead levels in Spain. Med Clínica (English Ed.) 147:481–487. https://doi.org/10.1016/j.medcle.2016.12.024

Sen Gupta S, Bhattacharyya KG (2011) Kinetics of adsorption of metal ions on inorganic materials: a review. Adv Colloid Interf Sci 162:39–58. https://doi.org/10.1016/j.cis.2010.12.004

Sen Gupta S, Bhattacharyya KG (2014) Adsorption of metal ions by clays and inorganic solids. RSC Adv 4:28537–28586. https://doi.org/10.1039/c4ra03673e

Sepúlveda P, Rubio MA, Baltazar SE, Rojas-Nunez J, Sánchez Llamazares JL, Garcia AG, Arancibia-Miranda N (2018) As(V) removal capacity of FeCu bimetallic nanoparticles in aqueous solutions: the influence of Cu content and morphologic changes in bimetallic nanoparticles. J Colloid Interface Sci 524:177–187. https://doi.org/10.1016/j.jcis.2018.03.113

Shahwan T (2014) Sorption kinetics: obtaining a pseudo-second order rate equation based on a mass balance approach. J Environ Chem Eng 2:1001–1006. https://doi.org/10.1016/j.jece.2014.03.020

Sheldon RA (2014) Green and sustainable manufacture of chemicals from biomass: state of the art. Green Chem 16:950–963. https://doi.org/10.1039/c3gc41935e

Shen G, Jiang DM, Lin F, Shi WZ, Ma XM (2005) Mössbauer study on the disordered Fe60Cr40 alloys prepared by mechanical alloying. Phys B Condens Matter 367:137–141. https://doi.org/10.1016/j.physb.2005.06.008

Shih YH, Hsu CY, Su YF (2011) Reduction of hexachlorobenzene by nanoscale zero-valent iron: kinetics, pH effect, and degradation mechanism. Sep Purif Technol 76:268–274. https://doi.org/10.1016/j.seppur.2010.10.015

Smarr MM, Grantz KL, Zhang C, Sundaram R, Maisog JM, Barr DB, Louis GMB (2016) Persistent organic pollutants and pregnancy complications. Sci Total Environ 551–552:285–291. https://doi.org/10.1016/j.scitotenv.2016.02.030

Song X, Zhang Y, Luo X, Chen P, Liu J (2019) 2D magnetic scallion sheathing-based biochar composites design and application for effective removal of arsenite in aqueous solutions. Chem Eng Res Des 152:384–392. https://doi.org/10.1016/j.cherd.2019.10.007

Sorensen JPR, Lapworth DJ, Nkhuwa DCW, Stuart ME, Gooddy DC, Bell RA, Chirwa M, Kabika J, Liemisa M, Chibesa M, Pedley S (2015) Emerging contaminants in urban groundwater sources in Africa. Water Res 72:51–63. https://doi.org/10.1016/j.watres.2014.08.002

Souza BS, Moreira FC, Dezotti MWC, Vilar VJP, Boaventura RAR (2013) Application of biological oxidation and solar driven advanced oxidation processes to remediation of winery wastewater. Catal Today 209:201–208. https://doi.org/10.1016/j.cattod.2012.08.037

Suazo-Hernández J, Sepúlveda P, Manquián-Cerda K, Ramírez-Tagle R, Rubio MA, Bolan N, Sarkar B, Arancibia-Miranda N (2019) Synthesis and characterization of zeolite-based composites functionalized with nanoscale zero-valent iron for removing arsenic in the presence of selenium from water. J Hazard Mater 373:810–819. https://doi.org/10.1016/j.jhazmat.2019.03.125

Sun Y-P, Li X, Cao J, Zhang W, Wang HP (2006) Characterization of zero-valent iron nanoparticles. Adv Colloid Interf Sci 120:47–56. https://doi.org/10.1016/j.cis.2006.03.001

Taylor P, Li X, Elliott DW, Zhang W, Li X, Elliott DW, Zhang W (2006) Zero-valent iron nanoparticles for abatement of environmental pollutants: materials and engineering aspects zero-valent iron nanoparticles for abatement of environmental pollutants: materials and engineering aspects. Crit Rev Solid State Materials Sci:37–41. https://doi.org/10.1080/10408430601057611

Teodosiu C, Gilca A, Barjoveanu G, Fiore S (2018) Emerging pollutants removal through advanced drinking water treatment: a review on processes and environmental performances assessment. J Clean Prod 197:1210–1221. https://doi.org/10.1016/j.jclepro.2018.06.247

Torres-Blancas T, Roa-Morales G, Ureña-Núñez F, Barrera-Díaz C, Dorazco-González A, Natividad R (2017) Ozonation enhancement by Fe–Cu biometallic particles. J Taiwan Inst Chem Eng 74:225–232. https://doi.org/10.1016/j.jtice.2017.02.025

Truex MJ, Mac Beth TW, Vermeul VR, Fritz BG, Mendoza DP, Mac Kley RD, Wietsma TW, Sandberg G, Powell T, Powers J, Pitre E, Michalsen M, Ballock-Dixon SJ, Zhong L, Oostrom M (2011) Demonstration of combined zero-valent iron and electrical resistance heating for in situ trichloroethene remediation. Environ Sci Technol 45:5346–5351. https://doi.org/10.1021/es104266a

Vergis BR, Kottam N, Hari Krishna R, Nagabhushana BM (2019) Removal of Evans Blue dye from aqueous solution using magnetic spinel ZnFe 2 O 4 nanomaterial: adsorption isotherms and kinetics. Nano-Struct Nano-Objects 18:100290. https://doi.org/10.1016/j.nanoso.2019.100290

Vidal J, Saez C, Cañizares P, Navarro V, Salazar R, Rodrigo MA (2018) ZVI–reactive barriers for the remediation of soils polluted with clopyralid: are they really Worth? Chem Eng J 350:100–107. https://doi.org/10.1016/j.cej.2018.05.142

Wang J, Zheng S, Shao Y, Liu J, Xu Z, Zhu D (2010) Amino-functionalized Fe3O4@SiO2 core-shell magnetic nanomaterial as a novel adsorbent for aqueous heavy metals removal. J Colloid Interface Sci 349:293–299. https://doi.org/10.1016/j.jcis.2010.05.010

Wang C, Luo H, Zhang Z, Wu Y, Zhang J, Chen S (2014) Removal of As(III) and As(V) from aqueous solutions using nanoscale zero valent iron-reduced graphite oxide modified composites. J Hazard Mater 268. https://doi.org/10.1016/j.jhazmat.2014.01.009

Wang J, Zhang Q, Shao X, Ma J, Tian G (2018) Properties of magnetic carbon nanomaterials and application in removal organic dyes. Chemosphere 207:377–384. https://doi.org/10.1016/j.chemosphere.2018.05.109

Wang Y, Song H, Chen J, Chai S, Shi L, Chen C, Wang Y, He C (2020) A novel solar photo-Fenton system with self-synthesizing H2O2: enhanced photo-induced catalytic performances and mechanism insights. Appl Surf Sci 512:145650. https://doi.org/10.1016/j.apsusc.2020.145650

Wanjala BN, Luo J, Fang B, Mott D, Zhong C (2011) Gold-platinum nanoparticles: alloying and phase segregation. J Mater Chem 21:4012–4020. https://doi.org/10.1039/c0jm02682d

Wen Z, Zhang Y, Cheng G, Wang Y, Chen R (2019) Simultaneous removal of As(V)/Cr(VI) and acid orange 7 (AO7) by nanosized ordered magnetic mesoporous Fe-Ce bimetal oxides: behavior and mechanism. Chemosphere 218:1002–1013. https://doi.org/10.1016/j.chemosphere.2018.11.208

Wu H, Wu Q, Zhang J, Gu Q, Wei L, Guo W, He M (2019) Chromium ion removal from raw water by magnetic iron composites and Shewanella oneidensis MR-1. Sci Rep 9:1–16. https://doi.org/10.1038/s41598-018-37470-1

Xia S, Gu Z, Zhang Z, Zhang J, Hermanowicz SW (2014) Removal of chloramphenicol from aqueous solution by nanoscale zero-valent iron particles. Chem Eng J 257:98–104. https://doi.org/10.1016/j.cej.2014.06.106

Xu L, Wang J (2011) A heterogeneous Fenton-like system with nanoparticulate zero-valent iron for removal of 4-chloro-3-methyl phenol. J Hazard Mater 186:256–264. https://doi.org/10.1016/j.jhazmat.2010.10.116

Yan W, Herzing AA, Kiely CJ, Zhang WX (2010) Nanoscale zero-valent iron (nZVI): aspects of the core-shell structure and reactions with inorganic species in water. J Contam Hydrol 118:96–104. https://doi.org/10.1016/j.jconhyd.2010.09.003

Yan W, Ramos MAV, Koel BE, Zhang WX (2012) As(III) sequestration by iron nanoparticles: study of solid-phase redox transformations with X-ray photoelectron spectroscopy. J Phys Chem C 116:5303–5311. https://doi.org/10.1021/jp208600n

Yang Z, Ji S, Gao W, Zhang C, Ren L, Tjiu WW, Zhang Z, Pan J, Liu T (2013) Magnetic nanomaterial derived from graphene oxide/layered double hydroxide hybrid for efficient removal of methyl orange from aqueous solution. J Colloid Interface Sci 408:25–32. https://doi.org/10.1016/j.jcis.2013.07.011

Ye H, Crooks RM (2007) Effect of elemental composition of PtPd bimetallic nanoparticles containing an average of 180 atoms on the kinetics of the electrochemical oxygen reduction reaction. J Am Chem Soc 129:3627–3633. https://doi.org/10.1021/ja068078o

Yousef RI, El-Eswed B, Al-Muhtaseb AH (2011) Adsorption characteristics of natural zeolites as solid adsorbents for phenol removal from aqueous solutions: kinetics, mechanism, and thermodynamics studies. Chem Eng J 171:1143–1149. https://doi.org/10.1016/j.cej.2011.05.012

Zamzow MJ, Eichbaum BR, Sandgren KR, Shanks DE (1990) Removal of heavy metals and other cations from wastewater using zeolites. Sep Sci Technol 25(13–15):1555–1569. https://doi.org/10.1080/01496399008050409

Zhang S, Niu H, Cai Y, Zhao X, Shi Y (2010) Arsenite and arsenate adsorption on coprecipitated bimetal oxide magnetic nanomaterials: MnFe2O4 and CoFe2O4. Chem Eng J 158:599–607. https://doi.org/10.1016/j.cej.2010.02.013

Zhang Y, Chen W, Dai C, Zhou C, Zhou X (2015) Structural evolution of nanoscale zero-valent iron (nZVI) in anoxic Co2+ solution: interactional performance and mechanism. Sci Rep 5:13966. https://doi.org/10.1038/srep13966

Zhang Z, Xia K, Pan Z, Yang C, Wang X, Zhang G, Guo Y, Bai R (2019) Removal of mercury by magnetic nanomaterial with bifunctional groups and core-shell structure: synthesis, characterization and optimization of adsorption parameters. Appl Surf Sci 500:143970. https://doi.org/10.1016/j.apsusc.2019.143970

Zhou C, Zhu H, Wang Q, Wang J, Cheng J, Guo Y, Zhou X, Bai R (2017) Adsorption of mercury (ii) with an Fe3O4 magnetic polypyrrole-graphene oxide nanocomposite. RSC Adv 7:18466–18479. https://doi.org/10.1039/c7ra01147d

Zhou CC, Gao ZY, Wang J, Wu MQ, Hu S, Chen F, Liu JX, Pan H, Yan CH (2018) Lead exposure induces Alzheimers's disease (AD)-like pathology and disturbes cholesterol metabolism in the young rat brain. Toxicol Lett 296:173–183. https://doi.org/10.1016/j.toxlet.2018.06.1065

Chapter 13
Advanced Magnetic Adsorbents Prepared from Emulsion Template for Water Treatment

Yongfeng Zhu, Hui Yu, Bin Mu, and Aiqin Wang

Contents

Abstract Adsorption technology regarded as an ideal method to remove water contaminants has been widely applied in practical applications, as the merit of the wide suitability and low cost. Among various adsorbents, magnetic recyclable adsorbents have gained more and more attention in recent years, which not only decrease the risk of secondary pollution but also realize the cyclic use of the

Y. Zhu · B. Mu · A. Wang (✉)
Key Laboratory of Clay Mineral Applied Research of Gansu Province, Center of Eco-material and Green Chemistry, Lanzhou Institute of Chemical Physics, Chinese Academy of Sciences, Lanzhou, People's Republic of China
e-mail: zhuyf@licp.cas.cn; mubin@licp.cas.cn; aqwang@licp.cas.cn

H. Yu
Key Laboratory of Clay Mineral Applied Research of Gansu Province, Center of Eco-material and Green Chemistry, Lanzhou Institute of Chemical Physics, Chinese Academy of Sciences, Lanzhou, People's Republic of China

Center of Materials Science and Optoelectronics Engineering, University of Chinese Academy of Sciences, Beijing, People's Republic of China

L. Meili, G. L. Dotto (eds.), *Advanced Magnetic Adsorbents for Water Treatment*, Environmental Chemistry for a Sustainable World 61,
https://doi.org/10.1007/978-3-030-64092-7_13

adsorbent after being regenerated and even increase significantly the additional value for enriching the scattered metals and precious metals. Recently, more and more studies have concerned the morphological control, homogeneous size, and tuned porous structure of magnetic adsorbents except for the adsorption performance. Hence, the emulsion template technique is applied to construct the magnetic adsorbents based on the advantages of facilely controlling the size distribution, crystallinity, and porous structure of magnetic materials.

Here, recent studies on the preparation of magnetic materials based on the emulsion template are reviewed, including magnetic nanoparticles, magnetic microspheres, and magnetic porous materials, and then the applications of the magnetic adsorbents for water treatment are summarized and discussed. The major points include the following aspects: (1) the emulsion template for preparation of the magnetic materials presents several advantages such as the confined reaction in the "microreactor," controlled shape, particle size and distribution, high polymerization degree, high productivity, low reaction temperature, and sufficient and tuned porous structure. (2) The obtained magnetic adsorbents exhibit excellent adsorption performance to the various pollutants, including heavy metals, dyes, and other organic contaminants, as well as the oil-water separation. It is expected that this review could be regarded as an important reference for the design and fabrication of novel adsorbents.

Keywords Emulsion template · Adsorption · Magnetic · Heavy metals · Organic pollutants · Imprinted polymer · Emulsion polymerization · Spinel ferrites · Porous materials · Nanoparticle

13.1 Introduction

Adsorption technology has been widely studied and used to remove the coexisting water contaminants in practical applications, due to its low cost and wide suitability. Especially, the adsorbents with recyclability always attract much attention in practical, which not only decreases the risk of secondary pollution (Rydin et al. 2000; Yin et al. 2018) but also increases significantly the additional value (Xue et al. 2019; Hashem et al. 2020). Generally, the strategies for designing the recycled adsorbents involve the large volume (Dlamini et al. 2020; Ren et al. 2019) and incorporation of magnetic particles (Yu et al. 2019). Compared with the former, the magnetic adsorbents have got more and more concerns, due to the diversified design, flexible operation, and excellent separating effect (Hua et al. 2012).

Magnetite (Fe_3O_4) and maghemite (γ-Fe_2O_3) are the most popular and widely used magnetic materials and could be directly used as adsorbents to eliminate various pollutants (Patel et al. 2019; Tsedenbal et al. 2020; Liu et al. 2021). However, the unavoidable problems are generally encountered, including magnetism loss and decreased adsorption performance, which might be related to the oxidization or decomposition of the naked magnetic particles in water (Zhu et al. 2013). Hence, most of the magnetic adsorbents are fabricated by incorporating Fe_3O_4 or γ-Fe_2O_3 into the adsorbent matrix (Ji et al. 2020; Fahimirad et al. 2018; Dehghani et al. 2021; Jung et al. 2019; Nuryono et al. 2020; Maleki et al. 2019; Tang et al. 2019a; Huang et al. 2020). Although magnetic adsorbents prepared by this strategy exhibit better adsorption performance and higher stability against acid or alkaline, their dispersibility and morphologies are ignored, which actually have a significant effect on the adsorption application. For example, nano-adsorbents generally displayed excellent adsorption performance due to the large specific surface area and the amount of active sites. But agglomeration, which could reduce the adsorption capacity, is still an important problem in large-scale applications. In addition, since many adsorbents possess sufficient adsorption sites, the sites located in the adsorbents interior usually fail to play role in the removal of pollutants. What is important is the recycling of the adsorbent is difficult in most of actual situation, and the risk of secondary pollution still exists. Therefore, the research to increase the dispersity of nano-adsorbent and realize the sufficient utilization of the adsorption sites has become the new hotspot.

Among the various strategies, the emulsion template is regarded as one of the classical and effective pathways for the preparation of particles with homogeneous size, controlled shape, or regular pore structure (Weng et al. 2020; Mokadem et al. 2020). With the development of the emulsion technique, novel hybrid materials are designed and prepared using emulsion template, and it indicated three attractive traits compared with other methods. First, the materials could be shaped with various morphologies, including spherical, hollow, and porous (Wang et al. 2020a, b; Stubenrauch et al. 2018; Thompson et al. 2019). Second, the morphologies of novel materials are tuned conveniently by changing the emulsion factors. Last but not least, the obtained materials will be endowed with some new function after incorporation of specific particles, such as photocatalytic or magnetic property derived from TiO_2 or Fe_3O_4 (Li et al. 2014a, b).

In this chapter, we review the studies related to the preparation of magnetic nanoparticle, magnetic microsphere, and magnetic porous material from the emulsion template and the application in water treatment (Fig. 13.1). It is expected that this review will be regarded as an important reference for the design and fabrication of other novel magnetic adsorbents.

13.2 Preparation of Magnetic Materials from Emulsion Template

13.2.1 Magnetic Nanoparticles

Magnetic particles have been widely applied in various fields, such as soil remediation, mineral processing industry, water purification plant, and so on (Anjali et al.

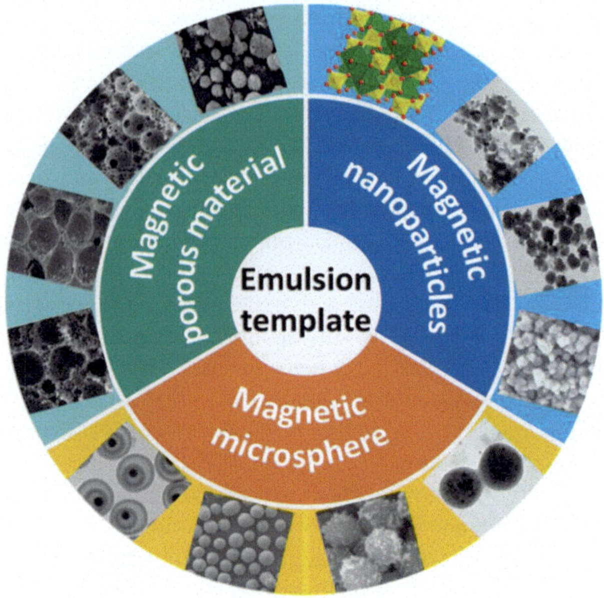

Fig. 13.1 The magnetic adsorbent prepared from the emulsion template

2019; Alhadidi et al. 2021; Li et al. 2021), and can be synthesized from many strategies, including coprecipitation (Aylar et al. 2020; Kavitha and Kurian 2020), thermal decomposition (Jesus et al. 2020), solvothermal (Fotukian et al. 2020), and microemulsion (Yousuf et al. 2019) (Table 13.1). Among these approaches, coprecipitation is most widely used, due to the product exhibited excellent dispersibility in water and convenient production process. And the size, shape, and magnetic property are affected by various parameters, including the type of ferric salts, the ratio of Fe^{2+} to Fe^{3+}, temperature, pH, etc. Even so, coprecipitation has several defects, including the large size and broad size distribution, which derived from the particles nucleation and subsequent growth up (Chen et al. 2016). In comparison, the magnetic nanoparticles prepared from the thermal decomposition at the presence of various stabilizing surfactants have monodisperse nanocrystals (Wu et al. 2008), while this process also has obvious drawbacks, including the complicated preparation process and expensive/toxic raw materials used (Xiao et al. 2016). Moreover, the obtained particles are hydrophobic and present a weak dispersibility in water. Solvothermal and hydrothermal techniques are good at preparing monodisperse magnetic particles with the controllable shape and narrow size distribution, but it is still limited in article due to the long synthetic time and high pressure.

Compared with other methods, the emulsion template, especially the microemulsion, exhibits better superiority in the preparation of magnetic nanoparticle. Because the reaction is limited in the "microreactor," the size, shape, and uniformity of particles could be controlled effectively. The general method is to mix two types of microemulsions, containing a salt or a complex of metal and a

Table 13.1 Comparison of the synthesis methods of iron oxide magnetic particles (Pang et al. 2016)

Methods	Reaction condition	Characteristic of the obtained products
Coprecipitation	Temperature: 20–90 °C	Shape control: Not good
	Duration: Minutes	Size distribution: Broad
	Solvent: Water	Crystallinity: Poor polydispersity
		Magnetization value: 20–80 emu/g
Thermal decomposition	Temperature: 100–320 °C	Shape control: Very good
	Duration: Hours–days	Size distribution: Very narrow
	Solvent: Organic compound	Crystallinity: High monodispersity
		Magnetization value: Up to 91 emu/g
Solvothermal	Temperature: 140–260 °C	Shape control: Good
	Duration: Hours	Size distribution: Narrow broad
	Solvent: Organic solvent or polyglycol	Crystallinity: High monodispersity
		Magnetization value: Up to 93 emu/g
Hydrothermal	Temperature: 150–220 °C	Shape control: Very good
	Solvent: Organic compound	Shape control: Good
		Magnetization value: Up to 113 emu/g

precipitating agent, respectively. Then the droplets take place collision and coalescence, and the magnetic nanoparticles nucleate and grow in the new droplets (Fig. 13.2).

For the fabrication and regulation of the morphology of magnetic nanoparticles, Pileni et al. conducted many pioneering works via emulsion template, including $CoFe_2O_4$ (Moumen et al. 1995a, b; Moumen and Pileni 1996a, b), Fe_3O_4 (Feltin and Pileni 1997), and cobalt-zinc ferrite magnetic nanoparticles (Hochepied and Pileni 2000). The size of the obtained magnetic particle could be adjusted in 2–11.6 nm. Soon afterwards, the magnetic spinel ferrites (SFs) prepared from the microemulsion become one of the hotspots following the relevant research of Pileni et al. SFs are the metal oxides which with the spinel structure, and the chemical constitution can be marked as AB_2O_4, where A and B represented various metal cations situated at tetrahedral (A site) and octahedral (B site), respectively. SFs can be classified as normal, inverse, and mixed based on the distribution of cations in tetrahedral and octahedral sites (Reddy and Yun 2016). The normal spinel includes $ZnFe_2O_4$, while $CdFe_2O_4$, $Fe[MFe]O_4$, $MgFe_2O_4$, $NiFe_2O_4$, $CoFe_2O_4$, and $CuFe_2O_4$ belong to the inverse spinel (Fröhlich et al. 2019; Masunga et al. 2019). The spinel with mixed structures is relatively rare, and $MnFe_2O_4$ is a typical example.

SFs could be synthesized by different methods, like calcination of the precursor (Popkov et al. 2020), sol-gel (Batoo and El-sadek 2013), hydrothermal (Ghahfarokhi and Shobegar 2020), ceramic method (Hilczer et al. 2016), coprecipitation (Ghone et al. 2018), and so on. However, the size of most SFs prepared from these methods is large and uncontrolled. In comparison, emulsion template is in favor of controlling

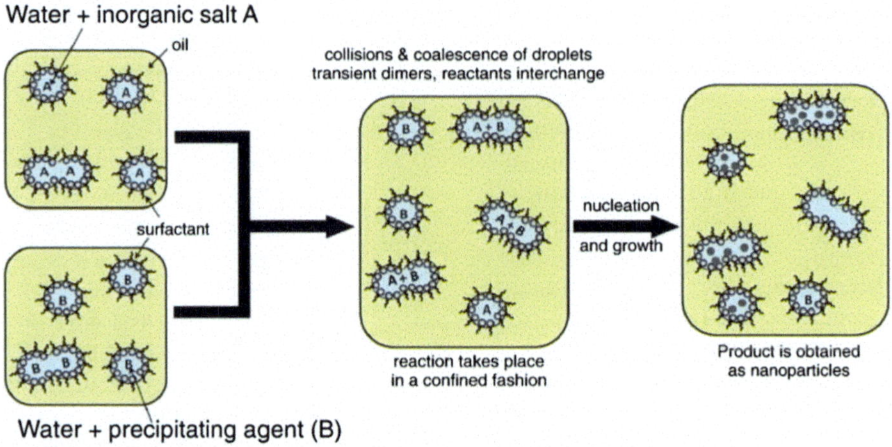

Fig. 13.2 The mechanism for the formation of metal particles by the microemulsion approach. (Reproduced with permission from Sanchez-Dominguez et al. 2012)

the size, morphology, shape and/or geometry, surface area, and homogeneity of magnetic particles (Ghone et al. 2018; Wang et al. 2012; Baig et al. 2019; Yousuf et al. 2019; Rafiq et al. 2020). Therefore, the microemulsion template is developed for the preparation of SFs with the controllable structure. For example, a series of $Mg_{1-x}Ca_xNi_yFe_{2-y}O_4$, $Zn_{1-x}Tb_xFe_2O_4$ were prepared via the microemulsion template, and the crystallite size of the synthesized samples could be facilely adjusted in the range of 15–45 nm.

A critical question is the interrelationship between the size of the obtained particles and the microemulsion characters. Many researches revealed that there is an almost linear correlation between them in few cases, but hardly find any correlation in most studies. So a range of experimental findings can be summarized as follows (Palmqvist 2003): (1) increasing reactant concentration will produce the increased particle size; (2) if the concentration of one of the reactants increases far beyond the other reactants, the particle size decreases; (3) the particle size might increase with the increase in the size of microemulsion droplet. It showed that the particles grown in the microemulsion droplet still have some complicated factors without control.

13.2.2 The Magnetic Microspheres

The introduction of the magnetic particle into the composite is very popular and has been widely reported every year (Fan et al. 2016; Xiao et al. 2016; Duan 2017). The purpose of these works is divided into three types. The first is to protect the magnetic particles from resisting the etching of acid or alkali, especially Fe_3O_4 and γ-Fe_2O_3 (Rott et al. 2018; Zhou et al. 2018). Fe_3O_4 and γ-Fe_2O_3 are the most widely used magnetic nanoparticles, but their physical properties are susceptible to change under different conditions. They are very unstable and easily transformed to other oxide forms at low pH, which affected their magnetic properties. So the coating or capsulation of magnetic particles is widely used in many studies (Lobato et al. 2019; Lobato et al. 2020). The second is to improve the dispersibility of magnetic particles. The agglomeration and formation the large clusters of magnetic particles in water is very common due to the hydrophobic surface (Lima and Feng 2012). The last reason of coated particles surface is to realize the functionalization of the nanoparticles by incorporation of various organic molecules or polymers (Wu et al. 2008; Ma et al. 2020; Kim et al. 2020).

Actually, the microfluidics technology may be the best method for the preparation of the material with near-perfect spherical shape (Zhang et al. 2018a, b; Kang et al. 2018). But the tedious process and the poor yield limit its large-scale production. In addition, in situ polymerization, pendant drop method, and emulsion template are developed for the preparation of the magnetic microsphere (Wang et al. 2010; Fang et al. 2019), and emulsion template technique is superior due to the high polymerization degree, high yield, and low reaction temperature.

Based on the preparation strategy, magnetic microsphere with four types of morphologies can be obtained by the emulsion template (Gervald et al. 2010) (Fig. 13.3): (a) the core-shell structure (magnetic particles as core and small molecule or polymer as shell), (b) the magnetic particle is embedded into the polymer matrix, (c) polymeric core with a surface layer of magnetic nanoparticles, and (d) the polymer further coats onto the magnetic particle supported polymeric core. Among them, the first three morphologies are very popular, but the reports involved the fourth structure is relatively rare. For example, Wang et al. (Wang et al. 2020a, b) fabricated triethylenetetramine-modified hollow Fe_3O_4/SiO_2/chitosan magnetic nanocomposite (Fe_3O_4/SiO_2/CS-TETA) with high specific surface by the emulsion polymerization. The carboxyl-functionalized polystyrene (PS) nanospheres were formed firstly by copolymerization of styrene and acrylic acid via the emulsion polymerization, and then the Fe_3O_4 nanoparticles were loaded onto. Later, the coating of silica onto the PS/Fe_3O_4 nanospheres and the calcinations at 500°C were carried out, and the hollow Fe_3O_4/SiO_2/CS nanocomposites were obtained finally after the chitosan modification (Fig. 13.4).

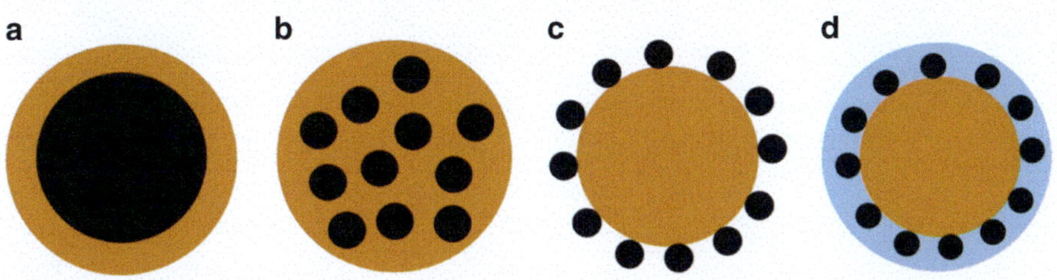

Fig. 13.3 Morphology types of magnetic polymer microspheres. (Reproduced with permission from Gervald et al. 2010)

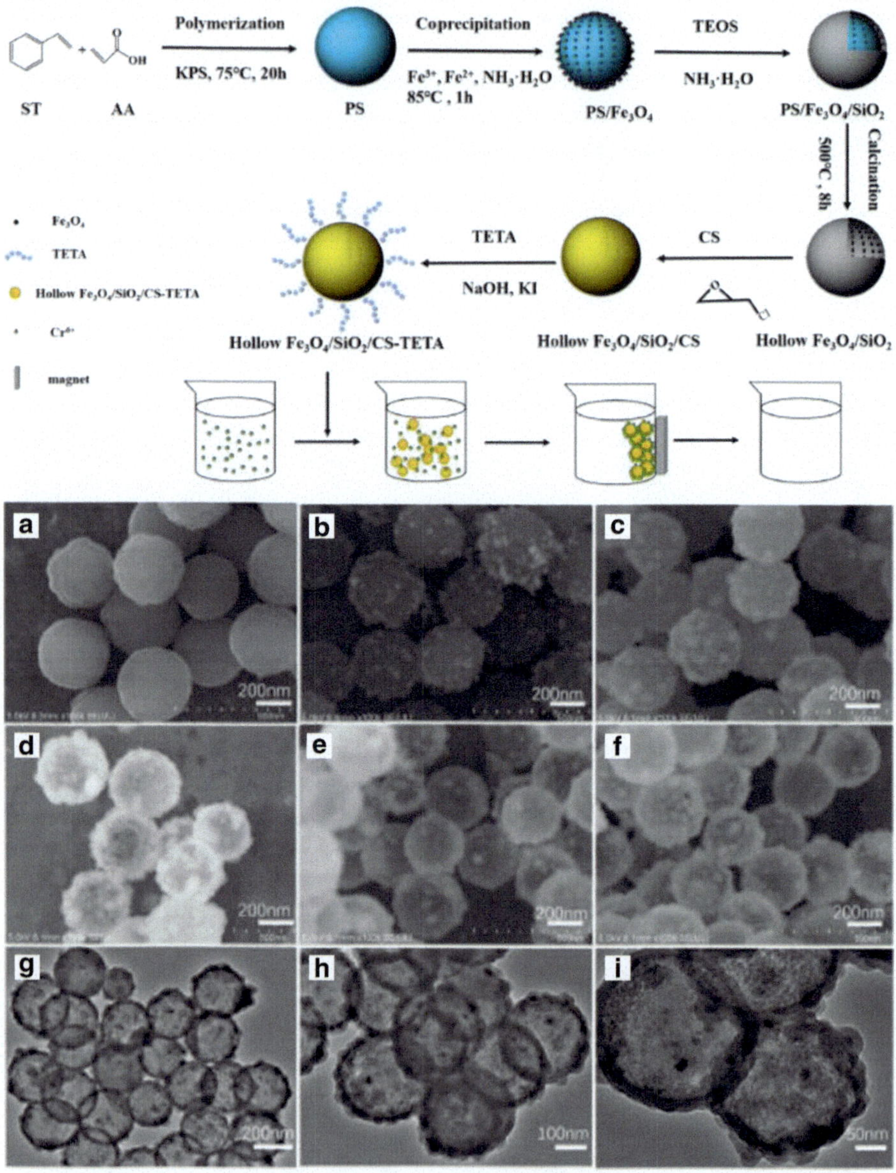

Fig. 13.4 Schematic illustration of the synthesis of hollow Fe_3O_4/SiO_2/CS-TETA nanocomposites and their application in recycle removal of Cr(VI). (Reproduced with permission from Wang et al. 2020a, b)

In order to achieve the coating, embedding, or supporting the magnetic particles, polymerization techniques such as emulsion polymerization, microemulsion polymerization, miniemulsion polymerization, dispersion polymerization, etc. have been used. Among them, emulsion polymerization is the most frequently adopted. However, due to the formation of the polymeric particles and oligomer occurs in the micelles and the aqueous phase simultaneously, the morphology and size of the obtained microspheres are very difficult to control via the emulsion polymerization. Compared with the emulsion polymerization, the structure, morphology, and size of microspheres could be efficiently regulated by microemulsion polymerization, miniemulsion polymerization, and nanoemulsion polymerization (Solans et al. 2005) (Fig. 13.5). But the problem is, if the microemulsion polymerization and miniemulsion polymerization are initiated with the free radical, partial magnetization might be lost due to the oxidizing initiator fragments (Zheng et al. 2005). In fact, the initiator types, concentrations of stabilizer, and the monomers dose also influence the

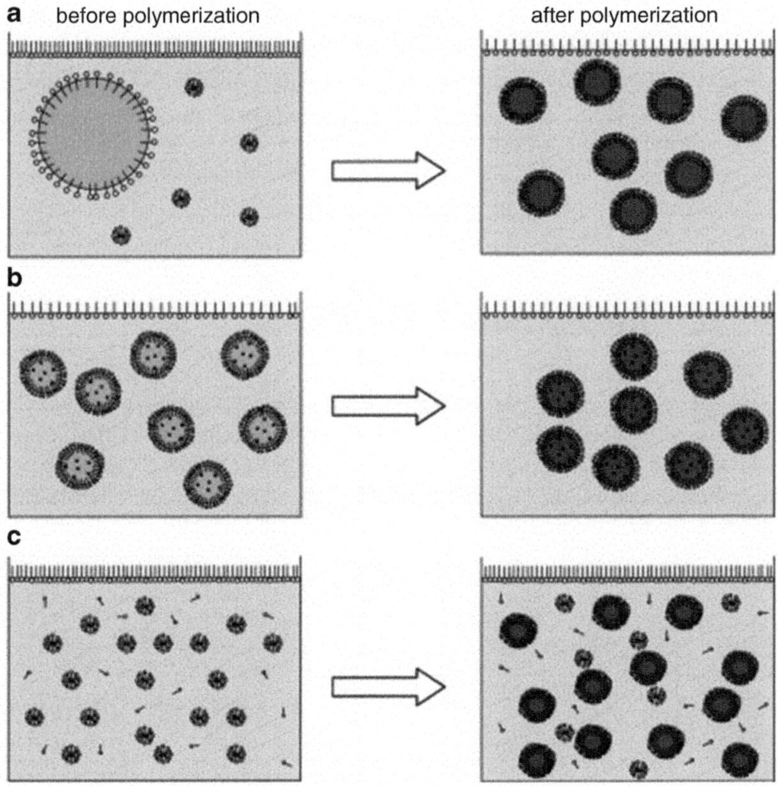

Fig. 13.5 Schematic representation of heterophase polymerization processes: (**a**) emulsion polymerization, (**b**) nanoemulsion polymerization, and (**c**) microemulsion polymerization. (Reproduced with permission from Solans et al. 2005)

morphologies and properties of magnetic nanocomposite (Hu et al. 2011; Feuser et al. 2015).

It is well known that the surfactants play a vital role in the microemulsion polymerization, miniemulsion polymerization, and nanoemulsion polymerization processes, which provide the droplets with colloidal stability against coalescence. But the inevitable migration of surfactants at the interface significantly affects the size and morphology of the obtained magnetic microspheres (Gharieh et al. 2019). In order to avoid this issue, the emulsifier-free miniemulsion polymerization might be a more wise choice (Zhang et al. 2016a, b). However, the stability of miniemulsion would be affected with ionic strength in the aqueous; hence, the hydrophilic monomer copolymerize with the hydrophobic monomer to keep the emulsion stability.

The dispersion of inorganic nanoparticles is another key problem to prepare polymer/inorganic nanocomposites by emulsion polymerization. Due to the nanoparticles have the high surface energy and easy to agglomerate together, the miniemulsion polymerization should be integrated with ultrasonic induction (Qiu et al. 2007). Compared with the conventional miniemulsion polymerization, ultrasound-induced miniemulsion polymerization possesses several advantages, such as no chemical initiators, low reaction temperatures, fast polymerization rate, higher monomer conversion and molecular weight. For instance, Teo et al. (Teo et al. 2009) prepared a novel poly(n-butyl methacrylate) latex bead with strong magnetism via one-pot method (Fig. 13.6). The O/W emulsion was prepared by dispersing the Fe_3O_4 nanoparticles into n-butyl methacrylate first and integrated with the high stirring and sonication under argon atmosphere. The polymerization reaction was preceded via continuous sonication without using any initiator.

13.2.3 The Magnetic Porous Materials

The porous material could be prepared from many approaches, such as hydrothermal synthesis (Kozyatnyk et al. 2019), freeze-drying (Anoshkin et al. 2018), porogenic solvent (Jiang and Kim 2013), or sacrificial hard template (Estevez et al. 2017). The as-prepared materials might have high porosity by these strategies, while the pore

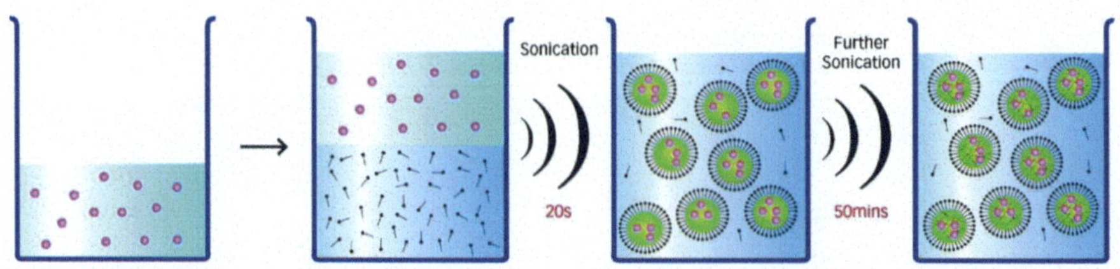

Fig. 13.6 A schematic of the process for magnetite nanocomposite spheres preparation by the sonochemically driven miniemulsion polymerization. (Reproduced with permission from Teo et al. 2009)

structure is not easy to control and tune. In comparison, the soft templates including the block copolymer template and colloidal template have been recognized to be more effective to synthesize ordered and disordered porous matrices (Wright et al. 2017). Especially, the emulsion template method is recognized to be an effective and versatile pathway for the preparation of polymeric materials with a well-defined porous structure, which is known as "polyHIPEs" (Chen et al. 2017, Zhang et al. 2018a, b, Gui et al. 2019). A polyHIPE is usually formed after finishing the polymerization reaction in the continuous phase of high internal phase emulsions (HIPEs), which has the large internal phase volume exceeded 74%, and then removing the dispersed phase. The interconnected pore will be formed as the thin membranes between the adjacent droplets are broke (Fig. 13.7) (Tan et al. 2018).

There are several important differences between the emulsion template used for the formation of the porous materials and the emulsion polymerization described above. First, the internal phase contents of HIPEs are above 74%, but the internal phase volume of emulsion polymerization and microencapsulation is significantly smaller. Second, the polymerization of HIPEs occurs in the continuous phase, while emulsion polymerization takes place in the dispersed phase. Moreover, HIPEs typically generate the monolithic material, but microspheres are obtained through the emulsion polymerization.

Both the surfactants and the amphiphilic solid particles are used to stabilize the HIPEs, but different characters of the surfactants and the amphiphilic particles generate different porous structures (Zhang et al. 2017). Generally speaking, surfactants or surfactant-like molecules are used to stabilize the emulsion in the

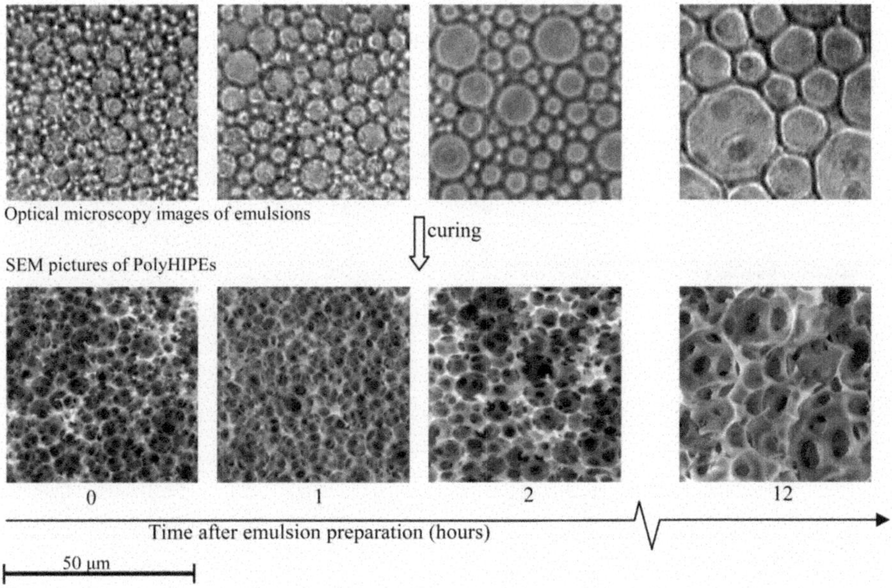

Fig. 13.7 Influence of time-dependent droplet coalescence on the morphology of polyHIPEs. (Reproduced with permission from Kovačič et al. 2007)

micromolecules; the porous materials with interconnected porous structure are obtained. On the contrary, the polymerization of the dispersed phase towards the amphiphilic particle-stabilized HIPEs always results in the closed-cell polymers with poor interconnectivity. Specifically, it has obvious positive effect of the stable particles modified by the surfactants. When the particles and the surfactants are synergistically stabilized in the emulsion, the stability of emulsion would be improved significantly and thus polyHIPEs attain excellent homogeneity. Meanwhile, the surfactants lead to the formation of the interconnected porous structure (Zheng et al. 2013).

At present, there are two approaches to obtain the magnetic porous materials from the emulsion template. The first is to disperse directly the magnetic particles into the emulsion continuous phase and then polymerization (Seeharaj et al. 2019). Due to the aggregation of the magnetic particles in the continuous phase, more researches are focused on the stabilization of the Pickering emulsion template with the magnetic particles. The magnetic particles should be modified with the organic molecules taken into account the inherent hydrophilicity of magnetic particles, such as surfactant, oleic acid, and so on (Zhang et al. 2019a, b, c, d). For example, Zhu et al. applied the amine-functionalized Fe_3O_4 nanoparticles ($Fe_3O_4 - NH_2$) to stabilize the HIPEs and fabricated novel magnetic porous polymers with a surface area of 5.532 m^2/g (Fig. 13.8). Our group also prepared the magnetic porous materials with sufficient interconnected porous structure from the HIPEs, which was stabilized with the amine-functionalized Fe_3O_4 (Zhu et al. 2016b; Lu et al. 2018a). Furthermore, we also developed another new type of magnetic porous adsorbent via the magnetic yeast and chitosan synergistically stabilized Pickering medium internal phase emulsions (Pickering MIPEs). As the droplet size and the stability of Pickering MIPEs could be adjusted by changing the synergistic effect between magnetic yeast

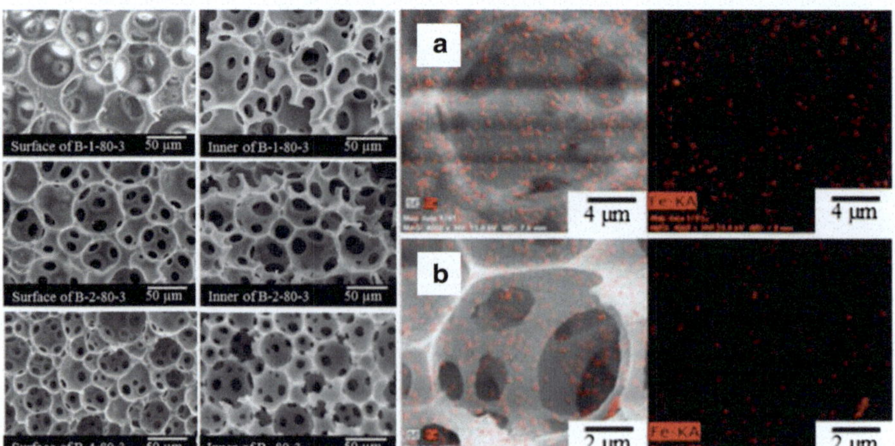

Fig. 13.8 SEM images of the surface and inner morphology of the emulsion-templated beads with different feeding amounts of FeNPs. (Reproduced with permission from Zhang et al. 2019a, b, c, d)

and chitosan, the pore structure of the as-prepared magnetic adsorbent could be flexibly tuned correspondingly (Lu et al. 2019a).

The shapes of the porous materials prepared from the emulsion template could be monolithic, microspherical, and even rod-like (Gokmen et al. 2009). The porous microspheres are formed easily by integrating the emulsion template with precipitation polymerization, but the formation of rod-like porous material still needs the microfluidic setup. We prepared the novel recyclable magnetic porous spheres by dropping the Pickering HIPEs into the hot liquid paraffin. The grafting polymerization reaction occurred between the hydroxypropyl cellulose and acrylic acid in the continuous phase of the Pickering HIPEs when the emulsion droplet fall (Zhu et al. 2017a; Zhu et al. 2017b). The size of as-prepared magnetic porous spheres was about 1.5 mm, and the sufficient porous structure existed in the spheres. The magnetic microsphere also attracts much attention due to the integrated advantages of nanoparticles and porous materials. In general, porous microspheres are formed by using the pore-foaming agent, and the strategy is classified as hard templates or soft templates. The uniform porous structure could be created after removing solid particles via etching in hard template, but the obvious flaws are the complicated removal process of hard templates and the harsh conditions.

In comparison, the post-processing of soft templates is more convenient, as the template removal could be achieved by a simple extraction or evaporation process. The microfluidics technology and the double-emulsion technique are the most representative soft template methods to prepare porous microspheres with interconnected porous structure. But the microfluidics technology possesses the gingerly preparation process, while the double-emulsion technique is simpler. In general, the formation of double emulsions needs a two-step emulsification process and also requires two kinds of surfactants to stabilize the oil-water (O/W) and water-oil (W/O) interfaces, respectively. The preparation process is flexible and suitable for the large-scale production (Fig. 13.9). Our group (Zhu et al. 2016c) prepared a series of magnetic porous microspheres via $(O_1/W)/O_2$ double emulsion. The silane-modified Fe_3O_4 particles and the surfactant of polyglycerol polyricinoleate were used to stabilize the internal O_1/W Pickering emulsion and the $(O_1/W)/O_2$ double emulsion, respectively. The results indicated that the magnetic microspheres presented a mean diameter of about 10 μm and interconnected porous structure.

13.3 The Application of the Magnetic Materials Prepared by Emulsion Template in Water Treatment

13.3.1 Removal of Heavy Metal Ions

Heavy metal ions, such as Pb^{2+}, Cr^{6+}, Cu^{2+}, Ni^{2+}, Cd^{2+}, Hg^{2+}, etc., are extremely noxious water pollutants and imposed serious side effects in living organisms. In addition, the prolonged excessive intake of heavy metal ions could damage the

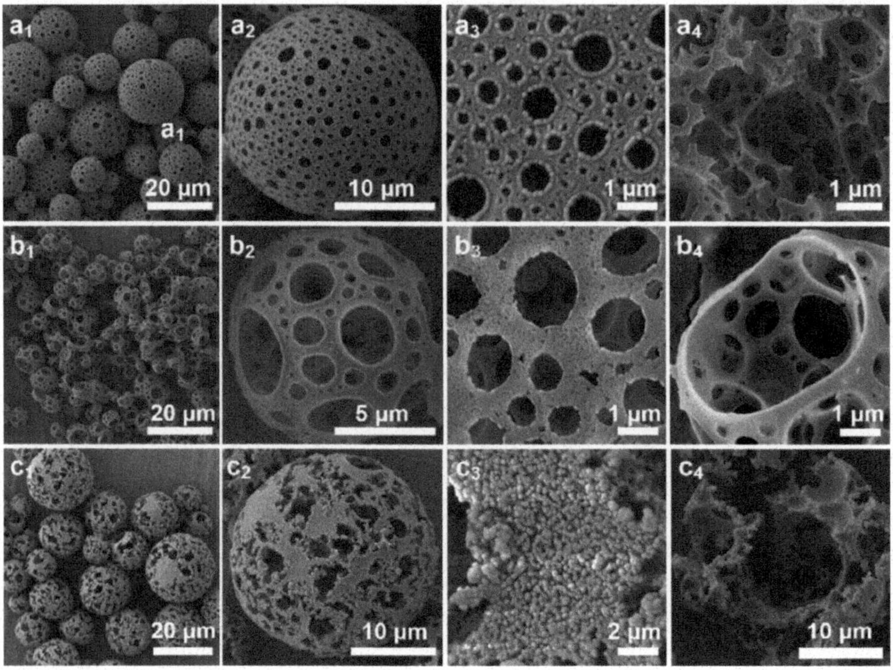

Fig. 13.9 SEM images of microspheres prepared from the double emulsions stabilized by a single anionic surfactant. (Reproduced with permission from Li et al. 2014a, b)

kidney, liver, brain function, and nervous system (Wadhawan et al. 2020). Adsorption is recognized as an efficient method for the removal of heavy metals, and magnetic adsorbents display distinct advantages including magnetic nanoparticles, magnetic composites with the micro−/nanospherical structure, and magnetic porous materials.

Magnetic Nanoparticles

Spinel ferrites (SFs) possess the superior chemical stability, enhanced magnetic properties, large surface area, vast of active sites at the corners, edges, and steps, so SFs applied in water treatment attract much attention and display enormous potential. For example, the adsorption performance of $MnFe_2O_4$ to Cu^{2+} and Pb^{2+} was reported about 197 mg/g and 21.64 mg/g (Ren et al. 2012). The magnetic MFe_2O_4 (M = Co, Ni, Cu, and Zn) nanoparticles had enhanced adsorption capacities of 69.4 and 47.1 mg/g for Cd^{2+} and Pb^{2+}, respectively (Yaqoob et al. 2019). The $MnFe_2O_4$ and $CoFe_2O_4$ prepared by Asadi et al. had the understanding adsorption capacities of 454.5 and 384.6 mg/g for Zn^{2+} (Asadi et al. 2020). The magnetic $Co_{0.6}Fe_{2.4}O_4$ microparticles with a uniform pore size of about 7.432 nm showed a high specific surface area of 97.155 m^2/g, and 80.32 mg/g of adsorption capacities towards Pb^{2+} (Kaur et al. 2015). The SFs prepared from emulsion template also present good adsorption performance for heavy metals. For instance, magnetic $Ni_{0.6}Fe_{2.4}O_4$ and $Co_{0.6}Fe_{2.4}O_4$ prepared from microemulsion template by Duan et al. had the maximum adsorption capacities of 189.04 mg/g and 80.32 mg/g for U(VI) or Pb(II) (Duan et al. 2015; Duan et al. 2016).

It is difficult to correlate the adsorption performance and the preparation method towards SFs, because the adsorption performances of SFs are affected by many factors, e.g., size and shape, metal ion doping, calcination temperature, and so on. Generally, SF nanoparticles with high surface area have superior adsorption performance. For example, Hu et al. (Hu et al. 2007) compared the adsorption capacities of $MnFe_2O_4$, $MgFe_2O_4$, $ZnFe_2O_4$, $CuFe_2O_4$, $NiFe_2O_4$, and $CoFe_2O_4$ to Cr(VI). The adsorption capacities followed the order $MnFe_2O_4 > MgFe_2O_4 > ZnFe_2O_4 > CuFe_2O_4 > NiFe_2O_4 > CoFe_2O_4$. The $MnFe_2O_4$ nanoparticles with a high surface area of 180 m^2/g showed shorter equilibrium time compared with other SFs. Besides, the chemical doping and calcination temperature also significantly affect the adsorption properties to heavy metals. The chemical doping could tune the adsorption properties of MFe_2O_4 by varying the particle sizes, morphologies, and functionalization, as well as variation of the adsorption characteristics. Especially, the adsorption performance of SFs is enhanced by introducing rare earth metal ions (Jacobo et al. 2004; Kuai et al. 2013). It is attributed to the structural disorders of SFs, caused by the doping of rare earth ions, is beneficial to increase the surface area and active binding sites. In addition, the calcination temperature may change the particle size, morphology, and surface area of SFs, resulting in the changed adsorption capacities (Ahalya et al. 2014).

The Magnetic Microsphere

Although the nano-adsorbents such as Fe_3O_4, $SrFe_2O_4$, and $Ni_{0.6}Fe_{2.4}O_4$ are conveniently prepared and recycled from the water, the adsorption performance still needs to enhance. Because the inherent physical and chemical property and the serious aggregation of nano-magnetic particles in water. Thus, many organic small molecules are used to modify the naked magnetic particles for increasing the adsorption sites, including ascorbic acid, oxalic acid, and so on (Feng et al. 2012). However, the increase in the adsorption performance is still limited. In addition, the organic molecule might diffuse into the water and cause the secondary pollution. Hence, introducing and immobilizing the magnetic particle into the matrix of polymeric adsorbent become the main direction of the current research, and the magnetic polymeric adsorbents with spherical structure are prepared and widely studied.

In this field, incorporation of the natural polymer into the adsorbent via the inverse emulsion is employed to prepare the spherical adsorbents. The natural polymers included carboxymethylcellulose and sodium alginate, especially chitosan and its derivatives have been widely applied, due to low cost, nontoxic, renewable, biodegradable, inherent adsorption performance, and high activity of the amino and hydroxyl. The simplest method for the preparation of the magnetic adsorbent based on chitosan is to disperse the magnetic particle into chitosan solution via the inverse emulsion and then cross-link with glutaraldehyde and epichlorohydrin or adjust the solution pH from acid to alkaline. The obtained magnetic spherical adsorbent could be used for the removal of many pollutants, including heavy metal, antibiotic, and dyes (Lian et al. 2015). Despite the magnetic adsorbents based on the chitosan and the magnetic particles are easy to prepare in mild condition, the obtained adsorbents usually show the weak adsorption performance for pollutants. For instance, Podzus et al. (Podzus et al. 2009) investigated the adsorption performance of magnetic chitosan composite for Cu^{2+}, the adsorption capacity was only about 19.4 mg/g. Zhang et al. (Zhang et al. 2019a, b, c, d) immobilized the Aspergillus onto the sodium tripolyphosphate crosslinked magnetic chitosan microspheres, the adsorption capacity of Cu^{2+} increased to 119.21 mg/g. It was attributed to the fact that the most of amino group and hydroxyl group, which played a critical role in the adsorption process participated in the cross-link reaction, especially for the chemical cross-link by formaldehyde, glutaraldehyde, epichlorohydrin, tripolyphosphate, ethylene glycol diglycidyl ether, and dimethyloldihydroxy ethylene urea. Hence, two strategies are applied to increase the adsorption performance of this type of adsorbent, that are: using of chitosan derivatives to replace chitosan during preparation of adsorbent, or modification the magnetic chitosan microspheres with others polymer.

For the first strategy, various chitosan derivatives have been used in the preparation of magnetic adsorbents, including quaternized chitosan, carboxylated chitosan, N-acyl chitosan, and so on. For example, Song et al. (Song et al. 2017) replaced the chitosan with derivatives of N-(2-hydroxyl)propyl-3-trimethyl ammonium chitosan chloride (HTCC) to prepare the As(III) imprinted magnetic adsorbent in microemulsions. The magnetic adsorbent showed excellent selectivity and recyclability for As(III) over a wide pH range. Moreover, the adsorption efficiency still maintained above 75% after 10 recycles. Tao et al. (Tao et al. 2016) modified chitosan with glutamine and fabricated a magnetic composite microsphere in the inverse emulsion for adsorbing Hg^{2+} and acid green 25 (AG25). The Hg^{2+} and AG25 all could be efficiently removed in weak acidic conditions, as the effective interactions between Hg^{2+} and the carboxyl, amide groups, as well as the hydrogen bonding between secondary amine of AG25 and carboxyl groups (Fig. 13.10).

Compared with the first strategy, the modification of magnetic chitosan microspheres with other polymers or introducing the other inorganic/organic adsorbents is more widely adopted. It is well known that the magnetic chitosan microspheres could be modified easily based on the high activity of chitosan's amino groups, including grafting polymerization, esterification, and acylation. For instance, the carboxylated chitosan magnetic spherical adsorbents with micro−/submicron size were fabricated by Xu et al. via the microemulsion method for Pb^{2+} removal (Xu et al. 2015). The chitosan magnetic microspheres with different sizes were formed in the microemulsion and then modified with ethylene diamine tetra acetic acid. The favorable recycle of both adsorbents displayed and 94% of elimination capacity could be kept after fifth cycle. Sun et al. (Sun et al. 2016a) grafted the quaternary ammonium groups onto the magnetic chitosan microspheres for removal of the Cr^{6+} under a high acid environment. The adsorption capacity could be reached

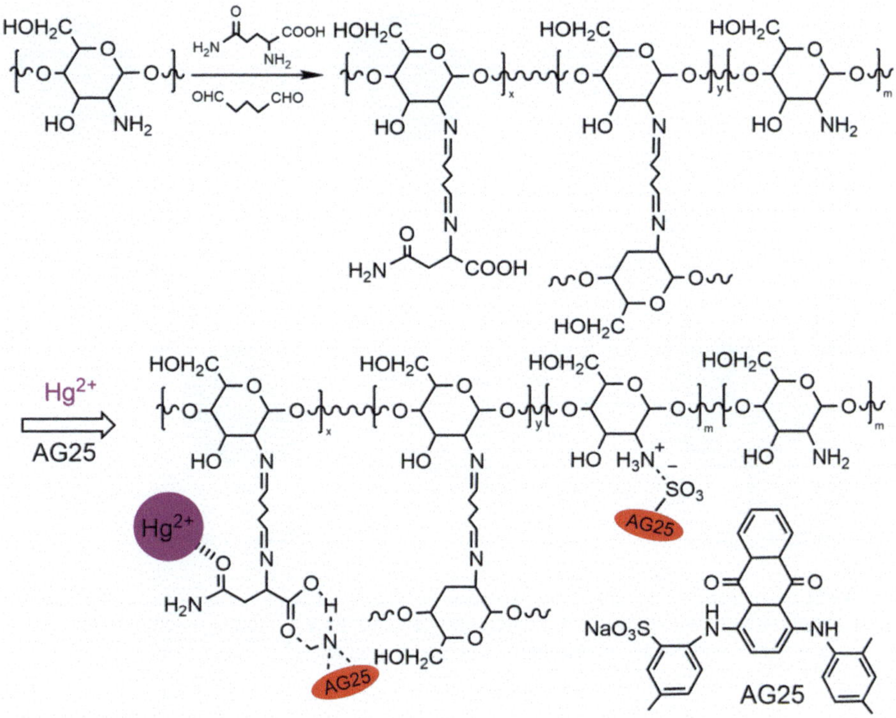

Fig. 13.10 Chemical cross-linking reaction of chitosan modified with glutamine and brief description for available adsorption mechanism of CS-Gln-MCM for removal of AG25 and Hg^{2+}. (Reproduced with permission from Tao et al. 2016)

to 233.1 mg/g for Cr^{6+} at pH 2.5 and 25 °C, depending on the initial Cr^{6+} concentration. Zheng et al. (Zheng et al. 2019) modified the magnetic chitosan microspheres with poly(4-vinyl pyridine) and the poly([2-(methacryloxy)ethyl] trimethylammonium chloride), the maximum adsorption capacities of two adsorbents for Cr^{6+} were 344.83 mg/g and 153.85 mg/g, respectively (Zheng et al. 2018).

Sun et al. (Sun et al. 2016b) adopted a large number of amino groups for modification of magnetic chitosan microspheres to increase the adsorption performance of Cr^{6+}. The adsorbent of polyethylenimine-modified magnetic chitosan microspheres (Fe_3O_4–SiO_2–CTS-PEI) exhibited high acid resistance and magnetic responsiveness, and the maximum adsorption capacity was 236.4 mg/g at 25°C, which was approximately 2.5 times for the unmodified magnetic microspheres. Xiao et al. (Xiao et al. 2017) introduced the amino groups and carboxyl groups into the spherical magnetic chitosan adsorbent for adsorption of Cu^{2+}. The Fe_3O_4 nanoparticles were supported onto the carboxyl-functionalized polystyrene particles (PS) by integrating the emulsifier-free emulsion polymerization and the in situ coprecipitation and then coated with cross-linked chitosan thin film. Finally, branch polyethylenimine (PEI) was grafted on the surface of PS/Fe_3O_4/CS via Michael

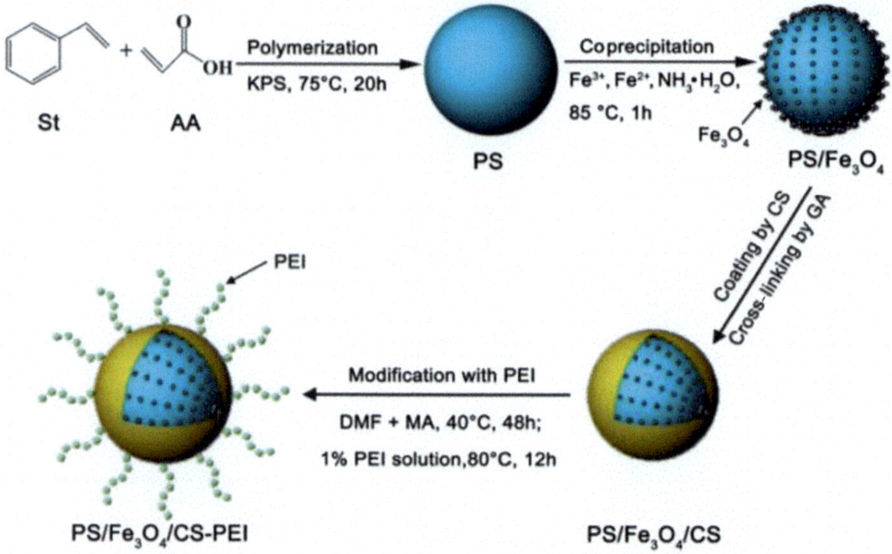

Fig. 13.11 Schematic illustration of the synthesis of PS/Fe_3O_4/CS-PEI composites and the photos of TEM (**a**) and SEM (**b**) of PS, TEM (**c**) and SEM (**d**) of PS/Fe_3O_4, TEM (**e**) of PS/Fe_3O_4/CS, and TEM (**f**) of PS/Fe_3O_4/CS-PEI. (Reproduced with permission from Xiao et al. 2017)

addition reaction and an amidation reaction. The adsorption capacity to Cu^{2+} reached 204.6 mg/g within 15 min (Fig. 13.11). Except for the chemical modification with polymers, some inorganic adsorbents also increased the adsorption capacity of the spherical magnetic chitosan adsorbent. Wang et al. (Wang et al. 2019a, b) introduced zinc oxide into the spherical magnetic chitosan adsorbent to eliminate the arsenic from groundwater, a high As(V) adsorption capacity achieved with 63.69 mg/g.

Due to the high activity of acrylate monomer and the strong affinity of carboxyl and acylamino for heavy metals, the spherical magnetic adsorbents prepared with acrylate monomers of acrylic acid and acrylamide via the emulsion template, especially the inverse emulsion is widely reported. The obtained adsorbents exhibit excellent adsorption performance and favorable reusability. For example, beadlike magnetic nanocomposite microgel adsorbent was prepared by polymerizing and cross-linking the poly(acrylic acid) (PAA) onto the silane-modified Fe_3O_4 particles to remove Pb^{2+} (Jiang et al. 2017) (Fig. 13.12). Due to the plentiful carboxyl groups, high wettability, and high swelling of the Fe_3O_4/PAA microgel adsorbent, the adsorption capacity towards the targeted metal ions of Pb^{2+} can be reached to 123.3 mg/g. Xie et al. (Xie et al. 2017) produced the magnetic microspherical adsorbent by polymerization of acrylic acid and acrylamide onto the cassava residue by an inverse emulsion method. The Cu(II) adsorption capacity of the adsorbents reached 110.5 mg/g when the pH was 6.4. Wanna et al. (Wanna et al. 2016) reported a magnetic adsorbent based on poly(methyl methacrylate) by the emulsion polymerization technique for heavy metal removal. The polyethylene glycol bis(amine)

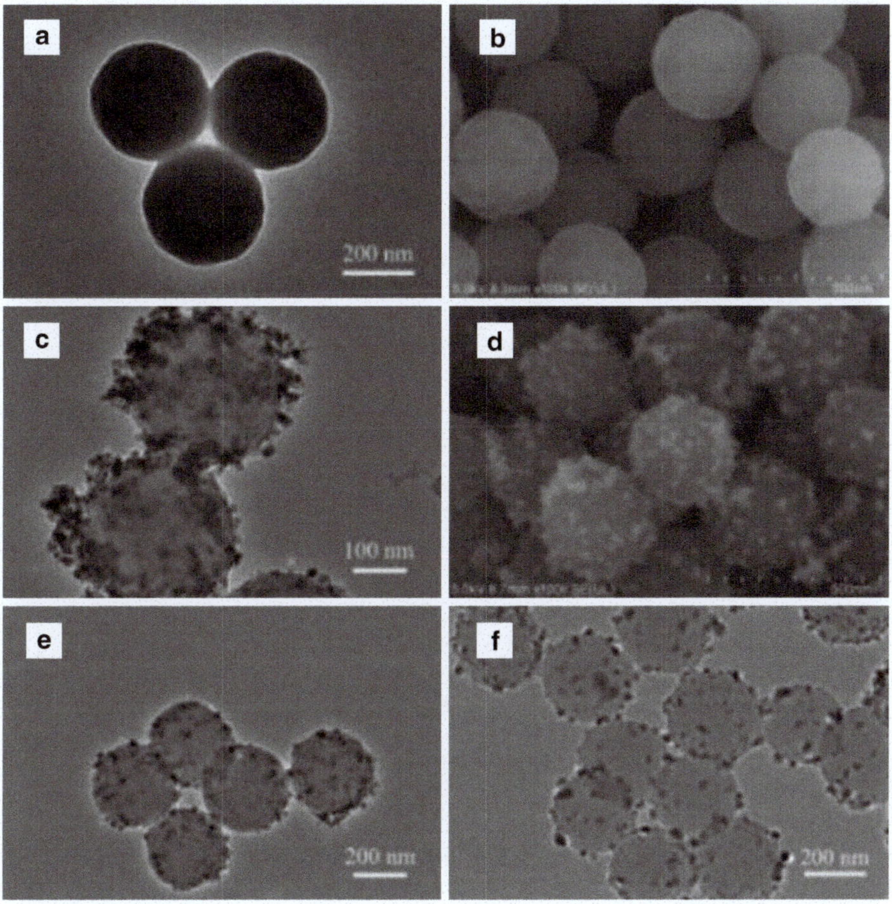

Fig. 13.11 (continued)

(PEG-bis(amine)) was grafted onto the magnetic nanoparticles after being modified with the poly(methyl methacrylate) by the reaction between the carboxyl groups derived from the hydrolysis of PMMA and the amino groups of polyethylene glycol bis(amine). The results indicated that the heavy metal uptake ratios of the adsorbents were 0.08, 0.04, 0.03, and 0.01 mmol/g for Pb^{2+}, Hg^{2+}, Cu^{2+}, and Co^{2+}, respectively. The cation radius of the heavy metal is the main effect factor for affecting the removal efficiency.

The conductive polymers including polypyrrole (PPY), polyaniline (PANI), polyindole (PIn), polythiophene (PTh), etc. have excellent adsorption performance for heavy metals, as the remarkable chelating property derived from the abundant of N-containing heterocyclic group. Due to the weak solubility but high activity of pyrrole, indole, and thiophene, the conductive polymer adsorbent is directly prepared from the O/W emulsion with mild condition. Chavez-Guajardo et al. coated

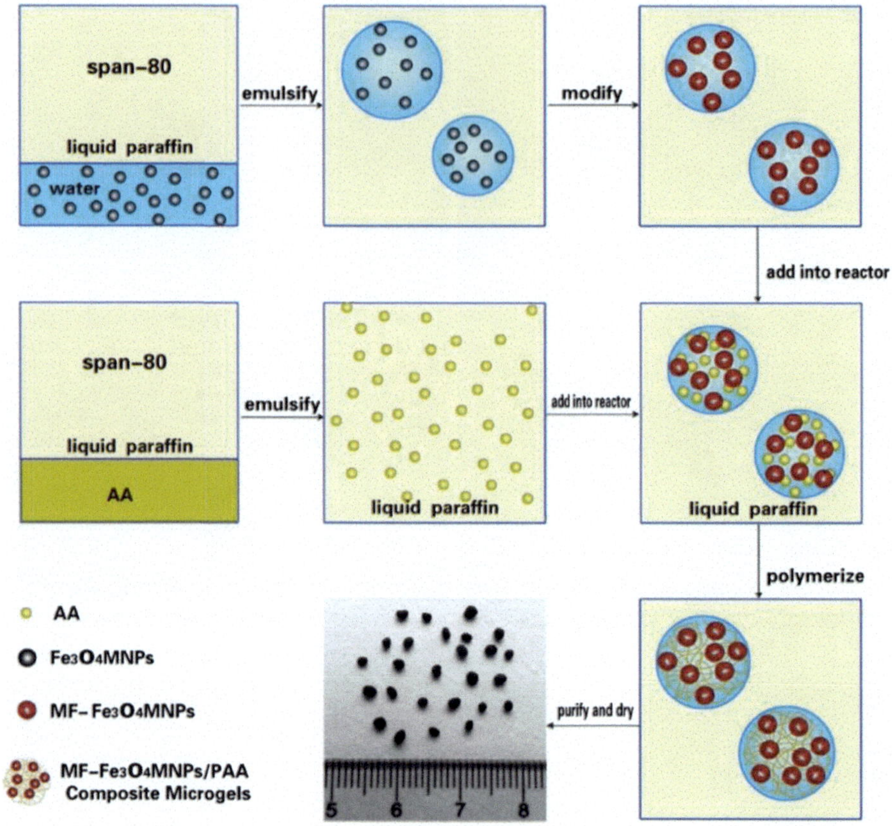

Fig. 13.12 Synthesis procedures of multi-functionalized Fe_3O_4 magnetite nanoparticles/ polyacrylic acid (MF-Fe_3O_4MNPs/PAA) composite microgels. (Reproduced with permission from Jiang et al. 2017)

the PPY and PANI onto the γ-Fe_2O_3 (PPY/γ-Fe_2O_3 and PANI/γ-Fe_2O_3) through the emulsion polymerization at room temperature. The maximum adsorption capacities of PPY/γ-Fe_2O_3 and PANI/γ-Fe_2O_3 were 209 and 196 mg/g and 171 and 107 mg/g for Cr^{6+} and Cu^{2+} (Chávez-Guajardo et al. 2015). Ebrahimpour et al. (Ebrahimpour et al. 2017) prepared three magnetic conductive polymers of PIn@Fe_3O_4, PTh@Fe_3O_4, and PIn-co-PTh@Fe_3O_4 by modification of Fe_3O_4 nanoparticles with polyindole (PIn), polythiophene (PTh), and poly(indole-co-thiophene) via in situ emulsion polymerization. The magnetic conductive polymers were used to pre-concentrate and determinate the aromatic amines in different real samples, and the PIn-co-PTh@Fe_3O_4 nanocomposite sorbent displayed higher extraction efficiency.

The Magnetic Porous Material

The powder adsorbent presents excellent adsorption performance in the treatment of wastewater, but the adsorption performance easily reduced as the inevitably aggravation in water. Although the millimeter-sized spherical adsorbent overcomes this shortcoming, most of the microspherical adsorbents have dense surface, and thus ions and organic molecules are difficult to diffuse into the matrix of the adsorbent, which limited the adsorption performance. Interestingly, the porous adsorbent could be good at resolving this problem. It possesses stable physical-chemical property, large specific surface area, and substantial exposed adsorption sites inside the adsorbent and high porosity, which could reduce mass transfer resistance. Hence, more and more studies are focused on the porous adsorbents for removal of pollutants. Among various methods, emulsion template technology might be the more effective approach for successful synthesis of porous materials with ordered porous structure, especially HIPEs. Up to now, the porous materials prepared from the HIPEs have been widely acted as the adsorbent for removal of various pollutants, including metal ions, dyes, antibiotic, and so on (Han et al. 2015; Pan et al. 2016; Zhang et al. 2019a, b, c, d). Due to the high porosity, the adsorbent prepared from HIPEs presents excellent removal efficiency for heavy metal ions. For example, Mert et al. used the humic acid-modified Fe_3O_4 (Fe_3O_4@HA) to stabilize HIPEs and formed magnetic polyHIPEs using styrene/divinylbenzene as monomer. Magnetic polyHIPEs were tested to remove Hg^{2+}, and the maximum adsorption capacity of 20.44 mmol/g was achieved (Mert et al. 2013). Zhu et al. prepared magnetic porous adsorbent of Pb^{2+} and Cd^{2+} from HIPEs, which was stabilized with amine-functionalized Fe_3O_4. The surface of the magnetic porous adsorbent possessed abundant benzene rings and was peculiarly prone to attach with the cation by π-bond, and the removal capacities of Pb^{2+} and Cd^{2+} were 257 and 129 mg/g at pH 5.5 (Zhu et al. 2018).

Our group devotes to study the porous adsorbent prepared from emulsion template for elimination of heavy metal (Zhu et al. 2016a; Zhu et al. 2016d; Zhu et al. 2017b). And the macroporous magnetic adsorbent of chitosan-g-poly(acrylic acid) was produced using the Fe_3O_4 nanoparticle-stabilized Pickering HIPE template. The porous adsorbent showed the high adsorption capacities of 308.84 mg/g and 695.22 mg/g, as well as a fast adsorption rate of 40 min for Cd^{2+} and Pb^{2+}, respectively (Zhu et al. 2016b; Lu et al. 2018a). Moreover, the favorable adsorption capacity of 88.95 mg/g for Sr^{2+} was reached by coating the magnetic porous materials with polyaniline (PANI), which was better than most of the other adsorbents (Lu et al. 2018b).

However, high consumption of organic phase and the addition of large amounts of surfactants restrict the application of conventional HIPE templates in the construction of porous adsorbents. To address these problems, we replaced the synthetic surfactant with the magnetic yeast (P-Yeast) to stabilize the HIPEs and developed a series of magnetic porous adsorbents from the HIPE template (Lu et al. 2019a). The stability of Pickering HIPEs and the corresponding porous structure could be controlled with the interaction between P-Yeast and acrylic acid. The open-cell

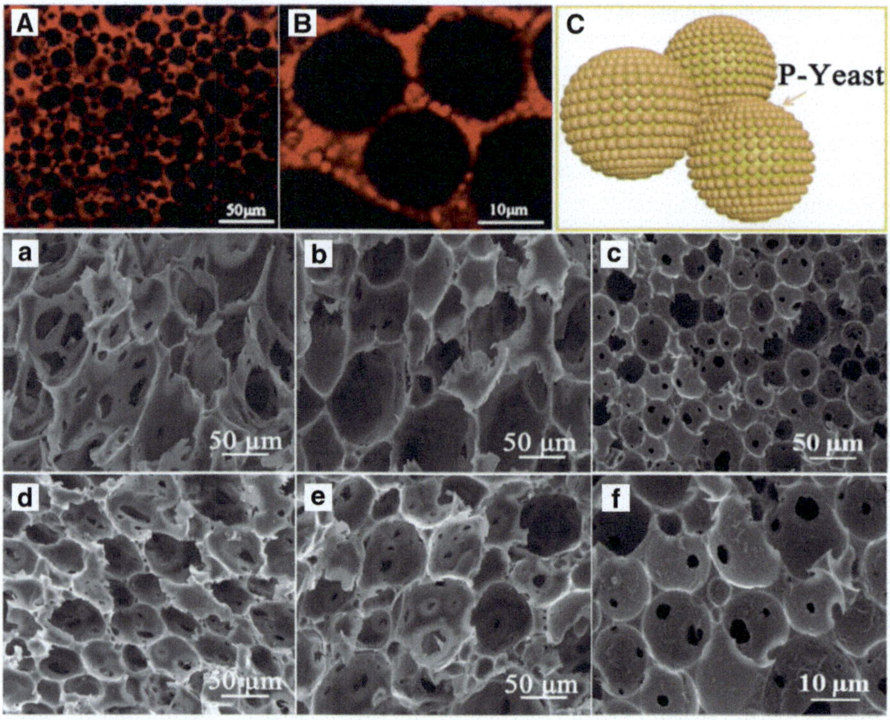

Fig. 13.13 CLSM images and the schematic diagram of P-Yeast-stabilized Pickering emulsions at 75% oil fraction. SEM images of the superporous adsorbent P-Yeast-PAA prepared with different amounts of AA. (Reproduced with permission from Lu et al. 2019a)

superporous adsorbent showed the fast and strong adsorption performance of 179.69 mg/g, 229.52 mg/g, and 166.81 mg/g for scattered metals of Rb^+, Cs^+, and Sr^{2+} (Fig. 13.13). Furthermore, another novel magnetic porous adsorbent was fabricated from the surfactant-free Pickering emulsion template stabilized with magnetic yeast (MY) and chitosan (Lu et al. 2019b). The Pickering emulsion had high stability at the middle phase emulsion (MIPEs) level, and the droplet size could be adjusted easily by controlling the interaction between yeast and chitosan via the pH varying. The microporous magnetic adsorbents with sufficient porous structure also exhibited excellent adsorption performance for Rb^+ and Sr^{2+}, and the saturation adsorption capacities of 168.98 and 151.91 mg/g for Rb^+ and Sr^{2+} were achieved within 25 or 10 min, respectively.

Despite the porous adsorbents with order porous structure are prepared from the emulsion template, but the monolithic adsorbent is needed to smash before using in some case, and the drastic process might destroy thoroughly the porous structure. So how to integrate the advantages of the spherical adsorbent and the porous structure is the research hotspot (Pan et al. 2015). The exciting finding is the porous spherical adsorbent could be directly formed in the multiphase emulsion, including the water-in-oil-in-water (W/O/W) emulsion or the oil-in-water-in-oil (O/W/O) emulsion. It should be pointed out the emulsion integrated with the microfluidic technique could obtain the porous spheres with perfect structure (Cao et al. 2016). For example, Cao constructed a kind of three-dimensional magnetic porous multi-walled carbon nanotube bead via the multiphase emulsion using a modified microfluidic device. The magnetic porous multi-walled carbon nanotube beads had good adsorption capability to oils and organic solvents with six times recyclability. However, it is difficult to realize the practical application of the porous microspherical adsorbent prepared by the microfluidic technique due to the complicated preparation process, the high production cost, and the toxic organic solvents.

Mudassir et al. (Mudassir et al. 2019) reported a magnetic microporous adsorbent by loading the Fe_3O_4 nanoparticles onto the macroporous polymeric beads, which prepared via the O/W/O emulsion, and finally modified with the poly(acrylic acid) for removal of Pb(II) and crystal violet. The adsorption capacities of 290.69 and 80.20 mg/g for Pb(II) and crystal violet were derived from the sufficient porous structure and abundant acrylic acid. The introduced Fe_3O_4 NPs not only endowed the magnetic property to the microsphere but also improved the BET surface area. Furthermore, the introduced Fe_3O_4 nanoparticles provided the auxiliary crosslinking point to enhance the mechanical strength of the adsorbent (Fig. 13.14).

A magnetic spherical porous adsorbent was synthesized through the integrated process of Pickering emulsion and precipitation polymerization (Zhu et al. 2016e; Zhu et al. 2017a; Zhu et al. 2017b). The Rb^+ and Cs^+ could be effectively removed within 15 and 30 min with the remarkable adsorption capacities of 310 and 448 mg/g, respectively (Fig. 13.15). In addition, the diameter of the spherical adsorbent was reduced from millimeter-level to micron order by adopting the O/W/O double emulsion. The microspherical adsorbent displayed the significant adsorption performance boost, and the removal for Cu^{2+} and Pb^{2+} could be achieved only within 3 min or 5 min, respectively, regardless of high (400 mg/L) or low (100 mg/L) initial concentrations (Zhu et al. 2016d).

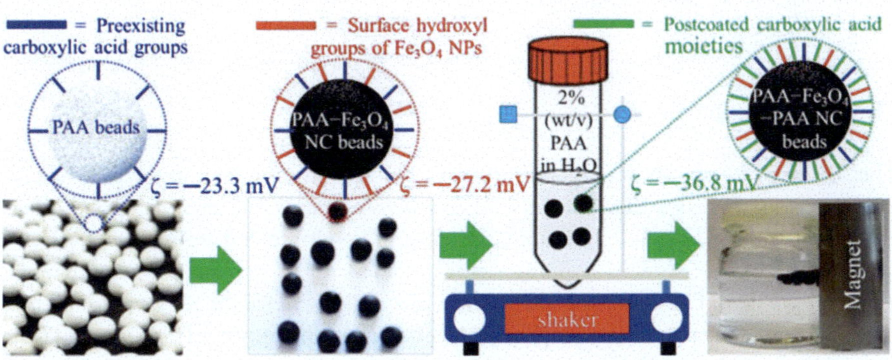

Fig. 13.14 Schematic view of the preparation of PAA (2.28–2.22 mm), PAA-Fe_3O_4 NC (2.26–2.20 mm), and PAA-Fe_3O_4-PAA NC (2.26–2.21 mm) beads. (Reproduced with permission from Mudassir et al. 2019)

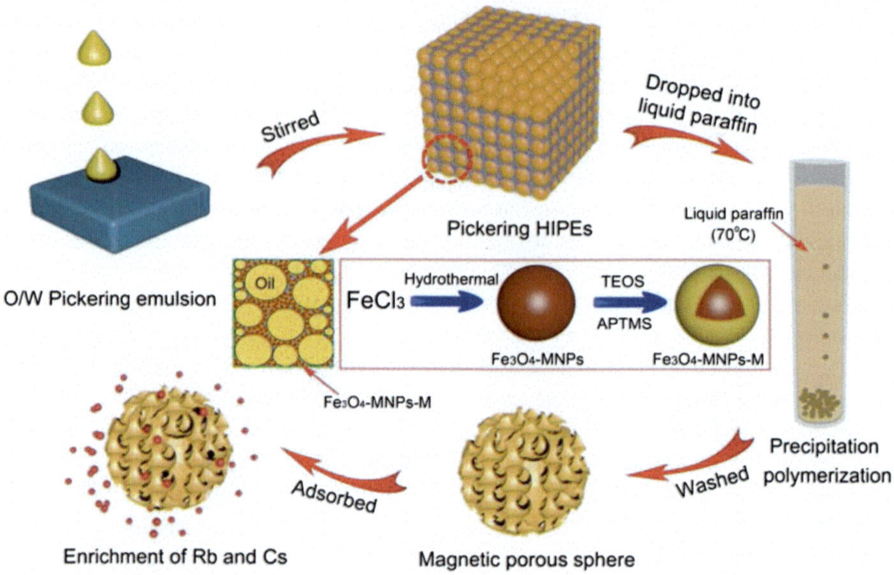

Fig. 13.15 Synthetic route of the interconnected magnetic porous spheres for enrichment of Rb^+ and Cs^+. (Reproduced with permission from Zhu et al. 2017a)

13.3.2 Removal of Organic Pollutant

Organic pollutants have become one of the most critical environmental issues besides heavy metals in water, as their durability and toxicity in the environment. The scope of organic pollutants is very broad, including endocrine-disrupting chemicals, pharmaceuticals, detergents, organic dyes, personal care products, pesticides, and common industrial organic chemicals (Lu and Astruc 2020; Routoula and Patwardhan 2020). And the adsorption technique plays an important role in the elimination of organic contaminants.

Magnetic Nanoparticles

The SFs used to adsorb organic pollutants are more common than in the removal of heavy metals, because SFs not only adsorb heavy metals and cationic dyes, other negatively charged organic pollutants also could be eliminated from aqueous solution (Konicki et al. 2013; Ding et al. 2015). For instance, the magnetic nanocomposite of $CaFe_2O_4$ and $MnFe_2O_4$ was synthesized for the removal of methyl orange, and the maximum capacity reached 344.83 mg/g. The $Ni_{0.6}Fe_{2.4}O_4$ nanoparticles fabricated by emulsion template for adsorption of Congo red, and 92.04% of Congo red could be removed within 9 min (Zeng et al. 2014). The adsorption mechanism of SFs to various cationic or anionic species included ion-exchange, electrostatic interactions, hydrogen bonding, and π-π interactions surface complexation (Zhang et al. 2010; Wang et al. 2012; Zhou et al. 2014). Particularly, the hydroxyl groups derived from M-OH and Fe-OH play an important role. More importantly, the charge of the hydroxyl groups would change with the variation of solution pH. Generally, SFs possess positive charge at low pH but will convert to the negative charge in the alkaline environment, due to the deprotonation of hydroxyl groups (Zafar et al. 2018) (Fig. 13.16). Except the surface charge, the recent reports revealed that the microstructure, particle size, and surface morphologies of SFs also affect the adsorption performance (Ding et al. 2015).

Except for the adsorption mechanism, the role of SFs for the removal of the organic pollutants also includes the catalytic degradation. SFs generate oxygen free radicals in the presence of strong oxidizing agents. For example, the peroxymonosulfate could be activated with the $CoFe_2O_4$ and then generated the sulfate radicals for the degradation of organic pollutants, such as diclofenac (Deng et al. 2013), methylene blue (Salami et al. 2019), and so on. In fact, many SFs possess the photocatalytic performance under visible light, such as $NiFe_2O_4$, $CuFe_2O_4$, and $ZnFe_2O_4$ (Mahmoodi 2013). Therefore, the integrated performances of the adsorption and the catalysis contribute to enhancing the adsorption properties of SFs to many pollutants.

The Magnetic Microsphere

The magnetic chitosan microspheres exhibit excellent adsorption performance for many organic pollutants. The adsorption capacities of chitosan-coated Fe_3O_4 for patulin and methylene blue (MB) were determined about 6.67 mg/g and 122 mg/g (Luo et al. 2017; Liu et al. 2018). And the performance was affected by the amount of the chitosan, magnetic particles, and cross-linking density. It was crucial to

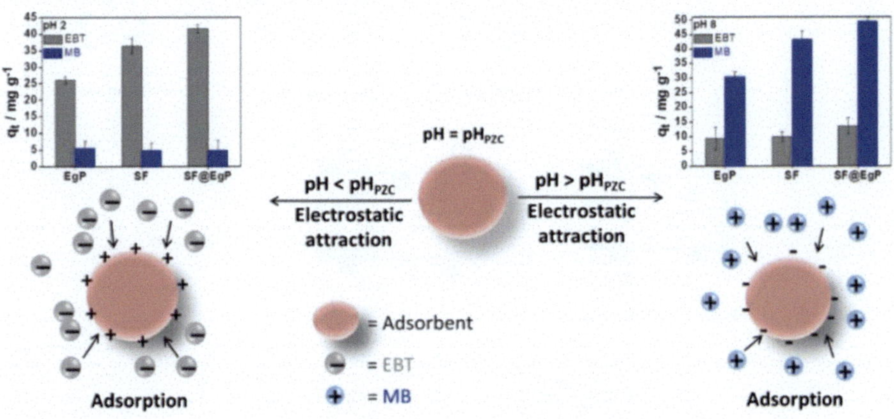

Fig. 13.16 Role of pH on $SrFe_2O_4$ (SF) for tunable adsorption of anionic dye Eriochrome Black T (EBT) and cationic dye methylene blue (MB). (Reproduced with permission from Zafar et al. 2018)

prevent from the inevitable agglomeration of magnetic particles in chitosan solution for the preparation of the magnetic chitosan adsorbent, which caused the heterogeneous magnetism to magnetic adsorbents. It confirms that this problem could be solved completely by supporting the magnetic particles onto inorganic materials, e.g., clay minerals and carbon nanotube. For example, the magnetic particles of chitosan/organic rectorite-Fe_3O_4 were prepared for removal of methylene blue (MB) and methyl orange (MO), and the maximum adsorption capacities for MB and MO were 24.69 mg/g and 5.56 mg/g, respectively (Fig. 13.17). The Fe_3O_4 was supported onto the rectorite first and then obtained the magnetic adsorbent of chitosan/organic rectorite-Fe_3O_4 microspheres (CS/Mt-OREC microspheres) by dispersing the rectorite-Fe_3O_4 into chitosan solution and cross-linked with the formaldehyde and epichlorohydrin in reversed-phase microemulsion (Zeng et al. 2015). Ma et al. prepared a chitosan/kaolin/Fe_3O_4 magnetic microsphere by supporting the Fe_3O_4 onto kaolin and emulsion cross-linking (Ma et al. 2014). The obtained microspheres showed stable adsorption performance for ciprofloxacin removal at least four adsorption-desorption cycles. Except the increased dispersity after incorporation of inorganic materials, it is also in favor of enhancing the mechanical strength of magnetic adsorbent, due to the additional cross-linking point of inorganic materials in the polymeric structure.

Above studies are focused on preparation with the magnetic spherical adsorbent based on the chitosan solution via the common inverse emulsion, which is stabilized

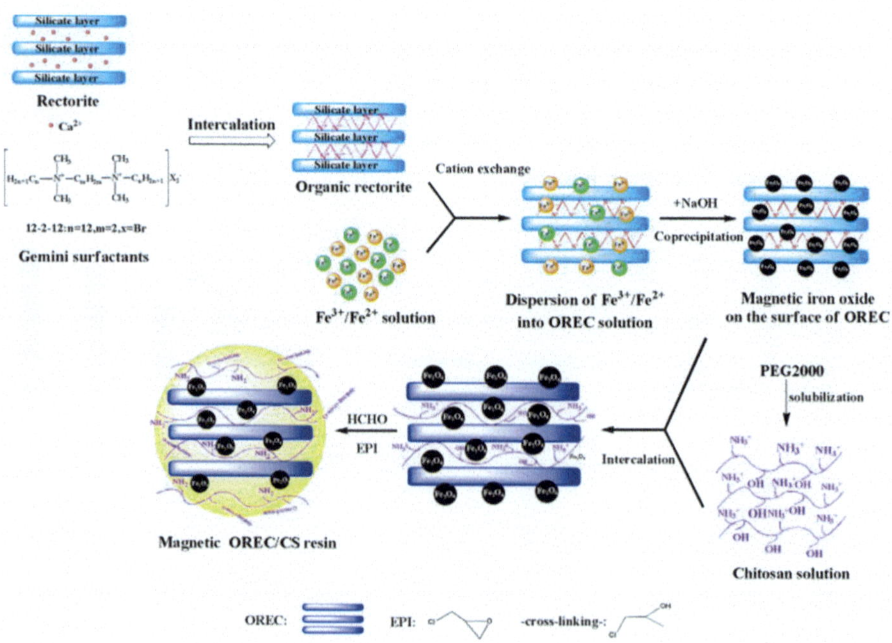

Fig. 13.17 The formation process of chitosan/organic rectorite-Fe_3O_4 microspheres. (Reproduced with permission from Zeng et al. 2015)

with surfactant. However, the residual surfactant might lead to the risk of secondary pollution for water. Recently, Pickering emulsion has been studied widely, due to the low usage levels and the high stability of the obtained emulsion, in which the surfactant is replaced with particles for stabilizing emulsion (Murray 2019). In addition, natural particles as the stabilized particle of Pickering emulsion become the study trend in recent years. In fact, chitosan also could be served as the stabilized particles for the formation of Pickering emulsion as its pH sensitivity (Li et al. 2019). Ou et al. prepared an imprint polymeric adsorbent by using the Pickering emulsion stabilized with chitosan nanoparticles (Fig. 13.18). Fe_3O_4 embedded into the matrix of adsorbent by directly dispersing the hydrophobic Fe_3O_4 into the dispersed phase. The erythromycin (ERY) adsorption capacity of magnetic adsorbent was about 52.32 μmol/g at 15 °C (Ou et al. 2015).

The functional groups of carboxyl, acylamino, amino, etc. have been widely incorporated into the adsorbent for removal of organic pollutants. For instance, Dai et al. (Dai et al. 2012) fabricated the Fe_3O_4/PAA microgel adsorbent by similar method for selective adsorption of tetracycline. The Fe_3O_4/PAA microgels possessed the molecular recognition ability by adopting the molecular imprinting technique, and the estimated adsorption capacity towards tetracycline was about 6.33 times higher than that of magnetic adsorbent without imprinting. Mao et al. (Mao et al. 2016) synthesized a pH-sensitive magnetic molecularly imprinted polymer via Pickering emulsion polymerization of methacrylic acid for selective adsorption of bifenthrin. The magnetic adsorbent displayed the outstanding adsorptive selectivity for bifenthrin, and the adsorption-desorption cycle could be easily operated by changing the pH of the solution.

Resin is also applied to remove pollutants due to favorable mechanical strength and abundant adsorption groups (Ming et al. 2015). The emulsion template is often adopted during the synthesis process at the aim of obtaining monodisperse resin microspheres. For example, iron-oxide nanoparticles were first coated with γ-methacryloxypropyl-trimethoxysilane and then polymerized with styrene and divinylbenzene in an O/W emulsion (Sehlleier et al. 2016). The MB adsorption capacity of the obtained adsorbents was about 298 mg/g. Lu et al. (Lu et al. 2017) fabricated magnetic hollow carbon microspheres (MHCMs) to remove rhodamine B by alternation of surfactant-free emulsion polymerization and microwave-assistant polycondensation (Fig. 13.19). The magnetic adsorbent with the multilayer core-shell structure was obtained through the emulsion polymerization and microwave-assistant hydrothermal method. The magnetic hollow carbon microspheres had uniform morphologies and high surface area. And the removal efficiency of rhodamine B (RB) reached to 99.5% and the adsorption capacity was 300 mg/g.

Wang et al. (Wang et al. 2019a, b) prepared a novel core-shell microspherical resin adsorbent of Fe_3O_4@lignosulfonate/phenolic through the emulsion polymerization to adsorb dyes. The maximum capacity was 283.6 mg/g in 40 min, which was much higher than those of most lignins and lignin-rich biomass. Zhu et al. (Zhu et al. 2016) synthesized magnetic resin polymer adsorbent with molecularly imprinted structure by Pickering emulsion, stabilized with magnetic eggshells. The

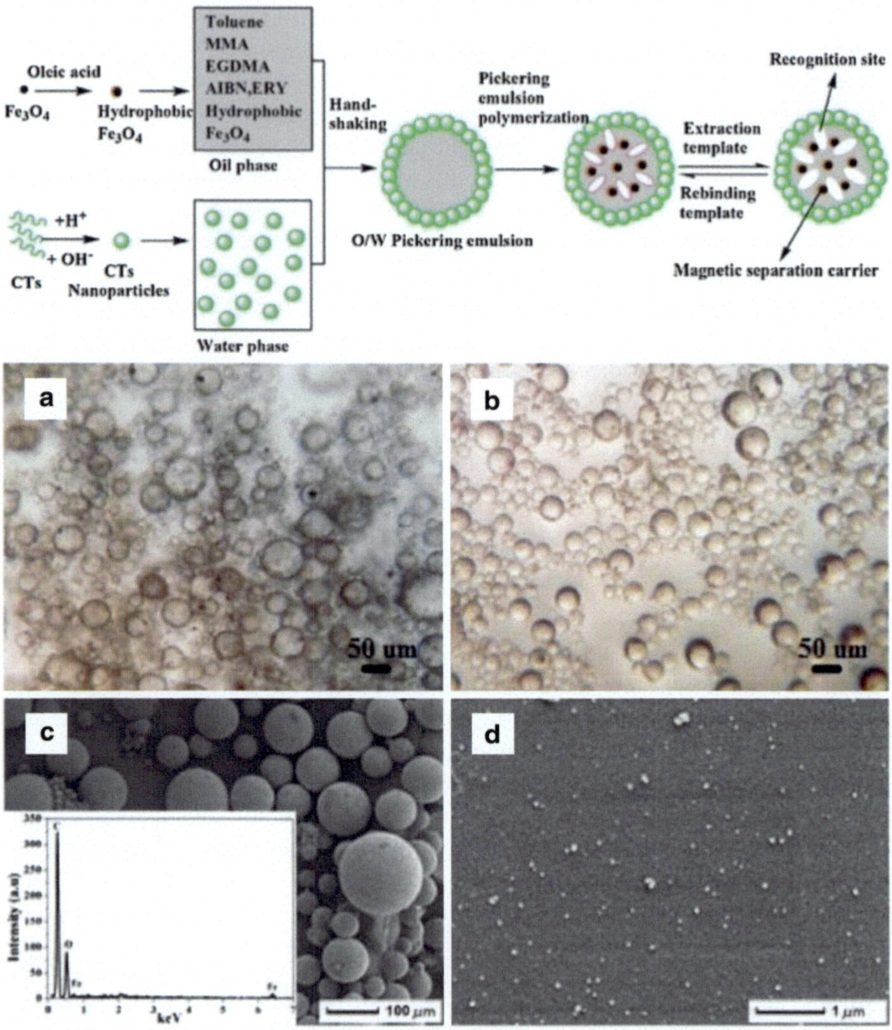

Fig. 13.18 Formation of magnetic imprinted polymers (MIPs) from O/W Pickering emulsion polymerization and the optical micrographs and SEM images of the Pickering emulsion and imprint polymeric adsorbent. (Reproduced with permission from Ou et al. 2015)

molecularly imprinted adsorbent with spherical and wrinkled morphology was obtained in the dispersed phase by polymerization of the monomer of methyl methacrylate. Adsorption experiments showed that the as-prepared molecularly imprinted adsorbent could selectively adsorb erythromycin, but it presented a low adsorption capacity of 47.393 mg/g, which might be due to the dense coating of the magnetic eggshell.

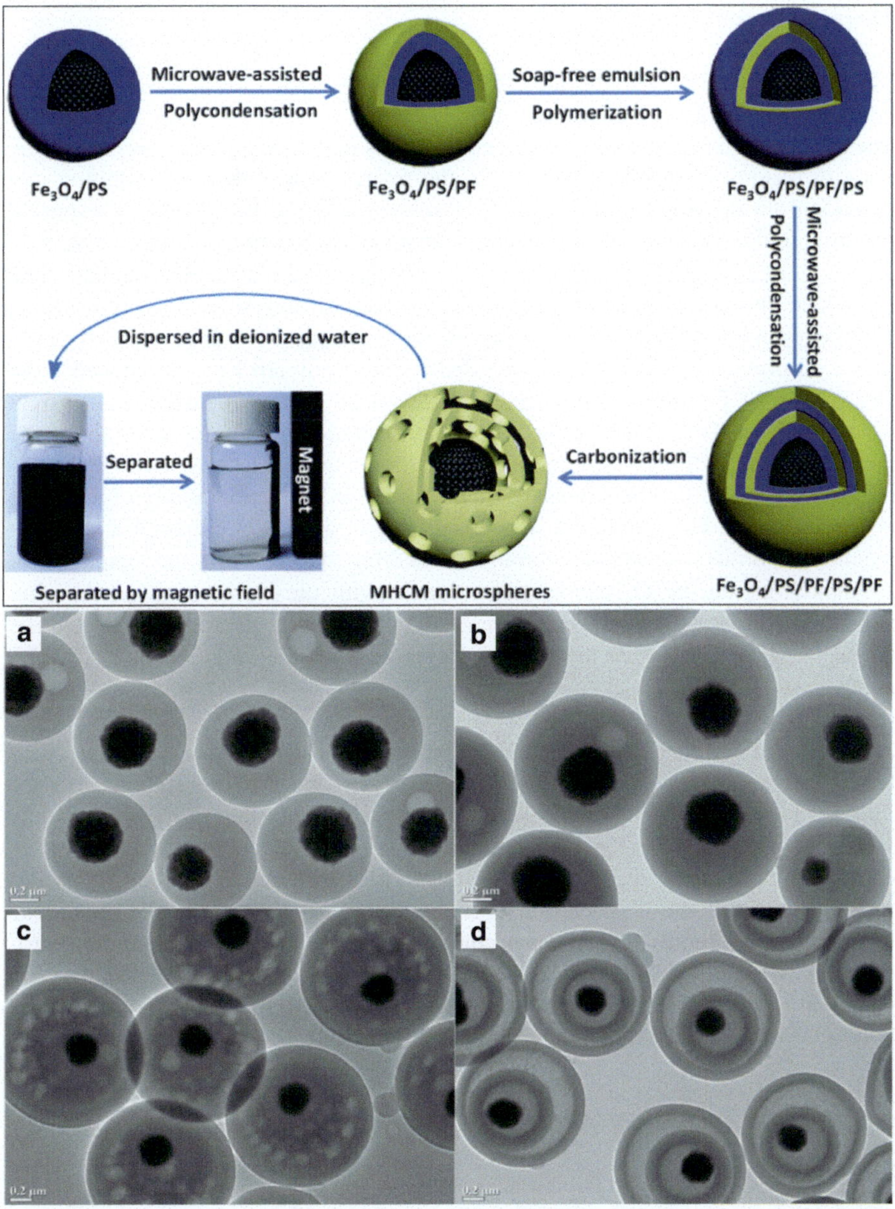

Fig. 13.19 Synthetic procedure of the MHCMs and the TEM images of the multilayer core-shell structure. (Reproduced with permission from Lu et al. 2017)

Magnetic Porous Material

More and more researches on the use of porous adsorbents for removal of organic pollutants are published in recent years (Wright et al. 2017; Kovačič et al. 2018). The sufficient porous structure of magnetic porous materials is conductive to fast the mass diffusion of organic pollutants in the matrix of adsorbent. In addition, the functional groups sited in the interior of adsorbent could be adequately exposed, resulting in the increased adsorption performance. For example, Du et al. (Du et al. 2019) fabricated Fe_3O_4@$Cu_3(btc)_2$ (Fe_3O_4@HKUST-1) magnetic particles and embedded into polyHIPEs, which were synthesized by ethylamine, divinylbenzene, and methyl methacrylate to form a polyHIPE composite by in situ polymerization. The adsorption experiment revealed that polyHIPEs introduced the Fe_3O_4@HKUST-1 displayed the higher removal efficiency for antibiotics of oxytetracycline (OTC), tetracycline, duomycin, and chlortetracycline than the unmodified polyHIPEs. Multiple actions including π-π interactions, hydrogen bonding, and electrostatic interactions resulted in the high extraction ability of magnetic polyHIPEs cake for antibiotics. Wu et al. (Wu et al. 2017) prepared a series of magnetic porous adsorbents through Pickering HIPEs for the removal of λ-cyhalothrin; the Fe_3O_4 nanoparticles coated with oleic acid (Fe_3O_4-OA) were applied to stabilize the emulsion. Because the irreversible adsorption of Fe_3O_4-OA at the oil-water interfaces, the throats decreased with the variation of Fe_3O_4-OA content. The maximum λ-cyhalothrin adsorption capacity at 298 K was 404.4 μmol/g (Fig. 13.20).

Azhar et al. (Azhar et al. 2019) obtained novel porous materials from the HIPE template stabilized with humic acid-modified Fe_3O_4 (HA-Fe_3O_4) and cationic fluorosurfactant (CFS). The HIPEs had increased stability than the emulsion only stabilized with CFS. The porous structure of as-prepared polyHIPEs was easily controlled by altering the concentrations of HA-Fe_3O_4 and/or CFS. The porous materials showed the high capacity for the raised oil absorption and methylene blue. More importantly, the foams adsorbent could be recycled by a simple centrifugation at least 10 cycles without obvious decrease in adsorption capacity. Zhu et al. (Zhu et al. 2015) fabricated multihollow magnetic imprinted microspheres by polymerization of Pickering double emulsion. The hydrophobic Fe_3O_4 nanoparticles

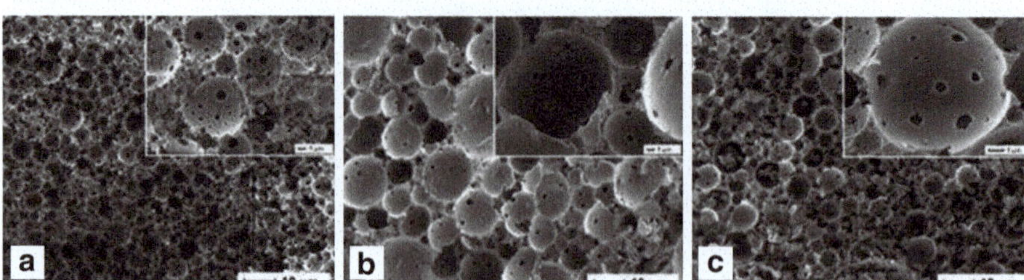

Fig. 13.20 SEM images of magnetic porous adsorbent with different amounts of Fe_3O_4 particles. (Reproduced with permission from Wu et al. 2017)

and hydrophilic cellulose nanocrystals were used to stabilize the W/O interface and the O/W interface, respectively. The selective recognition capability of the as-prepared microspheres for bifenthrin was proved to be more effective. Wang et al. (Wang et al. 2018a, b) also reported a molecularly imprinted multicore rattle-type microsphere for selective adsorption of bisphenol A through a facile Pickering emulsion polymerization, using silica nanoparticles as the stabilizer. Our group (Lu et al. 2018a) also prepared a novel magnetic porous adsorbent of chitosan-g-poly(2-acrylamide-2-methylpropanesulfonic acid) (CTS-g-AMPS) by grafting AMPS onto CTS in the Fe_3O_4-stabilized Pickering HIPEs; the as-prepared porous adsorbents could be employed to eliminate tetracycline and chlortetracycline. The adsorption capacities for tetracycline and chlortetracycline were 806.60 and 876.60 mg/g in a wide pH range of 3.0–11.0, respectively.

13.3.3 The Oil-Water Separation

Besides the soluble pollutants such as heavy metal ions, dyes, and antibiotics, many insoluble or weakly soluble pollutants are also contained in the wastewater. These pollutants may be originated from industrial oily wastewater or oil spill accidents. These types of insoluble or weakly soluble pollutants are also one of the most serious problems for the water environment (Zhang et al. 2019a, b, c, d). Generally, the separation of oil/water mixture can be classified into three main categories: oil removal, water removal, and controllable separation of oil and water. Among them, oil removal is the most attractive as compared to the other two, because of its simplicity and easiness.

Magnetic Nanoparticles

The magnetic nanoparticles could be served as the oil adsorbent, but the premise is the magnetic nanoparticles should be coated with the organic compounds. The functionalized magnetic nanoparticles have strong affinity for oil and thus could be to adsorb oil effectively. Oleic acid is widely used, as oleic acid has a high affinity to the Fe atoms of magnetic nanoparticles. Osama et al. (Osama et al. 2015) functionalized the Fe_3O_4 with oleic acid in the miniemulsion and then to remove oil. The result indicated that 95 wt. % of crude oil could be removed from the water surface. Zhu et al. also modified the Fe_3O_4 with sodium oleate; the obtained Fe_3O_4/sodium oleate showed excellent performance for the elimination of engine oil from the water surface (Zhu et al. 2012).

Magnetic Microsphere

Although the magnetic nanoparticles have the adsorption performance for oil after the coated with the organic small molecules, the adsorption capacity is relatively low, and thus the magnetic nanoparticles are incorporated into natural or synthetic polymers, including starch, alginic acid, chitosan, and so on. Among of them, the relevant studies of application chitosan to modify magnetic nanoparticles take the most part. Lü et al. (Lü et al. 2017) fabricated a class of chitosan-grafted magnetic nanoparticles by grafting the chitosan onto the silica-functionalized Fe_3O_4 via the Schiff base reaction. The chitosan-grafted magnetic nanoparticles could efficiently flocculate oil droplets at different pH conditions. The electrostatic attraction is dominant in acidic and neutral condition, but hydrophobic interaction plays a vital role in the alkaline condition.

The materials with superhydrophobic and superoleophilic properties could selectively collect oils or organic chemicals from water, which provide a novel strategy for the water-oil separation techniques (Chen et al. 2013). So many oil-adsorbed adsorbents are prepared with the hydrophobic monomer. For instance, Fe_3O_4/PS microspheres prepared through emulsion polymerization exhibited the fast rate for adsorption oil and the best oil absorbency, which was up to 2.492 times of their weight (Yu et al. 2015a, b). Chen also used the Fe_3O_4/PS microsphere to adsorb lubricating oil, and the adsorption amount was three times as the particles' weight (Chen et al. 2013). Another example involved the microspheres for adsorption of oil was prepared by coating the methyl methacrylate onto the Fe_3O_4/PS nanoparticles through secondary polymerization. The high hydrophobicity of the microspheres maintained in the wide pH range of 1–13. After 10 cycles, the nanoparticles still had a high oil absorption capacity of 3.22 g/g (Gu et al. 2014).

Magnetic Porous Material

The limited oil storage capacity of traditional oil/water separation materials (e.g., active carbon, zeolites, and other adsorbents) might restrict their practical applications. In comparison, the monolithic porous materials such as aerogels, sponges, and foams possessed sufficient and interconnected porous structure, which have the great potential in oil absorption, as their features of high oil adsorption capacities and easily recycling and reusing. The porous materials prepared from the HIPE template could be served as excellent oil adsorbents (Zheng et al. 2013; Yu et al. 2015a, b; Wang et al. 2018a, b; Zhang et al. 2019a, b, c, d).

Zhang et al. (Zhang et al. 2016a, b) prepared a poly(styrene-divinylbenzene) foam by the Pickering HIPEs through a one-step reaction process. The materials with different hierarchical pore structures were obtained by various Pickering emulsion stabilized with different types of Fe_3O_4. The adsorption capacity of the monolithic foam for chloroform was as high as 57.00 g/g. Zhou et al. prepared a hierarchical porous resin for removal of oily substance through the HIPE template stabilized with phenolic resin precursor and Tween 80. And then the dopamine hydrochloride, 1-dodecanethiol, and Fe_3O_4 particles grafted onto the interface of porous resin via adhesion of dopamine and Markel addition reaction. The as-synthesized hierarchical porous resin possessed a typical hierarchical porous structure, and the porous structure could be adjusted by varying the emulsion factors. The oil adsorption rate and the oil retention rate for toluene were 11.765 g/g and 86.43%, respectively (Zhou et al. 2019).

Zhang et al. employed the Span 20 together with Fe_3O_4 to synergistically stabilize styrene-based HIPEs and produced magnetic solid foam for removal of oil. The interconnected porous structure was constructed by varying the surfactant content, the amount of Fe_3O_4 particles, and other emulsion factors. The resulting magnetic solid foam exhibited excellent thermal stability. The oil adsorption capacity of the solid foam was 16 times its own mass even after 10 cycles of oil/water separation (Fig. 13.21) (Zhang et al. 2017).

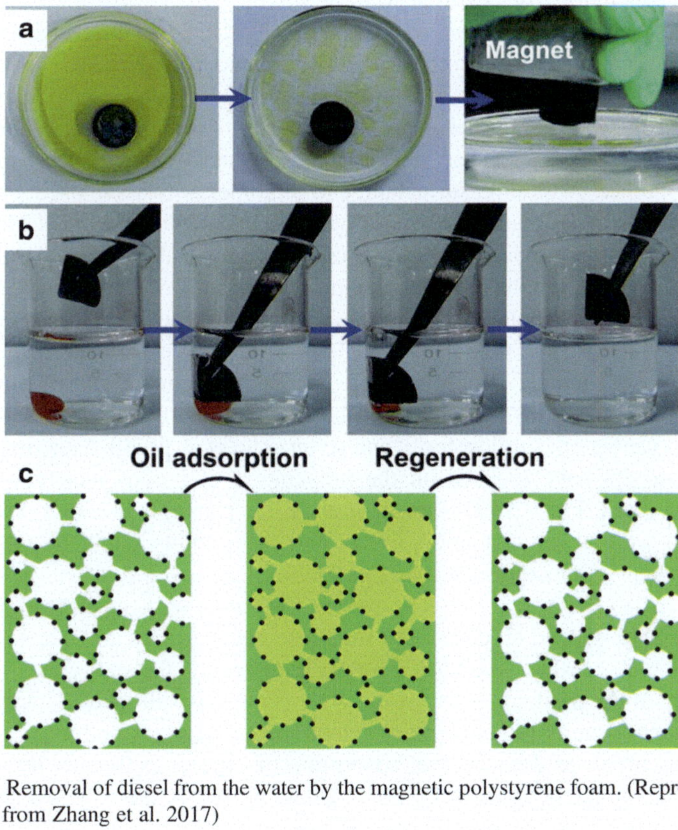

Fig. 13.21 Removal of diesel from the water by the magnetic polystyrene foam. (Reproduced with permission from Zhang et al. 2017)

13.4 Conclusions and Future Prospects

Magnetic adsorbents prepared from emulsion template have been attracting much attention in recent years. The magnetic materials prepared by emulsion template for water treatment are reviewed, including the magnetic nano-adsorbent, the spherical adsorbent, and the magnetic porous adsorbent. Magnetic nano-adsorbents of spinel ferrite exhibit high stability towards acid and excellent removal performance for various pollutants containing heavy metal and organic pollutant. Magnetic spherical adsorbent might be the most wide studies due to the flexible preparation method, various functionalization ways, and sufficient functional groups. Moreover, the molecular imprinting technique could be conveniently integrated with the preparation process to realize the adsorption selectivity. By contrast, the porous magnetic materials prepared from the HIPEs as adsorbent have increasingly been recognized as one of the research hotspots, which exhibit high porosity and excellent adsorption performance, tunable pore structure, pore size distribution, and mechanical strength. Although magnetic adsorbents prepared from the emulsion template present many advantages, some drawbacks still need to be solved.

First, the preparation process of magnetic adsorbent needs a large of organic phase. Although magnetic nanoparticles, magnetic microspheres, or magnetic porous materials prepared from the emulsion template display excellent adsorption performance in water treatment, it is unavoidable to consume highly the organic solvent and surfactant. Because the organic solvent contained much metal salt, residual monomer, cross-link, surfactant, and oligomer after being used, the attempt to cyclic utilization of organic solvent is not found in the relative studies. However, it might be the most important issue for realizing the practical application of magnetic adsorbent prepared from the emulsion template. In order to resolve this problem, the emulsion template is developed for the preparation of porous material by decreasing progressively the organic phase from 75% to 50%, even to 25% (Fresco-Cala and Cárdenas 2019; Kavousi and Nikfarjam 2019). Furthermore, the environmental harm of the toxic organic solvents could be alleviated by replacing with the edible oils, such as canola oil, sunflower oil, and so on (Zhu et al. 2020). More importantly, many researchers have focused on the preparation of porous materials from the water-based foam (Fig. 13.22) (Cervin et al. 2013; Huang et al. 2018). The water-based foam without any organic solvent, and the stability could be significantly increased when stabilization the interface between air and liquid with amphiphilic particles. Due to the green and easy preparation process, it must be the research hotspot in future beyond all doubt. For magnetic nanoparticles and magnetic microspheres, the ratio of water to oil has a significant effect on the size, crystallinity, and morphology of the product, and thus the green synthesis still takes a lot of efforts.

The second issue is the adsorption selectivity to different pollutants. Although many works involved the selective adsorption of organic pollutants by molecular imprinting technique (Cyganowski, 2020; Liu et al. 2020), the recycled and refined of valuable metal from the wastewater might be more meaningful. At present, a few works report the selective adsorption of metal ions based on the magnetic spherical

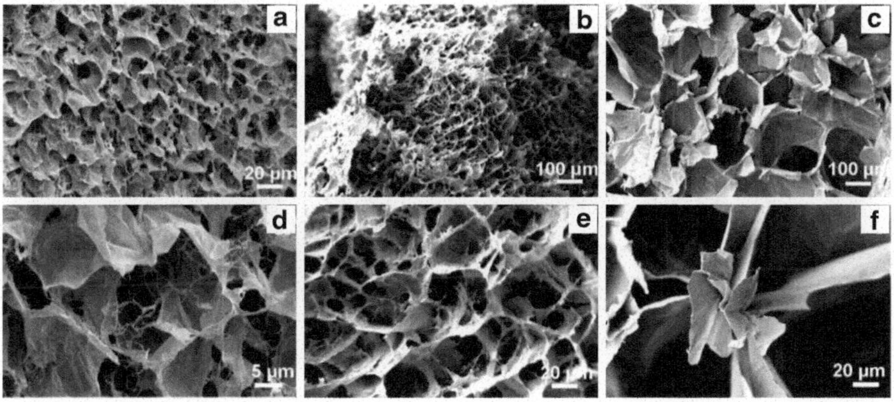

Fig. 13.22 SEM images of the porous structure template from water-based foam. (Reproduced with permission from Huang et al. 2018)

adsorbents, but the low adsorption efficiency limits the potential applications in practice. Therefore, many efforts should be paid to design and construct the functional magnetic adsorbents with excellent adsorption selectivity for rare and precious metals.

Last but not least, the weak reusability of most of the reported magnetic adsorbents should be resolved. The separation and reuse are the most important advantages of magnetic adsorbents, and there are many approaches to realize their cyclic utilization. The most common regeneration of the spent adsorbents is using the organic solvent (methyl alcohol, ethyl alcohol, etc.), acid (hydrochloric acid, sulfuric acid, etc.), or alkaline (sodium hydroxide, potassium hydroxide, etc.) (Ye et al. 2020; Zhao et al. 2020). Despite the adsorbents display excellent desorption and reusability under the evaluation condition, but it is urgent to explore the feasible and green approach to prevent from the secondary pollution, which derived from the desorbing agents, regenerating agents, and the desorbed pollutants from the spent adsorbents. This problem might be not important for the recycle and enrichment of the value metal ions, but it is crucial for adsorption of organic pollutants. In order to actually realize the recycle of the spent magnetic adsorbents, the carbonization technique might be a promising strategy for the regeneration of the spent adsorbent after adsorption organic pollutants.

Recently, the carbonaceous adsorbents have been applied in water treatment (Xiao et al. 2018; Dai et al. 2020). Because it exhibits excellent adsorption performance to organic or inorganic pollutants based on various adsorption mechanisms, including H-bond, π-π stacking, polar interaction to organic pollutant and coprecipitation, and electrostatic interaction to inorganic pollutants (Fig. 13.23). Moreover, the carbonization strategy as the potential approach to realize the recycle of adsorbent has been verified with our group's work (Tang et al. 2016). The cost-effective carbon/attapulgite composites were developed using waste hot-pot oil as a carbon precursor through a facile one-step calcination process (Tang et al. 2017;

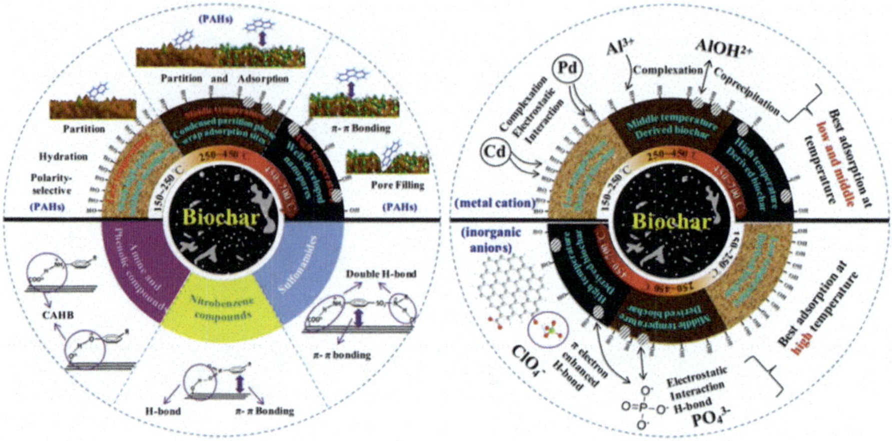

Fig. 13.23 Sorption mechanisms of organic (left) and inorganic pollutants (right) to biochars prepared under different pyrolytic temperatures. PAHs: polycyclic aromatic hydrocarbons. CAHB: charge-assisted H-bond, including negative charge-assisted H-bond (−) CAHB and positive charge-assisted H-bond (+) CAHB. (Reproduced with permission from Xiao et al. 2018)

Tang et al. 2018a; Tang et al. 2018b). The removal ratios to methyl violet and tetracycline still remained 77.6% and 60.2%, after ten times cycles of adsorption-regeneration via a facile thermal regeneration strategy, respectively (Tang et al. 2019b). Furthermore, a series of carbon/attapulgite composite adsorbents were successfully fabricated by a one-step in situ carbonization process using natural starch as the carbon source to decolorate the crude palm oil (Tian et al. 2018). And then the spent bleaching earth was further continuously transformed into carbon/attapulgite composite adsorbents after cyclic adsorption-thermal regeneration for the removal of dyes from wastewater. Therefore, the spent magnetic adsorbent after adsorption organic pollutants could be repeatedly regenerated and finally to be applied into soil for the remediation of heavy metal-polluted soil.

Acknowledgments The authors thank the funds by the National Natural Science Foundation of China (21706267), the Major Projects of the Natural Science Foundation of Gansu, China (No. 18JR4RA001), and the Youth Innovation Promotion Association CAS (2016370).

References

Ahalya K, Suriyanarayanan N, Sangeetha S (2014) Effect of pH and annealing temperatures on structural, magnetic, electrical, dielectric and adsorption properties of manganese ferrite nano particles. Mat Sci Semicon Proc 27:672–681. https://doi.org/10.1016/j.mssp.2014.08.009

Alhadidi QA, Zhou Z, Quiñones Deliz KY, Greenslet HY, Bonzongo JCJ (2021) Removal of type-A, type-B, and borderline metals from contaminated soils using zero valent iron and magnetic separation technology: a predictive approach for metal resources recovery. Chemosphere 274:129980. https://doi.org/10.1016/j.chemosphere.2021.129980

Ali R, Khan MA, Mahmood A, Chughtai AH, Sultan A, Shahid M, Ishaq M, Warsi MF (2014a) Structural, magnetic and dielectric behavior of $Mg_{1-x}Ca_xNiyFe_{2-y}O_4$ nano-ferrites synthesized by the micro-emulsion method. Ceram Int 40:3841–3846. https://doi.org/10.1016/j.ceramint.2013.08.024

Ali R, Mahmood A, Khan MA, Chughtai AH, Shahid M, Shakir I, Warsi MF (2014b) Impacts of Ni–Co substitution on the structural, magnetic and dielectric properties of magnesium nano-ferrites fabricated by micro-emulsion method. J Alloy Compd 584:363–368. https://doi.org/10.1016/j.jallcom.2013.08.114

Anjali C, Devendra S, Shekhar A (2019) Removal of pharmaceutical contaminants in wastewater using nanomaterials: a comprehensive review. Curr Drug Metab 20:483–505. https://doi.org/10.2174/1389200220666181127104812

Anoshkin VI, Campion J, Lioubtchenko DV, Oberhammer J (2018) Freeze-dried carbon nanotube aerogels for high-frequency absorber applications. ACS Appl Mater Int 10:19806–19811. https://doi.org/10.1021/acsami.8b03983

Asadi R, Abdollahi H, Gharabaghi M, Boroumand Z (2020) Effective removal of Zn (II) ions from aqueous solution by the magnetic $MnFe_2O_4$ and $CoFe_2O_4$ spinel ferrite nanoparticles with focuses on synthesis, characterization, adsorption, and desorption. Adv Powder Technol. https://doi.org/10.1016/j.apt.2020.01.028

Aylar N, Ali Reza M, Afshin F (2020) Alternating magnetic field and ultrasound waves as size controlling parameters in preparation of superparamagnetic Fe_3O_4 nanoparticles. J Nanosci Nanotechnol 20:871–877. https://doi.org/10.1166/jnn.2020.16904

Azhar U, Huyan C, Wan X, Zong C, Xu A, Liu J, Ma J, Zhang S, Geng B (2019) Porous multifunctional fluoropolymer composite foams prepared via humic acid modified Fe_3O_4 nanoparticles stabilized Pickering high internal phase emulsion using cationic fluorosurfactant as co-stabilizer. Arab J Chem 12:559–572. https://doi.org/10.1016/j.arabjc.2018.04.003

Baig MM, Yousuf MA, Agboola PO, Khan MA, Shakir I, Warsi MF (2019) Optimization of different wet chemical routes and phase evolution studies of $MnFe_2O_4$ nanoparticles. Ceram Int 45:12682–12690. https://doi.org/10.1016/j.ceramint.2019.03.114

Batoo KM, El-sadek MSA (2013) Electrical and magnetic transport properties of Ni–Cu–Mg ferrite nanoparticles prepared by sol–gel method. J Alloy Comp 566:112–119. https://doi.org/10.1016/j.jallcom.2013.02.129

Cao X, Zang L, Bu Z, Sun L, Guo D, Wang C (2016) Microfluidic fabrication of magnetic porous multi-walled carbon nanotube beads for oil and organic solvent adsorption. J Mater Chem A 4:10479–10485. https://doi.org/10.1039/C6TA01179A

Cervin NT, Andersson L, Sing Ng JB, Olin P, Bergström L, Wågberg L (2013) Lightweight and strong cellulose materials made from aqueous foams stabilized by nanofibrillated cellulose. Biomacromolecules 14:503–511. https://doi.org/10.1021/bm301755u

Chávez-Guajardo AE, Medina-Llamas JC, Maqueira L, Andrade CAS, Alves KGB, Melo CP (2015) Efficient removal of Cr (VI) and Cu (II) ions from aqueous media by use of polypyrrole/maghemite and polyaniline/maghemite magnetic nanocomposites. Chem Eng J 281:826–836. https://doi.org/10.1016/j.cej.2015.07.008

Chen M, Jiang W, Wang F, Shen P, Ma P, Gu J, Mao J, Li F (2013) Synthesis of highly hydrophobic floating magnetic polymer nanocomposites for the removal of oils from water surface. Appl Surf Sci 286:249–256. https://doi.org/10.1016/j.apsusc.2013.09.059

Chen L, Zhou C, Fiore S, Tong D, Zhang H, Li C, Ji S, Yu W (2016) Functional magnetic nanoparticle/clay mineral nanocomposites: preparation, magnetism and versatile applications. Appl Clay Sci 127–128:143–163. https://doi.org/10.1016/j.clay.2016.04.009

Chen X, Song X, Huang J, Wu C, Ma D, Tian M, Jiang H, Huang P (2017) Phase behavior of Pickering emulsions stabilized by graphene oxide sheets and resins. Energ Fuel 31:13439–13447. https://doi.org/10.1021/acs.energyfuels.7b02672

Cyganowski P (2020) Synthesis of adsorbents with anion exchange and chelating properties for separation and recovery of precious metals-a review. Sol Extr Ion Exc 38:143–165. https://doi.org/10.1080/07366299.2020.1720117

Dai J, Pan J, Xu L, Li X, Zhou Z, Zhang R, Yan Y (2012) Preparation of molecularly imprinted nanoparticles with superparamagnetic susceptibility through atom transfer radical emulsion polymerization for the selective recognition of tetracycline from aqueous medium. J Hazard Mater 205–206:179–188. https://doi.org/10.1016/j.jhazmat.2011.12.056

Dai Y, Zhang N, Xing C, Cui Q, Sun Q (2020) The adsorption, regeneration and engineering applications of biochar for removal organic pollutants: a review. Chemosphere 223:12–27. https://doi.org/10.1016/j.chemosphere.2019.01.161

Dehghani MH, Gholami S, Karri RR, Lima EC, Mahvi AH, Nazmara S, Fazlzadeh M (2021) Process modeling, characterization, optimization, and mechanisms of fluoride adsorption using magnetic agro-based adsorbent. J Environ Manag 286:112173. https://doi.org/10.1016/j.jenvman.2021.112173

Deng J, Shao Y, Gao N, Tan C, Zhou S, Hu X (2013) $CoFe_2O_4$ magnetic nanoparticles as a highly active heterogeneous catalyst of oxone for the degradation of diclofenac in water. J Hazard Mater 262:836–844. https://doi.org/10.1016/j.jhazmat.2013.09.049

Ding Z, Wang W, Zhang Y, Li F, Liu JP (2015) Synthesis, characterization and adsorption capability for Congo red of $CoFe_2O_4$ ferrite nanoparticles. J Alloy Comp 640:362–370. https://doi.org/10.1016/j.jallcom.2015.04.020

Dlamini DS, Tesha JM, Vilakati GD, Mamba BB, Mishra AK, Thwala JM, Li J (2020) A critical review of selected membrane- and powder-based adsorbents for water treatment: sustainability and effectiveness. J Clean Prod 277:123497. https://doi.org/10.1016/j.jclepro.2020.123497

Du F, Sun L, Tan W, Wei Z, Nie H, Huang Z, Ruan G, Li J (2019) Magnetic stir cake sorptive extraction of trace tetracycline antibiotics in food samples: preparation of metal–organic framework-embedded polyHIPE monolithic composites, validation and application. Anal Bioanal Chem 411:2239–2248. https://doi.org/10.1007/s00216-019-01660-1

Duan Y (2017) Novel preparation of Fe_3O_4/styrene-co-butyl acrylate composite microspheres via a phase inversion emulsion process. Colloid Polym Sci 295:1757–1763. https://doi.org/10.1007/s00396-017-4154-1

Duan S, Tang R, Xue Z, Zhang X, Zhao Y, Zhang W, Zhang J, Wang B, Zeng S, Sun D (2015) Effective removal of Pb(II) using magnetic $Co_{0.6}Fe_{2.4}O_4$ micro-particles as the adsorbent: synthesis and study on the kinetic and thermodynamic behaviors for its adsorption. Colloid Surf A-Physicochem Eng Asp 469:211–223. https://doi.org/10.1016/j.colsurfa.2015.01.029

Duan S, Liu X, Wang Y, Shao D, Alharbi NS, Alsaedi A (2016) Li J highly efficient entrapment of U(VI) by using porous magnetic $Ni_{0.6}Fe_{2.4}O_4$ micro-particles as the adsorbent. J Taiwan Inst Chem Eng 65:367–377. https://doi.org/10.1016/j.jtice.2016.05.041

Ebrahimpour E, Amiri A, Baghayeri M, Rouhi M, MansourLakouraj M (2017) Poly (indole-co-thiophene)@Fe_3O_4 as novel adsorbents for the extraction of aniline derivatives from water samples. Microchem J 131:174–181. https://doi.org/10.1016/j.microc.2016.12.022

Estevez L, Prabhakaran V, Garcia LA, Shin Y, Tao J, Schwarz MA, Darsell J, Bhattacharya P, Shutthanandan V, Zhang J (2017) Hierarchically porous graphitic carbon with simultaneously high surface area and colossal pore volume engineered via ice templating. ACS Nano 11:11047–11055. https://doi.org/10.1021/acsnano.7b05085

Fahimirad B, Rajabi M, Elhampour A (2018) A rapid and simple extraction of anti-depressant drugs by effervescent salt-assisted dispersive magnetic micro solid-phase extraction method using new adsorbent Fe_3O_4@SiO_2@N_3. Anal Chim Acta 1047:275–284. https://doi.org/10.1016/j.aca.2018.10.028

Fan L, Zhang B, Zhang H, Jia X, Chen X, Zhang Q (2016) Preparation of light core/shell magnetic composite microspheres and their application for lipase immobilization. RSC Adv 6:65911–65920. https://doi.org/10.1039/C6RA12764A

Fang Q, Zhang J, Bai L, Duan J, Xu H, Leung KCF, Xuan S (2019) In situ redox-oxidation polymerization for magnetic core-shell nanostructure with polydopamine-encapsulated-Au hybrid shell. J Hazard Mater 367:15–25. https://doi.org/10.1016/j.colsurfa.2010.04.016

Feltin N, Pileni MP (1997) New technique for synthesizing iron ferrite magnetic nanosized particles. Langmuir 13:3927–3933. https://doi.org/10.1021/la960854q

Feng L, Cao M, Ma X, Zhu Y, Hu C (2012) Superparamagnetic high-surface-area Fe_3O_4 nanoparticles as adsorbents for arsenic removal. J Hazard Mater 217–218:439–446. https://doi.org/10.1016/j.jhazmat.2012.03.073

Feuser PE, BubniakL S, Silva MCS, Viegas AC, Fernandes AC, Ricci-Junior E, Nele M, Tedesco AC, Sayer C, Araújo PHH (2015) Encapsulation of magnetic nanoparticles in poly(methyl methacrylate) by miniemulsion and evaluation of hyperthermia in U87MG cells. Eur Polym J 68:355–365. https://doi.org/10.1016/j.eurpolymj.2015.04.029

Fotukian SM, Barati A, Soleymani M, Alizadeh AM (2020) Solvothermal synthesis of $CuFe_2O_4$ and Fe_3O_4 nanoparticles with high heating efficiency for magnetic hyperthermia application. J Alloy Compd 816:152548. https://doi.org/10.1016/j.jallcom.2019.152548

Fresco-Cala B, Cárdenas S (2019) Preparation of macroscopic carbon nanohorn-based monoliths in polypropylene tips by medium internal phase emulsion for the determination of parabens in urine samples. Talanta 1981:295–301. https://doi.org/10.1016/j.talanta.2019.02.029

Fröhlich AC, Foletto EL, Dotto GL (2019) Preparation and characterization of $NiFe_2O_4$/activated carbon composite as potential magnetic adsorbent for removal of ibuprofen and ketoprofen pharmaceuticals from aqueous solutions. J Clean Prod 229:828–837. https://doi.org/10.1016/j.jclepro.2019.05.037

Gervald AY, Gritskova IA, Prokopov NI (2010) Synthesis of magnetic polymeric microspheres. Russ Chem Rev 79:219–229. https://doi.org/10.1070/RC2010v079n03ABEH004068

Ghahfarokhi SEM, Shobegar EM (2020) An investigation of the ethylene glycol surfactant on the structural, microstructure, magnetic and optical properties of $SrFe_2O_4$ nanoparticles. J Magn Magn Mater 495:165866. https://doi.org/10.1016/j.jmmm.2019.165866

Gharieh A, Khoee S, Mahdavian RA (2019) Emulsion and miniemulsion techniques in preparation of polymer nanoparticles with versatile characteristics. Adv Colloid Int Sci 269:152–186. https://doi.org/10.1016/j.cis.2019.04.010

Ghone DM, Mathe VL, Patankar KK, Kaushik SD (2018) Microstructure, lattice strain, magnetic and magnetostriction properties of holmium substituted cobalt ferrites obtained by co-precipitation method. J Alloy Comp 739:52–61. https://doi.org/10.1016/j.jallcom.2017.12.219

Gilani ZA, Warsi MF, Khan MA, Shakir I, Shahid M, Anjum MN (2015) Impacts of neodymium on structural, spectral and dielectric properties of $LiNi_{0.5}Fe_2O_4$ nanocrystalline ferrites fabricated via micro-emulsion technique. Phys E 73:169–174. https://doi.org/10.1016/j.physe.2015.06.001

Gokmen MT, Van Camp W, Colver PJ, Bon SAF, Du Prez FE (2009) Fabrication of porous, "clickable", polymer beads and rods through generation of high internal phase emulsion (HIPE) droplets in a simple microfluidic device. Macromolecules 42:9289–9294. https://doi.org/10.1021/ma9018679

Gu J, Jiang W, Wang F, Chen M, Mao J, Xie T (2014) Facile removal of oils from water surfaces through highly hydrophobic and magnetic polymer nanocomposites. Appl Surf Sci 301:492–499. https://doi.org/10.1016/j.apsusc.2014.02.112

Gui HG, Guan GW, Zhang T, Guo QP (2019) Crophase-separated, hierarchical macroporous polyurethane from a nonaqueous emulsion-templated reactive block copolymer. Chem Eng J 365:369–377. https://doi.org/10.1016/j.cej.2019.02.015

Han J, Du Z, Zou W, Li H, Zhang C (2015) In-situ improved phenol adsorption at ions-enrichment interface of porous adsorbent for simultaneous removal of copper ions and phenol. Chem Eng J 262:571–578. https://doi.org/10.1016/j.cej.2014.10.018

Hashem MA, Elnagar MM, Kenawy IM, Ismail MA, Qu W (2020) Synthesis and application of hydrazono-imidazoline modified cellulose for selective separation of precious metals from geological samples. Carbohydr Polym 237:116177. https://doi.org/10.1016/j.carbpol.2020.116177

Hilczer A, Kowalska K, Markiewicz E, Pietraszko A, Andrzejewski B (2016) Dielectric and magnetic response of $SrFe_{12}O_{19}$–$CoFe_2O_4$ composites obtained by solid state reaction. Mater Sci Eng B 207:47–55. https://doi.org/10.1016/j.mseb.2016.02.003

Hochepied JF, Pileni MP (2000) Magnetic properties of mixed cobalt–zinc ferrite nanoparticles. J Appl Phys 87:2472–2478. https://doi.org/10.1063/1.372205

Hu J, Lo IMC, Chen G (2007) Comparative study of various magnetic nanoparticles for Cr (VI) removal. Sep Purif Technol 56:249–256. https://doi.org/10.1016/j.seppur.2007.02.009

Hu J, Chen M, Wu L (2011) Organic-inorganic nanocomposites synthesized via miniemulsion polymerization. Polym Chem 2:760–772. https://doi.org/10.1039/C0PY00284D

Hua M, Zhang S, Pan B, Zhang W, Lv L, Zhang Q (2012) Heavy metal removal from water/wastewater by nanosized metal oxides: a review. J Hazard Mater 211–212:317–331. https://doi.org/10.1016/j.jhazmat.2011.10.016

Huang Y, Yang J, Chen L, Zhang L (2018) Chitin nanofibrils to stabilize long-life pickering foams and their application for lightweight porous materials. ACS Sustain Chem Eng 6:10552–10561. https://doi.org/10.1021/acssuschemeng.8b01883

Huang Y, Zhang W, Bai M, Huang X (2020) One-pot fabrication of magnetic fluorinated carbon nanotubes adsorbent for efficient extraction of perfluoroalkyl carboxylic acids and perfluoroalkyl sulfonic acids in environmental water samples. Chem Eng J 380:122392. https://doi.org/10.1016/j.cej.2019.122392

Jacobo SE, Duhalde S, Bertorello HR (2004) Rare earth influence on the structural and magnetic properties of NiZn ferrites. J Magn Magn Mater 272–276:2253–2254. https://doi.org/10.1016/j.jmmm.2003.12.564

Jesus ABC, Jesus JR, Lima RJS, Moura KO, Almeida JMA, Duque JGS, Meneses CT (2020) Synthesis and magnetic interaction on concentrated Fe_3O_4 nanoparticles obtained by the co-precipitation and hydrothermal chemical methods. Ceram Int in press. https://doi.org/10.1016/j.ceramint.2020.01.135

Ji J, Chen G, Zhao J, Wei Y (2020) Efficient removal of Pb (II) by inexpensive magnetic adsorbents prepared from one-pot pyrolysis of waste tyres involved magnetic nanoparticles. Fuel 282:118715. https://doi.org/10.1016/j.fuel.2020.118715

Jiang Y, Kim D (2013) Synthesis and selective adsorption behavior of Pd (II)-imprinted porous polymer particles. Chem Eng J 232:503–509. https://doi.org/10.1016/j.cej.2013.08.008

Jiang L, Chai F, Chen Q (2017) Soft magnetic nanocomposite microgels by in-situ crosslinking of poly acrylic acid onto superparamagnetic magnetite nanoparticles and their applications for the removal of Pb(II) ion. Eur Polym J 89:468–481. https://doi.org/10.1016/j.eurpolymj.2017.02.045

Jung Y, Ko YG, Do T, Chu Y, Choi US, C. Kim CH (2019) Core/shell hybrid fiber with aminated PAN and Fe_2O_3 as a high-capacity adsorbent for phosphate ions. J Hazard Mater 378:120726. doi:https://doi.org/10.1016/j.jhazmat.2019.06.003

Kang S, Rethinasabapathy M, Hwang SK, Lee GW, Jang SC, Kwak CH, Choe SR, Huh YS (2018) Microfluidic generation of Prussian blue-laden magnetic micro-adsorbents for cesium removal. Chem Eng J 341:218–226. https://doi.org/10.1016/j.cej.2018.02.025

Kaur M, Kaur N, Jeet K, Kaur P (2015) Effective removal of Pb(II) using magnetic $Co_{0.6}Fe_{2.4}O_4$ micro-particles as the adsorbent: synthesis and study on the kinetic and thermodynamic behaviors for its adsorption. Colloid Surf A 469:211–223. https://doi.org/10.1016/j.colsurfa.2015.01.029

Kavitha S, Kurian M (2020) Role of doped nitrogen and Sulphur in cobalt/cobalt-zirconium nanoferrites synthesized by facile methods. Ceram Int 46:4423–4434. https://doi.org/10.1016/j.ceramint.2019.10.168

Kavousi F, Nikfarjam N (2019) Highly interconnected macroporous structures made from starch nanoparticle-stabilized medium internal phase emulsion polymerization for use in cell culture. Polymer 18010:121744. https://doi.org/10.1016/j.polymer.2019.121744

Khan MA, Sabir M, Mahmood A, Asghar M, Mahmood K, Afzal Khan M, Ahmad I, Sher M, Warsi MF (2014) High frequency dielectric response and magnetic studies of $Zn_{1-x}Tb_xFe_2O_4$ nanocrystalline ferrites synthesized via micro-emulsion technique. J Magn Magn Mater 360:188–192. https://doi.org/10.1016/j.jmmm.2014.02.059

Kim JH, Kim SM, Yoon IH, Choi SJ, Kim I (2020) Selective separation of Cs-contaminated clay from soil using polyethylenimine-coated magnetic nanoparticles. Sci Total Environ 706:130020. https://doi.org/10.1016/j.scitotenv.2019.136020

Konicki W, Sibera D, Mijowska E, Lendzion-Bieluń Z, Narkiewicz U (2013) Equilibrium and kinetic studies on acid dye acid red 88 adsorption by magnetic $ZnFe_2O_4$ spinel ferrite nanoparticles. J Colloid Interf Sci 398:152–160. https://doi.org/10.1016/j.jcis.2013.02.021

Kovačič S, Štefanec D, Krajnc P (2007) Highly porous open-cellular monoliths from 2-hydroxyethyl methacrylate based high internal phase emulsions (HIPEs): preparation and void size tuning. Macromolecules 2007:8056–8060. https://doi.org/10.1021/ma071380c

Kovačič S, Drašinac N, Pintar A, Žagar E (2018) Highly porous cationic polyelectrolytes via oil-in-water concentrated emulsions: synthesis and adsorption kinetic study. Langmuir 34:10353–10362

Kozyatnyk I, Latham GK, Jansson S (2019) Valorization of humic acids by hydrothermal conversion into carbonaceous materials: physical and functional properties. ACS Sustain Chem Eng 7:2585–2592. https://doi.org/10.1021/acssuschemeng.8b05614

Kuai S, Zhang Z, Nan Z (2013) Synthesis of Ce^{3+} doped $ZnFe_2O_4$ self-assembled clusters and adsorption of chromium (VI). J Hazard Mater 250–251:229–237. https://doi.org/10.1016/j.jhazmat.2013.01.074

Li X, Sun G, Li Y, Yu JC, Wu J, Ma GH, Ngai T (2014a) Porous TiO_2 materials through Pickering high-internal phase emulsion templating. Langmuir 30:2676–2683. https://doi.org/10.1021/la404930h

Li Z, Liu H, Zeng L, Liu H, Yang S, Wang Y (2014b) Preparation of high internal water-phase double emulsions stabilized by a single anionic surfactant for fabricating interconnecting porous polymer microspheres. Langmuir 2014:12154–12163. https://doi.org/10.1021/la502564r

Li Q, Mao Q, Yang C, Zhang SJ, He GH, Zhang XJ, Zhang WJ (2019) Hydrophobic-modified montmorillonite coating onto crosslinked chitosan as the core-shell micro-sorbent for iodide adsorptive removal via Pickering emulsion polymerization. Inter J Biol Macromol 141:987–996. https://doi.org/10.1016/j.ijbiomac.2019.09.065

Li Y, Zhang Q, Yuan S, Yin H (2021) High-efficiency extraction of iron from early iron tailings via the suspension roasting-magnetic separation. Powder Technol 379:466–477. https://doi.org/10.1016/j.powtec.2020.10.005

Lian Q, Cui Y, Zheng X, Wu H (2015) Preparation and adsorption of magnetic $Co_{0.5}Ni_{0.5}Fe_2O_4$-chitosan nanoparticles. Russ J Appl Chem 88:1877−1883. https://doi.org/10.1134/S10704272150110208

Lima EWC, Feng R (2012) Agglomeration of magnetic nanoparticles. J Chem Phys 136:124109. https://doi.org/10.1063/1.3697865

Liu S, Ge H, Cheng S, Zou Y (2018) Green synthesis of magnetic 3D bio-adsorbent by corn straw core and chitosan for methylene blue removal. Environ Technol. https://doi.org/10.1080/09593330.2018.1556345

Liu Q, ZhouY LJ, Zho YB (2020) Novel cyclodextrin-based adsorbents for removing pollutants from wastewater: a critical review. Chemosphere 241:125043. https://doi.org/10.1016/j.chemosphere.2019.125043

Liu J, Zhang J, Xing L, Wang D, Wang L, Xiao H, Ke J (2021) Magnetic Fe_3O_4/attapulgite hybrids for cd(II) adsorption: performance, mechanism and recovery. J Hazard Mater 412:125237. https://doi.org/10.1016/j.jhazmat.2021.125237

Lobato NCC, Ferreira AD, Weidler PG, Franzreb M, Silva GC, Mansur MB (2019) Improvement of magnetic solvent extraction using functionalized silica coated Fe_3O_4 nanoparticles. Sep Purif Technol 229:115839. https://doi.org/10.1016/j.seppur.2019.115839

Lobato NCC, Ferreira AD, Weidler PG, Franzreb M, Silva GC, Mansur MB (2020) Microstructure and chemical stability analysis of magnetic core coated with SILICA and functionalized with silane OTS. Appl Surf Sci 505:144565–144571. https://doi.org/10.1016/j.apsusc.2019.144565

Lu F, Astruc D (2020) Nanocatalysts and other nanomaterials for water remediation from organic pollutants. Coordin Chem Rev 408:213180. https://doi.org/10.1016/j.ccr.2020.213180

Lu F, Huang C, You L, Wang J, Zhang Q (2017) Magnetic hollow carbon microspheres as a reusable adsorbent for rhodamine B removal. RSC Adv 7:23255–23264. https://doi.org/10.1039/C7RA03045B

Lü T, Chen Y, Qi DM, Cao ZH, Zhang D, Zhao HT (2017) Treatment of emulsified oil wastewaters by using chitosan grafted magnetic nanoparticles. J Alloy Compd 696:1205–1212. https://doi.org/10.1016/j.jallcom.2016.12.118

Lu T, Zhu Y, Qi Y, Wang W, Wang A (2018a) Magnetic chitosan-based adsorbent prepared via Pickering high internal phase emulsion for high-efficient removal of antibiotics. Int J Biol Macromol 106:870–877. https://doi.org/10.1016/j.ijbiomac.2017.08.092

Lu T, Zhu Y, Wang W, Qi Y, Wang A (2018b) Polyaniline-functionalized porous adsorbent for Sr^{2+} adsorption. J Radioanal Nucl Chem 317:907–917. https://doi.org/10.1007/s10967-018-5935-9

Lu T, Zhu Y, Qi Y, Kang Y, Wang A (2019a) Tunable superporous magnetic adsorbent prepared via eco-friendly Pickering MIPEs for high-efficiency adsorption of Rb^+ and Sr^{2+}. Chem Eng J 368:988–998. https://doi.org/10.1016/j.cej.2019.03.040

Lu T, Zhu Y, Wang W, Qi Y, Wang A (2019b) Interconnected superporous adsorbent prepared via yeast-based Pickering HIPEs for high-efficiency adsorption of Rb^+, Cs^+ and Sr^{2+}. Chem Eng J 361:1411–1422. https://doi.org/10.1016/j.cej.2018.11.006

Luo Y, Zhou Z, Yue T (2017) Synthesis and characterization of nontoxic chitosan-coated Fe_3O_4 particles for patulin adsorption in a juice-pH simulation aqueous. Food Chem 221:317–323. https://doi.org/10.1016/j.foodchem.2016.09.008

Ma W, Dai J, Dai X, Yan Y (2014) Preparation and characterization of chitosan/kaolin/Fe_3O_4 magnetic microspheres and their application for the removal of ciprofloxacin. Adsorpt Sci Technol 32:775–790. https://doi.org/10.1260/0263-6174.32.10.775

Ma J, Xia W, Fu X, Ding L, Kong Y, Zhang H, Fu K (2020) Magnetic flocculation of algae-laden raw water and removal of extracellular organic matter by using composite flocculant of Fe_3O_4/cationic polyacrylamide. J Clean Prod 248:119276. https://doi.org/10.1016/j.jclepro.2019.119276

Mahmoodi NM (2013) Zinc ferrite nanoparticle as a magnetic catalyst: synthesis and dye degradation. Mater Res Bull 48:4255–4260. https://doi.org/10.1016/j.materresbull.2013.06.070

Maleki A, Hajizadeh Z, Sharifi V, Emdadi Z (2019) A green, porous and eco-friendly magnetic geopolymer adsorbent for heavy metals removal from aqueous solutions. J Clean Prod 215:1233–1245. https://doi.org/10.1016/j.jclepro.2019.01.084

Mao Y, Cui J, Zhao J, Wu Y, Wang C, Lu J, Lin X, Yan Y (2016) Selective separation of bifenthrin by pH-sensitive/magnetic molecularly imprinted polymers prepared by Pickering emulsion polymerization. Fiber Polym 17:1531–1539. https://doi.org/10.1007/s12221-016-6570-0

Masunga N, Mmelesi OK, Kefeni KK, Mamba BB (2019) Recent advances in copper ferrite nanoparticles and nanocomposites synthesis, magnetic properties and application in water treatment: review. J Environ Chem Eng 7:103179. https://doi.org/10.1016/j.jece.2019.103179

Mert EH, Yıldırım H, Üzümcü AT, Kavas H (2013) Synthesis and characterization of magnetic polyHIPEs with humic acid surface modified magnetic iron oxide nanoparticles. React Funct Polym 73:175–181. https://doi.org/10.1016/j.reactfunctpolym.2012.09.005

Ming G, Duan H, Meng X, Sun G, Sun W, Liu Y, Lucia L (2015) A novel fabrication of monodisperse melamine-formaldehyde resin microspheres to adsorb lead (II). Chem Eng J 288:745–757. https://doi.org/10.1016/j.cej.2015.12.007

Mokadem Z, Saidi-Besbes S, Lebaz N, Elaissari A (2020) Magnetic monolithic polymers prepared from high internal phase emulsions and Fe_3O_4 triazole-functionalized nanoparticles for Pb^{2+}, Cu^{2+} and Zn^{2+} removal. React Funct Polym 155:104693. https://doi.org/10.1016/j.reactfunctpolym.2020.104693

Moumen N, Pileni MP (1996a) Control of the size of cobalt ferrite magnetic fluid. J Phys Chem 100:1867–1873. https://doi.org/10.1021/jp9524136

Moumen N, Pileni MP (1996b) New syntheses of cobalt ferrite particles in the range 2–5 nm: comparison of the magnetic properties of the nanosized particles in dispersed fluid or in powder form. Chem Mater 8:1128–1134. https://doi.org/10.1021/cm950556z

Moumen N, Lisiecki I, Briois V, Pileni MP (1995a) Micellar factors which play a role in the control of the nanosize particles of cobalt ferrite. Supramol Sci 2:161–168. https://doi.org/10.1016/0968-5677(96)89671-4

Moumen N, Veillet P, Pileni MP (1995b) Controlled preparation of nanosize cobalt ferrite magnetic particles. J Magn Magn Mater 149:67–71. https://doi.org/10.1016/0304-8853(95)00340-1

Mudassir MA, Hussain SZ, Jilani A, Zhang H, Ansari TM, Hussain I (2019) Magnetic hierarchically macroporous emulsion-templated poly(acrylic acid)–iron oxide nanocomposite beads for water remediation. Langmuir 35:8996–9003. https://doi.org/10.1021/acs.langmuir.9b01121

Murray BS (2019) Pickering emulsions for food and drinks. Curr Opin Food Sci 27:57–63. https://doi.org/10.1016/j.cofs.2019.05.004

Nuryono N, Miswanda D, Sakti SCW, Rusdiarso B, Krisbiantoro PA, Utami N, Otomo R, Kamiya Y (2020) Chitosan-functionalized natural magnetic particle@silica modified with (3-chloropropyl)trimethoxysilane as a highly stable magnetic adsorbent for gold(III) ion. Mater Chem Phys. https://doi.org/10.1016/j.matchemphys.2020.123507

Osama S, Nermen MH, Abdullah A (2015) Synthesis of magnetic nanoparticles and nanosheets for oil spill removal. Bentham Sci Pub 1:32–43. https://doi.org/10.2174/2210681205666150601215445

Ou H, Chen Q, Pan J, Zhang Y, Huang Y, Qi X (2015) Selective removal of erythromycin by magnetic imprinted polymers synthesized from chitosan-stabilized Pickering emulsion. J Hazard Mater 289:28–37. https://doi.org/10.1016/j.jhazmat.2015.02.030

Palmqvist AEC (2003) Synthesis of ordered mesoporous materials using surfactant liquid crystals or micellar solutions. Curr Opin Colloid Inter Face Sci 8:145–155. https://doi.org/10.1016/S1359-0294(03)00020-7

Pan J, Yin Y, Gan M, Meng M, Dai X, Wu R, Shi W, Yan Y (2015) Fabrication and evaluation of molecularly imprinted multi-hollow microspheres adsorbents with tunable inner pore structures derived from templating Pickering double emulsions. Chem Eng J 266:299–308. https://doi.org/10.1016/j.cej.2014.11.126

Pan J, Zeng J, Cao Q, Gao H, Gen Y, Peng Y, Dai X, Yan Y (2016) Hierarchical macro and mesoporous foams synthesized by HIPEs template and interface grafted route for simultaneous removal of λ-cyhalothrin and copper ions. Chem Eng J 284:1361–1372. https://doi.org/10.1016/j.cej.2015.09.023

Pang YL, Lim S, Ong HC, Chong WT (2016) Research progress on iron oxide-based magnetic materials: synthesis techniques and photocatalytic applications. Ceram Int 42:9–34. https://doi.org/10.1016/j.ceramint.2015.08.144

Patel RK, Chawla AK, Loulergue P, Teychene B, Pandey JK (2019) 3D printed microchannel loaded with hematite nanoadsorbent for fluoride removal from water. Mater Lett 254:190–193. https://doi.org/10.1016/j.matlet.2019.07.061

Podzus PE, Daraio ME, Jacobo SE (2009) Chitosan magnetic microspheres for technological applications: preparation and characterization. Physica B 404:2710–2712. https://doi.org/10.1016/j.physb.2009.06.093

Popkov V, Tolstoy VP, Semenov VG (2020) Synthesis of phase-pure superparamagnetic nanoparticles of $ZnFe_2O_4$ via thermal decomposition of zinc-iron layered double hydroxysulphate. J Alloy Comp 813:152179. https://doi.org/10.1016/j.jallcom.2019.152179

Qiu G, Wang Q, Wang C, Lau W, Guo Y (2007) Polystyrene/Fe_3O_4 magnetic emulsion and nanocomposite prepared by ultrasonically initiated miniemulsion polymerization. Ultrason Sonochem 14:55–61. https://doi.org/10.1016/j.ultsonch.2006.03.001

Rafiq MA, Javed A, Rasul MN, Khan MA, Hussain A (2020) Understanding the structural, electronic, magnetic and optical properties of spinel MFe_2O_4 (M = Mn, Co, Ni) ferrites. Ceram Int 46:4976–4983. https://doi.org/10.1016/j.ceramint.2019.10.237

Reddy DHK, Yun YS (2016) Spinel ferrite magnetic adsorbents: alternative future materials for water purification? Coordin Chem Rev 315:90–111. https://doi.org/10.1016/j.ccr.2016.01.012

Ren Y, Li N, Feng J, Luan T, Wen Q, Li Z, Zhang M (2012) Adsorption of Pb(II) and Cu(II) from aqueous solution on magnetic porous ferrospinel $MnFe_2O_4$. J Colloid Interface Sci 367:415–421. https://doi.org/10.1016/j.jcis.2011.10.022

Ren H, Hao J, Kang W, Wang G, Ju J, Li L, Cheng B (2019) Waste spunlaced facial puff derived monolithic flexible carbon framework (WCF): an ultralow-cost, recyclable and eco-friendly sorbent for oils and organic solvents. RSC Adv 9:31255–31263. https://doi.org/10.1039/C9RA05681E

Rott E, Nouri M, Meyer C, Minke R, Schneider M, Mandel K, Drenkova-Tuhtan A (2018) Removal of phosphonates from synthetic and industrial wastewater with reusable magnetic adsorbent particles. Water Res 145:608–617. https://doi.org/10.1016/j.watres.2018.08.067

Routoula E, Patwardhan SV (2020) Degradation of anthraquinone dyes from effluents: a review focusing on enzymatic dye degradation with industrial potential. Environ Sci Technol 54:647–664. https://doi.org/10.1021/acs.est.9b03737

Rydin E, Huser B, Welch EB (2000) Amount of phosphorus inactivated by alum treatments in Washington lakes. Limnol Oceanogr 45:226–230. https://doi.org/10.4319/lo.2000.45.1.0226

Salami R, Amini M, Bagherzadeh M, Hosseini H (2019) Vanadium supported on spinel cobalt ferrite nanoparticles as an efficient and magnetically recoverable catalyst for oxidative degradation of methylene blue. Appl Organomet Chem 33:e5127. https://doi.org/10.1002/aoc.5127

Sanchez-Dominguez M, Pemartin K, Boutonnet M (2012) Preparation of inorganic nanoparticles in oil-in-water microemulsions: a soft and versatile approach. Curr Opin Collid Int Sci 17:297–305. https://doi.org/10.1016/j.cocis.2012.06.007

Seeharaj P, Thasirisap E, Tridech C, Jindasuwan S (2019) Magnetic PolyHIPE composites with activated carbon and Iron oxide nanoparticles. IEEE T Magn 55:1–4. https://doi.org/10.1109/TMAG.2018.2862429

Sehlleier YH, Hardt S, Schulz C, Wiggers H (2016) A novel magnetically-separable porous iron-oxide nanocomposite as an adsorbent for methylene blue (MB) dye. J Environ Chem Eng 4:3779–3787. https://doi.org/10.1016/j.jece.2016.08.018

Solans C, Izquierdo P, Nolla N, Azemar MJ, Garcia-Celma (2005) Nano-emulsions. Curr Opin Colloid Interface Sci 10(3–4):102–110

Song X, Li L, Geng Z, Zhou L, Ji L (2017) Effective and selective adsorption of As(III) *via* imprinted magnetic Fe_3O_4/HTCC composite nanoparticles. J Environ Chem Eng 5:16–25. https://doi.org/10.1016/j.jece.2016.11.016

Stubenrauch C, Menner A, Bismarck A, Drenckhan W (2018) Emulsion and foam templating-promising routes to tailor-made porous polymers. Angew Chem Int Edit 57:10024–10032. https://doi.org/10.1002/anie.201801466

Sun X, Li Q, Yang L, Liu H (2016a) Chemically modified magnetic chitosan microspheres for Cr (VI) removal from acidic aqueous solution. Particuology 26:79–86. https://doi.org/10.1016/j.partic.2015.11.003

Sun X, Yang L, Dong T, Liu Z, Liu H (2016b) Removal of Cr(VI) from aqueous solution using amino-modified Fe_3O_4–SiO_2–chitosan magnetic microspheres with high acid resistance and adsorption capacity. J Appl Polym Sci 133:1–11. https://doi.org/10.1002/app.43078

Tan H, Tu Z, Jia H, Gou X, Ngai T (2018) Hierarchical porous protein scaffold templated from high internal phase emulsion costabilized by gelatin and gelatin nanoparticles. Langmuir 34:4820–4829. https://doi.org/10.1021/acs.langmuir.7b04047

Tang J, Mu B, Wang W, Zheng M, Wang A (2016) Fabrication of manganese dioxide/carbon/attapulgite composites derived from spent bleaching earth for adsorption of Pb(II) and brilliant green. RSC Adv 6:36534–36543. https://doi.org/10.1039/C5RA26362J

Tang J, Mu B, Zong L, Zheng M, Wang A (2017) Facile and green fabrication of magnetically recyclable carboxyl-functionalized attapulgite/carbon nanocomposites derived from spent bleaching earth for wastewater treatment. Chem Eng J 322:102–114. https://doi.org/10.1016/j.cej.2017.03.116

Tang J, Zong L, Mu B, Kang Y, Wang A (2018a) Attapulgite/carbon composites as a recyclable adsorbent for antibiotics removal. Korean J Chem Eng 35:1650–1661. https://doi.org/10.1007/s11814-018-0066-0

Tang J, Zong L, Mu B, Zhu Y, Wang A (2018b) Preparation and cyclic utilization assessment of palygorskite/carbon composites for sustainable efficient removal of methyl violet. Appl Clay Sci 161:317–325. https://doi.org/10.1016/j.clay.2018.04.039

Tang J, Wu I, Yu L, Fan X, Liu G, Yu Y (2019a) Study on adsorption properties and mechanism of thallium onto titanium-iron magnetic adsorbent. Sci Total Environ 694:133625. https://doi.org/10.1016/j.scitotenv.2019.133625

Tang J, Mu B, Zong L, Wang A (2019b) From waste hot-pot oil as carbon precursor to development of recyclable attapulgite/carbon composites for wastewater treatment. J Environ Sci 75:346–358. https://doi.org/10.1016/j.jes.2018.05.014

Tao X, Li K, Yan H, Yang H, Li A (2016) Simultaneous removal of acid green 25 and mercury ions from aqueous solutions using glutamine modified chitosan magnetic composite microspheres. Environ Pollut 209:21–29. https://doi.org/10.1016/j.envpol.2015.11.020

Teo BM, Chen F, Hatton TA, Grieser F, Ashokkumar M (2009) Novel one-pot synthesis of magnetite latex nanoparticles by ultrasound irradiation. Langmuir 25:2593–2595. https://doi.org/10.1021/la804278w

Thompson BR, Horozov TS, Stoyanov SD, Paunov VN (2019) Hierarchically structured composites and porous materials from soft templates: fabrication and applications. J Mater Chem A 14:8030–8049. https://xs.scihub.ltd/10.1039/C8TA09750J

Tian G, Wang W, Zhu Y, Zong L, Kang Y, Wang A (2018) Carbon/Attapulgite composites as recycled palm oil-Decoloring and dye adsorbents. Materials 11:86. https://doi.org/10.3390/ma11010086

Tsedenbal B, Hussain I, Lee JE, Koo BH (2020) Removal of Lead contaminants with gamma-Fe_2O_3 nanocrystals. Sci Adv Mater 12:422–426. https://doi.org/10.1166/sam.2020.3654

Wadhawan S, Jain A, Nayyar J, Mehta SK (2020) Role of nanomaterials as adsorbents in heavy metal ion removal from waste water: a review. J Water Pro Eng 33:101038. https://doi.org/10.1016/j.jwpe.2019.101038

Wang C, Yan J, Cui X, Cong D, Wang H (2010) Preparation and characterization of magnetic hollow PMMA nanospheres via in situ emulsion polymerization. Colloid Surf A: Physicochem Eng Asp 363:71–77. https://doi.org/10.1016/j.colsurfa.2010.04.016

Wang L, Li J, Wang Y, Zhao L, Jiang Q (2012) Adsorption capability for Congo red on nanocrystalline MFe_2O_4 (M = Mn, Fe, Co, Ni) spinel ferrites. Chem Eng J 181–182:72–79. https://doi.org/10.1016/j.cej.2011.10.088

Wang F, Yu X, Yang Z, Duan H, Zhang Z, Liu H (2018a) Dual pH- and light-responsive amphiphilic random copolymer nanomicelles as particulate emulsifiers to stabilize the oil/water interface. J Phys Chem C 122:18995–19003. https://doi.org/10.1021/acs.jpcc.8b05065

Wang X, Azhar U, Wang Y, Chen J, Xu A, Zhang S, Geng B (2018b) Highly porous and chemical resistive P(TFEMA-DVB) monolith with tunable morphology for rapid oil/water separation. RSC Adv 8:8355–8364. https://doi.org/10.1039/C8RA00501J

Wang G, Liu Q, Chang M, Jang J, Sui W, Si C, Ni Y (2019a) Novel Fe_3O_4@lignosulfonate/phenolic core-shell microspheres for highly efficient removal of cationic dyes from aqueous solution. Ind Crop Prod 127:110–118. https://doi.org/10.1016/j.indcrop.2018.10.056

Wang S, Lu Y, Ouyang X, Liang X, Yu D, Yang L, Huang F (2019b) Fabrication of chitosan-based MCS/ZnO@Alg gel microspheres for efficient adsorption of As(V). Inter J Biol Macromol 139:886–895. https://doi.org/10.1016/j.ijbiomac.2019.08.070

Wang R, Li W, Lu R, Peng J, Liu X, Liu K, Peng H (2020a) A facile synthesis of cationic and super-hydrophobic polyHIPEs as precursors to carbon foam and adsorbents for removal of non-aqueous-phase dye. Colloid Surf A 605:125334. https://doi.org/10.1016/j.colsurfa.2020.125334

Wang X, Liu X, Xiao C, Zhao H, Zhang M, Zheng N, Kong W, Zhang L, Yuan H, Zhang L, Lu J (2020b) Triethylenetetramine-modified hollow Fe_3O_4/SiO_2/chitosan magnetic nanocomposites for removal of Cr(VI) ions with high adsorption capacity and rapid rate. Micropor Mesopor Mater 297:110041. https://doi.org/10.1016/j.micromeso.2020.110041

Wanna Y, Chindaduang A, Tumcharern G, Phromyothin D, Porntheerapat S, Nukeaw J, Hofmann H, Pratontep S (2016) Efficiency of SPIONs functionalized with polyethylene glycol bis(amine) for heavy metal removal. J Magn Magn Mater 414:32–37. https://doi.org/10.1016/j.jmmm.2016.04.064

Weng S, Jin M, Wan D (2020) An emulsion-templated and amino diol-dictated porous material as an efficient and well recyclable boric acid scavenger. Colloid Surf A 611:125873. https://doi.org/10.1016/j.colsurfa.2020.125873

Wright AJ, Main MJ, Cooper NJ, Blight BA, Holder SJ (2017) Poly high internal phase emulsion for the immobilization of chemical warfare agents. ACS Appl Mater Int 9:31335–31339. https://doi.org/10.1021/acsami.7b09188

Wu W, He Q, Jiang C (2008) Magnetic Iron oxide nanoparticles: synthesis and surface functionalization strategies. Nanos Res Lett 3:397–415. https://doi.org/10.1007/s11671-008-9174-9

Wu Y, Ma Y, Pan J, Gu R, Luo J (2017) Porous and magnetic molecularly imprinted polymers via Pickering high internal phase emulsions polymerization for selective adsorption of λ-cyhalothrin. Front Chem 5:1–10. https://doi.org/10.3389/fchem.2017.00018

Xiao D, Lu T, Zeng R, Bi Y (2016) Preparation and highlighted applications of magnetic microparticles and nanoparticles: a review on recent advances. Microchim Acta 183:2655–2675. https://doi.org/10.1007/s00604-016-1928-y

Xiao C, Liu X, Mao S, Zhang L, Lu J (2017) Sub-micron-sized polyethylenimine-modified polystyrene/Fe_3O_4/chitosan magnetic composites for the efficient and recyclable adsorption of Cu(II) ions. Appl Surf Sci 394:378–385. https://doi.org/10.1016/j.apsusc.2016.10.116

Xiao X, Chen B, Chen Z, Zhu L, Schnoor JL (2018) Insight into multiple and multilevel structures of biochars and their potential environmental applications: a critical review. Environ Sci Technol 52:5027–5047. https://doi.org/10.1021/acs.est.7b06487

Xie X, Xiong H, Zhang Y, Tong Z, Liao A, Qin Z (2017) Preparation magnetic cassava residue microspheres and its application for Cu(II) adsorption. J Environ Chem Eng 5:2800–2806. https://doi.org/10.1016/j.jece.2017.05.024

Xu Y, Dang Q, Liu C, Yan J, Fan B, Cai J, Li J (2015) Preparation and characterization of carboxyl-functionalized chitosan magnetic microspheres and submicrospheres for Pb2+ removal. Colloid Surf A Physicochem Eng Asp 482:353–364. https://doi.org/10.1016/j.colsurfa.2015.06.028

Xue D, Li T, Liu Y, Yang Y, Zhang Y, Cui J, Guo D (2019) Selective adsorption and recovery of precious metal ions from water and metallurgical slag by polymer brush graphene–polyurethane composite. React Funct Polym 136:138–152. https://doi.org/10.1016/j.reactfunctpolym.2018.12.026

Yaqoob KA, Bououdina M, Najar BA AMS, Vijaya JJ (2019) Selectivity and efficient Pb and Cd ions removal by magnetic MFe_2O_4 (M=Co, Ni, Cu and Zn) nanoparticles. Mater Chem Phys 232:254–264. https://doi.org/10.1016/j.matchemphys.2019.04.077

Ye J, Nyobe D, Tang B, Bin L, Pi L, Huang S, Fu F, Cai Y, Guan G, Hao X (2020) Facilely synthesized recyclable mesoporous magnetic silica composite for highly efficient and fast adsorption of methylene blue from wastewater: thermodynamic mechanism and kinetics study. J Mol Liq 303:112656. https://doi.org/10.1016/j.molliq.2020.112656

Yin H, Ren C, Li W (2018) Introducing hydrate aluminum into porous thermally-treated calcium-rich attapulgite to enhance its phosphorus sorption capacity for sediment internal loading management. Chem Eng J 348:704–712. https://doi.org/10.1016/j.cej.2018.05.065

Yousuf MA, Baig MM, Waseem M, Haider S, Shakir I, Khand SUD, Warsi MF (2019) Low cost micro-emulsion route synthesis of Cr-substituted $MnFe_2O_4$ nanoparticles. Ceram Int 45:22316–22323. https://doi.org/10.1016/j.ceramint.2019.07.259

Yu L, Hao G, Gu J, Zhou S, Zhang N, Jiang W (2015a) Fe_3O_4/PS magnetic nanoparticles: synthesis, characterization and their application as sorbents of oil from waste water. J Magn Magn Mater 394:4–21. https://doi.org/10.1016/j.jmmm.2015.06.045

Yu S, Tan H, Wang J, Liu X, Zhou K (2015b) High porosity supermacroporous polystyrene materials with excellent oil–water separation and gas permeability properties. ACS Appl Mater Interfaces 7:6745–6753. https://doi.org/10.1021/acsami.5b00196

Yu M, Wang L, Hu L, Li Y, Luo D, Mei S (2019) Recent applications of magnetic composites as extraction adsorbents for determination of environmental pollutants. TrAC Trends Anal Chem 11:115611. https://doi.org/10.1016/j.trac.2019.07.022

Zafar MN, Amjad M, Tabassum M, Ahmad I, Zubair M (2018) $SrFe_2O_4$ nanoferrites and $SrFe_2O_4$/ground eggshell nanocomposites: fast and efficient adsorbents for dyes removal. J Clean Prod 199:983–994. https://doi.org/10.1016/j.jclepro.2018.07.204

Zeng S, Duan S, Tang R, Li L, Liu C, D. Sun (2014) Magnetically separable $Ni_{0.6}Fe_{2.4}O_4$ nanoparticles as an effective adsorbent for dye removal: synthesis and study on the kinetic and thermodynamic behaviors for dye adsorption. Chem Eng J 258:218–228. doi:https://doi.org/10.1016/j.cej.2014.07.093

Zeng L, Xie M, Zhang Q, Kang Y, Guo X, Xiao H, Peng Y, Luo J (2015) Chitosan/organic rectorite composite for the magnetic uptake of methylene blue and methyl orange. Carbohydr Polym 123:89–98. https://doi.org/10.1016/j.carbpol.2015.01.021

Zhang S, Niu H, Cai Y, Zhao X, Shi Y (2010) Arsenite and arsenate adsorption on coprecipitated bimetal oxide magnetic nanomaterials: $MnFe_2O_4$ and $CoFe_2O_4$. Chem Eng J 158:599–607. https://doi.org/10.1016/j.cej.2010.02.013

Zhang L, Liang S, Liu R, Yuan T, Zhang S, Xu Z, Xu H (2016a) Facile preparation of multifunctional uniform magnetic microspheres for T1-T2 dual modal magnetic resonance and optical imaging. Colloid Surf B: Physicochem Eng Asp 144:344–354. https://doi.org/10.1016/j.colsurfb.2016.04.014

Zhang N, Zhong S, Zhou X, Jiang W, Wang T, Fu J (2016b) Superhydrophobic P (St-DVB) foam prepared by the high internal phase emulsion technique for oil spill recovery. Chem Eng J 298:117–124. https://doi.org/10.1016/j.cej.2016.03.151

Zhang N, Zhong S, Chen T, Zhou Y, Jiang W (2017) Emulsion-derived hierarchically porous polystyrene solid foam for oil removal from aqueous environment. RSC Adv 7:22946–22953. https://doi.org/10.1039/C7RA02953E

Zhang B, Huyan Y, Wang J, Chen X, Zhang H, Zhang Q (2018a) Fe_3O_4@SiO_2@CCS porous magnetic microspheres as adsorbent for removal of organic dyes in aqueous phase. J Alloy Compd 735:1986–1996. https://doi.org/10.1016/j.jallcom.2017.11.349

Zhang S, Fan X, Zhang F, Zhu Y, Chen J (2018b) Synthesis of emulsion-templated magnetic porous hydrogel bead and their application for catalyst of fenton reaction. Langmuir 34:3669–3677. https://doi.org/10.1021/acs.langmuir.8b00009

Zhang C, Liu S, Li S, Tao Y, Wang P, Ma X, Chen L (2019a) Enahanced biosorption of Cu(II) by magnetic chitosan microspheres immobilized Aspergillus sydowii (MCMAs) from aqueous solution. Colloid Surface A 581:123813. https://doi.org/10.1016/j.colsurfa.2019.123813

Zhang H, Zhao R, Pan M, Deng J, Wu Y (2019b) Biobased, porous poly(high internal phase emulsions): prepared from biomass-derived vanillin and laurinol and applied as an oil adsorbent. Ind Eng Chem Res 58:5533–5542. https://doi.org/10.1021/acs.iecr.9b00515

Zhang T, Sanguramath AR, Israel S, Silverstein SM (2019c) Emulsion templating: porous polymers and beyond. Macromolecules 52:5445–5479. https://doi.org/10.1021/acs.macromol.8b02576

Zhang W, Ruan G, Li X, Jiang X, Huang Y, Du F, Li J (2019d) Novel porous carbon composites derived from a graphene-modified high-internal-phase emulsion for highly efficient separation and enrichment of triazine herbicides. Anal Chim Acta 1071:17–24. https://doi.org/10.1016/j.aca.2019.04.041

Zhao L, Lv W, Hou J, Li Y, Duan J, Ai S (2020) Synthesis of magnetically recyclable g-C_3N_4/Fe_3O_4/ZIF-8 nanocomposites for excellent adsorption of malachite green. Microchem J 152:104425. https://doi.org/10.1016/j.microc.2019.104425

Zheng W, Gao F, Gu H (2005) Magnetic polymer nanospheres with high and uniform magnetite content. J Magn Magn Mater 288:403–410. https://doi.org/10.1016/j.jmmm.2004.09.125

Zheng Z, Zheng X, Wang H, Du Q (2013) Macroporous graphene oxide-polymer composite prepared through Pickering high internal phase emulsions. ACS Appl Mater Interf 5:7974–7982. https://doi.org/10.1021/am4020549

Zheng C, Zheng H, Wang Y, Wang Y, Qu W, An Q (2018) Synthesis of novel modified magnetic chitosan particles and their adsorption performance toward Cr (VI). Bioresour Technol 267:1–8. https://doi.org/10.1016/j.biortech.2018.06.113C

Zheng C, Zheng H, Sun Y, Xu B, Wang Y, Zheng X, Wang Y (2019) Simultaneous adsorption and reduction of hexavalent chromium on the poly(4-vinyl pyridine) decorated magnetic chitosan biopolymer in aqueous solution. Bioresour Technol 293:122038. https://doi.org/10.1016/j.biortech.2019.122038

Zhou L, Ji L, Ma P, Shao Y, Zhang H, Gao W, Li Y (2014) Development of carbon nanotubes/$CoFe_2O_4$ magnetic hybrid material for removal of tetrabromobisphenol A and Pb(II). J Hazard Mater 265:104–114. https://doi.org/10.1016/j.jhazmat.2013.11.058

Zhou J, Liu Y, Zhou X, Ren J, Zhong C (2018) Magnetic multi-porous bio-adsorbent modified with amino siloxane for fast removal of Pb(II) from aqueous solution. Appl Surf Sci 427:976–985. https://doi.org/10.1016/j.apsusc.2017.08.110

Zhou X, Zhou W, Ju D, Peng Y, Zhou L, Tang S, Wang J (2019) A research on the preparation of oil-adsorbing hydrophobic porous resins by high internal phase emulsions (HIPEs) template. Sci Eng Compos Mater 26:261–269. https://doi.org/10.1515/secm-2019-0015

Zhu L, Li C, Wang J, Zhang H, Zhang J, Shen Y, Li C, Wang C, Xie A (2012) A simple method to synthesize modified Fe_3O_4 for the removal of organic pollutants on water surface. Appl Surf Sci 258:6326–6330. https://doi.org/10.1016/j.apsusc.2012.03.031

Zhu J, Wei J, Chen M, Gu H, Rapole SB, Pallavkar S, Ho TC, Hopper J, Guo Z (2013) Magnetic nanocomposites for environmental remediation. Adv Powder Technol 24:459–467. https://doi.org/10.1016/j.apt.2012.10.012

Zhu W, Ma W, Li C, Pan J, Dai X (2015) Well-designed multihollow magnetic imprinted microspheres based on cellulose nanocrystals (CNCs) stabilized Pickering double emulsion polymerization for selective adsorption of bifenthrin. Chem Eng J 276:249–260. https://doi.org/10.1016/j.cej.2015.04.084

Zhu Y, Jiang D, Sun D, Yan Y, Li C (2016) Fabrication of magnetic imprinted sorbents prepared by Pickering emulsion polymerization for adsorption of erythromycin from aqueous solution. J Environ Chem Eng 4:3570–3579. https://doi.org/10.1016/j.jece.2016.07.036

Zhu Y, Wang F, Zheng Y, Wang F, Wang A (2016a) Rapid enrichment of rare-earth metals by carboxymethyl cellulose-based open-cellular hydrogel adsorbent from HIPEs template. Carbohydr Polym 140:51–58. https://doi.org/10.1016/j.carbpol.2015.12.003

Zhu Y, Zheng Y, Wang F, Wang A (2016b) Fabrication of magnetic macroporous chitosan-*g*-poly (acrylic acid) hydrogel for removal of Cd^{2+} and Pb^{2+}. Inter J Biol Macromol 93:483–492. https://doi.org/10.1016/j.ijbiomac.2016.09.005

Zhu Y, Zheng Y, Wang F, Wang A (2016c) Fabrication of magnetic porous microspheres via (O_1/W)/O_2 double emulsion for fast removal of Cu^{2+} and Pb^{2+}. J Taiwan Inst Chem E 67:505–510. https://doi.org/10.1016/j.jtice.2016.08.006

Zhu Y, Zheng Y, Wang F, Wang A (2016d) Monolithic supermacroporous hydrogel prepared from high internal phase emulsions (HIPEs) for fast removal of Cu^{2+} and Pb^{2+}. Chem Eng J 284:422–430. https://doi.org/10.1016/j.cej.2015.08.157

Zhu Y, Zheng Y, Zong L, Wang F, Wang A (2016e) Fabrication of magnetic hydroxypropyl cellulose-g-poly(acrylic acid) porous spheres via Pickering high internal phase emulsion for removal of Cu^{2+} and Cd^{2+}. Carbohydr Polym 149:242–250. https://doi.org/10.1016/j.carbpol.2016.04.107

Zhu Y, Wang W, Zhang H, Ye X, Wu Z, Wang A (2017a) Fast and high-capacity adsorption of Rb^+ and Cs^+ onto recyclable magnetic porous spheres. Chem Eng J 327:982–991. https://doi.org/10.1016/j.cej.2017.06.169

Zhu Y, Zhang H, Wang W, Ye X, Wu Z, Wang A (2017b) Fabrication of a magnetic porous hydrogel sphere for efficient enrichment of Rb^+ and Cs^+ from aqueous solution. Chem Eng Res Des 125:214–225. https://doi.org/10.1016/j.cherd.2017.07.021

Zhu H, Tan X, Tan L, Zhang H, Liu H, Fang M, Hayat T, Wang X (2018) Magnetic porous polymers prepared via high internal phase emulsions for efficient removal of Pb^{2+} and Cd^{2+}. ACS Sustainable Chem Eng 6:5206–5213. https://doi.org/10.1021/acssuschemeng.7b04868

Zhu Y, Wang W, Yu H, Wang A (2020) Preparation of porous adsorbent via Pickering emulsion template for water treatment: a review. J Environ Sci 88:217–236. https://doi.org/10.1016/j.jes.2019.09.001

Chapter 14
Chitosan-Based Magnetic Adsorbents

Juliana M. N. dos Santos and Guilherme L. Dotto

Contents

Abstract Chitosan-based magnetic adsorbents are promising materials in the adsorption field due to their high specific surface area, selectivity, chemical and physical stability, and cost-effectiveness. These remarkable properties combined with chemical/physical adsorbent modifications, but mostly with its affordable regeneration, provided by solid–liquid phase separation through the application of an external magnetic field, lead magnetic chitosan to become a suitable adsorbent candidate for environmental protection purposes. In this regard, this chapter writing has relied on several published documents, carefully addressed and reviewed, about chitosan-based magnetic adsorbents applied to environmental remediation. Basically, these adsorbents consist of a magnetic core coated with chitosan. In this material combination, the magnetic particles and the polymer share their particular characteristics and mutually contribute to overcome their individual limitations. The magnetic characteristics are mostly provided by particles of iron oxides such as

J. M. N. dos Santos (✉) · G. L. Dotto
Chemical Engineering Department, Federal University of Santa Maria–UFSM, Santa Maria, RS, Brazil

L. Meili, G. L. Dotto (eds.), *Advanced Magnetic Adsorbents for Water Treatment*, Environmental Chemistry for a Sustainable World 61,
https://doi.org/10.1007/978-3-030-64092-7_14

Fe_3O_4, even though alternatives less susceptible to chemical oxidation already exist. Chitosan as the major component of the magnetic adsorbent surface, thus mainly responsible for the adsorbate–adsorbent interactions, undergoes modifications in order to obtain higher adsorption capacities and better mechanical and chemical properties. The crosslinking procedure provides greater chemical and mechanical resistance in acidic media as it allows the achievement of a three-dimensional network. The functional groups' adding, known as the functionalization step, increases the number and diversity of active sites available for adsorption, enabling the contaminants' removal with little or no affinity for unmodified chitosan. Although chitosan has already assumed shapes of spheres, fibers, powders, films, and flakes in previous reports, not all of these formats have their magnetic version. For now, only beads, films, and powders have received the attention of researchers. When applied to effluent treatment, chitosan-based adsorbents remove mostly dyes, medicines, and ion metals by adsorption. In case of dyes and medicines, the adsorption mechanism may occur by ion exchange and ion pair formation. The metal ion adsorption occurs by chelation, as NH_2 and OH functional groups act as chelators to bind metal ions.

Keywords Magnetic chitosan · Magnetic adsorbent · Crosslinking · Functionalization · Dye · Pharmaceutical · Metal · Adsorption · Environmental remediation · Adsorbent shaping

14.1 Introduction

The potable water demand, due to intensified population growth and industrial activities, is still rising while natural reservoirs remain constant. It is not possible to create water, but it is possible and vital to preserve it. Industries are considered the main polluting agents of the aquatic environment, especially those that use or synthesize nonbiodegradable molecules in their processes and do not properly treat the effluents before dumping them into the water-receiving bodies. Dyes, medicines, and some metals are examples of pollutants that are difficult to remove and degrade (Hessel et al. 2007; Lasheen et al. 2016; Shalla et al. 2018).

Conventional treatments are designed to remove suspended particles and solids but are not able to efficiently deal with contaminated effluents by micropollutants. Therefore, technologies such as neutralization, suspended solids' removal by physical or chemical separation, biological treatments, and physicochemical treatments via adsorption, ion exchange, chemical oxidation, and separation membranes have already been developed and used as tertiary treatment in some processes. However, adsorption has proved to be the alternative with greater flexibility and operation simplicity (Perry et al. 1997; Geankoplis 1998; Liang et al. 2019b).

Due to using solids as adsorbents, adsorptive processes must comprise a solid–liquid separation step after adsorption, industrially performed by the unitary operations of centrifugation and filtration. That is the context in which magnetic adsorbents acquire prominence, as they can be separated from the liquid phase with the simple magnetic field application, thus enabling operation cost reduction and adsorbent shelf life increase (Liu et al. 2011; Reddy and Lee 2013; Li et al. 2018).

Magnetic particles, usually iron oxides, with a large specific surface area and high saturation of magnetization, can be used as adsorbents in an isolated or "neat" form. However, associating these particles to other materials can promote the attainment of an adsorbent with greater applicability and efficiency, as each of the formative elements begins sharing its characteristics and, at the same time, they mutually contribute to overcome their individual limitations. It is like a two-way street (Reddy and Yun 2016; dos Santos et al. 2019b).

Combining iron oxide and chitosan, for example, may provide the attainment of a versatile and efficient magnetic adsorbent for removing pollutants as an outcome. The amine and hydroxyl active sites, originally chitosan exclusive, become the main adsorbate–adsorbent interaction points in the generated composite, which also exhibits typical magnetic properties from magnetic iron oxides. While chitosan is able to reduce magnetic particle agglomeration and oxidation tendency, the presence of the oxide minimizes the structural polymer fragility when exposed to acidic media (Muzzarelli et al. 2012; Ranjbar Bandforuzi and Hadjmohammadi 2019).

Chitosan is a widely studied polymer in pollutant removal and is also considered a low-cost material. From the industrial point of view, chitosan is mainly obtained from shrimp and crab shells rich in chitin. Before undergoing the alkaline deacetylation process, the peels are subjected to acid treatment to remove calcium carbonates, followed by basic treatment to eliminate remaining proteins. Depending on the application and required purity, the peels still go through the discoloration and deodorization steps (El Knidri et al. 2018).

Although hydroxyl groups also make part of the chitosan polymer chain, amino groups are more reactive and provide chitosan with higher cationicity and better adsorption capacity, especially for negatively charged molecules. It should be mentioned that this polymer is very versatile and in addition to assuming various shapes, with some structure modifications, chitosan can adsorb other pollutant species, including cationic molecules. These modifications, also known as functionalization steps, include the addition of functional groups and association of different materials (Muzzarelli et al. 2012).

Whereas magnetic chitosan is formed by a magnetic core coated with the polymer, it can be inferred that the adsorbent surface is essentially composed by chitosan. This may explain why studies specifically focus on chitosan modifying to achieve more efficient magnetic adsorbents in pollutant removal through adsorption. By reading this chapter, it will be possible to get an overview about attainment of chitosan-based magnetic adsorbents, understanding the magnetic particles value as key elements, the adsorbent improving methods, the existing shapes, and the characteristics of main effluents treated using chitosan-based adsorbents. Figure 14.1 presents a scheme that illustrates, in a very didactic way, all the aspects to be contemplated in this chapter, from the synthesis of chitosan magnetic adsorbent to its application.

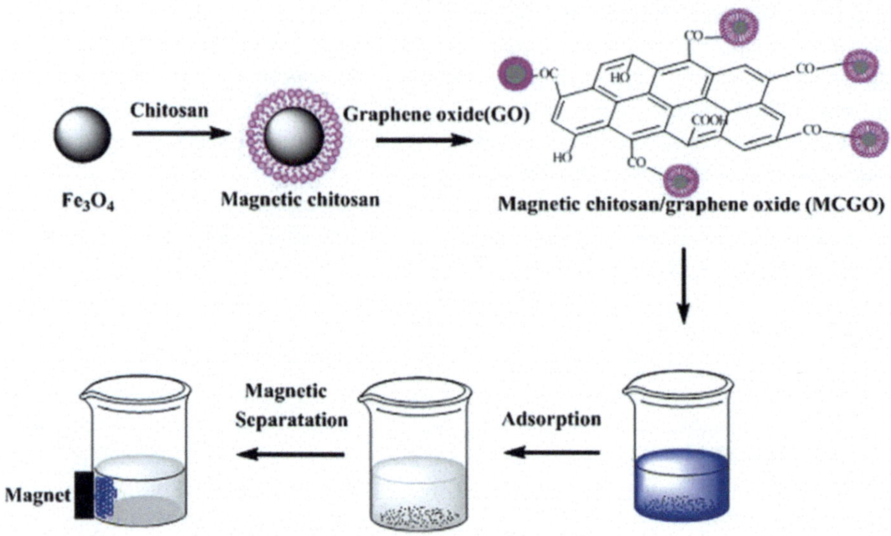

Fig. 14.1 Synthesis scheme of a magnetic chitosan adsorbent with Fe_3O_4 and graphene oxide (MCGO), applied in Methylene Blue adsorption and subsequently separated from the liquid phase through the magnet approximation. (Reprinted with permission of Bioresource Technology (Elsevier) from Fan et al. 2012b)

14.2 General Aspects Involved in Synthesis of Magnetic Chitosan Adsorbents

The magnetic adsorbents employed in wastewater treatment are basically formed through the association between magnetic particles and some other polymeric, carbonaceous, or mineral material that can be chemically or physically modified. The chitosan-based magnetic adsorbent development is no different. In this case, the addition of chitosan polymer chains, with amine and hydroxyl functional groups, also promotes chemical stability to magnetic particles while achieving a composite with better adsorbent characteristics than those presented by constituent elements individually (Muzzarelli et al. 2012; Reddy and Yun 2016; Rahmi et al. 2019b).

In general, there are two ways to synthesize magnetic chitosan adsorbents, which differ from each other in the obtaining step of magnetic particles: one-step and multistep. The one-step synthesis, also named in literature as the in situ method, consists of magnetic component coprecipitation directly into the polymer matrix previously dissolved in acid medium. Thus, magnetic particles and the magnetic composite itself are simultaneously obtained. Otherwise, when magnetic elements are formed prior to the composite and/or its modifications, then the synthesis method is identified as a multistep (Cesano et al. 2015; Wu et al. 2016; Fan et al. 2018; Pandi et al. 2019).

Kloster et al. (2019) compared the two aforementioned procedures in chitosan/ iron oxide nanocomposite film formation. In in situ films, the magnetite precursor salts were added to an acidic chitosan solution and, after the film-molding procedure, they were immersed in NaOH solution to promote magnetite precipitation. When applied in Congo Red dye removal, the in situ films maintained their integrity, but the strong interactions developed between iron oxide particles and the chitosan matrix triggered the decrease of adsorption ability, as active sites availability has been restricted (Kloster et al. 2015b, 2019).

The films obtained via the multistep method employed ultrasonic agitation. Magnetite was synthesized via coprecipitation using NH_4OH as the precipitating agent and, only after drying, was added to the chitosan matrix. After ultrasound and manual shaking, the resulting suspension got the film shape. When compared to those obtained via the in situ route, the multistep films achieved higher adsorption capacity mainly due to the presence of the remaining acetic acid from the chitosan solubilization step. Upon effluent contact, the acid changed the pH of the colored solution from basic to acid, favoring the adsorption of Congo Red anionic molecules. Moreover, the film fracture during adsorption increased the surface contact, enhancing the mass transfer (Kloster et al. 2015a, 2019).

Although significant, the one-step and multistep methodologies applied for acquiring magnetic particles are not alone in the feature determination of chitosan magnetic adsorbents. Just as sol-gel transition and casting are methodologies directly related to some composites' final shape (Sects. 14.3.2 and 14.3.3, respectively), the magnetic particle type used, the crosslinking, and functionalization procedures influence this material attainment as well. In order to facilitate the readers' understanding about this matter, the equally relevant aspects previously mentioned will be detailed in the following sections.

14.2.1 Magnetic Particles and Their Main Features

The particles that make up chitosan adsorbents and share with them their magnetic properties contain iron or iron oxide. In wastewater treatment, they act as adsorbents in the neat or associated form, but may also play the catalyst role if oxy-reduction processes are employed. When it comes to chitosan-based composites, oxides are most commonly used, especially magnetite (Fe_3O_4). However hematite (α-Fe_2O_3), maghemite (γ-Fe_2O_3), ferrite, and zero iron (nZVI) also have already been reported as elements of magnetic composites (Ngomsik et al. 2005; Song and Zhang 2006; Sable et al. 2018; Liang et al. 2019a).

The iron oxide synthesis may occur via the physical or chemical route. Physical methods include mechanical milling and pulsed laser ablation, while chemical methods entail hydrothermal synthesis, coprecipitation, sol-gel, thermal decomposition, electrodeposition, chemical decomposition, etc. Coprecipitation, however, presents the large-scale producing of iron oxide as its main advantage, therefore being more profitable for industries (Wang et al. 2014; Chen et al. 2015; Alonso et al. 2018).

The most common precursor reagents used for chitosan magnetic adsorbents synthesis are inorganic salts purchased from supplier companies. However, there are reports in published literature about magnetic materials obtained from fly ash, produced during coal combustion in thermal power plants, and iron sand. Rahmi et al. (2019a) isolated Fe_3O_4 from iron sand widely available in Aceh, Indonesia, by HCl leaching and subsequent NH_4OH basic precipitation. Mu et al. (2020) obtained magnetic cohenite (Fe_3C) when the carbon, deposited on fly ash surface through vapor deposition, reacted with the iron oxides present (Rahmi et al. 2019a; Mu et al. 2020).

The small particle size, high crystallinity, and tendency for agglomeration are some of the inherent characteristics of magnetic materials and may spoil their applicability. Depending on the precursor elements and the adopted synthesis route, these features' manifestation may be greater or lesser. Magnetite, for example, is an easily obtainable oxide with a high associated magnetic saturation value. Despite its widespread use as a magnetic element of composites, if it is not under an inert atmosphere, Fe_3O_4 particles oxidize very easily, even coated (Chen et al. 2015; Alonso et al. 2018; Wu et al. 2019b; Ranjbar Bandforuzi and Hadjmohammadi 2019).

Trying to overcome these limitations, some authors employed other iron oxides in the attainment of adsorbents. Spinel ferrites (MFe_2O_4, where M is a metal cation), used as an alternative, are metal oxides, with a normal, inverse, or mixed structure, most commonly obtained via the chemical route. When compared with aforementioned iron oxides, ferrite is more stable, presenting a high magnetic permeability, larger surface area, mechanical and chemical resistance to oxidation, not requiring the coating, and operating under a controlled atmosphere (Muzzarelli et al. 2012; Reddy and Yun 2016; Alonso et al. 2018; Mohammadzadeh Pakdel and Peighambardoust 2018).

The magnetic saturation and particles' magnetic behavior are also important features and shall be regarded if a magnetic adsorbent, separable from the liquid phase by magnet approximation, is a desired outcome. In studies related to wastewater treatment, magnetic particles are mostly ferromagnetic or superparamagnetic. This indicates the absence of spontaneous magnetization, i.e., their magnetic domains are naturally randomly oriented. When applying an external field, these domains become oriented along the field axis, reaching a certain magnetization volume (M_S) (Laurent et al. 2017).

In ferromagnetism behavior, there is a coercive force (H_C) that causes M_S to slowly reduce from the remaining magnetization (M_R) to zero magnetization when the external field is removed, causing hysteresis. In superparamagnetism, a characteristic behavior of particles smaller than 20 nm, hysteresis does not occur. At the same moment the field is removed, the domains will be randomly oriented again. Both parameters, M_R and H_C, depend on the nature and the size of materials. Magnetization saturation, however, is influenced only by chemical composition. This explains, for example, the different M_S values obtained by the synthesized ferrites in Santos et al. (2019a, b) studies (Laurent et al. 2017).

Santos et al. (2019a, b) synthesized chitosan magnetic adsorbents with cobalt ($CoFe_2O_4$) and zinc ferrite ($ZnFe_2O_4$) as core material. While $CoFe_2O_4$ and its respective composite, applied in Indigotine Blue dye adsorption, showed ferromagnetic behavior and higher M_S values (50.34 and 16.45 emu g^{-1}, respectively), the behavior of $ZnFe_2O_4$ and its composite, used to remove Diclofenac, was characterized as superparamagnetic, with lower M_S values (26.16 and 9.75 emu g^{-1}, respectively). It is important to mention that $CoFe_2O_4$ and $ZnFe_2O_4$ were synthesized in the same way, only varying the precursor salts (dos Santos et al. 2019b, a).

In this backdrop, it is noteworthy that adsorbent composites obtained M_S lower values than those presented by isolated magnetic particles. This was visible not only in the previously cited studies but also in many others reported in literature. The magnetic saturation reduction always occurs and is not related to synthesis methodologies. In chitosan-coated adsorbents, this is due to the polymer amorphous structure that changes the core magnetic moments when adsorbents are subjected to external magnetic fields (Wang et al. 2014; Hui et al. 2018; Rahmi et al. 2019a).

14.2.2 Crosslinking of Chitosan Polymeric Chains

Chitosan ((1-4)-2-amine-2-deoxy-β-D-glucan) is a naturally occurring amino polysaccharide, derived from chitin ((1-4)-2-acetamide-2-deoxy-β-D-glucan). The chitin conversion to chitosan is most commonly performed via alkaline deacetylation, but may also occur through enzymatic route. In alkaline deacetylation, the chitin acetyl group (C_2H_3O) is removed by hydrolysis, leading to amine group formation (NH_2). The resulting copolymer with N-acetyl glucosamines and glucosamines may be called chitosan if glucosamines represent at least 50% of the constituent units (Sivashankari and Prabaharan 2017; Roman et al. 2019).

After cellulose, chitosan is the most available biopolymer in nature. Even widely studied for contaminants' removal via adsorption, chitosan's inherent characteristics such as low porosity and mechanical fragility when exposed to acidic media and high temperatures limit its applicability. As most of the contaminated effluents to be treated by adsorption present acid behavior, chitosan crosslinking is a constant procedure adopted in studies related to wastewater matter (Rong et al. 2017; Mohammadzadeh Pakdel and Peighambardoust 2018).

Crosslinking of chitosan polymer chains is accomplished by a polymer reacting with a crosslinking chemical agent. In literature, glutaraldehyde, epichlorohydrin, ethylene glycol, and tripolyphosphate are the most common crosslinking agents used. Their constituent molecules present at least two reactive functional groups that allow bridging between polymer chains and amino groups by introducing covalent bonds (Crini and Badot 2008).

Figure 14.2 shows chitosan before and after crosslinking using glutaraldehyde. It is possible to observe that chitosan modification through crosslinking does not chemically affect the polymer; it only creates new interchain bonds, allowing a three-dimensional network conformation. The lower mobility and high molecular

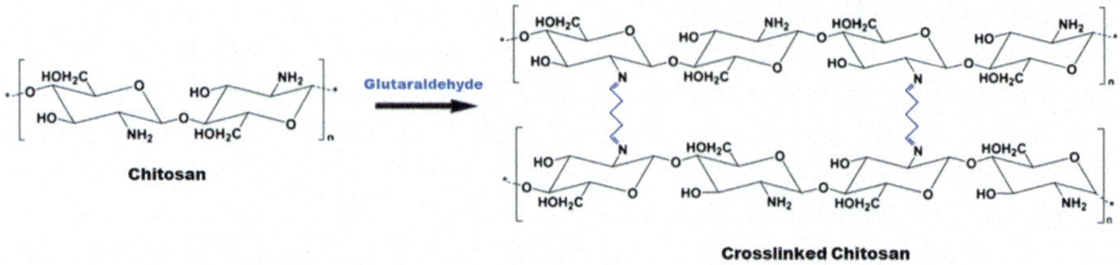

Fig. 14.2 Chitosan polymer chains before and after crosslinking using glutaraldehyde as the crosslinker. It is possible to observe that chitosan modification through crosslinking does not chemically affect the polymer; it only creates new interchain bonds, allowing a three-dimensional network conformation

mass provide a higher mechanical resistance and also enable the composite formation when some particles get trapped among polymer chains (Mourya and Inamdar 2008; Bée et al. 2017).

Although crosslinking is fast and efficient to improve chitosan characteristics, especially mechanical ones, this procedure is highly dependent on the reagent concentration employed and is responsible for a significant reduction in adsorption capacity as active sites may be crosslinked. Glutaraldehyde, for example, is a crosslinking agent that directly interacts with chitosan amino groups, while epichlorohydrin primarily interacts with hydroxyl groups, leaving amines, known as the main polymer reactive points, free to interact with adsorbate (Obeid et al. 2013; Jawad et al. 2019; Rahmi et al. 2019a).

In light of the crosslinking methodology constraints, especially the already proven environmental toxicity of crosslinking agents such as glutaraldehyde and epichlorohydrin, many authors had evaluated the performance of some other chemicals in promoting polymeric chains crosslinking. Mi et al. (2015) used citric acid as an alternative crosslinker to tripolyphosphate. Besides being used in the food industry, this polycarboxylic acid may provide lone pairs similar to those present in sodium tripolyphosphate, but preventing the release of phosphates responsible for water sources' eutrophication (Mi et al. 2015; Mirabedini et al. 2017).

In 2016 and 2017, Mirabedini et al. published documents using glyoxal in chitosan crosslinking to form the Fe_3O_4NPs/CS/Glyoxal/PVA and Fe_3O_4NPs/CS/Glyoxal hydrogel films, respectively. Due to its low cost and biocompatibility, glyoxal is widely studied in the biomedical field as a crosslinking agent of naturally occurring polymers. In Cr (VI) adsorption, glyoxal promoted the production of chemically and mechanically stable adsorbent films at acidic pH, between 3 and 4, when Cr(VI) is predominantly present in its ionic and highly toxic form ($HCrO_4^-$) (Gupta and Jabrail 2006; Mirabedini and Kassaee 2016; Mirabedini et al. 2017; Jawad et al. 2019).

Sulfuric acid (H_2SO_4) also played the crosslinker role in Marques et al. (2019) and Rahmi et al. (2019b) reports. In the first paper, Marques et al. added sulfuric acid to chitosan, in different proportions of SO_4^{2-}/NH^{3+}, seeking to produce crosslinked nonmagnetic membranes to remove Cu^{2+} by filtration. Rahmi et al. (2019b) reported the synthesis of magnetic chitosan nanocomposite beads to adsorb Methylene Blue dye. If compared to non-crosslinked spheres, H_2SO_4 crosslinked magnetic chitosan beads showed rough and amorphous surface, making the adsorption process easier (Marques et al. 2019; Rahmi et al. 2019b).

14.2.3 *Functionalization*

In adsorbents composed by a chitosan-coated magnetic core, functionalization is considered a differential. This procedure enables the extending of adsorbent application possibilities as it improves structural characteristics and makes the active sites available to interact with target molecules more diverse and numerous. Functionalization consists of adding functional groups to those characteristic of chitosan, allowing the adsorption capacity increase and, in many cases, enabling efficient molecules adsorption with little or no affinity for chitosan (Zhao et al. 2016; Danalıoğlu et al. 2017).

In those composites in which the chitosan polymer chains have been crosslinked, the simple adding of functional groups or functional groups holder materials is particularly important. As already mentioned in Sect. 14.2.2, the crosslinking procedure is responsible for significantly reducing available adsorbent sites. It is important to emphasize the lack of standard functionalization processes. Functionalization may be performed at any synthesis stage through chitosan copolymerization, grafting, and polymer blending reactions (Zhang et al. 2016; Abou El-Reash 2016; Iordache et al. 2018; Wu et al. 2019a).

Chitosan holds both amine (NH_2) and hydroxyl (OH) functional groups. However, the amino groups are the most reactive, which protonate and thus favor polymer interaction with molecules negatively charged in acidic medium, especially anionic dyes and medicines. Therefore, if the cationic molecules or metal cations' adsorption is a goal, the grafting of anionic clusters on chitosan magnetic adsorbents surface is required. To this end, chitosan has already been modified by xanthate groups' insertion, with sulfur atoms and carboxyl groups able to interact with cationic species (Xu et al. 2015; Tanhaei et al. 2019).

Chitosan modification may also occur by introducing a quaternary ammonium salt into the primary amino group, at the C-2 position to produce cationic hybrid multifunctional particles, grafting the amidoxime, carrier of amine and C–O groups. The Schiff base formation and the addition of chelating agents such as ethylenediaminetetraacetic acid (EDTA), diethylenetriaminepentaacetic acid (DTPA), or nitrilotriacetic acid (NTA) particularly contribute to enhancing the interaction between magnetic chitosan and metal ions. Anyway, there are countless ways to achieve functionalized chitosan. To better understand the relationship between magnetic chitosan functionalization and effluent treatment using adsorption, some scientific works found in the literature will be highlighted as follows (Hamza et al. 2018; Soares et al. 2019; Wu et al. 2019a; Anush and Vishalakshi 2019).

Zheng et al. (2019) synthesized modified magnetic chitosan microparticles to remove food dyes. After the synthesis via crosslinking, [2-(methacryloxy) ethyl] trimethylammonium polychloride polymer was grafted to the surface by free radical polymerization. According to the authors, the high adsorption capacities obtained for Food Yellow (833.33 mg g^{-1}) and Acid Yellow 23 (666.67 mg g^{-1}) dyes over a wide pH range (2–10) are due to the abundance of grafted quaternary ammonium groups, since unmodified microparticles are more susceptible to pH changes (Zheng et al. 2019).

Mahdavinia and Mosallanezhad (2016), unlike most of the scientific papers found, developed a magnetic chitosan adsorbent to adsorb a cationic dye, Methylene Blue. Initially, the authors performed magnetic particles synthesis with the k-carrageenan polymer and coprecipitated Fe_3O_4. Then, they added chitosan by crosslinking. The k-carrageenan is a natural polymer with anionic sulfate in its structure (—SO^{3-}) that provides high affinity for cationic molecules. Regardless of magnetization degree and the added chitosan amount, Methylene Blue adsorption was favored at a more basic pH, following the Langmuir equilibrium model (Mahdavinia and Mosallanezhad 2016).

Zhang et al. (2016) applied grafting copolymerization to form three different magnetic brush-shaped particles. The nucleus was made up of magnetic chitosan particles, while the polymeric branches shaped the brush-like outer layer to promote the increasing of surface area and the number of functional groups available to interact with diclofenac and tetracycline hydrochloride molecules, whose particle branches, composed by 2-methyl-acryloyl-oxyethyl-trimethyl ammonium chloride monomer, presented a positively charged surface throughout the studied pH range, indicating a great potential for adsorption of anionic contaminants (Zhang et al. 2016).

Soares et al. (2019) synthesized magnetic hybrid nanoadsorbents with a magnetite core encapsulated by a silicon network, highly enriched in quaternary chitosan. Encapsulation of Fe_3O_4 particles occurred by alkaline hydrolysis. Quaternary chitosan formation was basically achieved through the introducing of a quaternary ammonium salt into its primary amine. Surface charge tests showed the cationic nature of magnetic nanoparticles over a wide pH range (2–7), facilitating interaction with anionic molecules. The maximum adsorption capacity for diclofenac was 240 mg g^{-1} at pH 6, which is considerably higher than other adsorbents except for some nonmagnetic materials (Soares et al. 2019).

14.3 Common Shapes of Chitosan-Based Magnetic Adsorbents

Chitosan is a very versatile polymer and has been used as the adsorbent in various forms. Chitosan beads, flakes, films, powders, and fibers were synthesized to suit different applications, but also to efficiently remove pollutants by adsorption.

However, when it comes specifically to chitosan-based magnetic adsorbents, not all shapes aforementioned own their magnetic version. Thereafter, information about magnetic chitosan shapes already studied will be better developed, with emphasis on the structural characteristics and on synthesis specificities (Vakili et al. 2014; Xu et al. 2015).

14.3.1 Powders

Powders are particulate materials whose constituent particles may assume nanometric dimensions. Among all shapes of chitosan-based magnetic adsorbent already studied, powders stand out for synthesis simplicity, although further modification/functionalization steps are considered, especially because the material assumes a powder form just after elimination of any remaining liquid from synthesis step. The powders or chitosan nanocomposites enjoy a higher specific surface area and lower resistance to intraparticle diffusion due to the small size of constituent elements, thus greatly favoring pollutant adsorption (Hamza et al. 2018; Zeraatkar Moghaddam et al. 2018; Subedi et al. 2019).

On the other hand, particle size may limit the powder application to fixed-bed columns as they provoke hydrodynamic blockage and, consequently, pressure drop increase. In batch adsorption systems, the use of adsorbent powders such as chitosan powder is also quite restricted, thus considering the difficulty in retrieving very small particles without filtration and high-speed centrifugation installed. However, the magnetic powders are able to shift this situation and become a viable alternative mainly in adsorption batch systems. In Fig. 14.3, it is possible to observe the way chitosan magnetic powders are easily attracted by the external magnet (Van Hoa et al. 2016; Hamza et al. 2018; Liang et al. 2019a).

Fig. 14.3 Mesoporous graphene/Fe_3O_4/chitosan nanocomposite synthesized in the Van Hoa et al. (2016) study; (**a**) spread in solution and (**b**) under external magnetic field. (Reprinted with permission of Journal of Water Process Engineering (Elsevier) from Van Hoa et al. 2016)

14.3.2 Beads

Beads or spheres also fit into the particulate adsorbent category. They can be synthesized over a wide diameter range, depending on assessed application and the absence or presence of associated material. In published studies, when measured, chitosan beads diameters with magnetic properties ranged from 0.002 to 3.4 mm. Unlike powders, beads require simple storage conditions and can be more easily employed in large-scale operations as adsorbent column filling without material loss (Zhu et al. 2010, 2012; Wang et al. 2014; Bée et al. 2017; Malwal and Gopinath 2017; Zheng et al. 2018; Rahmi et al. 2019b; Heydaripour et al. 2019; Mu et al. 2020).

Chitosan is a polymer that could be presented in gel form. Therefore, it is widely used to encapsulate different materials with nanometer proportions as well as iron oxide particles. Chitosan bead gelling is performed through a sudden pH change, when chitosan dissolved in acidic medium is added dropwise into a basic coagulation solution, with NaOH being the most commonly used base. The sphere diameter obtained is then directly dependent on the drop size dispensed (Wang et al. 2018; Mu et al. 2020).

This sphere shaping process is named in literature as sol-gel transition by direct but controlled exposure to alkali medium, allowing to mold materials in various formats, including in fiber shape. However, chitosan gelling can also occur through crosslinking with ionic agents and complexation. Regardless of whether the final product is a physical or chemical gel, the sol-gel transition procedure is always highly dependent on pH conditions, chitosan deacetylation degree, temperature, and concentration of the employed reagents (Rwei et al. 2005; Nilsen-Nygaard et al. 2015; Wu et al. 2019b).

Figure 14.4, presented by authors Malwal and Gopinath (2017), illustrates the synthesis scheme of magnetic chitosan beads applied to adsorption. Except for glutaraldehyde treatment and external silica layer addition, this scheme represents the synthesis methodology adopted in most studies published on this subject. The silica outer layer addition, to form stabilized magnetic-chitosan beads for arsenic removal (SiO_2@Fe_3O_4–chitosan hybrid spheres), provided greater material stability in different media besides As(V) adsorptive capacity increase (about 20 times higher), with M_S equal to 9.1 emu g^{-1} (Malwal and Gopinath 2017).

Chitosan magnetic spheres, Fe_3O_4@$Zr(OH)_4$-impregnated (MICB), were also applied to remove arsenic in the Wang et al. (2014) study, but from an underground real water. The authors also evaluated the As (III) species adsorption using column system and demonstrated the decreasing of As (III) concentration from 0.103 to 0.05 mg L^{-1} in just 60 min of operation, without previous effluent treatment. After 5 h, the As (III) concentration was already less than 0.01 mg L^{-1}, using 1 g L^{-1} as adsorbent dosage. By presenting approximately 96% of contaminant removal, the studied operation became an alternative to ion exchange membrane system, widely employed in arsenic removal, without any previous adding of oxidizing agents to convert As (III) to As (V) (Wang et al. 2014).

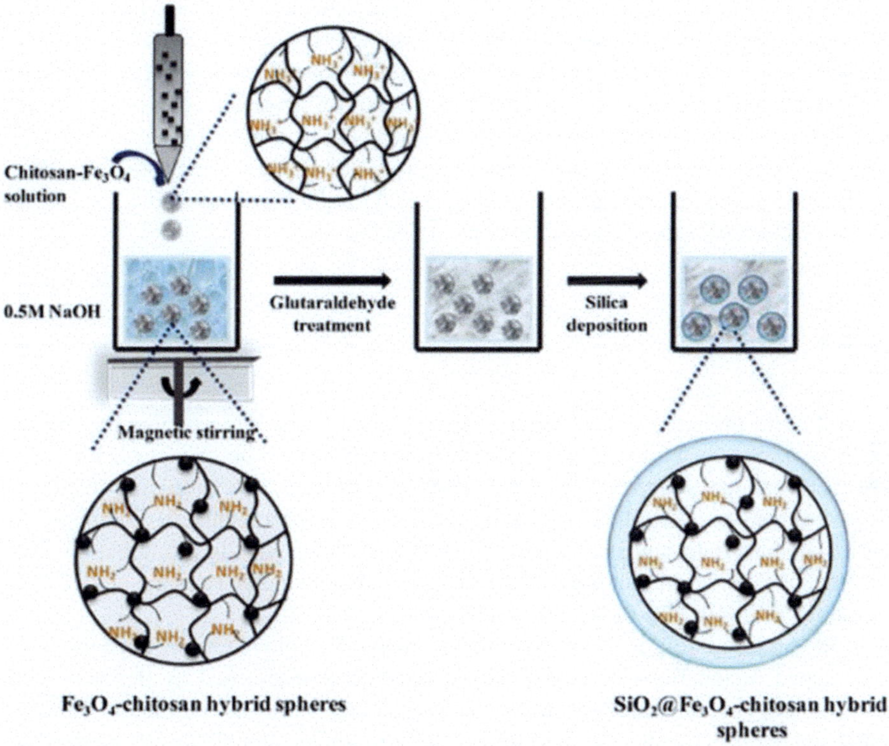

Fig. 14.4 Scheme to demonstrate the magnetic chitosan beads' synthesis (SiO_2@Fe_3O_4-chitosan hybrid spheres) adopted by Malwal and Gopinath (2017). The chitosan sphere shaping was performed when chitosan and Fe_3O_4 acid suspension was dropwise poured into a basic coagulation solution. In this study, the beads also were silica-coated. (Reprinted with permission of Colloid and Interface Science Communications (Elsevier) from Malwal and Gopinath 2017)

The literature indicates that chitosan beads drying methodology may alter this material's structure, thus directly influencing their adsorbent potential. When Malwal and Gopinath (2017) compared lyophilization and simple air drying, the authors found that chitosan spheres presented an aerogel structure with higher porosity after undergoing lyophilization procedure, due to shrinkage force reduction. The porous and fibrillar morphology of chitosan magnetic beads may be highly beneficial for pollutant removal by adsorption (Malwal and Gopinath 2017).

In addition to estimating the effect of chitosan and PVA association in magnetic beads production, Wang et al. (2018) also evaluated the adsorptive performance after consecutive freeze–thaw cycles, where water was eliminated by sublimation. The freeze–thaw step did not change the gel's original conformation, but made beads insoluble as gel crosslinking points were intensified. Thus, there were the desirable reduction in chitosan swelling degree and beads' surface area increase from 36.8 to 60.1 $m^2\ g^{-1}$ (Wang et al. 2018).

The sol-gel transition procedure to obtain adsorbent beads does not preclude crosslinking methodology use or functionalization. Beyond studies already mentioned, Vakili et al. (2019) synthesized magnetic chitosan spheres (CS-DEO-SP) crosslinked with diepoxyoctane (DEO) and amino-functionalized with spermine (SP). Crosslinking ensured better stability of beads in acid medium required for Cr (VI) adsorption, while active sites' increase, promoted by functionalization, resulted in adsorption capacity growth from 49.8 to 112.41 mg g^{-1}, with 0 and 1.5 g L^{-1} as SP concentrations, respectively (Vakili et al. 2019).

14.3.3 Films

In addition to particulate shape, chitosan magnetic composites can also be synthesized as films. Generally, film-shaped adsorbents, even nonmagnetics, facilitate adsorbent–effluent separation after the adsorption process. When holding magnetic properties, the films applied to adsorption lose some prominence if compared to particulate materials, since solid–liquid separation can be equally promoted through magnet approximation, but with a larger surface area available (Cesano et al. 2015; Kloster et al. 2019).

In literature, casting is the most common method employed to produce chitosan film-shaped, magnetic or not. The casting method is performed by deposition of chitosan solution on a flat substrate. Over time and with the combined acting of radial and hydrodynamic (viscous) forces, the solution become flattened and the film is then conceived through evaporation or solvent solidification (Kloster et al. 2015a; Danglad-Flores et al. 2018).

Naturally, chitosan may exhibit some fragility in film-shaping due to its crystallinity. Characteristics such as stiffness and rupture tendency can be intensified in magnetic film manufacturing as iron oxide addition induces a rise in crystallinity. That is the context in which the use (or not) of plasticizers is an important aspect to be considered in such film synthesis. Plasticizer agents are able to reduce intermolecular forces, providing greater mobility of polymer chains and, consequently, the formation of more flexible and soft films (Kloster et al. 2015a; Chen et al. 2018).

There is a long list of plasticizing agents and glycerol, polyols, fatty acids, hydrated salts, triacetin, citrate, polyethylene, and water are some of them. In chitosan magnetic films, the most commonly employed agent is glycerol as it owns hydrogen bonding and hydrophobic group, both suitable for this polymer. By glycerol employing, Kloster et al. (2015a, b) achieved increasing of magnetite mass percentage in chitosan films, maintaining its flexibility. While nonplasticized films became brittle with 7% weigh of magnetic material, films containing between 20% and 30% of glycerol remained resistant by adding up to 10% weight of magnetite (Kloster et al. 2015a; Chen et al. 2018).

For effective chitosan plasticization, the plasticizing agent must be able to break its intermolecular bonding network without creating new crosslinked intermolecular networks by hydrogen bonding. Chen et al. (2018) evaluated glycerol and ionic liquids' performance in chitosan plasticization. Glycerol, by presenting a single hydrogen bonding site, was more efficient in promoting molecular mobility than ionic liquids, carriers of multiple hydrogen bonding sites, which increase the possibility of crosslinking between polymer chains (Chen et al. 2018).

The authors Kloster et al. (2019) evaluated Congo Red anionic dye adsorption by chitosan magnetic films synthesized via the in situ and the multistep routes (Sect. 14.2), with or without glycerol plasticizer addition. The use of plasticizer ensured greater adsorption capacities for both films as it increased the access of active adsorption sites by dye molecules. This may be explained for free volume increasing caused by the greater spacing between polymer chains or the film free spaces, left by glycerol molecules which solubilized in aqueous solution (Kloster et al. 2019).

The material combinations for chitosan magnetic films' production also occur just as crosslinking. Castrejón-Parga et al. (2014) and Marques Neto et al. (2019) synthesized films containing multiwalled carbon nanotubes (MWCNTs) apart from chitosan and iron oxide. Both studies adopted, as the method, the MWCNTs' homogenization in a polymer matrix of chitosan solubilized in acetic acid. The MWCNTs were already decorated with magnetic particles.

Castrejón-Parga et al. (2014) referred to the addition of MWCNTs, decorated with magnetite, to chitosan films as material reinforcement. The polymer chitosan chains interact with decorated nanotubes without losing film characteristic behavior, with Ms. = 0.004 emu g^{-1}. The authors did not evaluate the obtained material as adsorbent. However, when discussing the structural analyzes performed, they concluded that this materials' combination was able to promote higher thermal stability and greater dispersion of magnetic particle, reducing clusters formation (Castrejón-Parga et al. 2014).

Marques Neto et al. (2019) developed an adsorbent film (CLCh/MWCNT/Fe) containing chitosan crosslinked with glutaraldehyde (CLCh) and MWCNTs decorated with iron oxide (Fig. 14.5). In this study, besides causing surface area increase

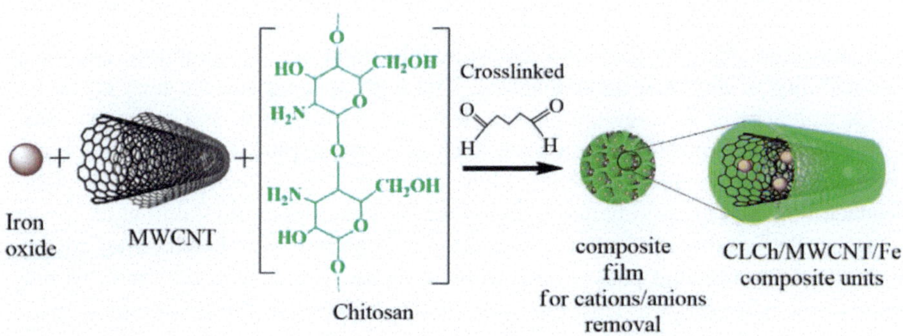

Fig. 14.5 Scheme to illustrate an adsorbent film-shaped (CLCh/MWCNT/Fe), containing glutaraldehyde crosslinked chitosan (CLCh) and multiwalled carbon nanotubes (MWCNTs) decorated with iron oxides. (Reprinted with permission of Chemosphere (Elsevier) from Marques Neto et al. 2019)

of chitosan films (from 25.20 to 49.68 m^2 g^{-1}), carbon nanotubes significantly contributed to adsorption of Cr (III) and Cr (VI), present in synthetic solution, but also in real samples of electroplating effluent. The Cr (VI) species was able to interact with the positively charged film surface, but a steric effect reduced Cr (VI) removal by adsorption. This effect could be generated by Cr (III) coadsorption through electrostatic attraction which prevented new Cr species adsorption (Marques Neto et al. 2019).

The Cr (VI) adsorption by chitosan magnetic films have been also evaluated in two other studies. In 2016, Mirabedini and Kassaee produced an adsorbent containing magnetite and chitosan crosslinked with glyoxal and polyvinyl alcohol (Fe_3O_4NPs/CS/glyoxal/PVA hydrogel film), with $q_m = 33.78$ mg g^{-1} at pH 3, where q_m means maximum adsorption capacity. In 2017, Mirabedini et al., using the same synthesis route, assessed an adsorbent only containing magnetite and chitosan crosslinked with glyoxal (Fe_3O_4NPs/CS/Glyoxal), with $q_m = 27.25$ mg g^{-1} at pH 4. Although the authors did not compare nor evaluate the magnetic properties, it is possible to observe that PVA addition led to increase of adsorption capacity, possibly due to high hydroxyl groups' availability in PVA (Mirabedini and Kassaee 2016; Mirabedini et al. 2017).

Metal adsorption was also evaluated by Lasheen et al. (2012). The authors synthesized films of chitosan and nanomagnetite (NMag-CS) to adsorb Cu^{2+}, Pb^{2+}, Cd^{2+}, Cr (VI), and Ni^{2+} metal ions. As adsorbent, the films showed good characteristics such as high specific surface area (640.59 m^2 g^{-1}), total volume of 0.18 cm^3 g^{-1}, and superparamagnetic behavior (Ms = 21 emu g^{-1}). The chitosan affinity order with each ion (Cu^{2+} > Cr (VI) > Pb^{2+} > Cd^{2+} > Ni^{2+}) directly reflected on adsorption capacities experimentally obtained, indicating that the adsorption occurred by complexation, based on Lewis acid and base theory (Lasheen et al. 2016).

14.4 Application in Wastewater Treatment

The desirable characteristics in a magnetic adsorbent are the same that guarantee applicability of any adsorbent material. The high specific surface area ensures a larger contact surface among phases, favoring adsorbate access to active sites available for adsorption. Other aspects such as adequate pore diameter and volume, mechanical and chemical stability, high selectivity, cost-effectiveness, affordable regeneration, and environmental viability are no less important (Vakili et al. 2014).

Regardless of the adsorption system adopted (continuous or batch), after adsorption, the fluid and solid phases need to be separated from each other and this may involve the addition of at least one more unitary operation. Filtration and centrifugation are the most used to enable adsorbents recovery/regeneration, mostly particulate solids. These steps, which also entail costs to the final process, may be suppressed by using materials that can be quickly and efficiently separated from the liquid phase through simple approximation of an external magnetic field. It is mainly in this context that feasibility of any magnetic adsorbent takes place (Liu et al. 2011; De Gisi et al. 2016; Basheer 2018).

In view of its biodegradable and nontoxic nature and relatively low cost due to its high bioavailability, chitosan owns a wide range of applications, including drug release, tissue engineering, and wastewater treatment. With respect to magnetic chitosan, the situation is repeated, especially the empowerment of this material in the treatment of effluents contaminated by dyes, drugs, and metals. Many countries, usually developed ones, are committed to keeping water bodies free of these contaminants either by strengthening corporate oversight or by checking specified limits, but above all by supporting the development of new technologies. All these efforts are directly related to the large number of published documents on this topic (Hessel et al. 2007; Dotto et al. 2017; Shariatinia and Jalali 2018).

Among the effluents that must undergo treatment before returning to the environment, colored wastewater is more difficult to treat. In most cases, in addition to the presence of dyes, these effluents present high alkalinity, chemical and biological oxygen demand, and total dissolved solids. The complex aromatic structures of dyes afford chemical stability and make the biodegradation difficult. If dyes are not removed, upon reaching water bodies, they can significantly affect the photosynthetic activity of aquatic plants as there is reduced light penetration through water. The organic nature of molecules also stimulates the growth of lethal microorganisms in the environment as whole, including human life (Baban et al. 2010; Srinivasan and Viraraghavan 2010; Rauf and Salman Ashraf 2012; Katheresan et al. 2018).

Medicine-contaminated effluents are no less harmful when discharged into the environment without proper treatment. Pharmaceutical industries, hospitals, and inadequate disposal of animal waste and expired drugs are the main sources of marine water, groundwater, and surface water contamination by medicines in many countries around the world. The bioaccumulation potential of aquatic and terrestrial organisms makes pharmaceutical products an imminent threat to fauna, flora, and human health as people are exposed to these molecules through the food chain or contaminated water (Doble and Kumar 2005; Tiwari et al. 2017; Courtier et al. 2019).

The metal presence in effluents of mining, electroplating, tannery, steel works, automobile, battery, and semiconductor industries is also recurrent. Unlike organic contaminants, metals cannot be degraded until they become environmentally harmless. In the human body, metals such as lead, chromium, copper, zinc, arsenic, and cadmium exhibit a high degree of bioaccumulation and reactivity, which may induce genetic mutations and inhibit essential enzymes' activity. In the environment, they also interfere with plant growth and development. In addition to dye and medicine molecules, efficient metal removal requires additional treatment implementation to those conventionally adopted at wastewater treatment plants (Doble and Kumar 2005; Mi et al. 2015; Kongarapu et al. 2018).

Although a chitosan-based magnetic adsorbent is a composite and both components contribute significantly to material final structure, it is the polymeric matrix and its modifications generally determine the considered pollutants' adsorption mechanism. In case of dyes and medicines, the adsorption mechanism may occur by ion exchange and ion pair formation. Metal adsorption occurs by chelation, as NH_2 and OH functional groups act as a chelator to bind metal ions (Kongarapu et al. 2018; Liang et al. 2019a).

Table 14.1 shows a compilation from existing documents about using chitosan-based magnetic adsorbents in target pollutant removal, highlighting features such as magnetic saturation (M_S, emu g^{-1}), adsorption capacity (q, mg g^{-1}), and the viable number of adsorption–desorption cycles for adsorbent magnetic chitosan.

14.5 Conclusions

Chitosan-based magnetic adsorbents are promising materials in the adsorption field due to their high specific surface area, selectivity, chemical and physical stability, and cost-effectiveness. These remarkable properties combined with chemical/physical adsorbent modifications, but mostly with its affordable regeneration, provided by solid–liquid phase separation through the application of an external magnetic field, lead magnetic chitosan to become a suitable adsorbent candidate for environmental protection purposes. In this regard, this chapter writing was completed relying on several published documents, carefully addressed and reviewed, about chitosan-based magnetic adsorbents applied to environmental remediation.

The attainment of chitosan magnetic adsorbents essentially occurs through the combination of a magnetic material and a polymer, which share their particular characteristics and mutually contribute to overcoming their individual limitations. Chitosan is the material that externally coats the core composed by magnetic particles. Therefore, this polymer is expected to be mainly responsible for adsorbate–adsorbent interactions when chitosan-based magnetic adsorbents are employed, and also the main material to improve when researchers aim better adsorption efficiency and structural characteristics.

The development of these adsorbents involves many aspects that vary depending on the sort of magnetic particle used, the improvement methodologies considered, the desired material's final shape, and the pollutant species to be adsorbed. Among the iron oxides widely used to afford magnetic characteristics to chitosan adsorbents, the magnetite (Fe_3O_4) obtained via coprecipitation stands out, even though alternatives less susceptible to oxidation already exist, such as ferrite. In general, magnetic particles exhibit ferromagnetic or superparamagnetic behavior and high magnetization saturation, which is always reduced after polymeric coating.

As aforementioned, the methodologies able to improve the performance of chitosan-based magnetic adsorbents on pollutant removal by adsorption, are mostly related to polymer modification, either by crosslinking or functionalization. The first one provides greater chemical and mechanical resistance in acidic media as it allows the achievement of a three-dimensional network. The functionalization procedures, in its turn, through functional groups adding, is able to increase the number and diversity of active sites available for adsorption, enabling the contaminants removal with little or none affinity for unmodified chitosan.

Table 14.1 Compilation of published studies about chitosan-based magnetic adsorbents applied to wastewater treatment by adsorption

Material	Adsorbate	M_s (emu g^{-1})*	q (mg g^{-1})	Adsorption–desorption cycles	References
Mγ-Fe_2O_3/CSCs	Methyl Orange	1.44	29.37	–	Zhu et al. (2010)
γ-Fe_2O_3/chitosan composite film	Methyl Orange	–	29.41	–	Jiang et al. (2012)
Magnetic chitosan/poly (vinyl alcohol) hydrogel beads (MCPHBs)	Methyl Orange	3.83	68.86	12	Wang et al. (2018)
Chitosan/maghemite composite (MagCHbead)	Methyl Orange	–	779	4	Obeid et al. (2013)
Glutaraldehyde - grafted chitosan coated with $NiFe_2O_4$ (G-g-C@NF nanocomposite)	Methyl Orange Congo Red	25	551.2 274.7	–	Zeraatkar Moghaddam et al. (2018)
Chitosan/clay/nanomagnetite composite (chitosan/heulandite/Fe_3O_4)	Methyl Orange Methylene Blue	–	149.2 45.1	–	Cho et al. (2015)
Magnetic chitosan and graphene oxide (MCGO)	Methylene Blue	–	180.83	–	Fan et al. (2012b)
Magnetic β-cyclodextrin–chitosan/graphene oxide materials (MCCG)	Methylene Blue	55.13	84.32	5	Fan et al. (2013)
Magnetic chitosan-crosslinked κ-carrageenan bionanocomposite	Methylene Blue	12.2	123.1	5	Mahdavinia and Mosallanezhad (2016)
Mesoporous graphene/Fe_3O_4/chitosan nanocomposite (RGO@Fe_3O_4@chitosan)	Methylene Blue	–	249.25	5	Van Hoa et al. (2016)

(continued)

Table 14.1 (continued)

Material	Adsorbate	M_s (emu g^{-1})*	q (mg g^{-1})	Adsorption–desorption cycles	References
Magnetic polyethyleneimine nanoparticles coated by sulfonated chitosan (PEI@MNP/SCS)	Methylene Blue	Over 52	–	1	Zhao et al. (2016)
Magnetic chitosan/clay beads	Methylene Blue	–	82.00	–	Bée et al. (2017)
H_2SO_4 crosslinked magnetic chitosan nanocomposite bead	Methylene Blue	–	20.41	–	Rahmi et al. (2019b)
Chitosan/active charcoal magnetic composite (CTN/AC-Fe_3O_4)	Methylene Blue Reactive Blue 4	–	500.0 250.0	–	Karaer and Kaya (2016)
Xanthated chitosan@Al_2O_3/iron oxide (XCAF)	Methylene Blue Safranin O	–	197.8 169.8	–	Tanhaei et al. (2019)
Magnetic chitosan "fluid" (MCF)	Congo Red	25.0	1724.1	5	Hui et al. (2018)
Chitosan/iron oxide nanocomposite films	Congo red	–	700.00	–	Kloster et al. (2019)
Magnetic chitosan grafted with graphene oxide (MCGO)	Methyl Blue	–	95.16	5	Fan et al. (2012a)
$CoFe_2O_4$/chitosan magnetic composite	Blue Indigotine	16.45	380.88	5	dos Santos et al. (2019b)
Magnetic graphene/chitosan (MGCh)	Acid Orange 7	29.87	42.7	–	Sheshmani et al. (2014)
Glutaraldehyde crosslinked magnetic chitosan nanoparticles (GMCNs)	FD&C Blue 1 D&C Yellow 5	–	475.61 292.07	3	Zhou et al. (2014)

Microsphere containing glutamine modified chitosan and silica-coated Fe_3O_4 nanoparticles (CS-Gln-MCM)	Acid Green 25 Hg(II)	3.48	698.95199.23	5	Tao et al. (2016)
Modified magnetic chitosan microparticles (MCDs)	Food Yellow 3 Acid Yellow 23	10.32	833.33 666.67	5	Zheng et al. (2019)
Magnetic quaternary chitosan hybrid nanoparticles (Fe_3O_4@SiO_2/SiHTCC)	Diclofenac	–	240.4	–	(Soares et al. 2019)
$ZnFe_2O_4$/chitosan magnetic particles	Diclofenac	9.75	188.00	4	dos Santos et al. (2019a)
Amine-functionalized chitosan (AmCS@Fe_3O_4)	Diclofenac	17.5	469.48	4	Liang et al. (2019a)
Chitosan-based magnetic composite particles with core-brush topology (CD-MCP)	Diclofenac Sodium Tetracycline Hydrochloride	12.5	1640 40.2	3	Zhang et al. (2016)
Magnetic chitosan-grafted graphene oxide nanocomposite (MCGO)	Ciprofloxacin	14.76	282.9	4	Wang et al. (2016)
Fe_3O_4/activated carbon/chitosan (MACC: Magnetic activated carbon/chitosan)	Ciprofloxacin Erythromycin Amoxicillin	5.78	90.10 178.57 526.31	–	Danalıoğlu et al. (2017)
Genipin-crosslinked chitosan/graphene oxide-SO_3H (GC/MGO-SO_3H) composite	Ibuprofen Tetracycline	55.4	473.25 556.28	5	Liu et al. (2019)
Ethylenediamine-modified crosslinked magnetic chitosan resin (EMCMCR)	Cr(VI)	–	51.813	10	Hu et al. (2011)
Magnetic chitosan nanoparticles	Cr(VI)	–	55.80	–	Thinh et al. (2013)
Magnetic chitosan and graphene oxide-ionic liquid (MCGO-IL)	Cr(VI)	–	145.35	4	Li et al. (2014)
Chitosan/MWCNT/Fe_3O_4 composite nanofibers	Cr(VI)	–	358	5	Beheshti et al. (2016)

(continued)

Table 14.1 (continued)

Material	Adsorbate	M_s (emu g^{-1})*	q (mg g^{-1})	Adsorption–desorption cycles	References
Fe_3O_4NPs/CS/glyoxal/PVA hydrogel film	Cr(VI)	–	33.78	3	Mirabedini and Kassaee (2016)
Fe_3O_4 NPs/CS/Glyoxal Hydrogel film	Cr(VI)	–	27.25	3	Mirabedini et al. (2017)
Magnetic chitosan (Chi@Fe_3O_4) and graphene oxide modified magnetic chitosan (Chi@Fe_3O_4GO) nanocomposites	Cr(VI)	–	142.32 and 100.51	4	Subedi et al. (2019)
Thiourea-modified magnetic chitosan (Fe_3O_4@Al_2O_3-CS)	Cr(VI)	9.2	327.8	5	Cai et al. (2019)
Aminated crosslinked chitosan beads (CS-DEO-SP)	Cr(VI)	–	358.10	5	Vakili et al. (2019)
Magnetic crosslinked chitosan on graphene oxide (MCC@GO) and nitrogen-doped graphene oxide (MCC@NGO)	Cr(VI)	11.1 10.2	15.9 30.2	5	Zeraatkar Moghaddam et al. (2019)
CLCh/MWCNT/Fe film	Cr(VI) Cr(III)	–	449.30 66.25	10	Marques Neto et al. (2019)
Chi-CG Magnetic beads	Cr(VI) Cu(II)	1.15	138.53 156.49	3	Abou El-Reash (2016)
Epichlorohydrin crosslinked chitosan Schiff Base-Fe_3O_4 nanocomposite (ECSBNC)	Cr(VI) Cu (II)	–	83.33 90.90	–	Anush and Vishalakshi (2019)
Chitosan combined with magnetic Loofah biochar (CMLB)	Cr(VI) Cu(II)	–	30.14 54.68	3	Xiao et al. (2019)
Combination of carboxymethyl chitosan-coated magnetic nanoparticles and chitosan-citrate complex gel beads (NOCC-MNPs/CCGBs)	Cu(II)	71.99	294.11	–	Mi et al. (2015)

Chitosan-modified Mn ferrite nanoparticles	Cu(II)	12.0	65.1	–	Meng et al. (2015)
Chitosan/polyvinyl alcohol (CTN/PVA-Fe_3O_4) magnetic	Cu(II)	–	143.0	–	(Karaer and Kaya 2017)
Magnetic chitosan composite adsorbent (CsFeAC)	Cu(II)	28.32	216.60	5	(Li et al. 2017)
Magnetic bentonite/carboxymethyl chitosan/sodium alginate (Mag-Ben/CCS/Alg)	Cu(II)	7.05	56.79	4	Zhang et al. (2019a)
Chitosan – cellulose enwrapped magnetic carbon foam (CCMF)	Cu(II)	–	115.65	–	Zhang et al. (2019b)
Triethylene – tetramine-grafted chitosan(TETA-CS)	Pb(II)	–	370.63	4	Kuang et al. (2013)
Carboxylated chitosan magnetic microspheres (EDCMM) and submicrospheres (EDCMSM)	Pb(II)	14.117 66.563	141.83 164.81	5	Xu et al. (2015)
Magnetic chitosan-(D-glucosimine methyl) benzaldehyde (MCS-Sch)	Pb(II)	51.2	121.95	7	Shahraki and Samareh Delarami (2018)
Phosphorylated chitosan/$CoFe_2O_4$ composite (P-MCS)	Pb(II) Cd(II)	20.67	151.06 71.53	5	Wu et al. (2019a)
APTS-Fe_3O_4/APT@CS	Pb(II)	14.96	625.34	5	Liang et al. (2019b)
NMag-CS film	Cu^{2+}, Pb^{2+}, Cr (VI), $Cd^{2+}$and Ni^{2+}	21.00	123.40, 116.20, 114.90, 112.30 and 109.80, respectively.	5	Lasheen et al. (2016)
Magnetic chitosan grafted with alkyl acrylate (Mag-CSg-HA) (Mag-CSg- BMA) (Mag-CSg-BA)	Ni(II)	32.3 30.1 33.7	121.96 104.17 101.01	5	Iordache et al. (2018)

(continued)

Table 14.1 (continued)

Material	Adsorbate	M_s (emu g^{-1})*	q (mg g^{-1})	Adsorption–desorption cycles	References
Silica stabilized magnetic-chitosan beads (SiO_2 @Fe_3O_4- chitosan hybrid spheres)	As(V)	9.1	1.69	–	Malwal and Gopinath (2017)
Magnetic nanoparticles impregnated chitosan beads (MICB)	As(V) As(III)	17.14	35.70 35.30	5	Wang et al. (2014)
Chitosan/polydopamine@C@magnetic fly ash adsorbent beads (CPCFMA)	Ag(I)	5.57	57.02	5	Mu et al. (2020)
Amidoxime-functionalized magnetic chitosan microparticles (AM-MG-CH)	U(VI) Eu(III)	–	1.5 mmol 2.47 mmol	5	Hamza et al. (2018)
Magnetic chitosan–cystamine composite (CDF-CS)	Au(III)	9.5	478.47	4	Zhao et al. (2019)

In addition to being the major surface constituent of chitosan-based magnetic adsorbents, the chitosan polymer matrix also determines its final shape. Although chitosan has already assumed shapes of spheres, fibers, powders, films, and flakes in previous reports, not all of these formats have their magnetic version. For now, only beads, films, and powders have received the attention of researchers. Powders are the most studied, but beads and films molded through sol-gel transition and casting methods, respectively, have also had their applicability highlighted.

In the prospect of wastewater treatment, chitosan-based adsorbents have been considered as interesting sorbents for effective removal of organic molecules as well as metal ions discharged from industrial and municipal wastes into water sources. In case of dyes and medicines, the adsorption mechanism may occur by ion exchange and ion pair formation. The metal ions' adsorption occurs by chelation, as NH_2 and OH functional groups act as a chelator to bind metal ions.

Although chitosan-based magnetic adsorbents have already been widely studied in the context of environmental remediation, there is still opportunity to develop different shapes of adsorbent such as fibers, foams, and aerogels with a larger specific surface area. In the future, studies should focus on alternative magnetic particles to magnetite, wherein adsorbents with better chemical stability will be attained and maybe with higher M_S values. Furthermore, by using functionalization, chitosan-based magnetic adsorbents are able to develop affinities with other sort of pollutants, such as phenols and radionuclides present in the effluents of oil industry and nuclear power plants, respectively.

References

Abou El-Reash YG (2016) Magnetic chitosan modified with cysteine-glutaraldehyde as adsorbent for removal of heavy metals from water. J Environ Chem Eng 4:3835–3847. https://doi.org/10.1016/j.jece.2016.08.014

Alonso J, Barandiarán JM, Fernández Barquín L, García-Arribas A (2018) Magnetic nanoparticles, synthesis, properties, and applications. In: Magnetic nanostructured materials. Elsevier, pp 1–40

Anush SM, Vishalakshi B (2019) Modified chitosan gel incorporated with magnetic nanoparticle for removal of cu(II) and Cr(VI) from aqueous solution. Int J Biol Macromol 133:1051–1062. https://doi.org/10.1016/j.ijbiomac.2019.04.179

Baban A, Yediler A, Ciliz NK (2010) Integrated water management and CP implementation for wool and textile blend processes. Clean (Weinh) 38:84–90. https://doi.org/10.1002/clen.200900102

Basheer AA (2018) New generation nano-adsorbents for the removal of emerging contaminants in water. J Mol Liq 261:583–593. https://doi.org/10.1016/j.molliq.2018.04.021

Bée A, Obeid L, Mbolantenaina R et al (2017) Magnetic chitosan/clay beads: a magsorbent for the removal of cationic dye from water. J Magn Magn Mater 421:59–64. https://doi.org/10.1016/j.jmmm.2016.07.022

Beheshti H, Irani M, Hosseini L et al (2016) Removal of Cr (VI) from aqueous solutions using chitosan/MWCNT/Fe 3 O 4 composite nanofibers-batch and column studies. Chem Eng J 284:557–564. https://doi.org/10.1016/j.cej.2015.08.158

Cai W, Zhu F, Liang H et al (2019) Preparation of thiourea-modified magnetic chitosan composite with efficient removal efficiency for Cr(VI). Chem Eng Res Des 144:150–158. https://doi.org/10.1016/j.cherd.2019.01.031

Castrejón-Parga KY, Camacho-Montes H, Rodríguez-González CA et al (2014) Chitosan–starch film reinforced with magnetite-decorated carbon nanotubes. J Alloys Compd 615:S505–S510. https://doi.org/10.1016/j.jallcom.2013.12.269

Cesano F, Fenoglio G, Carlos L, Nisticò R (2015) One-step synthesis of magnetic chitosan polymer composite films. Appl Surf Sci 345:175–181. https://doi.org/10.1016/j.apsusc.2015.03.154

Chen F, Chen M, Yang C et al (2015) Terbium-doped gadolinium oxide nanoparticles prepared by laser ablation in liquid for use as a fluorescence and magnetic resonance imaging dual-modal contrast agent. Phys Chem Chem Phys 17:1189–1196. https://doi.org/10.1039/C4CP04380D

Chen M, Runge T, Wang L et al (2018) Hydrogen bonding impact on chitosan plasticization. Carbohydr Polym 200:115–121. https://doi.org/10.1016/j.carbpol.2018.07.062

Cho DW, Jeon BH, Chon CM et al (2015) Magnetic chitosan composite for adsorption of cationic and anionic dyes in aqueous solution. J Ind Eng Chem 28:60–66. https://doi.org/10.1016/j.jiec.2015.01.023

Courtier A, Cadiere A, Roig B (2019) Human pharmaceuticals: why and how to reduce their presence in the environment. Curr Opin Green Sustain Chem 15:77–82. https://doi.org/10.1016/j.cogsc.2018.11.001

Crini G, Badot P-M (2008) Application of chitosan, a natural aminopolysaccharide, for dye removal from aqueous solutions by adsorption processes using batch studies: a review of recent literature. Prog Polym Sci 33:399–447. https://doi.org/10.1016/j.progpolymsci.2007.11.001

Danalıoğlu ST, Bayazit ŞS, Kerkez Kuyumcu Ö, Salam MA (2017) Efficient removal of antibiotics by a novel magnetic adsorbent: magnetic activated carbon/chitosan (MACC) nanocomposite. J Mol Liq 240:589–596. https://doi.org/10.1016/j.molliq.2017.05.131

Danglad-Flores J, Eickelmann S, Riegler H (2018) Deposition of polymer films by spin casting: a quantitative analysis. Chem Eng Sci 179:257–264. https://doi.org/10.1016/j.ces.2018.01.012

De Gisi S, Lofrano G, Grassi M, Notarnicola M (2016) Characteristics and adsorption capacities of low-cost sorbents for wastewater treatment: a review. Sustain Mater Technol 9:10–40. https://doi.org/10.1016/j.susmat.2016.06.002

Doble M, Kumar A (2005) Biotreatment of industrial effluents. Elsevier. https://doi.org/10.1016/B978-0-7506-7838-4.X5000-3

dos Santos JMN, Pereira CR, Foletto EL, Dotto GL (2019a) Alternative synthesis for ZnFe 2 O 4 / chitosan magnetic particles to remove diclofenac from water by adsorption. Int J Biol Macromol 131:301–308. https://doi.org/10.1016/j.ijbiomac.2019.03.079

dos Santos JMN, Pereira CR, Pinto LAA et al (2019b) Synthesis of a novel CoFe2O4/chitosan magnetic composite for fast adsorption of indigotine blue dye. Carbohydr Polym 217:6–14. https://doi.org/10.1016/j.carbpol.2019.04.054

Dotto GL, Santos JMN, Tanabe EH et al (2017) Chitosan/polyamide nanofibers prepared by Forcespinning ® technology: a new adsorbent to remove anionic dyes from aqueous solutions. J Clean Prod 144:120–129. https://doi.org/10.1016/j.jclepro.2017.01.004

El Knidri H, Belaabed R, Addaou A et al (2018) Extraction, chemical modification and characterization of chitin and chitosan. Int J Biol Macromol 120:1181–1189. https://doi.org/10.1016/j.ijbiomac.2018.08.139

Fan L, Luo C, Li X et al (2012a) Fabrication of novel magnetic chitosan grafted with graphene oxide to enhance adsorption properties for methyl blue. J Hazard Mater 215–216:272–279. https://doi.org/10.1016/j.jhazmat.2012.02.068

Fan L, Luo C, Sun M et al (2012b) Preparation of novel magnetic chitosan/graphene oxide composite as effective adsorbents toward methylene blue. Bioresour Technol 114:703–706. https://doi.org/10.1016/j.biortech.2012.02.067

Fan L, Luo C, Sun M et al (2013) Synthesis of magnetic β-cyclodextrin-chitosan/graphene oxide as nanoadsorbent and its application in dye adsorption and removal. Colloids Surf B Biointerf 103:601–607. https://doi.org/10.1016/j.colsurfb.2012.11.023

Fan C, Li K, He Y et al (2018) Evaluation of magnetic chitosan beads for adsorption of heavy metal ions. Sci Total Environ 627:1396–1403. https://doi.org/10.1016/j.scitotenv.2018.02.033

Geankoplis CJ (1998) Processos de Transporte y Operaciones Unitarias, 3rd edn. Compañía Editorial Continental

Gupta KC, Jabrail FH (2006) Glutaraldehyde and glyoxal cross-linked chitosan microspheres for controlled delivery of centchroman. Carbohydr Res 341:744–756. https://doi.org/10.1016/j.carres.2006.02.003

Hamza MF, Roux JC, Guibal E (2018) Uranium and europium sorption on amidoxime-functionalized magnetic chitosan micro-particles. Chem Eng J 344:124–137. https://doi.org/10.1016/j.cej.2018.03.029

Hessel C, Allegre C, Maisseu M et al (2007) Guidelines and legislation for dye house effluents. J Environ Manag 83:171–180. https://doi.org/10.1016/j.jenvman.2006.02.012

Heydaripour J, Gazi M, Oladipo AA, Gulcan HO (2019) Porous magnetic resin-g-chitosan beads for adsorptive removal of phenolic compounds. Int J Biol Macromol 123:1125–1131. https://doi.org/10.1016/j.ijbiomac.2018.11.168

Hu X-j, Wang J-s, Liu Y-g et al (2011) Adsorption of chromium (VI) by ethylenediamine-modified cross-linked magnetic chitosan resin: isotherms, kinetics and thermodynamics. J Hazard Mater 185:306–314. https://doi.org/10.1016/j.jhazmat.2010.09.034

Hui M, Shengyan P, Yaqi H et al (2018) A highly efficient magnetic chitosan "fluid" adsorbent with a high capacity and fast adsorption kinetics for dyeing wastewater purification. Chem Eng J 345:556–565. https://doi.org/10.1016/j.cej.2018.03.115

Iordache ML, Dodi G, Hritcu D et al (2018) Magnetic chitosan grafted (alkyl acrylate) composite particles: synthesis, characterization and evaluation as adsorbents. Arab J Chem 11:1032–1043. https://doi.org/10.1016/j.arabjc.2015.12.010

Jawad AH, Norrahma SSA, Hameed BH, Ismail K (2019) Chitosan-glyoxal film as a superior adsorbent for two structurally different reactive and acid dyes: adsorption and mechanism study. Int J Biol Macromol 135:569–581. https://doi.org/10.1016/j.ijbiomac.2019.05.127

Jiang R, Fu Y-Q, Zhu H-Y et al (2012) Removal of methyl orange from aqueous solutions by magnetic maghemite/chitosan nanocomposite films: adsorption kinetics and equilibrium. J Appl Polym Sci 125:E540–E549. https://doi.org/10.1002/app.37003

Karaer H, Kaya I (2016) Synthesis, characterization of magnetic chitosan/active charcoal composite and using at the adsorption of methylene blue and reactive blue4. Microporous Mesoporous Mater 232:26–38. https://doi.org/10.1016/j.micromeso.2016.06.006

Karaer H, Kaya İ (2017) Synthesis, characterization and using at the copper adsorption of chitosan/polyvinyl alcohol magnetic composite. J Mol Liq 230:152–162. https://doi.org/10.1016/j.molliq.2017.01.030

Katheresan V, Kansedo J, Lau SY (2018) Efficiency of various recent wastewater dye removal methods: a review. J Environ Chem Eng 6:4676–4697. https://doi.org/10.1016/j.jece.2018.06.060

Kloster GA, Marcovich NE, Mosiewicki MA (2015a) Composite films based on chitosan and nanomagnetite. Eur Polym J 66:386–396. https://doi.org/10.1016/j.eurpolymj.2015.02.042

Kloster GA, Muraca D, Meiorin C et al (2015b) Magnetic characterization of chitosan–magnetite nanocomposite films. Eur Polym J 72:202–211. https://doi.org/10.1016/j.eurpolymj.2015.09.014

Kloster GA, Mosiewicki MA, Marcovich NE (2019) Chitosan/iron oxide nanocomposite films: effect of the composition and preparation methods on the adsorption of Congo red. Carbohydr Polym 221:186–194. https://doi.org/10.1016/j.carbpol.2019.05.089

Kongarapu RJ, Nayak AK, Khobragade MU, Pal A (2018) Surfactant bilayer on chitosan bead surface for enhanced Ni(II) adsorption. Sustain Mater Technol 18. https://doi.org/10.1016/j.susmat.2018.e00077

Kuang SP, Wang ZZ, Liu J, Wu ZC (2013) Preparation of triethylene-tetramine grafted magnetic chitosan for adsorption of Pb(II) ion from aqueous solutions. J Hazard Mater 260:210–219. https://doi.org/10.1016/j.jhazmat.2013.05.019

Lasheen MR, El-Sherif IY, Tawfik ME et al (2016) Preparation and adsorption properties of nano magnetite chitosan films for heavy metal ions from aqueous solution. Mater Res Bull 80:344–350. https://doi.org/10.1016/j.materresbull.2016.04.011

Laurent S, Henoumont C, Stanicki D, et al (2017) Magnetic properties. In: MRI contrast agents from molecules to particles no title. pp 5–11

Li L, Luo C, Li X et al (2014) Preparation of magnetic ionic liquid/chitosan/graphene oxide composite and application for water treatment. Int J Biol Macromol 66:172–178. https://doi.org/10.1016/j.ijbiomac.2014.02.031

Li J, Jiang B, Liu Y et al (2017) Preparation and adsorption properties of magnetic chitosan composite adsorbent for Cu2+ removal. J Clean Prod 158:51–58. https://doi.org/10.1016/j.jclepro.2017.04.156

Li H, Li X, Chen Y et al (2018) Removal and recovery of thallium from aqueous solutions via a magnetite-mediated reversible adsorption-desorption process. J Clean Prod 199:705–715. https://doi.org/10.1016/j.jclepro.2018.07.178

Liang XX, Omer AM, Hu Z et al (2019a) Efficient adsorption of diclofenac sodium from aqueous solutions using magnetic amine-functionalized chitosan. Chemosphere 217:270–278. https://doi.org/10.1016/j.chemosphere.2018.11.023

Liang XX, Ouyang X, Wang S et al (2019b) Efficient adsorption of Pb(II) from aqueous solutions using aminopropyltriethoxysilane-modified magnetic attapulgite@chitosan (APTS-Fe3O4/APT@CS) composite hydrogel beads. Int J Biol Macromol 137:741–750. https://doi.org/10.1016/j.ijbiomac.2019.06.244

Liu JJ, Wang XC, Fan B (2011) Characteristics of PAHs adsorption on inorganic particles and activated sludge in domestic wastewater treatment. Bioresour Technol 102:5305–5311. https://doi.org/10.1016/j.biortech.2010.12.063

Liu Y, Liu R, Li M et al (2019) Removal of pharmaceuticals by novel magnetic genipin-crosslinked chitosan/graphene oxide-SO3H composite. Carbohydr Polym 220:141–148. https://doi.org/10.1016/j.carbpol.2019.05.060

Mahdavinia GR, Mosallanezhad A (2016) Facile and green rout to prepare magnetic and chitosan-crosslinked κ-carrageenan bionanocomposites for removal of methylene blue. J Water Process Eng 10:143–155. https://doi.org/10.1016/j.jwpe.2016.02.010

Malwal D, Gopinath P (2017) Silica stabilized magnetic-chitosan beads for removal of arsenic from water. Colloid Interf Sci Commun 19:14–19. https://doi.org/10.1016/j.colcom.2017.06.003

Marques Neto J d O, Bellato CR, DDC S (2019) Iron oxide/carbon nanotubes/chitosan magnetic composite film for chromium species removal. Chemosphere 218:391–401. https://doi.org/10.1016/j.chemosphere.2018.11.080

Marques JS, Pereira MR, Sotto A, Arsuaga JM (2019) Removal of aqueous copper(II) by using crosslinked chitosan films. React Funct Polym 134:31–39. https://doi.org/10.1016/j.reactfunctpolym.2018.10.009

Meng Y, Chen D, Sun Y et al (2015) Adsorption of Cu 2+ ions using chitosan-modified magnetic Mn ferrite nanoparticles synthesized by microwave-assisted hydrothermal method. Appl Surf Sci 324:745–750. https://doi.org/10.1016/j.apsusc.2014.11.028

Mi FL, Wu SJ, Chen YC (2015) Combination of carboxymethyl chitosan-coated magnetic nanoparticles and chitosan-citrate complex gel beads as a novel magnetic adsorbent. Carbohydr Polym 131:255–263. https://doi.org/10.1016/j.carbpol.2015.06.031

Mirabedini M, Kassaee MZ (2016) Removal of toxic Cr(VI) from water by a novel magnetic chitosan/glyoxal/PVA hydrogel film. Desalin Water Treat 57:14266–14279. https://doi.org/10.1080/19443994.2015.1065763

Mirabedini M, Kassaee MZ, Poorsadeghi S (2017) Novel magnetic chitosan hydrogel film, cross-linked with glyoxal as an efficient adsorbent for removal of toxic Cr(VI) from water. Arab J Sci Eng 42:115–124. https://doi.org/10.1007/s13369-016-2062-1

Mohammadzadeh Pakdel P, Peighambardoust SJ (2018) Review on recent progress in chitosan-based hydrogels for wastewater treatment application. Carbohydr Polym 201:264–279. https://doi.org/10.1016/j.carbpol.2018.08.070

Mourya VK, Inamdar NN (2008) Chitosan-modifications and applications: opportunities galore. React Funct Polym 68:1013–1051. https://doi.org/10.1016/j.reactfunctpolym.2008.03.002

Mu C, Zhang L, Zhang X et al (2020) Selective adsorption of Ag (I) from aqueous solutions using chitosan/polydopamine@C@magnetic fly ash adsorbent beads. J Hazard Mater 381:120943. https://doi.org/10.1016/j.jhazmat.2019.120943

Muzzarelli RAA, Boudrant J, Meyer D et al (2012) Current views on fungal chitin/chitosan, human chitinases, food preservation, glucans, pectins and inulin: a tribute to Henri Braconnot, precursor of the carbohydrate polymers science, on the chitin bicentennial. Carbohydr Polym 87:995–1012. https://doi.org/10.1016/j.carbpol.2011.09.063

Ngomsik A-F, Bee A, Draye M et al (2005) Magnetic nano- and microparticles for metal removal and environmental applications: a review. Comptes Rendus Chim 8:963–970. https://doi.org/10.1016/j.crci.2005.01.001

Nilsen-Nygaard J, Strand S, Vårum K et al (2015) Chitosan: gels and interfacial properties. Polymers (Basel) 7:552–579. https://doi.org/10.3390/polym7030552

Obeid L, Bée A, Talbot D et al (2013) Chitosan/maghemite composite: a magsorbent for the adsorption of methyl orange. J Colloid Interface Sci 410:52–58. https://doi.org/10.1016/j.jcis.2013.07.057

Pandi K, Viswanathan N, Meenakshi S (2019) Hydrothermal synthesis of magnetic iron oxide encrusted hydrocalumite-chitosan composite for defluoridation studies. Int J Biol Macromol 132:600–605. https://doi.org/10.1016/j.ijbiomac.2019.03.115

Perry RH, Green DW, Maloney JO (1997) Perry's chemical engineers' handbook, 7th edn. McGraw-Hill, New York

Rahmi, Fathurrahmi, Lelifajri, PurnamaWati F (2019a) Preparation of magnetic chitosan using local iron sand for mercury removal. Heliyon 5:1–8. https://doi.org/10.1016/j.heliyon.2019.e01731

Rahmi, Ishmaturrahmi, Mustafa I (2019b) Methylene blue removal from water using H2SO4 crosslinked magnetic chitosan nanocomposite beads. Microchem J 144:397–402. https://doi.org/10.1016/j.microc.2018.09.032

Ranjbar Bandforuzi S, Hadjmohammadi MR (2019) Modified magnetic chitosan nanoparticles based on mixed hemimicelle of sodium dodecyl sulfate for enhanced removal and trace determination of three organophosphorus pesticides from natural waters. Anal Chim Acta 1078:90–100. https://doi.org/10.1016/j.aca.2019.06.026

Rauf MA, Salman Ashraf S (2012) Survey of recent trends in biochemically assisted degradation of dyes. Chem Eng J 209:520–530. https://doi.org/10.1016/j.cej.2012.08.015

Reddy DHK, Lee SM (2013) Application of magnetic chitosan composites for the removal of toxic metal and dyes from aqueous solutions. Adv Colloid Interf Sci 201–202:68–93. https://doi.org/10.1016/j.cis.2013.10.002

Reddy DHK, Yun Y-S (2016) Spinel ferrite magnetic adsorbents: alternative future materials for water purification? Coord Chem Rev 315:90–111. https://doi.org/10.1016/j.ccr.2016.01.012

Roman DL, Roman M, Sletta H et al (2019) Assessment of the properties of chitin deacetylases showing different enzymatic action patterns. J Mol Graph Model 88:41–48. https://doi.org/10.1016/j.jmgm.2019.01.002

Rong SY, Mubarak NM, Tanjung FA (2017) Structure-property relationship of cellulose nanowhiskers reinforced chitosan biocomposite films. J Environ Chem Eng 5:6132–6136. https://doi.org/10.1016/j.jece.2017.11.054

Rwei SP, Chen TY, Cheng YY (2005) Sol/gel transition of chitosan solutions. J Biomater Sci Polym Ed 16:1433–1445. https://doi.org/10.1163/156856205774472290

Sable SS, Shah KJ, Chiang P-C, Lo S-L (2018) Catalytic oxidative degradation of phenol using iron oxide promoted sulfonated-ZrO2 by Advanced Oxidation Processes (AOPs). J Taiwan Inst Chem Eng 91:434–440. https://doi.org/10.1016/j.jtice.2018.06.030

Shahraki S, Samareh Delarami H (2018) Magnetic chitosan-(d-glucosimine methyl)benzaldehyde Schiff base for Pb+2 ion removal. Experimental and theoretical methods. Carbohydr Polym 200:211–220. https://doi.org/10.1016/j.carbpol.2018.07.081

Shalla AH, Bhat MA, Yaseen Z (2018) Hydrogels for removal of recalcitrant organic dyes: a conceptual overview. J Environ Chem Eng 6:5938–5949. https://doi.org/10.1016/j.jece.2018.08.063
Shariatinia Z, Jalali AM (2018) Chitosan-based hydrogels: preparation, properties and applications. Int J Biol Macromol 115:194–220. https://doi.org/10.1016/j.ijbiomac.2018.04.034
Sheshmani S, Ashori A, Hasanzadeh S (2014) Removal of acid Orange 7 from aqueous solution using magnetic graphene/chitosan: a promising nano-adsorbent. Int J Biol Macromol 68:218–224. https://doi.org/10.1016/j.ijbiomac.2014.04.057
Sivashankari PR, Prabaharan M (2017) Deacetylation modification techniques of chitin and chitosan. In: Chitosan based biomaterials, vol 1. Elsevier, pp 117–133
Soares SF, Fernandes T, Sacramento M et al (2019) Magnetic quaternary chitosan hybrid nanoparticles for the efficient uptake of diclofenac from water. Carbohydr Polym 203:35–44. https://doi.org/10.1016/j.carbpol.2018.09.030
Song Q, Zhang ZJ (2006) Correlation between spin−orbital coupling and the superparamagnetic properties in magnetite and cobalt ferrite spinel nanocrystals. J Phys Chem B 110:11205–11209. https://doi.org/10.1021/jp060577o
Srinivasan A, Viraraghavan T (2010) Decolorization of dye wastewaters by biosorbents: a review. J Environ Manag 91:1915–1929. https://doi.org/10.1016/j.jenvman.2010.05.003
Subedi N, Lähde A, Abu-Danso E et al (2019) A comparative study of magnetic chitosan (Chi@Fe3O4) and graphene oxide modified magnetic chitosan (Chi@Fe3O4GO) nanocomposites for efficient removal of Cr(VI) from water. Int J Biol Macromol 137:948–959. https://doi.org/10.1016/j.ijbiomac.2019.06.151
Tanhaei B, Ayati A, Sillanpää M (2019) Magnetic xanthate modified chitosan as an emerging adsorbent for cationic azo dyes removal: kinetic, thermodynamic and isothermal studies. Int J Biol Macromol 121:1126–1134. https://doi.org/10.1016/j.ijbiomac.2018.10.137
Tao X, Li K, Yan H et al (2016) Simultaneous removal of acid green 25 and mercury ions from aqueous solutions using glutamine modified chitosan magnetic composite microspheres. Environ Pollut 209:21–29. https://doi.org/10.1016/j.envpol.2015.11.020
Thinh NN, Hanh PTB, Ha LTT et al (2013) Magnetic chitosan nanoparticles for removal of Cr (VI) from aqueous solution. Mater Sci Eng C 33:1214–1218. https://doi.org/10.1016/j.msec.2012.12.013
Tiwari B, Sellamuthu B, Ouarda Y et al (2017) Review on fate and mechanism of removal of pharmaceutical pollutants from wastewater using biological approach. Bioresour Technol 224:1–12. https://doi.org/10.1016/j.biortech.2016.11.042
Vakili M, Rafatullah M, Salamatinia B et al (2014) Application of chitosan and its derivatives as adsorbents for dye removal from water and wastewater: a review. Carbohydr Polym 113:115–130. https://doi.org/10.1016/j.carbpol.2014.07.007
Vakili M, Deng S, Liu D et al (2019) Preparation of aminated cross-linked chitosan beads for efficient adsorption of hexavalent chromium. Int J Biol Macromol 139:352–360. https://doi.org/10.1016/j.ijbiomac.2019.07.207
Van Hoa N, Khong TT, Thi Hoang Quyen T, Si Trung T (2016) One-step facile synthesis of mesoporous graphene/Fe3O4/chitosan nanocomposite and its adsorption capacity for a textile dye. J Water Process Eng 9:170–178. https://doi.org/10.1016/j.jwpe.2015.12.005
Wang J, Xu W, Chen L et al (2014) Preparation and evaluation of magnetic nanoparticles impregnated chitosan beads for arsenic removal from water. Chem Eng J 251:25–34. https://doi.org/10.1016/j.cej.2014.04.061
Wang F, Yang B, Wang H et al (2016) Removal of ciprofloxacin from aqueous solution by a magnetic chitosan grafted graphene oxide composite. J Mol Liq 222:188–194. https://doi.org/10.1016/j.molliq.2016.07.037
Wang W, Zhang H, Shen J, Ye M (2018) Facile preparation of magnetic chitosan/poly (vinyl alcohol) hydrogel beads with excellent adsorption ability via freezing-thawing method. Colloids Surf A Physicochem Eng Asp 553:672–680. https://doi.org/10.1016/j.colsurfa.2018.05.094

Wu X, Huang M, Zhou T, Mao J (2016) Recognizing removal of norfloxacin by novel magnetic molecular imprinted chitosan/γ-Fe2O3 composites: selective adsorption mechanisms, practical application and regeneration. Sep Purif Technol 165:92–100. https://doi.org/10.1016/j.seppur.2016.03.041

Wu D, Wang Y, Li Y et al (2019a) Phosphorylated chitosan/CoFe2O4 composite for the efficient removal of Pb(II) and Cd(II) from aqueous solution: adsorption performance and mechanism studies. J Mol Liq 277:181–188. https://doi.org/10.1016/j.molliq.2018.12.098

Wu J, Cheng X, Yang G (2019b) Preparation of nanochitin-contained magnetic chitosan microfibers via continuous injection gelation method for removal of Ni(II) ion from aqueous solution. Int J Biol Macromol 125:404–413. https://doi.org/10.1016/j.ijbiomac.2018.11.212

Xiao F, Cheng J, Cao W et al (2019) Removal of heavy metals from aqueous solution using chitosan-combined magnetic biochars. J Colloid Interface Sci 540:579–584. https://doi.org/10.1016/j.jcis.2019.01.068

Xu Y, Dang Q, Liu C et al (2015) Preparation and characterization of carboxyl-functionalized chitosan magnetic microspheres and submicrospheres for Pb 2+ removal. Colloids Surf A Physicochem Eng Asp 482:353–364. https://doi.org/10.1016/j.colsurfa.2015.06.028

Zeraatkar Moghaddam A, Ghiamati E, Pourashuri A, Allahresani A (2018) Modified nickel ferrite nanocomposite/functionalized chitosan as a novel adsorbent for the removal of acidic dyes. Int J Biol Macromol 120:1714–1725. https://doi.org/10.1016/j.ijbiomac.2018.09.198

Zeraatkar Moghaddam A, Esmaeilkhanian E, Shakourian-Fard M (2019) Immobilizing magnetic glutaraldehyde cross-linked chitosan on graphene oxide and nitrogen-doped graphene oxide as well-dispersible adsorbents for chromate removal from aqueous solutions. Int J Biol Macromol 128:61–73. https://doi.org/10.1016/j.ijbiomac.2019.01.086

Zhang S, Dong Y, Yang Z et al (2016) Adsorption of pharmaceuticals on chitosan-based magnetic composite particles with core-brush topology. Chem Eng J 304:325–334. https://doi.org/10.1016/j.cej.2016.06.087

Zhang H, Omer AM, Hu Z et al (2019a) Fabrication of magnetic bentonite/carboxymethyl chitosan/sodium alginate hydrogel beads for cu (II)adsorption. Int J Biol Macromol 135:490–500. https://doi.org/10.1016/j.ijbiomac.2019.05.185

Zhang Z, Li H, Li J et al (2019b) A novel adsorbent of core-shell construction of chitosan-cellulose magnetic carbon foam: synthesis, characterization and application to remove copper in wastewater. Chem Phys Lett 731:1–7. https://doi.org/10.1016/j.cplett.2019.07.001

Zhao W, Huang X, Wang Y et al (2016) A recyclable and regenerable magnetic chitosan absorbent for dye uptake. Carbohydr Polym 150:201–208. https://doi.org/10.1016/j.carbpol.2016.05.037

Zhao M, Zhao J, Huang Z et al (2019) One pot preparation of magnetic chitosan-cystamine composites for selective recovery of Au(III) from the aqueous solution. Int J Biol Macromol 137:721–731. https://doi.org/10.1016/j.ijbiomac.2019.07.022

Zheng C, Zheng H, Wang Y et al (2018) Synthesis of novel modified magnetic chitosan particles and their adsorption performance toward Cr(VI). Bioresour Technol 267:1–8. https://doi.org/10.1016/j.biortech.2018.06.113

Zheng C, Zheng H, Wang Y et al (2019) Modified magnetic chitosan microparticles as novel superior adsorbents with huge "force field" for capturing food dyes. J Hazard Mater 367:492–503. https://doi.org/10.1016/j.jhazmat.2018.12.120

Zhou Z, Lin S, Yue T, Lee TC (2014) Adsorption of food dyes from aqueous solution by glutaraldehyde cross-linked magnetic chitosan nanoparticles. J Food Eng 126:133–141. https://doi.org/10.1016/j.jfoodeng.2013.11.014

Zhu HY, Jiang R, Xiao L, Li W (2010) A novel magnetically separable γ-Fe2O3/crosslinked chitosan adsorbent: preparation, characterization and adsorption application for removal of hazardous azo dye. J Hazard Mater 179:251–257. https://doi.org/10.1016/j.jhazmat.2010.02.087

Zhu HY, Fu YQ, Jiang R et al (2012) Novel magnetic chitosan/poly(vinyl alcohol) hydrogel beads: preparation, characterization and application for adsorption of dye from aqueous solution. Bioresour Technol 105:24–30. https://doi.org/10.1016/j.biortech.2011.11.057

Chapter 15
Methods Used for Performance Enhancement of Iron-Based Magnetic Adsorbents in Water Systems

Parmila Devi and Anil K. Saroha

Contents

Abstract Magnetic adsorbents are increasingly gaining attention as a focus of research for contaminant removal from water and wastewater due to their low cost, easy synthesis and separation from aqueous medium. However, the commercial applications of zero-valent iron (ZVI) and iron oxides are limited due to particle agglomeration and low oxidation stability. The agglomeration behavior of ZVI and particle oxidative stability can be enhanced by modification of magnetic adsorbents using an appropriate stabilization procedure such as surface coating or functionalization. This chapter systematically describes the recent advances in modification methods of magnetic adsorbents for contaminant removal from water

P. Devi
Department of Chemical and Biological Engineering, College of Engineering, University of Saskatchewan, Saskatoon, SK, Canada

A. K. Saroha (✉)
Department of Chemical Engineering, Indian Institute of Technology, Delhi, New Delhi, India
e-mail: aksaroha@chemical.iitd.ac.in

L. Meili, G. L. Dotto (eds.), *Advanced Magnetic Adsorbents for Water Treatment*, Environmental Chemistry for a Sustainable World 61,
https://doi.org/10.1007/978-3-030-64092-7_15

and wastewater. Various modification methods including carbon stabilization, polymer coating, inorganic oxide impregnation, functionalization, and miscellaneous stabilization procedures are summarized. These stabilization procedures can enhance the adsorption capacities of magnetic adsorbents by expanding the surface area and pore volume of the adsorbents and generating new active sites and/or functional groups on the adsorbent surface. The key process parameters and relevant published data along with the results of these studies have been discussed.

Keywords Magnetic adsorbents · Wastewater · Mechanism · Modification · Functionalization · Surfactants · Carbon support · Polymers · Iron oxides

15.1 Introduction

Adsorption is a widely used technique for the removal of organic and inorganic contaminants from water and wastewater. However, the tedious separation of spent adsorbents from water systems is the main limitation (Devi and Saroha 2016; Gupta et al. 2011). Recently, magnetic adsorbents are increasingly studied for contaminant removal due to their super paramagnetic nature, which enables their easy separation under magnetic conditions. Iron oxide-based adsorbents such as magnetite (Fe_3O_4, FeO, and Fe_2O_3) are considered as a cost-effective alternative for pollutant removal owing to their nontoxic nature, easy preparation procedure, and abundance in nature (Devi et al. 2019a, b). Apart from iron oxides, zero-valent iron (ZVI) particles have been widely used for the remediation of water (surface and groundwater) and removal of different types of pollutants including chlorophenols, hydrocarbons, and heavy metals. The easy separation from aqueous medium due to their magnetic properties and high surface areas for adsorption makes them green and a safe alternative for environmental remediation (Devi and Saroha 2015; Devi et al. 2019a, b). However, there are certain limitations that restrict the field-scale and wider applications of magnetic adsorbents. The proper dispersion of magnetic adsorbents in aquatic medium is hard to achieve as the magnetic particles tend to agglomerate due to strong interactions between the particles through magnetic and Van der Waals forces of attraction. Also, ZVI has low stability as iron particles in the zero-valent state are vulnerable to oxidation in the presence of air (Devi and Saroha 2015; Devi et al. 2019a, b). The agglomeration of particles and oxidative stability can be improved by modification of ZVI particles using an appropriate stabilization method like surface coating, incorporation of support material, and functionalization. Modification or functionalization of magnetic adsorbents can be effective in overcoming the agglomeration challenges as well as reducing their leachability and toxicity in aquatic medium. Also, these procedures have resulted in enhancing the sorption capacity of magnetic adsorbents. Among the various modification procedures, modification of magnetic adsorbents with carbon particles has resulted in an enhancement in particle stability, making them more compatible for further functionalization. The most common functional groups used for particle functionalization are amino, anhydride, and thiol groups. The prefunctionalization was usually performed using silanols (alkoxy silane) and postfunctionalization using surfactants, chelating ligands, and inorganic oxides (Karami and Zeynizadeh 2019).

15.2 Types of Magnetic Particles

Zero-valent iron and iron oxides are the commonly used magnetic particles employed for the preparation of various types of magnetic adsorbents. The composition and structural properties of magnetic particles are given in Table 15.1.

15.3 Methods for Preparation of Magnetic Adsorbents

Considerable efforts have been made by researchers to develop the methods for preparation of magnetic adsorbents. The commonly used methods include coprecipitation, reduction, thermal decomposition, and hydrothermal synthesis (Fig. 15.1). Coprecipitation involves the chemical reaction between iron salts in

Table 15.1 Types of iron nanoparticles

Composition	Type	Structure/magnetic properties	Common name
Fe	α-Fe	Crystalline phase with body-centered cubic (BCC) lattice	–
	γ-Fe	Crystalline phase with face-centered cubic (FCC) lattice	–
	Amorphous Fe	Absence of crystalline phase	–
Fe_2O_3	α-Fe_2O_3	In the α-Fe_2O_3 structure, all Fe^{3+} ions have an octahedral coordination	Hematite
	γ-Fe_2O_3	γ-Fe_2O_3 has a cation-deficient AB_2O_4 spinel, the metal atoms A and B occur in tetrahedral and octahedral environments, respectively	Maghemite
Fe_3O_4	–	The cubic spinel Fe_3O_4. Ferrimagnetic at temperatures below 858 K	Magnetite
FeO	–	Cubic Fe^{2+}oxide. Antiferromagnetic (TC = 185 K) in the bulk state	Wustite
$Fe_2O_3.H_2O$	α-FeOOH	Antiferromagnetic in the bulk state (TC = 393 K)	Goethite
	β-FeOOH	Paramagnetic at 300 K	Akaganeite
	γ-FeOOH	Paramagnetic at 300 K	Lepidocrocite
	δ-FeOOH	Ferrimagnetic.	Ferroxyhite

Reprinted with the permission from Kaur et al. (2014)

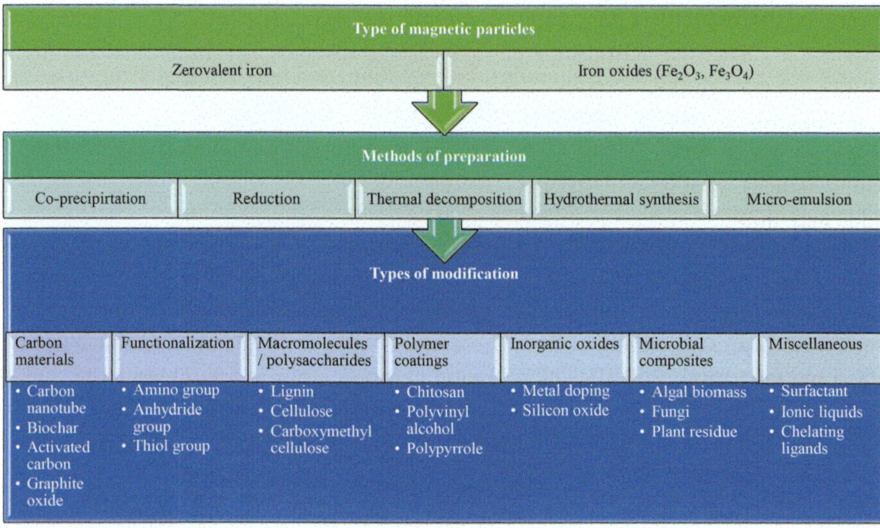

Fig. 15.1 Flow diagram showing methods of preparation of magnetic adsorbents and different types of modifications

alkaline solution under anoxic conditions. The limitations of this process are requirement of very high pH conditions and formation of nonuniform agglomerated particles. Further, thermal decomposition is reported to have high efficiency for the formation of monodispersed particles. In this process, organic metallic salts are dissolved in organic solvent containing surfactant and undergo thermal treatment at 200–300 °C. In hydrothermal synthesis, chemical reactions are performed in aqueous medium under high temperature (120–260 °C) and pressure (0.3–4 MPa) conditions (Qiao et al. 2019). Calcination/annealing temperature is the key parameter that can affect the adsorption properties of magnetic adsorbents as it controls particle size and morphology of spinal ferrite, carbon-based adsorbents, and magnetic composites. Particle size of magnetic adsorbents was found to decrease with an increase in calcination temperature, while surface area was found to increase with an increase in calcination temperature. The micro-emulsion process involves the synthesis of magnetic nanoparticles in a medium of two immiscible liquids (water and oil) and surfactants. The aqueous phase is dispersed as nano-droplets in the oil phase and these nano-droplets provide restricted reaction media for the shape and size distribution of particles. It is reported that the magnetic particles prepared by the micro-emulsion process had a smaller size and high saturation magnetization values compared to the magnetic particles prepared using other methods (Qiao et al. 2019). Among the synthesis procedures, chemical reduction is mostly used for the production of magnetic adsorbents and magnetic nanoparticles. It is a simple method and can be safely carried out at room temperature with readily available chemical reagents. Devi and Saroha (2014, 2015) performed chemical reduction of iron salt using sodium borohydride for the synthesis of ZVI.

15.4 Modifications of Magnetic Adsorbents

Modification of magnetic adsorbents is very effective in overcoming the problems of agglomeration, leaching, and toxicity. Modification of magnetic particles can be performed using stabilizers (carbon nanotubes, biochar, and activated carbon), coating (silica), impregnation (metal oxides), and functionalization (amino, anhydride, and thiol groups). Functionalization of magnetic adsorbents using amino, anhydride, and thiol groups yields higher sorption capacities (Gómez-Pastora et al. 2014). A large number of studies are reported in the literature on the application of magnetic adsorbents for contaminant removal from wastewater (Fig. 15.2), out of which 35% used bare magnetic nanoparticles (MNPs), 31% organic coated MNPs, 22% functionalized MNPs, and 12% inorganic coated MNPs (Gómez-Pastora et al. 2014).

15.4.1 Carbon-Based Materials as Support

Many studies are available in the literature on the usage of carbon materials as support material for magnetic particles (Table 15.2). The most commonly used carbon materials are (1) biochar; (2) carbon nanotubes; (3) activated carbon (AC); and (iv) graphene oxide. Biochar and AC are widely used in environmental applications due to their porous structures and strong affinity toward a wide range of organic compounds. High surface area and porosity of AC can be developed using thermal treatment followed by physical or chemical activation. Literature data suggest an enhancement in adsorption capacities of carbonaceous adsorbents due to an increase in specific surface area and pore volume after magnetic modification (Hao et al. 2018). After magnetization, the surface area of magnetic coconut shell-derived carbon material was found to increase to 952 m^2/g compared to raw coconut shell-derived carbon (416 m^2/g). Altintig et al. (2017) used zinc chloride AC for Fe impregnation and the resultant Fe-AC was used as an adsorbent for the removal of methylene blue. The Fe-AC and AC showed an adsorption capacity of 357 mg/g and 303 mg/g, respectively, in 30 min at solution pH 4. Devi and Saroha (2014) used paper mill sludge-based biochar for stabilization of ZVI particles and subsequently used as an adsorbent for pentachlorophenol removal from synthetic effluent. It was found that biochar and ZVI complement each other by resolving the shortcomings. Biochar serves as a support material and prevents the agglomeration of ZVI particles by screening the magnetic forces, whereas ZVI provides magnetic and reductive characteristics to the magnetic composites. Agricultural waste-based biochar and AC are increasingly used for magnetic adsorbent preparation (Devi et al. 2019a, b; Devi et al. 2020). The main constituents in agricultural wastes are usually lignin and cellulose. The properties of biochar or AC diverge depending on the amount of lignin or cellulose in precursor material. Lignin-derived biochar has a poorly defined structure; thus, metals are largely adsorbed by a combination of ion exchange and metal chelation (Noor et al. 2017).

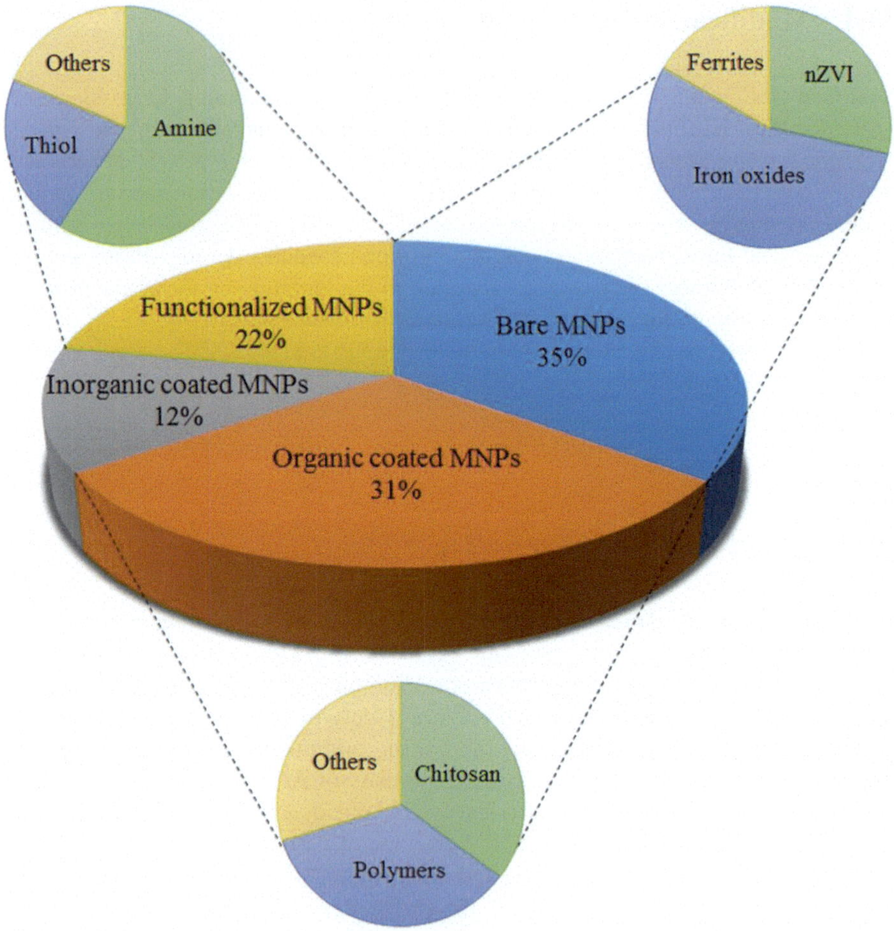

Fig. 15.2 Distribution of studies reported in literature according to the type of material used for contaminant removal. (Reprinted with permission from Gómez-Pastora et al. 2014)

Carbon nanotubes (CNTs) are pin-like hollow-layered structure with high thermal resistance, chemical stability, and high surface area. Combination of magnetic particles with carbon nanotubes produces a promising material with superior adsorption capacity for contaminant removal from water and wastewater. Pourzamani et al. (2017) synthesized SWCNT-MN hybrid adsorbent by combining magnetic nanoparticle with CNT for application as hybrid adsorbent for xylene removal.

Apart from biochar and AC, graphene and graphene oxide (GO) have been widely used as support material in adsorption as well as catalysis reactions owing to high surface area, conductivity, high chemical & physical stability, and high adsorption capacity (Zandi-Atashbar et al. 2018). Various researchers mixed GO with magnetic particles and used it for the removal of chromium, dyes, and organic contaminants. However, the commercial application of these materials is restricted

Table 15.2 Carbonaceous materials as support for magnetic adsorbents

Adsorbent	Modification	Surface area (m^2/g)	Pore volume (cm^3/g)	Pore diameter (nm)	Adsorbate	Adsorption capacity (mg/g)	Reaction conditions	References
Single-wall carbon nanotubes magnetic nanoparticle	Single-wall carbon nanotubes	187	11.9	0.6	Xylene	50	C_o—100 mg/L; pH 8; adsorbent dosage—2 g/L; time—0.33 h	Pourzamani et al. (2017)
Magnetic oxidized multiwalled carbon nanotube	Carbon nanotube	169	1.0	–	Methylene blue	1.1	C_o—3.0 × 10^{-5} Mol/L; pH 6.6; adsorbent dosage—0.01 g/25 mL; 25 °C	Duman et al. (2016)
Magnetic activated carbon	Coconut-derived activated carbon	952	0.4	0.27	Phenol	23.7	C_o—100 mg/L; adsorbent dosage—0.5 g/100 mL; time—72 h	Hao et al. (2018)
Magnetic biochar	Biochar	5.4	0.02	1.5	Phenol	3.5	C_o—100 mg/L; adsorbent dosage—0.5 g/100 mL; time—72 h	Hao et al. (2018)
Magnetic carbon composites (MCC)	Pinewood saw dust carbon	43	0.11	10	Mercury	167.2	C_o—200 mg/L; pH—6.5; adsorbent dosage—20 mg/10 mL; time—24 h; 25 °C	Wang et al. (2018)
Magnetic activated carbon	Activated carbon	940	0.27	2.9	Methylene blue	357.1	C_o—250 mg/L; pH—4; adsorbent dosage—0.5 g/100 mL; time—0.5 h	Altıntıg et al. (2017)

(continued)

Table 15.2 (continued)

Adsorbent	Modification	Surface area (m^2/g)	Pore volume (cm^3/g)	Pore diameter (nm)	Adsorbate	Adsorption capacity (mg/g)	Reaction conditions	References
Magnetic bio-adsorbent	Pistachio shell AC	–	–	–	Lead Arsenic Cadmium	147.05151.5119.04	C_o—20 mg/L; pH—5.5; adsorbent dosage—0.02 g/50 mL; 25 °C	Nejadshafiee and Islami (2019)
Nitrogen-doped reduced graphene oxide decorated with Fe_3O_4 nanoparticles	Nitrogen-doped reduced graphene oxide	–	–	–	Phosphate	135.5	C_o—50 mg/L; pH—5; adsorbent dosage—5 mg/25 mL; time—12 h	Akram et al. (2019)
Biochar-ZVI composites	Biochar	101	0.08	4.7	Pentachlorophenol	40.2	C_o—50 mg/L; pH—7; time—24 h; adsorbent dosage—10 mg/50 mL; 25 °C	Devi and Saroha (2014)
Magnetic graphene oxide composites	Graphene oxide	–	–	–	Lead Cadmium	128.2 385.1	C_o—100 mg/L; pH—5; adsorbent dosage—100 mg/100 mL; 25 °C	Bao et al. (2020)
Magnetic adsorbent from walnut shell	NaOH/activated carbon	398	0.10	–	Lead	28.6	C_o—50 mg/L; pH—5; adsorbent dosage—2 g/L; time—0.16 h	Safinejad et al. (2017)
Municipal surplus sludge magnetic biochar	–	39	0.06	3.7	Tetracycline	32.6	C_o—100 mg/L; pH—7; adsorbent dosage—20 mg/20 mL	Tang et al. (2018)

due to their minimal stability in aqueous medium. In order to solve this problem, researchers introduced the cross-coupling agent to link up magnetic nanomaterials to GO. Bao et al. (2020) synthesized Fe_3O_4/SiO_2/GO composites using n-propyltrimethoxysilane (NPTS) as a cross-coupling agent to connect Fe_3O_4/SiO_2 and GO. The synthesized adsorbent exhibited excellent magnetic and adsorption properties for cadmium and lead with maximum adsorption capacities of 128.2 and 385.1 mg/g, respectively. Although crosslinking process is effective in the preparation of stable magnetic adsorbents, these crosslinking reagents are very expensive. Therefore, it is challenging for researchers to formulate environmental-friendly and cost-attractive Fe_3O_4/GO composites.

15.4.2 Polymer Coatings

The use of polymer coating is a feasible method to improve the adsorption capacity of magnetic particles by adding or changing the functional groups. Various polymers, such as chitosan, poly-acrylic acid, polystyrene, polyethylenimine, poly (methyl methacrylate), polypyrrole, polyacrylamide, and polydopamine, have been used in the literature (Tables 15.3 and 15.4). Most of these polymers have great potential due to their low cost and nontoxic nature. Among these polymers, chitosan is widely used for the coating of magnetic particles and it acts as an excellent chelating agent for heavy metals, radionuclides, dyes, etc. However, the amine group of chitosan is bit unstable in acidic medium and loses its structural stability due to protonation reaction. The structural stability of chitosan can be improved by sulfonation, nitration, carboxymethylation, hydroxylation, etc. The affinity of carboxymethylated chitosan for Congo red dye is reported to be four times higher than that of pure chitosan due to the presence of active –COOH, –OH, and $-NH_2$ groups (Kim et al. 2016). Hence, various crosslinkers (such as glutaraldehyde and polydopamine) have been explored to enhance the chemical stability and adsorption capacity of chitosan-based magnetic adsorbents for the removal of dyes from wastewater. It is reported that crosslinked magnetic chitosan nanocomposites are effective in the adsorption of Acid Red 2, Methylene blue & Malachite green, and Congo red dyes from aqueous solutions (Kim et al. 2016; Liu et al. 2018).

Similarly, polyamidoamine (PAMAM) dendrimers are the commonly used polymers in many environmental applications due; to their highly branched three-dimensional shapes, ability to flocculate & adsorb, and the presence of reactive functional groups (Liu et al. 2018; Wu et al. 2019). The three-dimensional amide chains are abundant in amino groups, which enable them to interact with other compounds. Therefore, the grafting of amino groups on magnetic adsorbents can improve their adsorption selectivity toward different compounds. Sehlleier et al. (2016) synthesized iron oxide/polymer composite using PAMAM as a surface modifier for magnetic particles for the adsorption of flavonoids. Liu et al. (2018) synthesized a high surface area polyacryl amide (PAM)-coated magnetic adsorbent

Table 15.3 Types of polymer used for modification of magnetic nanoparticles

Polysaccharides/polymer	Structure	Functional groups	Interactions	References
Starch	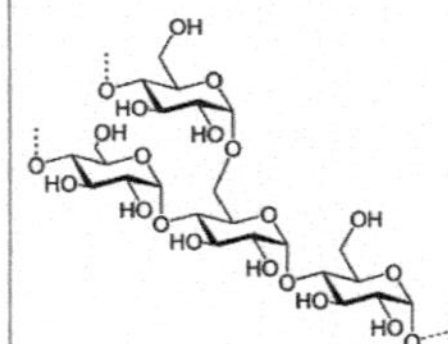	Hydroxyl	H-bonding	Xu et al. (2018)
Alginic acid	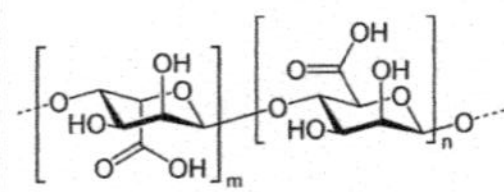	Hydroxyl, ketone	Protonation/deprotonation, electrostatic	Abdi et al. (2019)
Chitosan	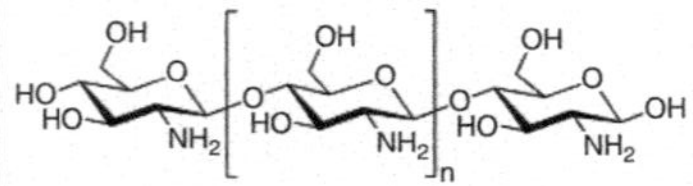	Hydroxyl, amine	Electrostatic	Son et al. (2018)
Cellulose	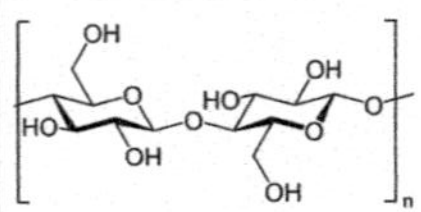	Hydroxyl, ether	Van der Waals forces	Karami and Zeynizadeh (2019)
Poly (N-isopropylacrylamide)	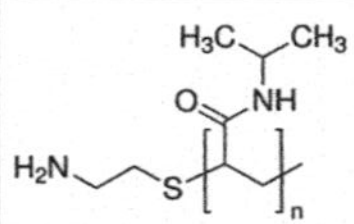	Amine	π–π interactions	Zhao et al. (2014)
Polyacrylic acid	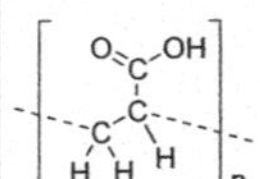	Carboxyl	π–π interactions	Zeng et al. (2015)
Polyvinyl alcohol	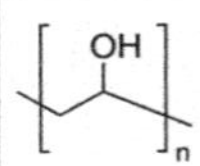	Hydroxyl	Electrostatic interactions	Lv et al. (2017)

Table 15.4 Modification of magnetic adsorbents using polymers for removal of pollutants from water and wastewater

Adsorbent	Modification	Surface area (m^2/g)	Pore volume (cm^3/g)	Pore diameter (nm)	Adsorbate	Adsorption capacity (mg/g)	Reaction conditions	References
Chitosan-modified magnetic kelp biochar (chi-KBm)	Chitosan	6.2	–	–	Copper	26.9	C_o—1000 mg/L; pH—6.9; adsorbent dosage—0.5 g/30 mL	Son et al. (2018)
Xanthate-modified magnetic chitosan/poly (vinyl alcohol) particles (XMMCP)	Xanthate-modified chitosan & polyvinyl alcohol	15.5	0.05	–	Lead	97	C_o—107 mg/L; pH—6.0; adsorbent dosage—0.75 g/250 mL; time—24 h	Lv et al. (2017)
Modified magnetic chitosan microparticles	Chitosan	1.5	–	3.38	Food yellow Acid yellow	833.3 666.7	C_o—500 mg/L; pH—2; adsorbent dosage—25 mg/50 mL	Zheng et al. (2019)
Magnetic chitosan composite adsorbent (CsFeAC)	Chitosan	28.0	–	–	Copper	216.6	C_o—500 mg/L; pH—5.5; adsorbent dosage—0.05 g/500 mL	Li et al. (2017)
Magnetic xanthate-modified chitosan	Chitosan	13.3	–	–	Methylene blue Safranin O	197.8 169.8	C_o—20 mg/L; pH—4; adsorbent dosage—0.8 mg/L	Tanhaei et al. (2019)
EDTA-functionalized magnetic chitosan composite adsorbent	EDTA-functionalized chitosan	2.6	7.6	–	Methylene blue (MB) Lead (Pb) Copper (cu)	459.9 220.0 225.0	C_o—400 mg/L (MB), 100 mg/L (Pb & cu); pH—6; 30 °C	Chen et al. (2019)
Calcon-imprinted magnetic chitosan (CIMC) nanoparticles (NPs)	Calcon-imprinted chitosan; glutaraldehyde as linker	–	–	–	Calcon dye	51.7	C_o—100 mg/L; pH—3; adsorbent dosage—0.05 g/20 mL, 25 °C	Fat'hi and Nasab (2018)
Calcon-imprinted magnetic chitosan (CIMC) nanoparticles (NPs)	Calcon-imprinted chitosan; Epichlorohydrin as linker	–	–	–	Calcon dye	39.2	C_o—100 mg/L; pH—3; adsorbent dosage—0.05 g/20 mL, 25 °C	Fat'hi and Nasab (2018)

(continued)

Table 15.4 (continued)

Adsorbent	Modification	Surface area (m^2/g)	Pore volume (cm^3/g)	Pore diameter (nm)	Adsorbate	Adsorption capacity (mg/g)	Reaction conditions	References
Magnetic waste tyre-activated carbon–chitosan composite	Activated carbon–chitosan	1281	0.54	4.05	Methylparaben propylparaben	85.9 90.0	C_o—63 mg/L; pH—6; adsorbent dosage—64 mg/50 mL; time—57 min; 25 °C	Mashile et al. (2020)
Fe_3O_4-contained polyvinyl alcohol/chitosan composite nanofibers	Polyvinyl alcohol/chitosan	–	–	–	Chromium	213.3	Co—100 mg/L; pH—3; adsorbent dosage—0.05 g/1 mL	Yan et al. (2017)
LDH@Fe_3O_4/crosslinked PVA NCs	Layered double hydroxides (LDH) & polyvinyl alcohol (PVA)	87	0.39	42	Methyl orange	19.7	C_o—30 mg/L; pH—6; adsorbent dosage—0.1 g/10 mL; time—7 h; 25 °C	Mallakpour and Hatami (2019)
Thiol-modified magnetic polypyrrole nanocomposite	Mercapto-acetic acid-modified polypyrrole	49	0.15	19.6	Silver	806.5	C_o—600 mg/L; pH—5.6; adsorbent dosage—0.5 g/L; 25 °C	Mahlangu et al. (2019)
Superparamagnetic Fe_3O_4@SiO_2 nanoparticles	Polyvinylpyrrolidone	89	–	–	Phenanthrene	18.8	C_o—1 mg/L; pH—6; adsorbent dosage—0.5 mg/100 mL; 27 °C	Wang et al. (2019)
Chitosan magnetic composite adsorbent	Quaternary ammonium salt-modified chitosan	28	–	–	Copper	216.6	C_o—500 mg/L; pH—5.5; time—2 h; 25 °C	Li et al. (2016)

for the removal of copper and cadmium from synthetic solution. The active sites of the adsorbent favored the binding of metal ions, and removal efficiencies of 96.9% and 78.1% were obtained for copper and cadmium, respectively. This adsorbent showed very good regeneration potential, and only a 6% reduction in the adsorption capacity was obtained after five consecutive adsorption–regeneration cycles.

Poly(vinyl alcohol) (PVA) is a commonly used polymer for the modification of magnetic adsorbents. It is biodegradable, nonexpensive, nontoxic, water soluble, and provides good chemical and mechanical resistance. However, high solubility and low thermal stability of PVA restrict its applications (Mallakpour and Hatami 2019). Therefore, Mallakpour and Hatami (2019) developed an adsorbent by combining layered double hydroxide–Fe_3O_4 in the PVA matrix and applied it for the remediation of dyes in water. The synthesized adsorbent showed remarkable features and the multicycling study indicated stable and extraordinary adsorption efficiency after three cycles. Polypyrrole (PPy) is an amine-based polymer, capable of effectively interacting with organic or inorganic analytes. Nascimento et al. (2019) used magnetic molecularly imprinted polypyrrole (MMIPPy) for selective removal of pharmaceutical active compounds and obtained an adsorption capacity of 1508 mg/g in 10 min.

15.4.3 Functionalization

Functionalization of magnetic adsorbents with various functional groups (e.g., hydroxyl, carboxyl, amine, anhydride, and thiol) is a very effective method to enhance selectivity and adsorptive capacity. Functionalization is basically performed either to boost the quantity of preexisting functional groups or to graft new functional groups onto the surface of carbon- or silica-coated materials. Such small molecules, bound on the surface of the particles, typically exist as a monolayer and thus do not have a large number of functional groups for adsorption. So, polymer coatings are effective in preventing the leaching of magnetic particles in aqueous medium, sometimes their wider application is limited due to the comparatively poor adsorption efficiency and water-swollen property.

Therefore, hydroxamic acid is increasingly used as a functionalization agent for modification of magnetic polyacrylamide-based adsorbents. It is well known that hydroxamic acid group has an ability to form stable chelates with heavy metal ions. Literature studies have shown that hydroxamic acid group can functionalize and enhance the adsorption capacities of several polymers such as polyvinyl alcohol, polyacrylonitrile, and polyacrylamide (Zhao et al. 2014). It is reported that acidic functional groups are necessary for the separation of metal ions from wastewater. Therefore, Nejadshafiee and Islami (2019) grafted 4-butane sultone ($C_4H_8SO_3H$; strong chelating agent) onto Fe_3O_4 NPs@AC and used it for the removal of heavy metal. The $-SO_3H$ and $-OH$ functional groups present on the surface of Fe_3O_4 NPs@AC@$C_4H_8SO_3H$ composites formed a complex with metal ion and facilitated their removal from aqueous solution. In another study, amine group-functionalized magnetic montmorillonite-based adsorbent was successfully synthesized using the solvothermal reaction. This one-pot solvothermal synthesis reaction was carried out in ethylene glycol ±198 °C for 6 h (Irawan et al. 2019).

The functionalization of polymeric material-coated magnetic adsorbents is a multiple-step and time-consuming process. These multicomponent systems are less stable and easily prone to reduction in their performance and activity with time. Thus, the stability, economic, and procedural factors primarily hinder the large-scale application of these N-functionalized polymeric materials. Therefore, Akram et al. (2019) applied a greener approach to synthesize nitrogen-functionalized graphene oxide Fe_3O_4 composites (N-GO/Fe_3O_4) without using any toxic reducing agents, organic linker, capping, or template agent. The resultant adsorbent effectively removed phosphate from aqueous solution. The phosphate molecules were separated from aqueous solution through electrostatic attraction between positively charged (N-functionalities and Fe_3O_4 nanoparticles) and negatively charged phosphate species. According to hard–soft acid–base principles, silver has a great affinity for sulfur-containing functional groups. Therefore, Mahlangu et al. (2019) prepared magnetic nanocomposites by coating Fe_3O_4 with thioglycolic acid-functionalized PPy. The resultant adsorbent showed a high affinity for silver and a maximum adsorption capacity of 806.5 mg/g was obtained at 25 °C. The adsorption phenomenon was well described by pseudo-second-order rate equation and Langmuir isotherm model. Table 15.5 summarizes various studies reported in the literature on the applications of different functionalized magnetic adsorbents for the removal of pollutants from water and wastewater systems.

15.4.4 Inorganic Oxides

Various metal ions including nonmagnetic ions have been impregnated on magnetic adsorbents to enhance the adsorption performance of magnetic adsorbents (Table 15.6). The most commonly used metal ions are nickel and rare metal ions (Devi and Saroha 2015). An impregnation with rare earth metals causes structural disorders in the lattice structure and grain size of spinal ferrite and carbon nanotubes. These changes can be favorable for adsorption capacity as structural disorders can increase the surface area, pore volume and pore diameter, and decrease particle size. The magnetization strength of magnetic adsorbents can increase or decrease due to these changes; however, this did not influence the magnetic separation properties of magnetic adsorbents (Liu et al. 2018). Devi and Saroha (2015) developed the Ni–ZVI magnetic biochar composites (Ni–ZVI-MBC) by impregnation of nickel on the ZVI biochar composites and used it for the removal of pentachlorophenol from wastewater. Higher pentachlorophenol removal efficiency has been achieved using these adsorbent owing to higher adsorption potential of biochar and reductive properties of Fe–Ni bimetals. It has been reported that the impregnation of nickel particles on ZVI is effective in preventing ZVI passivation due to corrosion thereby enhancing the removal rate of pentachlorophenol.

Table 15.5 Application of functionalized magnetic adsorbents for the removal of pollutants from water and wastewater systems

Adsorbent	Modification	Surface area (m^2/g)	Pore volume (cm^3/g)	Pore diameter (nm)	Adsorbate	Adsorption capacity (mg/g)	Reaction conditions	References
Thiol-modified magnetic polypyrrole nanocomposite	Mercapto-acetic acid-modified polypyrrole	49	0.15	19.6	Silver	806.5	C_o—600 mg/L; pH—5.6; adsorbent dosage—0.5 g/L; 25 °C	Mahlangu et al. (2019)
Superparamagnetic $Fe_3O_4@SiO_2$ nanoparticles	Polyvinylpyrrolidone	89	–	–	Phenanthrene	18.8	C_o—1 mg/L; pH—6; adsorbent dosage—0.5 mg/100 mL; 27 °C	Wang et al. (2019)
Magnetic nanoparticles	3-aminopropyl trimethoxysilane (APTMS) and dendrimer-like polyamidoamine	–	–	–	Cadmium Cobalt Zinc Lead Copper	100.8 95.4 89.6 108.8 87.8	C_o—160 mg/L; pH—4; adsorbent dosage—200 mg/L; time 5 min	Harinath et al. (2017)
Magnetic mesoporous silica nanospheres	Aminated magnetic mesoporous silica	148	0.09	3.9	Mercury	243.83	C_o—100 mg/L; pH—4; adsorbent dosage—0.02 g/50 mL; time—10 min	Fu et al. (2019)
Magnetic multiporous bio-adsorbent	Amino siloxane	45	–	101	Lead	133.3	C_o—60 mg/L; pH—5; adsorbent dosage—0.75 g/L	Zhou et al. (2018)
Crosslinked magnetic chitosan nanoparticles (Fe_3O_4@CS-BAL)	2,2′-(butane-1,4-diylbis (oxy))dibenzaldehyde	139	–	–	Reactive red 239	200.0	C_o—100 mg/L; pH—7; adsorbent dosage—20 mg/20 mL; time—2 min	Banaei et al. (2018)

(continued)

Table 15.5 (continued)

Adsorbent	Modification	Surface area (m^2/g)	Pore volume (cm^3/g)	Pore diameter (nm)	Adsorbate	Adsorption capacity (mg/g)	Reaction conditions	References
$Fe_3O_4@SiO_2@NH_2$	3-aminopropyltrimethoxy-silane	–	–	–	Methyl red	81.4	C_o—100 mg/L; pH—5.3; adsorbent dosage—0.5 g/L	Ghorbani and Kamari (2019)
Mercaptoamine-functionalized silica-coated magnetic nanoparticles	Mercaptoamine-functionalized silica	72	–	–	Mercury Lead	355 292	C_o—50 mg/L; pH—5–7; adsorbent dosage—10 mg/50 mL; time—2 h; 25 °C	Bao et al. (2017)
Municipal surplus sludge magnetic biochar	Glacial acetic acid	102	0.14	3.9	Tetracycline	60.4	C_o—100 mg/L; pH—7; adsorbent dosage—20 mg/20 mL	Tang et al. (2018)

Table 15.6 Application of metal ion-modified magnetic nanoparticles for removal of pollutants from water systems

Adsorbent	Modification	Surface area (m^2/g)	Pore volume (cm^3/g)	Pore diameter (nm)	Adsorbate	Adsorption capacity (mg/g)	Reaction conditions	References
Magnetic spinel Fe_2CuO_4/rGO nanocomposite	Graphene oxide (GO)	187	3.14	30.54	Mercury		C_o—0.51 mg/mL; pH—7; time—60 min; 24 °C	Zandi-Atashbar et al. (2018)
Titanium iron magnetic nano-sized adsorbent	Titanium oxide	289			Thallium	111.3	C_o—150 mg/L; pH—7; adsorbent dosage—0.1 g/L; 25 °C	Tang et al. (2019)
$MnFe_2O_4$-graphene oxide magnetic nanoparticles	Graphene oxide (GO)	98	0.23	12	Lanthanum Cerium	1001 982	C_o—100 mg/L; pH—7; adsorbent dosage—0.3 g/L; 25 °C	Ghobadi et al. (2018)
Ni-ZVI biochar composites	Nickel and biochar	168	0.09	2.3	Pentachlorophenol	50	C_o—50 mg/L; pH—6; time—24 h; adsorbent dosage—10 mg/50 mL; 25 °C	Devi and Saroha (2015)
Fe_3O_4@Al_2O_3@Zn–Fe LDH	Zn–Fe layer double hydroxide	157	0.34	1.29	Arsenic	67.57	C_o—65.6; pH—4; adsorbent dosage—28.9 mg; time 12.8 min	Adlnasab et al. (2019)
Calcium-modified iron oxide adsorbents	$CaCO_3$; polyvinyl alcohol	9	–	–	Phosphate	17.0	C_o—200 mg/L; pH—7; adsorbent dosage—1.5 g/100 mL	Han et al. (2017)

Manganese dioxide (MnO_2) nanoparticles are increasingly used for the synthesis of composite adsorbents due to their strong oxidizing abilities, higher adsorption capacities, and superior chemical stability. Magnetic composites prepared using a mixture of Fe_3O_4 and MnO_2 have demonstrated high sorption capacities for heavy metals (cadmium (Cd^{2+}) and arsenic (As^{3+} and As^{5+})) owing to their high stability and surface area. Dai et al. (2016) synthesized nano-Fe_3O_4/MnO_2 using hydrothermal procedure and reported 10.46 mg/g of adsorption capacity for Cd^{2+} removal. Similarly, Zhao et al. (2012) used Fe_3O_4/MnO_2 core–shell nanoplates and reported adsorption capacities of 72.8 mg/g and 32.1 mg/g for As^{3+} and As^{5+}, respectively.

The impregnation of metal oxides on iron particle has proved very effective in enhancing the adsorption capacity of the magnetic adsorbents. However, it had little impact in improving the oxidative stability and preventing their aggregation in the aqueous solution. Therefore, silica shells have been increasingly used for the immobilization of magnetic particles in order to enhance their chemical and oxidative stability. According to recent reports, 3-aminopropyl trimethoxysilane and chloromethyl thiirane have been used more frequently as surface modification reagents on silica-coated magnetic nanoparticles. These modified adsorbents contain sulfur and amine sites that can significantly enhance their interactions with metal particles like mercury (Hg) and lead (Pb). Amine sites interact with metal ions by donating a lone pair of electrons resulting in the formation of N–Hg(II) and N–Pb (II) complexes (Fig. 15.3). Bao et al. 2017 used XPS spectra to provide validation of metal ion binding to nitrogen atoms and the presence of nitrate ions on the surfaces of modified silica-coated magnetic nanoparticles.

15.4.5 *Macromolecules/Polysaccharides*

Biomolecules/polysaccharides (lignin, starch, cellulose, and carboxymethyl cellulose (CMC)) are widely used for modification of magnetic adsorbents as they are nontoxic, biodegradable, and have good compatibility with magnetic adsorbents. Cellulose is not a good stabilizer for magnetic adsorbents due to its insolubility in water. However, CMC aids in stabilizing the ZVI particles and then in improving their stability and dispersibility in aqueous medium (Devi and Dalai 2019). The degree of substitution of CMC plays an important role in the stabilization of ZVI

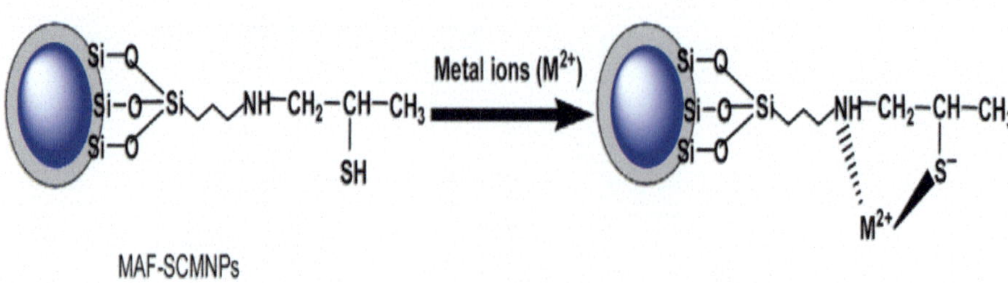

Fig. 15.3 Mechanism for the removal of Hg(II) and Pb(II) on MAF-SCMNPs. (Bao et al. 2017)

particles as CMC interacts with ZVI through hydroxyl (–OH) and carboxyl (–COO) functional groups resulting in better dispersion and less agglomeration of ZVI particles. It is reported that CMC with a higher degree of substitution causes dispersion of ZVI particles due to stronger repulsive forces and greater steric hindrance compared to CMC with a lower degree of substitution. Similarly, starch and lignin have low adsorption capacity and some sort of functionalization is required in order to increase the adsorption capacity. Therefore, chemical modification such as alkylation, hydroxylation, or amination is required to increase the chemical active sites. Meng et al. (2019) synthesized lignin derivate magnetic hydrogel microspheres (LDMHMs) via blending with Fe_3O_4 and used for heavy metals and dye removal. The adsorption capacity of 33 mg/g, 55 mg/g, and 23 mg/g were obtained for lead, mercury, and nickel, respectively. The spent adsorbents were separated by magnetic separation followed by regeneration using the hydrochloric acid and then reused for multiple adsorption–desorption cycles. Recycling studies showed that these adsorbents were able to retain more than 90% adsorption capacity after multiple cycles, thus indicating high reusability and adsorption potential. Jiang et al. (2019) synthesized calcium lignosulfonate-modified magnetic Fe_3O_4/C and used it as an adsorbent for the removal of azo dyes. High adsorption efficiencies were obtained, and calcium lignosulfonate-modified magnetic Fe_3O_4/C were able to remove 98%, 92%, and 99% of Congo red, Titan yellow, and Eriochrome blue black R dyes, respectively. Karami and Zeynizadeh (2019) prepared EDA-functionalized magnetic cellulose adsorbent for the removal of metal ions including copper, lead, cadmium, zinc, and nickel (Fig. 15.4). These adsorbents

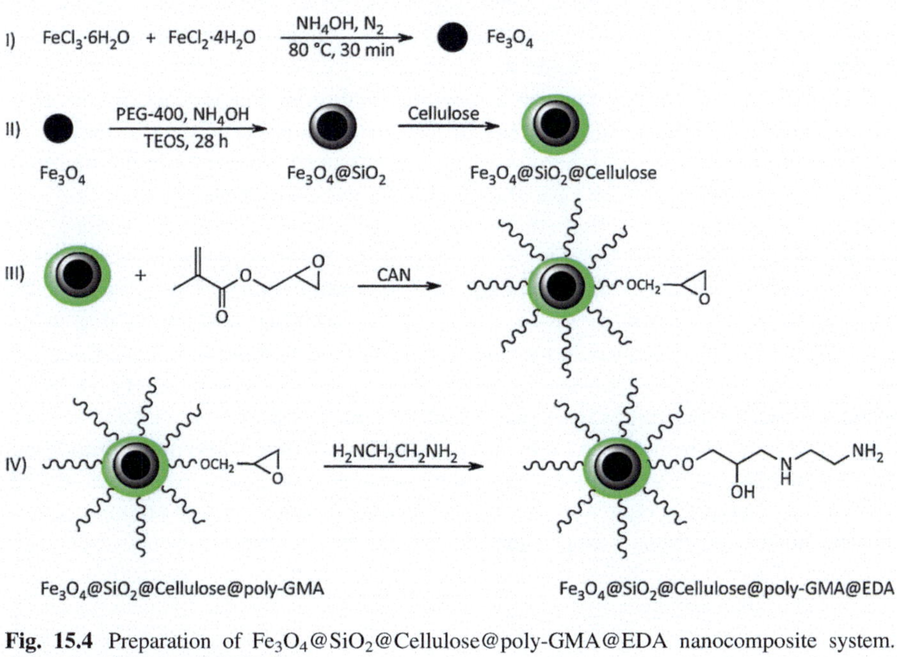

Fig. 15.4 Preparation of Fe_3O_4@SiO_2@Cellulose@poly-GMA@EDA nanocomposite system. (Karami and Zeynizadeh 2019)

showed higher affinity to adsorb lead (129.0 mg/g), cadmium (117.9 mg/g), zinc (77.4 mg/g), and nickel (70.1 mg/g) from aqueous solution. Similarly, a number of reports are available in the literature on the use of biomolecules/polysaccharides for magnetic adsorbents, which are shown in Table 15.7.

15.4.6 Miscellaneous

Layered double hydroxide (LDH) is a hydrophilic, two-dimensional nano-clay with a high charge density. The hydrophilic nature restricted their use in many industrial fields such as polymeric NPs. Therefore, the modification of LDH materials with bio-safe and organic molecules is a significant way to improve their interactions with the polymeric matrixes. Nowadays, the use of LDHs as a safe adsorbent for the elimination of toxic dyes from aqueous solution is a very interesting field for researchers (Mallakpour and Hatami 2019). One of the limitations of LDHs as an adsorbent comes from the platelet-like structure of these materials, which causes their easy dispersion in the aqueous solution. As a result, this would create difficulty in their separation from the water solution. Thus, for the easy separation, these materials are combined with Fe_3O_4 to synthesize a sorbent, which can be isolated from the solution using an external magnet (Li et al. 2017). In recent years, combined LDHs with magnetite nanoparticles (NPs) are increasingly used for the separation of noxious cations, degradation of dyes, drug delivery, and to develop the mechano-chemical properties of polymeric nanocomposites (NCs). For example, Koilraj and Sasaki (2016) used Fe_3O_4/MgAl–NO_3–LDH as a separable adsorbent to uptake phosphate from the synthetic solution.

Surfactant plays an important role in the modification of magnetic adsorbents as it allows the easy dispersion of magnetic particles on the support material as well as in aqueous medium. Devi and Saroha (2014, 2015) used cetyltrimethylammonium ammonium bromide (CTAB) for the dispersion of ZVI on the biochar surface. Literature reports show that the discrete surfactant aggregates formed by cetyltrimethylammonium ammonium bromide (CTAB) on the graphite surface had a high adsorption capacity for bisphenol A. The modification of $Fe_3O_4@SiO_2$ NPs with polyvinylpyrrolidone (PVP) is advantageous for polar adsorbents due to its easy dispersion in aqueous solution and affinity toward hydrophobic contaminants. Wang et al. (2019) synthesized Fe_3O_4/SiO_2 and PVP-modified Fe_3O_4/SiO_2 nanoparticles for the removal of phenanthrene and found that the surface area of the modified Fe_3O_4/SiO_2 was reduced to 60.82 m^2/g compared to nonmodified Fe_3O_4/SiO_2 (88.7 m^2/g). Contrary to this, the adsorption capacity of PVP-modified Fe_3O_4/SiO_2 nanoparticles was increased twofold (18.8 mg/g) compared to nonmodified Fe_3O_4/SiO_2 nanoparticles.

Table 15.7 Macro/biomolecule-modified magnetic nanoparticles for removal of pollutants from water systems

Adsorbent	Modification	Surface area (m^2/g)	Pore volume (cm^3/g)	Pore diameter (nm)	Adsorbate	Adsorption capacity (mg/g)/ removal efficiency (%)	Reaction conditions	References
Lignin derivate magnetic hydrogel microspheres	Lignin	–	–	–	Methylene blue Methyl orange Malachite green	43 39 155	C_o—50 mg/L (except MO; C_o—10 mg/L); pH—8; adsorbent dosage—10 mg/20 mL; time—24 h	Meng et al. (2019)
Lignin derivate magnetic hydrogel microspheres	Lignin	–	–	–	Lead Mercury Nickel	33 55 23	C_o—50 mg/L; pH—5; adsorbent dosage—10 mg/20 mL; time—24 h	Meng et al. (2019)
Carboxymethyl cellulose-coated ZVI	Carboxymethyl cellulose	36	0.09	10.8	Naphthenic acid	100%	C_o—10 mg/L; pH—7; time—2 h; adsorbent dosage—100 mg/50 mL; 25 °C	Devi and Dalai (2019)
EDA-functionalized magnetic cellulose nanocomposite	Cellulose with glycidyl methacrylate (GMA) and then ethylenediamine (EDA)	–	–	–	Copper	81.7	C_o—400 mg/L; pH—5.5; adsorbent dosage—0.05 g/25 mL; 25 °C	Karami and Zeynizadeh (2019)
Magnetic lignin-based adsorbent	Lignin	–	–	–	Congo red Titan yellow Eriochrome blue black R	303 263.2 308	C_o—40 mg/L; pH—7; adsorbent dosage—0.01 mg/50 mL; time—180 min; 25 °C	Jiang et al. (2019)
Egg-albumen-formaldehyde-based magnetic polymeric resin	Egg-albumen-formaldehyde	179	–	24	Cadmium	149.3	C_o—100 mg/L; pH—2; adsorbent dosage—100 mg/250 mL; time—6 h; 25 °C	Ahamad et al. (2019)

15.5 Mechanism of Contaminant Removal on Magnetic Adsorbents

The possible mechanism of contaminant removal on magnetic adsorbents has been proposed by various researchers (Bao et al. 2017; Tang et al. 2018, 2019; Yan et al. 2017). In most of the literature reports, the FTIR and XPS analyses were used to elucidate the chemical interactions between the contaminants and the adsorbent surface. Tang et al. (2019) investigated the adsorption mechanism of thallium removal on titanium-coated magnetic adsorbents and found that the –OH groups present on the surface of magnetic adsorbent were primarily involved in removal of thallium through protonation and deprotonation mechanisms. At an acidic pH, thallium appeared to form inner-surface complexes with adsorbent surface through electrostatic attractive forces. Yan et al. (2017) studied the effect of solution pH on the removal of chromium using PVA/chitosan-based magnetic adsorbents and found that chromium was primarily removed from the solution through electron transfer, electron sharing, and oxidation reduction reactions on the adsorbent surface. At neutral solution pH, chromium exists as CrO_4^{2-} and interacts with the active functional groups (–NH_2 and –OH) present on the adsorbent surface. While at acidic solution pH, chromium exists as $HCrO_4^-$ and the negatively charged $HCrO_4^-$ readily interacts with positively charged –OH_2^+ and NH_3^+ functional groups through electrostatic interactions (Yan et al. 2017). Similarly, Bao et al. (2017) reported that nitrogen atoms and nitrite ions present on magnetic nanoparticles' surface were responsible for the effective removal of mercury and lead from the aqueous solution. Tang et al. (2018) found that the aromatic contaminants like tetracycline interact with the adsorbent surface via strong π–π interactions and H-bonding mechanisms. The active functional groups like C=O and –NH_2 facilitate the strong interaction of aromatic contaminants with magnetic adsorbents through electron transfer mechanism.

15.6 Conclusions

This chapter discussed several modification methods used for the performance enhancement of the iron-based magnetic adsorbents. In most of the studies, the magnetic particles used were iron oxides due to their low cost, easy synthesis procedure, and relatively higher chemical stability. Among the modification procedures, carbon-based and polymer-modified magnetic adsorbents exhibited superior activity and oxidative stability. Polymer-modified magnetic adsorbents combine the functional properties of polymer and the magnetic & reductive properties of iron oxide particles. The use of modified magnetic adsorbents for the removal of contaminants has several advantages such as higher adsorption capacity, faster removal rates, and surface area compared to traditional adsorbents. The commercial applications of magnetic adsorbents in environmental technologies are still in the incipient stage and a great deal of work is still needed to establish modified magnetic adsorbents on commercial scale. Therefore, the objective of this chapter is to provide state of the art on the application of magnetic adsorbents into water treatment technologies and to suggest modifications for the implementation of these technologies.

References

Abdi G, Alizadeh A, Amirian J, Rezaei S, Sharma G (2019) Polyamine-modified magnetic graphene oxide surface: feasible adsorbent for removal of dyes. J Mol Liq 289:111118. https://doi.org/10.1016/j.molliq.2019.111118

Adlnasab L, Shekari N, Maghsodi A (2019) Optimization of arsenic removal with Fe_3O_4@Al_2O_3@Zn-Fe LDH as a new magnetic nano adsorbent using Box-Behnken design. J Environ Chem Eng 7. https://doi.org/10.1016/j.jece.2019.102974

Ahamad T, Naushad M, Eldesoky GE, Alqadami AA, Khan A (2019) Synthesis and characterization of egg-albumen-formaldehyde based magnetic polymeric resin (MPR): highly efficient adsorbent for Cd(II) ion removal from aqueous medium. J Mol Liq 286:110951. https://doi.org/10.1016/j.molliq.2019.110951

Akram MY, Ahmed S, Li L, Akhtar N, Ali S, Muhyodin G, Zhu XQ, Nie J (2019) N-doped reduced graphene oxide decorated with Fe_3O_4 composite: stable and magnetically separable adsorbent solution for high performance phosphate removal. J Environ Chem Eng 7:103137. https://doi.org/10.1016/j.jece.2019.103137

Altıntıg E, Altundag H, Tuzen M, Sarı A, Sarı A (2017) Effective removal of methylene blue from aqueous solutions using magnetic loaded activated carbon as novel adsorbent. Chem Eng Res Des 122:151–163. https://doi.org/10.1016/j.cherd.2017.03.035

Banaei A, Farokhi Yaychi M, Karimi S, Vojoudi H, Namazi H, Badiei A, Pourbasheer E (2018) 2,2′-(butane-1,4-diylbis(oxy))dibenzaldehyde cross-linked magnetic chitosan nanoparticles as a new adsorbent for the removal of Reactive Red 239 from aqueous solutions. Mater Chem Phys 212:1–11. https://doi.org/10.1016/j.matchemphys.2018.02.036

Bao S, Li K, Ning P, Peng J, Jin X, Tang L (2017) Highly effective removal of mercury and lead ions from wastewater by mercaptoamine-functionalised silica-coated magnetic nano-adsorbents: Behaviours and mechanisms. Appl Surf Sci 393:457–466. https://doi.org/10.1016/j.apsusc.2016.09.098

Bao S, Yang W, Wang Y, Yu Y, Sun Y (2020) One-pot synthesis of magnetic graphene oxide composites as an efficient and recoverable adsorbent for Cd(II) and Pb(II) removal from aqueous solution. J Hazard Mater 381:120914. https://doi.org/10.1016/j.jhazmat.2019.120914

Chen B, Zhao H, Chen S, Long F, Huang B, Yang B, Pan X (2019) A magnetically recyclable chitosan composite adsorbent functionalized with EDTA for simultaneous capture of anionic dye and heavy metals in complex wastewater. Chem Eng J 356:69–80. https://doi.org/10.1016/j.cej.2018.08.222

Dai HL, Xin YE, Xiang L, Zhan P, Xiu-Guo LU, Liu ZM (2016) Adsorption on cadmium by nano-Fe_3O_4/MnO_2 and its adsorption mechanism. China Water Wastewater 32:81–85

Devi P, Dalai AK (2019) Effects of carboxymethyl cellulose grafting on stability and reactivity of zerovalent iron in water systems. J Cleaner Prod 229:66–74. https://doi.org/10.1016/j.jclepro.2019.04.364

Devi P, Saroha AK (2014) Synthesis of the magnetic biochar composites for use as an adsorbent for the removal of pentachlorophenol from the effluent. Bioresour Technol 169:525–531. https://doi.org/10.1016/j.biortech.2014.07.062

Devi P, Saroha AK (2015) Simultaneous adsorption and dechlorination of pentachlorophenol from effluent by Ni–ZVI magnetic biochar composites synthesized from paper mill sludge. Chem Eng J 271:195–203. https://doi.org/10.1016/j.cej.2015.02.087

Devi P, Saroha AK (2016) Risk assessment and technical feasibility of usage of paper mill sludge biochar-based exhausted adsorbent for geopolymeric brick formation. Environ Sci Pollut Res 23:21641–21651. https://doi.org/10.1007/s11356-016-7343-4

Devi P, Kothari P, Dalai AK (2019a) Effects of carboxymethyl cellulose grafting on stability and reactivity of zerovalent iron in water systems. J Clean Prod 229:65–74. https://doi.org/10.1016/j.jclepro.2019.04.364

Devi P, Kothari P, Dalai AK (2019b) Stabilization and solidification of arsenic and iron contaminated canola meal biochar using chemically modified phosphate binders. J Hazard Mater (In press). https://doi.org/10.1016/j.jhazmat.2019.121559

Devi P, Dalai AK, Chaurasia SP (2020) Activity and stability of biochar in hydrogen peroxide based oxidation system for degradation of naphthenic acid. Chemosphere 241:125007. https://doi.org/10.1016/j.chemosphere.2019.125007

Duman O, Tunç S, Polat TG, Bozoğlan BKI (2016) Synthesis of magnetic oxidized multiwalled carbon nanotube-κ-carrageenan-Fe_3O_4 nanocomposite adsorbent and its application in cationic Methylene Blue dye adsorption. Carbohydr Polym 147:79–88. https://doi.org/10.1016/j.carbpol.2016.03.099

Fat'hi MR, Nasab SJH (2018) Synthesis of calcon-imprinted magnetic chitosan nanoparticles as a novel adsorbent and its application in selective removal of calcon dye from aqueous solutions. Int J Biol Macromol 114:1151–1160. https://doi.org/10.1016/j.ijbiomac.2018.03.103

Fu Y, Sun Y, Chen Z, Ying S, Wang J, Hu J (2019) Functionalized magnetic mesoporous silica/poly(m-aminothiophenol) nanocomposite for Hg(II) rapid uptake and high catalytic activity of spent Hg(II) adsorbent. Sci Total Environ 691:664–674. https://doi.org/10.1016/j.scitotenv.2019.07.153

Ghobadi M, Gharabaghi M, Abdollahi H, Boroumand Z, Moradian M (2018) MnFe2O4-graphene oxide magnetic nanoparticles as a high-performance adsorbent for rare earth elements: synthesis, isotherms, kinetics, thermodynamics and desorption. J Hazard Mater 351:308–316. https://doi.org/10.1016/j.jhazmat.2018.03.011

Ghorbani F, Kamari S (2019) Core–shell magnetic nanocomposite of Fe_3O_4 @SiO_2 @NH_2 as an efficient and highly recyclable adsorbent of methyl red dye from aqueous environments. Environ Technol Innov 14:100333. https://doi.org/10.1016/j.eti.2019.100333

Gómez-Pastora J, Bringas E, Ortiz I (2014) Recent progress and future challenges on the use of high performance magnetic nano-adsorbents in environmental applications. Chem Eng J 256:187–204. https://doi.org/10.1016/j.cej.2014.06.119

Gupta A, Yadav R, Devi P (2011) Removal of hexavalent chromium using activated coconut shell and activated coconut coir as low cost adsorbent. IIOAB J 2:8–12

Han C, Lalley J, Iyanna N, Nadagouda MN (2017) Removal of phosphate using calcium and magnesium-modified iron-based adsorbents. Mater Chem Phys 198:115–124. https://doi.org/10.1016/j.matchemphys.2017.05.038

Hao Z, Wang C, Yan Z, Jiang H, Xu H (2018) Magnetic particles modification of coconut shell-derived activated carbon and biochar for effective removal of phenol from water. Chemosphere 211:962–969. https://doi.org/10.1016/j.chemosphere.2018.08.038

Harinath Y, Reddy DHK, Sharma LS, Seshaiah K (2017) Development of hyperbranched polymer encapsulated magnetic adsorbent (Fe3O4@SiO2-NH2-PAA) and its application for decontamination of heavy metal ions. J Environ Chem Eng 5:4994–5001. https://doi.org/10.1016/j.jece.2017.09.031

Irawan C, Nata IF, Lee CK (2019) Removal of Pb(II) and As(V) using magnetic nanoparticles coated montmorillonite via one-pot solvothermal reaction as adsorbent. J Environ Chem Eng 7:103000. https://doi.org/10.1016/j.jece.2019.103000

Jiang C, Wang X, Qin D, Da W, Hou B, Hao C, Wu J (2019) Construction of magnetic lignin-based adsorbent and its adsorption properties for dyes. J Hazard Mater 369:50–61. https://doi.org/10.1016/j.jhazmat.2019.02.021

Karami S, Zeynizadeh B (2019) Reduction of 4-nitrophenol by a disused adsorbent: EDA-functionalized magnetic cellulose nanocomposite after the removal of Cu^{2+}. Carbohydr Polym 211:298–307. https://doi.org/10.1016/j.carbpol.2019.01.113

Kaur R, Hasan A, Iqbal N, Alam S, Saini M (2014) Synthesis and surface engineering of magnetic nanoparticles for environmental cleanup and pesticide residue analysis: a review. J Sep Sci 37:1805–1825. https://doi.org/10.1002/jssc.201400256

Kim HR, Jang JW, Park JW (2016) Carboxymethyl chitosan-modified magnetic-cored dendrimer as an amphoteric adsorbent. J Hazard Mater 317:108–116. https://doi.org/10.1016/j.jhazmat.2016.06.025

Koilraj P, Sasaki K (2016) Fe_3O_4/MgAl–NO_3 layered double hydroxide as a magnetically separable sorbent for the remediation of aqueous phosphate. J Environ Chem Eng 4:984–991. https://doi.org/10.1016/j.jece.2016.01.005

Li J, Jiang B, Liu Y, Qiu C, Hu J, Qian G, Guo W, Ngo HH (2017) Preparation and adsorption properties of magnetic chitosan composite adsorbent for Cu^{2+} removal. J Clean Prod 158:51–58. https://doi.org/10.1016/j.jclepro.2017.04.156

Li K, Li P, Cai J, Xiao S, Yang H, Li A (2016) Efficient adsorption of both methyl orange and chromium from their aqueous mixtures using a quaternary ammonium salt modified chitosan magnetic composite adsorbent. Chemosphere 154:310–318. https://doi.org/10.1016/j.chemosphere.2016.03.100

Liu Z, Li X, Zhan P, Hu F, Ye X (2018) Removal of cadmium and copper from water by a magnetic adsorbent of PFM: adsorption performance and micro-structural morphology. Sep Purif Technol 206:199–207. https://doi.org/10.1016/j.seppur.2018.06.007

Lv L, Chen N, Feng C, Gao Y, Li M (2017) Xanthate-modified magnetic chitosan/poly (vinyl alcohol) adsorbent: preparation, characterization, and performance of Pb(II) removal from aqueous solution. J Taiwan Inst Chem Eng 78:485–492. https://doi.org/10.1016/j.jtice.2017.06.009

Mahlangu T, Das R, Abia LK, Onyango M, Ray SS, Maity A (2019) Thiol-modified magnetic polypyrrole nanocomposite: An effective adsorbent for the adsorption of silver ions from aqueous solution and subsequent water disinfection by silver-laden nanocomposite. Chem Eng J 360:423–434. https://doi.org/10.1016/j.cej.2018.11.231

Mallakpour S, Hatami M (2019) An effective, low-cost and recyclable bio-adsorbent having amino acid intercalated LDH@Fe3O4/PVA magnetic nanocomposites for removal of methyl orange from aqueous solution. Appl Clay Sci 174:127–137. https://doi.org/10.1016/j.clay.2019.03.026

Mashile GP, Mpupa A, Nqombolo A, Dimpe M, Nomngongo PN (2020) Recyclable magnetic waste tyre activated carbon-chitosan composite as an effective adsorbent rapid and simultaneous removal of methylparaben and propylparaben from aqueous solution and wastewater. J Water Proc Eng 33:101011. https://doi.org/10.1016/j.jwpe.2019.101011

Meng Y, Li C, Liu X, Lu J, Cheng Y, Xiao LP, Wang H (2019) Preparation of magnetic hydrogel microspheres of lignin derivate for application in water. Sci Total Environ 685:847–855. https://doi.org/10.1016/j.scitotenv.2019.06.278

Nascimento TA, de Oliveira HL, Borges KB (2019) Magnetic molecularly imprinted polypyrrole as a new selective adsorbent for pharmaceutically active compounds. J Environ Chem Eng 7:103371. https://doi.org/10.1016/j.jece.2019.103371

Nejadshafiee V, Islami MR (2019) Adsorption capacity of heavy metal ions using sultone-modified magnetic activated carbon as a bio-adsorbent. Mater Sci Eng C 101:42–52. https://doi.org/10.1016/j.msec.2019.03.081

Noor NM, Othman R, Mubarak NM, Abdullah EC (2017) Agricultural biomass-derived magnetic adsorbents: preparation and application for heavy metals removal. J Taiwan Inst Chem Eng 78:168–177. https://doi.org/10.1016/j.jtice.2017.05.023

Pourzamani H, Parastar S, Hashemi M (2017) The elimination of xylene from aqueous solutions using single wall carbon nanotube and magnetic nanoparticle hybrid adsorbent. Process Saf Environ Prot 109:688–696. https://doi.org/10.1016/j.psep.2017.05.010

Qiao K, Tian W, Bai J, Wang L, Zhao J, Du Z, Gong X (2019) Application of magnetic adsorbents based on iron oxide nanoparticles for oil spill remediation: a review. J Taiwan Inst Chem Eng 97:227–236. https://doi.org/10.1016/j.jtice.2019.01.029

Safinejad A, Chamjangali MA, Goudarzi N, Bagherian G (2017) Synthesis and characterization of a new magnetic bio-adsorbent using walnut shell powder and its application in ultrasonic assisted removal of lead. J Environ Chem Eng 5:1429–1437. https://doi.org/10.1016/j.jece.2017.02.027

Sehlleier YH, Hardt S, Schulz C, Wiggers H (2016) A novel magnetically-separable porous iron-oxide nanocomposite as an adsorbent for methylene blue (MB) dye. J Environ Chem Eng 4:3779–3787. https://doi.org/10.1016/j.jece.2016.08.018

Son EB, Poo KM, Mohamed HO, Choi YJ, Cho WC, Chae KJ (2018) A novel approach to developing a reusable marine macro-algae adsorbent with chitosan and ferric oxide for simultaneous efficient heavy metal removal and easy magnetic separation. Bioresour Technol 259:381–387. https://doi.org/10.1016/j.biortech.2018.03.077

Tang J, Wu W, Yu L, Fan X, Liu G, Yu Y (2019) Study on adsorption properties and mechanism of thallium onto titanium iron magnetic adsorbent. Sci Total Environ 694:133625. https://doi.org/10.1016/j.scitotenv.2019.133625

Tang L, Yu J, Pang Y, Zeng G, Deng Y, Wang J, Ren X, Ye S, Peng B, Feng H (2018) Sustainable efficient adsorbent: alkali-acid modified magnetic biochar derived from sewage sludge for aqueous organic contaminant removal. Chem Eng J 336:160–169. https://doi.org/10.1016/j.cej.2017.11.048

Tanhaei B, Ayati A, Sillanpää M (2019) Magnetic xanthate modified chitosan as an emerging adsorbent for cationic azo dyes removal: kinetic, thermodynamic and isothermal studies. Int J Biol Macromol 121:1126–1134. https://doi.org/10.1016/j.ijbiomac.2018.10.137

Wang H, Liu Y, Ifthikar J, Shi L, Khan A, Chen Z, Chen Z (2018) Towards a better understanding on mercury adsorption by magnetic bio-adsorbents with Γ-Fe_2O_3 from pinewood sawdust derived hydrochar: influence of atmosphere in heat treatment. Bioresour Technol 256:269–276. https://doi.org/10.1016/j.biortech.2018.02.019

Wang L, Shen C, Cao Y (2019) PVP modified $Fe_3O_4@SiO_2$ nanoparticles as a new adsorbent for hydrophobic substances. J Phys Chem Solids 133:28–34. https://doi.org/10.1016/j.jpcs.2019.05.004

Wu Y, Chen C, Zhou Q, Li QX, Yuan Y, Tong Y, Wang H, Zhou X, Sun Y, Sheng X (2019) Polyamidoamine dendrimer decorated nanoparticles as an adsorbent for magnetic solid-phase extraction of tetrabromobisphenol A and 4-nonylphenol from environmental water samples. J Colloid Interface Sci 539:361–369. https://doi.org/10.1016/j.jcis.2018.12.064

Xu W, Hussain A, Liu Y (2018) A review on modification methods of adsorbents for elemental mercury from flue gas. Chem Eng J 346:692–711. https://doi.org/10.1016/j.cej.2018.03.049

Yan E, Cao M, Jiang J, Gao J, Jiang C, Ba X, Yang X, Zhang D (2017) A novel adsorbent based on magnetic Fe3O4contained polyvinyl alcohol/chitosan composite nanofibers for chromium (VI) removal. Solid State Sci 72:94–102. https://doi.org/10.1016/j.solidstatesciences.2017.08.014

Zandi-Atashbar N, Ensafi AA, Ahoor AH (2018) Magnetic Fe2CuO4/rGO nanocomposite as an efficient recyclable catalyst to convert discard tire into diesel fuel and as an effective mercury adsorbent from wastewater. J Clean Prod 172:68–80. https://doi.org/10.1016/j.jclepro.2017.10.146

Zeng G, Liu Y, Tang L, Yang G, Pang Y, Zhang Y, Zhou Y, Li Z, Li M, Lai M (2015) Enhancement of Cd (II) adsorption by polyacrylic acid modified magnetic meso- porous carbon. Chem Eng J 259:153–160. https://doi.org/10.1016/j.cej.2014.07.115

Zhao F, Tang WZ, Zhao D, Meng Y, Yin D, Sillanpää M (2014) Adsorption kinetics, isotherms and mechanisms of Cd(II), Pb(II), Co(II) and Ni(II) by a modified magnetic polyacrylamide microcomposite adsorbent. J Water Process Eng 4:47–57. https://doi.org/10.1016/j.jwpe.2014.09.003

Zhao Z, Liu J, Cui F, Feng H, Zhang L (2012) One pot synthesis of tunable Fe_3O_4–MnO_2 core–shell nanoplates and their applications for water purification. J Mater Chem 22:9052–9057. https://doi.org/10.1039/c2jm00153e

Zheng C, Zheng H, Wang Y, Sun Y, An Y, Liu H, Liu S (2019) Modified magnetic chitosan microparticles as novel superior adsorbents with huge "force field" for capturing food dyes. J Hazard Mater 367:492–503. https://doi.org/10.1016/j.jhazmat.2018.12.120

Zhou J, Liu Y, Zhou X, Ren J, Zhong C (2018) Magnetic multi-porous bio-adsorbent modified with amino siloxane for fast removal of Pb(II) from aqueous solution. Appl Surf Sci 427:976–985. https://doi.org/10.1016/j.apsusc.2017.08.110

Index

L. Meili, G. L. Dotto (eds.), *Advanced Magnetic Adsorbents for Water Treatment*, Environmental Chemistry for a Sustainable World 61,
https://doi.org/10.1007/978-3-030-64092-7

E

F

G

H

The manufacturer's authorised representative in the EU is Springer Nature Customer Service Centre GmbH, Europaplatz 3, 69115 Heidelberg, Germany. If you have any concerns regarding our products, please contact ProductSafety@springernature.com

Printed and bound by CPI Group (UK) Ltd, Croydon, CR0 4YY
30/09/2026
02225248-0018